Studying
Animal
Behavior

Studying Animal Behavior

Autobiographies of the Founders

Edited by
Donald A. Dewsbury

The University of Chicago Press
Chicago and London

The University of Chicago Press, Chicago 60637
The University of Chicago Press, Ltd., London

98 97 96 95 94 93 92 91 90 89   5 4 3 2 1

Originally published under the title, *Leaders in the Study of Animal
Behavior: Autobiographical Perspectives,* in the Animal Behavior
series edited by Douglas K. Candland.

Reprinted by arrangement with Associated University Presses, Inc.

Library of Congress Cataloging in Publication Data

Leaders in the study of animal behavior.
    Studying animal behavior : autobiographies of the founders /
edited by Donald A. Dewsbury.
        p.   cm.
    Reprint. Originally published: Leaders in the study of animal
behavior. Lewisburg : Bucknell University Press, c1985.
    Includes index.
    1. Ethologists—Biography.   2. Animal behavior.   I. Dewsbury,
Donald A., 1939-   . II. Title.
QL26.L43   1989   89-4978
591.51—dc20   CIP
ISBN 0-226-14410-0 (pbk.)

Contents

	Preface	7
1	Two Pillars of Wisdom GERARD P. BAERENDS	13
2	Curiosity, Milieu, and Era VINCENT G. DETHIER	43
3	"Fishy, Fishy, Fishy": Autobiographical Sketches IRENÄUS EIBL-EIBESFELDT	69
4	Of Dogs, Mice, People, and Me JOHN LANGWORTHY FULLER	93
5	Reflections of an Experimental Naturalist DONALD R. GRIFFIN	121
6	A Lifelong Attempt to Understand Animals HEINI HEDIGER	145
7	The Wild-Goose Chase ECKHARD H. HESS	183
8	Ethology in Relation to Other Disciplines ROBERT A. HINDE	193
9	Those Critical Periods of Social Reinforcement JOHN A. KING	205
10	The Cat Who Walks By Himself PAUL LEYHAUSEN	225
11	My Family and Other Animals KONRAD Z. LORENZ	259
12	The Ontogeny of an Ethologist AUBREY MANNING	289
13	Hark Ye to the Birds: Autobiographical Marginalia PETER MARLER	315
14	In Haldane's Footsteps JOHN MAYNARD SMITH	347
15	It's a Long Long Way to Tipperary, the Land of my Genes CURT P. RICHTER	357
16	Investigative Behavior: Toward a Science of Sociality JOHN PAUL SCOTT	389

17 Watching and Wondering   431
      NIKO TINBERGEN

18 In the Queendom of Ants: A Brief Autobiography   465
      EDWARD O. WILSON

19 Backstage and Upstage with "Animal Dispersion"   487
      VERO C. WYNNE-EDWARDS

Preface

This volume was conceived in a comparative psychology seminar I taught at the University of Florida several years ago. In contrast to the usual seminars, which are organized with respect to areas of research, this one was focused on the careers of individual scientists, in an effort to appreciate the continuity of their work over time and in different areas. We found that there was little biographical material available on many of the prominent figures in animal behavior. Autobiographical material was available for few of even the best-known animal behaviorists. This book was designed to fill the gap we discovered.

Autobiographical statements provide insight into the nature of the development of scientific careers and fields that are duplicated by no other material. There is no other way in which to get the firsthand perspective of the individuals who made the history—regardless of how much the objective historian may question the accuracy of that perspective. Volumes such as the *History of Psychology in Autobiography* (e.g. Boring and G. Lindzey, eds., vol. 5 [New York: Appleton-Century-Crofts, 1967]) and *Pioneers in Neuroendocrinology* (J. Meites, B. T. Donovan, and S. M. McCann, eds., [New York: Plenum, 1978]) have served this function.

The field of animal behavior was created by individuals from diverse backgrounds and perspectives and has matured in this century. Although some of the important pioneers are no longer alive, others are of such status and position as to provide irreplaceable perspective on this growth. I felt the time was ripe for a volume of autobiographies by prominent animal behaviorists. Although I would not contend that I was the most appropriate editor for such a volume, nobody else had undertaken the task. Therefore, I decided to launch such an effort.

In formulating plans for the volume I consulted with many colleagues, both at the University of Florida and elsewhere. My major concern was shared by others: how does one select the individuals for inclusion in such a volume? There are many more important animal behaviorists than can be included in such a book. In making choices one necessarily omits some of these excellent scientists. I received some advice that the whole project be dropped because it had to be inherently unfair and to hurt the feelings of some who were not invited. Although truly sensitive to this argument, I decided that the benefits of such a book were likely to outweigh the harm it might do and I decided to continue with it. My hope is that we shall be able to publish a second volume and right some of the wrongs promulgated with the first.

I did feel, however, that decisions such as these should not be undertaken by one individual. Therefore, I invited a group of animal behaviorists with diverse backgrounds to function as a selection committee that would make recommenda-

7

tions regarding invitations to contribute to the book. All whom I asked agreed to serve. These were: Edwin M. Banks of the University of Illinois; Robert Boice of the State University of New York, Albany; Gordon Burghardt of the University of Tennessee; Devra Kleiman of the U.S. National Zoological Park; Aubrey Manning of the University of Edinburgh; and Martin W. Schein of West Virginia University.

We developed a procedure that represents an honest effort to make fair choices. Although other methods could have been used, I believe ours to be at least one reasonable approach. I generated an initial list of potential contributors and circulated it to the committee members with a request for additions and deletions. The result was a list of sixty-nine potential invitees. We each rank ordered these sixty-nine names with the constraint that there be no self-rankings or rankings of close relatives. The primary basis for invitations was the mean rank of each individual. I decided to issue twenty-six invitations in the hope of generating twenty chapters and therefore issued invitations to those twenty-six individuals with the top mean rankings. My plan was to exercise some editorial judgment in issuing additional invitations so as to balance the content of the volume as letters of decline arrived. However, the response was so good that I am happy to say I never got that opportunity.

The list of invitees was truly the result of a group process and all of us ranked some individuals quite high only to find them not invited. I wonder if any two animal behaviorists would generate the same two lists of twenty-six names! Nevertheless, I believe the list to be a fair one and to not promulgate more injustices than necessary. I ran a correlational analysis of the rankings by the seven raters. Of the twenty-one correlations, all were positive, all were significant at at least the .05 level, and all but two were significant at the .01 level. Two raters generated a correlation of $+.70$. All raters generated correlations with the consensus of $+0.61$ or better. I cannot argue that these data indicate the correctness of our choices. However, they do indicate that we displayed considerable communality of choice—despite the fact that individuals were deliberately selected for the committee so as to reflect diverse approaches to animal behavior.

Several of those invited had to decline for various reasons. Harry F. Harlow was already seriously ill and passed away on December 6, 1981. Frank A. Beach felt that as he had already written two similar chapters ("Frank A. Beach," in G. Lindzey, ed., *A History of Psychology in Autobiography* [Englewood Cliffs, N.J.: Prentice-Hall, 1974]; "Confessions of an imposter," in J. Meites, B. T. Donovan, and S. M. McCann, eds., *Pioneers in Neuroendocrinology II* [New York: Plenum, 1978]), a third chapter would be inappropriate. Autobiographical books had already been written by invitees B. F. Skinner *Particulars of My Life* [New York: McGraw-Hill, 1976]) and Karl von Frisch *A Biologist Remembers* [New York: Springer-Verlag, 1980]). Professor von Frisch passed away in 1981. Lester R. Aronson, John T. Emlen, and William H. Thorpe were also unable to contribute chapters. The result is a collection of nineteen autobiographical chapters by a diverse collection of distinguished animal behaviorists.

I hope that this volume can serve two major functions. First, I hope it records for posterity the perspectives, lives, and accomplishments of some of those scientists who shaped the field of animal behavior as we know it today. Second, I hope the book will be useful to students and young scientists grappling with career decisions, the problems of selecting research problems, and the difficulties of life so admirably solved by the authors of these chapters. I hope that the volume gives the reader as much pleasure as it has given me in its assembly. It has been a pleasure to work with such a distinguished and cooperative group of individuals and I have greatly enjoyed reading their stories. If others enjoy them as much as I have, I hope that Volume Two will sometime see the light of day.

I believe that many books require a strong editor. With this one, however, I have deliberately used a light editorial hand. By doing so I hope that what is lost with some reduction in the uniformity of style is more than compensated for by permitting the personality and style of each contributor fuller expression.

I thank Laura Sley, Susan Fuentes, and Gerri Lennon for aid with typing parts of the manuscript and my family Joyce, Bryan, and Laura, for bearing with me during its production.

The people at Bucknell University Press deserve special thanks for supporting this project at a time when other publishers would not. Douglas Candland provided special support in many aspects of the production of the book; it might not have been completed without his efforts.

Studying
Animal
Behavior

Gerard P. Baerends (b. 30 March 1916). Photograph by Eddy de Jongh

1

Two Pillars of Wisdom

Gerard P. Baerends

How I Became a Biologist

I was born in 1916 in The Hague, the seat of the government of the Netherlands, and lived there for my first twenty-four years. This city was surrounded by areas that were interesting for a naturalist. To the northwest it had become fused with the fishing port and seaside resort of Scheveningen, interjecting into a broad zone of dunes with a lush vegetation and a wide sandy beach, stretched out along the west coast of Holland. Low polderland extended inland of the dunes and was protected by them against the sea. Although heavily used for pasture and for agricultural or horticultural purposes, this area contained many small lakes, pools, ditches, and marshes, with a richness of wildlife.

My parents liked to walk and bicycle in these areas, taking me—their only child—with them from an early age. My mother particularly was interested in natural history and taught me the names of many plants and birds in our surroundings and, with the use of illustrated books, of other parts of the world.

My father's influence was different. First, he set an example, stimulating me to develop manual skills and insight in solving all sorts of technical problems, that has proved to be of great use in my later experiments. Second, I learned from him much about dealing with people. One of his tasks as a civil servant was to find solutions for problems people individually experienced with civilian laws that did not adequately cover their particular cases. He liked to talk with me about his considerations when I had matured sufficiently to understand and respond. I have often felt that these discussions have been very effective in preparing me for the administrative aspects of the functions I had later to fulfill.

Among the books my mother introduced me to were the writings of two well-known Dutch schoolmasters from the beginning of this century, E. Heimans and J. P. Thijsse, who have contributed tremendously to opening the eyes of people in our country to the pleasures of making observations on the life of animals and plants, asking questions, and searching for the answers. Around the turn of the century, Heimans and Thijsse wrote a series of books introducing the opportunities for such observations in different landscapes. In 1896 they began to edit a

---

This title is paraphrased after T. E. Lawrence's "Seven Pillars of Wisdom," and refers to the recurrent influence of my split interest and involvement in both ethology and marine biology during my career.

monthly magazine called *De Levende Natuur*, in which the publication of such observations was encouraged.

Perhaps the greatest impact of all was made by a series of albums, published by a large biscuit firm that packed the illustrations with their products, and for which Heimans and Thijsse wrote the text. I remember that I had always been fascinated by a particular picture, in one of these albums, it was a representation of the digger wasp *Ammophila sabulosa,* paralyzing a big yellow caterpillar.

The enthusiasm raised by the activities of Heimans and Thijsse also led to the foundation of the Nederlandse Jeugdbond voor Natuurstudie (N.J.N.; Dutch Youths Association for the Study of Nature). This society, with branches in various high schools, still flourishes and is entirely run by its members, ranging in age from eleven to twenty-three, who organize lectures, excursions, camps, a regular magazine, and monthly exhibitions, mainly with the expertise available in their own group. I joined the N.J.N. when I entered high school in 1928 and this turned out to have been a decisive step for my future career.

In early autumn of that year, soon after the lessons had started, a schoolmate who was one year ahead of me took some of us newcomers out on the beach. Although I had often visited the beach and had superficially looked at shells and parts of other marine organisms, I was now for the first time actually introduced to the wealth of different kinds of invertebrates washed ashore with weeds, pieces of old ropes and netting, cork floats, and so on. I came home with a bucket full of nasty smelling treasures, the start of a biological collection that continued to grow during the next ten years.

I had fallen for biology and this was not going to change. During my high school years I devoted all my free time, and part of the time I was supposed to spend on other disciplines, to extending my knowledge of biology, preferably about life in the field. My biology teacher, Dr. A. Schierbeek, stimulated me by providing opportunities for self-study, for instance, through asking me to organize the extensive collection of shells and birds he had assembled in the school. The N.J.N. brought me in contact with more experienced naturalists. Among them were Lukas (Luuk) Tinbergen (the younger brother of Niko Tinbergen), and Joost ter Pelkwijk. Gifted with scientific as well as artistic capacities, they had an important stimulating influence on others. Niko was at that time studying biology in Leiden, but was still an active member of the N.J.N., contributing through lectures, guiding excursions, and writing in the magazine. Niko had been at the same high school as I, and we both now feel that Dr. Schierbeek must have influenced our interest in behavior. Although the history of biology was his speciality, he had also written a popular book on instinct and intelligence!

After finishing high school in spring 1934 I fervidly wished to study biology. However, at that time of economic recession, the prospects for biologists were abominable. This discipline was generally considered a luxury and of little or no practical use. Today I still feel grateful to my parents that they nevertheless yielded to my wish, in spite of the sacrifices that had to be made to make my study financially possible and of the warnings of well-meaning advisers.

Thus I began the study of biology at Leiden in September 1934 and so did my schoolmate Josina (Jos) van Roon, who was also an active N.J.N. member. Luuk and Joost had arrived in Leiden one year earlier. Together with some other students, a group was formed in the first two years with a strong common interest in field studies.

How My Biological Basis was Shaped

For a long time the study of biology in the Netherlands has consisted of essentially three parts. After three years of lectures and practical courses on various subjects of biology, in which botany and zoology were equally represented, and on fundamentals in other fields of science, a first examination, the *kandidaatsexamen*, had to be passed. Three more years followed in which the student had to work for periods varying from three to twelve months on research subjects in different fields of biology. This period had to be concluded by the *doctoraalexamen*, which was required for starting research for the Ph.D. degree. Among the different teachers contributing to my education in the first period, only those whose influence I repeatedly thought of in later years should be mentioned here. One of them was the professor of general botany, L. J. M. Baas Becking, a plant physiologist with a very wide interest. He did pioneer work on plant ecology, in particular on the influences of abiotic factors on the occurrence of aquatic plant life. He made us enthusiastic for the quantitative study of different factors and processes in the environment, and impressed on me the slogan that everything measurable should be measured, while special effort should be devoted to making quantifiable that which had eluded measurement so far.

The chair of general zoology at Leiden was occupied by C. J. van der Klaauw, a comparative anatomist and theoretical biologist. He made an effort to teach us to organize our thinking, and was for me an example of a determined administrator and a devoted and fair director of the laboratory. Under van der Klaauw's official supervision, H. P. Wolvekamp gave us a thorough introduction to animal physiology. At the end of my third year, Wolvekamp invited me to become his assistant, at first in giving a course in physiological chemistry (to my surprise and embarrassment, since I had never given more attention to the nonbiological subjects than strictly necessary), and later also in carrying out research. Wolvekamp tended to be a perfectionist in his experimental work and I have later often felt the benefit of having undergone his "drill." I also credit him with having drawn my attention to developmental biology, a subject that, in spite of its basic position in biology, was absent from Leiden's curriculum.

One year before I arrived at the University of Leiden, Niko Tinbergen had been appointed assistant in the Zoological Laboratory in the hope that he would further develop studies on animal behavior, similar to those he had started some years earlier with his thesis on the orientation of the bee-catching digger wasp *Philanthus triangulum* (Tinbergen, 1932). Niko also acted as a demonstrator in the anatomy courses, which I enjoyed tremendously. Although behavior was not programmed in the curriculum before the third year, my contacts with Niko became intensive as early as the second part of my first year of study, when he invited Jos and me to help him carry out a program on herring gull behavior that he had started in a gullery in the dunes between Leiden and The Hague. Later in that year he asked us to act as "slaves" in assisting George van Beusekom, who, aiming at a Ph.D., was continuing some aspects of the *Philanthus* orientation study (van Beusekom, 1948) in Leiden's field station at Hulshorst. We did this work on top of the regular study program, in the early mornings in the spring and on our summer vacation.

My first important experience with ethological methods, however, began just before I went to Leiden. In spring 1934, Jos and I had started to make regular observations on the pair formation behavior of the grey heron, profiting from the presence of a large heronry in the treetops of a park near our homes (Baerends & Baerends-Van Roon 1960; Baerends & Cingel, 1962). As a guide we used a paper published by Dr. Jan Verwey (1930) that can still be considered a classical example of how to describe behavior, the kind of questions that can be asked, and the ways of finding answers to them. Verwey, a pioneer in the study of animals in relation to their environment, held an assistantship in zoology at Leiden when Niko arrived there as a student, and had a significant influence on his development. In 1931 he became director of the Zoological Station of the Dutch Zoological Society at the port of Den Helder, a center for providing the Dutch universities with facilities for marine biological work. I first met Verwey when, in the winter of our first year, Jos and I had been recruited for carrying out at the Station, measurements on the great numbers of migrating starlings killed by colliding with lighthouses. By that time the emphasis of Verwey's work was in the field of ecology, his behavioral work being restricted to problems of the causation and orientation of animal migration. Like the botanist Baas Becking, Verwey put great effort into measuring factors in the environment, and during the earlier phases of my study we repeatedly volunteered to assist with such measurements on cruises in the coastal area. My fancy for marine life, initiated at the Scheveningen beach, became greatly enhanced through these contacts.

Before I had passed my *kandidaats-examen,* I was already able, unexpectedly, to start my first research subject for the *doctoraal-examen,* later extended for my Ph.D. When Jos and I arrived in the Hulshorst camp in the summer of 1936, to act again as "slaves" for Van Beusekom, we found that the latter had fallen ill and was consequently forced to give up his experiments for that season. Niko suggested that we should look for a suitable subject to start a project for ourselves. Stimulated by many articles in *De Levende Natur* (e.g., Bouwman) and the work of the French entomologists, such as Fabre, Ferton and others, I had for several years been interested in digger wasps. I have already mentioned the picture of *Ammophila sabulosa* that fascinated me at a very early age. I had found another *Ammophila* species at Hulshorst the year before, and I had discovered in the work of the Swedish entomologist Adler a suggestion that this species (called *Ammophila campestris* at that time) might provision (with paralyzed caterpillars) several of its larvae alternately and in different nests, each containing only one cell. Alternative provisioning would be an achievement never found in other representatives of the Sphegidae. We decided to test this suggestion and began continuous observations on the provisioning behavior of color-marked wasps during their entire daily activity period (0800–1800) and for as many consecutive days as the weather allowed us. Although Niko consented to this plan, he personally favored a study of the orientation of *Ammophila,* because these wasps, when returning to the previously dug nest with their prey, walk long distances—in contrast to *Philanthus,* which exclusively transports its prey in flight. We agreed to combine our study of the life history with one on orientation. If properly done, this work was to be honored with credits for our *doctoraal-examen.* As it turned out, it kept us busy for five consecutive summer seasons. In between we fulfilled the other requirements for our *doctoraal-examen,* after I had passed the *kandidaats-examen* in the spring of 1937. Later we added a third project, for which van der Klaauw accepted responsibility, to our *Ammophila* work. It was a microanatomical description of the morphogenesis of the larva, combined with a study of the influence of temperature and humidity on the speed of development. This work provided arguments for considering the complicated program our wasp was found to follow in provisioning its brood as an adaptation to the unstable Dutch climate. Publication of the results was postponed by the outbreak of the Second World War (Baerends & Baerends-Van Roon, 1950a).

The other research subjects I worked on for my *doctoraal examen* were all concerned with the sea. First, Baas Becking put me on a study of the possible role of silica as a limiting factor for the growth of diatoms. He further encouraged me to take oceanography as my second specialization, and arranged for me an extremely valuable period at the Marine Biological Laboratory at Plymouth. Here I came in contact with Atkins, Harvey, Cooper, and Russell, who together did pioneering work in the Channel on the influence of nutrients on plankton growth. Under the supervision of Wolvekamp I worked on the gas-binding properties of the blood of cephalopods (Wolvekamp et al., 1942), trying to relate species differences in this respect to habitat differences. I completed this experience in the biological and chemical aspects of sea research with chiefly theoretical work in physical oceanography. Baas Becking, who in the meantime had been charged with the reorganization of biological research in the Dutch East Indies had me in mind for a position at the Oceanographic Institute at Batavia (now Jakarta).

However, in 1939 when I was about to end that *Ammophila* season, the war broke out and I was called for military service. Although I had not formally completed all requirements for the *doctoraal-examen,* in October 1939 my professors generously decided that I should be given this degree. As long as our country was not directly involved in the war, I could find some time for biology; I completed the data analysis on the orientation of *Ammophila* and completed the first draft for that part of my thesis. In May 1940 the German forces launched their attack on the Netherlands, overrunning our defenses in the course of five days. Initially our entire army was taken prisoner, but the nonprofessional soldiers were soon released. For me this meant that I could add the season of 1940 to the *Ammophilia* work. After its completion in September I began the final analysis of the life-history data. I did this at our home in The Hague with little contact with the laboratory, since in November 1940 the German authorities had closed Leiden University in reaction to protests against the firing of Jewish professors. Then more advanced students tried to take over the lessons and help their younger collegues to prepare for examinations. My share was to continue the physiology lectures and it seemed safer to do this in The Hague, further away from the university.

The Control of a Behavioral Repertoire

It was not until the last writing of my thesis, in autumn 1940, that I detected, through handling the data, the rules of the program through which the wasps control the provisioning of their different nests. In the beginning of 1941 I had finished the draft for the life-history part and confronted Niko with it. After having read it, Niko told me that the work on the sticklebacks—of which I had been out of touch in the last couple of years—had brought him to a similar fundamental conclusion, viz., that behavior is causally controlled through the grouping of activities in different coordinating systems, which he called "instincts," connected with each other through hierarchical relationships. Since he was not quite satisfied with the way I had formulated the discussion of my conclusions, he proposed to rephrase this part together and so we did. The entire thesis was finished before the summer, and I was fortunate to obtain my degree on July 11, 1941 when the university had been reopened for examinations, but for a period that was not to last long (Baerends, 1941a,b).

The appearance in 1935 of Konrad Lorenz's paper on the social companion in the life of birds, and the personal contacts between Niko and Konrad shortly thereafter, had a tremendous influence on the behavior work in our group at Leiden. Thus far it had been especially inspired by von Frisch's work on sensory physiology and his techniques for making animals answer questions (Tinbergen & Kruijt, 1938). Another source of inspiration had been the more descriptive life-history work of the British ornithologists. Lorenz, however, proposed an attractive theoretical framework for general principles on how behavior is brought about that in addition seemed to give insight into its evolution. We felt a strong desire to test these hypotheses for cases other than birds and to extend, amend, or possibly correct them.

When Lorenz postulated his concept of "fixed action pattern," he considered this element to be an independent behavior unit, with an internal basis of its own, and released by an external situation specific for each activity. He rejected the idea of control of fixed action patterns by superimposed instincts, such as McDougall had proposed.

The stickleback data argued in favor of the idea that a hierarchical control mechanism would be underlying the behavior of this fish; Kortlandt (1940a,b) was using this principle to explain his observations on causation and ontogeny of the behavior of cormorants, and our data on the life history of *Ammophila campestris* provided experimental evidence proving the existence of internal factors controlling groups of activities in an insect. We found that our wasp provisioned each of its cells (distributed singly over an area of about 25 m^2) in three phases. In phase I the nest is dug and the first caterpillar is brought in and the egg then placed upon it. In phase II, provided the larva has hatched, a small quantity of caterpillars is added; in phase III a larger quantity is added. The phases in one cell are intercalated by phases in another nest.

Each phase begins with an inspection visit, without a caterpillar, in which the nest closure is removed and the wasp enters the cell. Experimental manipulation of the nest contents, before inspection as well as before provisioning visits, showed that the amount of provisioning during the ensuing phase is determined by the situation found during inspection visits only. During a provisioning visit, the further behavior of the wasp cannot be influenced by changing the nest contents, although it is then just as well in a position to perceive stimulation from the cell.

This finding meant that a particular external stimulus could, under specific circumstances, trigger in the animal a state that persists for an amount of time determined at the moment of its perception. Thus the inspection visit could be considered appetitive behavior subserving activation of a state controlling provisioning behavior with respect to a particular cell (Baerends, 1941a, b). Remarkably, the state persisted notwithstanding intercalated periods of feeding behavior and of resting during the nights or for one or more days of unfavorable weather. Furthermore, it was found that within the state underlying a phase, a number of substates could be distinguished, for instance a state in which the wasp was able to hunt, one in which it transported its prey, and a state for opening the cell and storing the caterpillar. In each substate the wasp had a specific set of action patterns at its disposal and showed increased sensitivity to a particular set of external stimuli.

The idea that a hierarchical structure of systems and subsystems underlies the behavior of an individual has remained at the basis of my later ethological research. It influenced the approach to my next ethological project, started a few months before I obtained my Ph.D., when I was appointed assistant in the Zoological Laboratory at Leiden to carry out ethological research with Tinbergen and physiological research with Wolvekamp. Following an earlier suggestion of Konrad Lorenz, Niko asked me to investigate the possibilities of cichlid fish as subjects for our further research.

Jos, who in the meantime had passed her *doctoraal examen* and had accepted a part-time job as van der Klaauw's secretary, joined me again in this new task. We observed and described two different types of reproductive behavior of representatives of this very large fish family; *Hemichromis bimaculatus* as an example of a substrate spawner and *Tilapia natalensis* (now known as *Sarotherodon mossambica*) as an example of an oral incubator. Through deductions from our data we tried to make tentative conclusions about how the behavior of these fish might be hierarchically organized. We tried to understand the form of different displays as based on conflicts between simultaneously activated incompatible states, and we began to wonder to what extent such interactions could be responsible for the extensive radiation found in the behavior patterns of different species of *Cichlidae*. The mechanism underlying the various color patterns most cichlids are able to assume were studied in the hope that it would be possible to use different patterns as indicators for particular motivational states. Another line of research was the experimental analysis of the relative effectiveness of different visual features of the parents for attracting their young (Baerends & Baerends-Van Roon, 1950b).

However, it became increasingly clear that work at the university should not be continued much longer. Attempts to Nazify it were made repeatedly. A possible way to counter such actions seemed to be to drain the university of its students and staff. Consequently many professors eased the way for students to move to other universities, encouraged and assisted their coworkers to find less exposed positions elsewhere, and prepared themselves for resignation. I was fortunate enough to be selected in the spring of 1942 for the post of fisheries biologist in the State Institute for Fisheries Research. This seemed the end of my ethological work; however, the new job fully met my marine biological interest, and allowed Jos and me to marry. In September 1942 several professors at Leiden were taken hostage; among them were van der Klaauw and Tinbergen.

Fishery Biologist

Before the war, fisheries research in the Netherlands had been treated in the proverbial stepmotherly fashion. The strong decline of the fish stocks since 1930, however, led to the wish to intensify and reorganize this research. Because of the war activities, fishing had become almost impossible and the stocks were expected to recover through this respite. My task was to make a study of the overfishing phenomenon and to design proposals to prevent its recurrence after the war ended, keeping the particular interests of our Dutch fishing industry in mind. My work was also expected to lead to a program for our fisheries research in the future.

First I tried to master the literature on population dynamics in general and on fish populations in particular. Then I studied the fishery statistics of the various countries operating in the North Sea. On the basis of my opinion of what overfishing actually implied, and of the historical picture of the efforts put into North Sea fishing and the yields obtained, I made an estimate of the size of the optimum catch of demersal fish the North Sea would be able to produce (Baerends, 1947). I argued that the only way to keep the North Sea populations at a level of maximum productivity would be to set, internationally, a limit on the total catch; this could probably best be achieved by keeping the fishing effort per country below an agreed level. In addition I considered auxilliary measures, such as sizes of gear and closed periods or closed zones, to ensure that the fish caught would have a good market value.

To be able to evaluate the statistics and to judge the feasibility and consequences of various protective measures, I had to become well acquainted with the various aspects of the fishing industry. Only after the war was over did it become possible for me to investigate the actual state of the stocks and to test several of the measures considered by making, in autumn and winter 1945–46, trips with some of the few ships of our fishing fleet that had survived the war.

My study served as a basis of the Dutch policy at an international meeting organized in 1946 in London for the purpose of reaching a convention that was to protect the North Sea from becoming overfished again. Unfortunately, because of too many conflicts between the immediate interests of the different countries and of insufficient appreciation for the reality of the threat, the resulting agreement was unsatisfactory. As a consequence, around 1955 signs of overfishing had again become apparent. Only recently, in the early seventies and in the context of the European Economic Community, was the principle of allocating catch quotas to the countries participating in the fisheries carried through.

I did not find convincing evidence that the stocks of pelagic fish (herring and mackerel) ever reached the overfishing state before 1940; however, this certainly happened after 1965. In the literature on the North Sea herring it was generally assumed that different races or populations existed, the individuals of which would mix during part of the year. I felt that in considerations of rational exploitation of the North Sea herring, more knowledge about these populations, especially about the degree to which they were in fact isolated units without an appreciable gene exchange, was indispensable.

Back to the University

In the midst of my efforts to get our sea fisheries research going, in the spring of 1946, I was offered the chair of zoology at the University of Groningen, as the successor to the physiologist E. H. Hazelhoff, who had died in 1945. Niko Tinbergen and Jan Verwey had been invited before me, but both had refused because they preferred to continue their work in Leiden and Den Helder, respectively. I found it difficult to decide. The marine environment fascinated me, I liked the work and life at sea, and enjoyed the contacts with fishermen and other people involved in the fishing industry. Moreover, I wanted to remain involved in the department of sea fisheries research in our country for which I had made the preparations in the past years.

The University of Groningen was at that time the smallest one in our country, although second in age only to Leiden. The number of students arriving per year to specialize in biology had for the past decade not surpassed three, and the laboratory consisted of an odd collection of old and queerly structured buildings. It had only a small scientific and technical staff. However, the trustees of the university told me explicitly that they wanted the zoological laboratory to develop into a research center, if possible with a character diverging from that of the zoological laboratories of our other universities. Education of students could be of secondary importance since their number was expected to remain relatively small. And because the expected extension of the neighboring telephone exchange was only feasible on our premises, the trustees were rather confident that a new laboratory would have to be built after a few years. It was clear that in Groningen I would be in a unique position to organize an institute entirely after my personal likings and insights.

I tried to find a solution that, if I chose to go to Groningen, would enable me to return to ethology but also to satisfactorily complete the fisheries work I had started and give continuous support to the development of our fishery institute.

This seemed to be possible if the research of the zoological laboratory would be directed toward problems concerning the activity of the animal as a whole, thus combining both ethology and ecology. The incorporation of ecology I also found very desirable with respect to finding jobs for my students; it was expected that the demand for ecologists would for the time being exceed that for ethologists.

On November 1, 1946 I began my task in Groningen. A considerable amount of time had to be devoted to adapting the curriculum of the students and the laboratory situation to the research I intended to promote. I was responsible for the entire training in zoology of students working for a degree in biology or taking zoology as a subsidiary subject while majoring in another study, such as human psychology, medicine, pharmacy, or geology. For the biology students, I designed a program suitable to preparing them for the kind of research I had in mind. Our program, however, also had to give them enough background so that they would be able to switch to different specializations taught in our other universities, if later in their study they might wish to do so. Lectures and courses had to be prepared, for help with which I could obtain to some degree from one senior lecturer, Dr. Sven Dijkgraaf, and a few assistants. The laboratory, initially used for comparative anatomy and later for research on respiratory physiology, had poor facilities for keeping animals under conditions appropriate for ethological work. While making the necessary administrative preparations, I rounded off my work at the fisheries institute and broke in my successor. In consultation with him, I also designed our future fisheries research program; since he preferred to restrict himself for the time being to the urgent problems of the demersal fish, in Groningen I could embark upon herring research.

Homing in Herring Races

The Dutch fishing fleet heavily relied upon summer- and autumn-spawning herring that seemed to be derived from spawning concentrations in different areas of the North Sea that were each visited by the herring shoals at somewhat different times of the season. For judging the chance of overfishing, it was of great importance to know whether these differences concerned different populations, each behaving as a separate, self-supporting unit within the common area of the North Sea. Different herring races used to be distinguished through quantification of meristic characters (numbers of vertebrae, of rays in particular fins, of scales in a particular body region, etc.). Since a considerable overlap occurred between the numbers found for these features in different populations, identification was only possible by means of comparing the averages for representative samples. Only Heincke had tried to identify the race of individual herrings through combination of different features, but without practical success.

The frequency distributions of the parameters in a sample taken from a catch often revealed that they consisted of mixtures of populations. Were such mixtures schooling together, and if so during the whole year, thus also when spawning on the sea bottom? Or had the mixing taken place during or after they were caught? I planned to tackle this question by sampling from spatially separated fish concentrations in drift nets separate from each side of these standing nets, as well as by trying to catch samples from a single school after having detected it with an echo sounder. It turned out that in this way the variation in the vertebrae frequency distributions of samples became largely reduced to a few distinct types.

We then tried to characterize these types by studying the ring patterns on otoliths and scales; this, as Hjortf had demonstrated for Norwegian herring, offered the possibility of reconstructing the time at which important events in the life of an individual fish such as spawning and periods of rapid growth, occurred. Two of my students who contributed most to the program, J. J. Zijlstra and K. H. Postuma, were assimilated in 1955 by the well-developing Fisheries Institute, and took the herring project with them to continue it further. I think that from all the evidence obtained it may be concluded that individual herrings return year after year for spawning on the grounds where they were born, just as has been found for salmon born upstream in rivers, and always at the same brief period of the year (Postuma, 1974; Postuma & Zijlstra, 1958). How this achievement is brought about by the behavioral organization of these fish and to what extent differences between the behavior of different races are due to genetic differences in the structure of the underlying behavioral organization or to differences in the external input processed, are interesting ethological questions following from the ecological work.

Behavioral Organization

In the planning of our ethological work in Groningen, the problem of how behavior is causally organized became the leading theme. On the one hand, I wanted to tackle the analysis and description of this organization for species as well as individuals, or rather more modestly, of parts of it. On the other hand, I hoped to become able to understand the divergences or radiation in behavior, found when comparing related species, in terms of differences between species in the selection pressures on modifiable elements in the structure of their behavioral organization. For this comparative work I chose the cichlid fish family and later I planned work with gallinaceous birds. I also considered it a challenge to try to understand the complicated display behavior of the ruff, a remarkable wader bird, on the basis of the notion of species-specific behavioral organization and the interaction between different motivational states (Hogan-Warburg, 1966; Van Rijn, 1973). Furthermore, as a contribution to Tinbergen's program of comparative studies of larid birds we made use of a suitable opportunity in our area to study the reproductive behavior of the black tern (Baerends et al., 1956).

For the analysis of behavioral organization, suitable methods had to be developed. One approach I wanted to try out was that of statistical analysis of temporal relationships between different elementary patterns, followed by experiments aimed at testing whether a coincident occurrence of different activities found could be attributed to the effect of common causal factors. A second way was by investigating for motivational state. The statistical/experimental approach was first used by Wiepkema in his analysis of the organization of the behavior by which males of the bitterling, a fish in which the females deposit eggs in mussels defend their territories around a mussel they have claimed and attract and guide females. Following a suggestion of Dr. Don Jensen, an American comparative psychologist working as a guest in our laboratory, Wiepkema was the first to apply factor analysis in causal ethological work (Wiepkema, 1961).

My coworkers Waterbolk and Brouwer successfully used different patterns of black markings to follow changes of motivation in courting male guppies (Baerends et al., 1955). In the cichlids we found the control of the color patterns more complicated because they not only subserve communication between conspecifics, but also camouflage. Consequently compromises occur in which it is more difficult to recognize the underlying basic patterns; today they are still a subject of our research.

From Herring to Herring Gulls

Among the theoretical concepts Lorenz has postulated, that of the "innate releasing mechanism" (IRM), a hypothetical afferent mechanism enabling an animal to react appropriately to objects vital to its survival without having had previous experience with it, has stimulated much of our experimental work. As part of the herring gull work started in the early thirties, Niko made me carry out dummy experiments to investigate the features a gull uses to recognize eggs. This work was left in a pioneering stage by the Leiden ethologists, but I turned to it again when, in 1950, I was looking for problems suitable for tackling in a field course on ethology. Until then the results of the tests had suggested that, in accordance with Lorenz's ideas, the gull was reacting to stimulation from only a few characters of an object, the "key stimuli," when accepting it for incubation. Moreover, by intensifying such key stimuli beyond their usual strength, the effectiveness of a dummy could be dramatically increased in spite of unnaturalistic appearance. In contrast I had incidentally found indications that the gulls were able to perceive differences between dummies and real eggs. In these cases they must have made use of many other features than key stimuli only. I suspected that the methods used thus far had been insufficiently sensitive for demonstrating susceptibility for details.

After improving the method (simultaneous presentation of two dummies on the nest rim and watching to see which of them was retrieved first) and with the collaboration of many students I continued the experiments during fifteen incubation seasons. The study expanded to include information processing and decision making in the bird (Baerends & Drent, 1982). In addition, it spawned another project concerning the gull's incubation behavior, again because we thought we had come across a discrepancy between a Lorenzian statement and our own observations.

According to Lorenz, an external stimulus situation would affect a fixed action pattern only by releasing and/or directing it. In contrast we found that the incubation act, after having been triggered by the eggs, was maintained only as long as the clutch produced feedback information corresponding with a full clutch of smooth rounded eggs at about 38°C. When we experimentally interfered with this feedback stimulation, the bird would soon show "interruptive behavior," which might consist of resettling, short bursts of nest building and/or feather preening, and even of leaving the nest. This feedback effect of the external situation demanded further investigation, directed in particular at understanding the way in which the interruptive behavior is caused and its composition determined. This last question was also important since, from observing the birds during the tests, we had obtained the impression that the kind of interruptive behavior shown during the rim-egg tests might give an indication about the bird's motivation when deciding between the two dummies. For an answer I considered more knowledge of the functional organization of the incubation behavior of the gulls necessary (Baerends & Drent, 1970).

The analysis of the way the behavior of a species is organized I consider an essential part in our efforts to understand how the nervous and endocrine systems manage to bring about behavior (Baerends, 1971; 1975b). It leads to description of the principles through which behavior is controlled and to the ways in which the performance of the various tasks is programmed. Physiology and endocrinology, supported by anatomy in the widest sense, attack the problem from the opposite side of the gap to be bridged. These disciplines have started by investigating the characteristics of the "gadgets" involved, and are making progress in understanding how different gadgets interact. Only with the aid of causal ethology, however can one find out how physiological activity leads to functional, integrated behavior. For this reason the Max Planck Society brought together in 1958 Lorenz, the ethologist, and Von Holst, the physiologist, at the Max Planck Institut für Verhaltensphysiologie at Seewiesen.

Structuring the Institute

Initially Dr. Dijkgraaf represented physiology in our laboratory. He had studied the lateral line organ of fish under von Frisch and had, working in Groningen during the war isolated from information from abroad, independently and with ethological methods, detected the capacity of bats to discriminate between objects with ultrasound. He left in 1948 when he was appointed professor of comparative physiology in Utrecht. Physiologists eligible for the lecturer post and with a sufficiently strong interest in ethology to fill the niche he had occupied, were extremely rare; instead I tried to find young investigators with experience in ethology but with a basic knowledge of physiology that would enable them to develop into behavioral physiologists.

One of them was Dr. Leendert de Ruiter, who joined our staff in 1953 and, in the course of years, built within the scope of our laboratory a subdepartment consisting of three groups: one group studying the processing of receptor messages, a second group dealing with the control of motor patterns, and a third group concentrating on central processes in the nervous system. Group 1 chose chemoreception of flies as the main subject for tackling the problems, group 2 the respiratory mechanism of fish, and group 3 initially chose nutritional homeostasis and later also extended its interest to the central mechanisms involved in aggression. All groups have developed successfully up to the present day.

Work on the endocrinology of behavior, using the reproductive behavior of the stickleback as a subject, was gradually developed under Dr. Bertha Baggerman, after she had taken her degree with me in 1957 on the timing of migration in this fish (Baggerman, 1957). In addition, endocrinology was incorporated in the neurophysiology groups. For support of all physiological work we attracted experts in fields of anatomy to our staff.

Meanwhile I chose to use the vacant lecturer post for reinforcing ecology. To my delight I found Luuk Tinbergen willing to join our group; in 1949 as lecturer, but later this post was changed into a second full professorship in zoology at Groningen University. Since 1946 Luuk had been carrying out sophisticated fieldwork on the food selection behavior of insectivorous birds and its population-ecological consequences. This project suited the general program of our laboratory excellently. Based at Groningen, Luuk went on with this work until his untimely death in 1955 (Tinbergen, 1960). Then this type of ecological work had to be discontinued. Instead, Dr. de Ruiter, who also had a great interest in population ecology and evolution, initiated an experimental project on the influence of various environmental factors on the population dynamics of the polymorphic snail *Cepaea nemoralis*. This study was undertaken by H. Wolda, who gradually took over Luuk Tinbergen's task.

The second zoology chair was given in 1956 to de Ruiter; a new lecturer's post for an ecologist was created in 1956; Wolda occupied it until he left us in 1971 for the Smithsonian Tropical Research Institute in Panama. His successor, Dr. R. H. Drent, who for several years worked with me on the herring gull project, is again a behavioral ecologist; he is nowadays directing several projects in our laboratory concerning optimal foraging theory.

The International Contacts among Ethologists

Before the war, most of our international contacts had been with German-speaking ethologists; also most of the literature on the experimental analysis of behavior was written in German. As a result of our experiences during the German occupation of our country, the relations with German and Austrian colleagues were only reestablished gradually and with care after the war. In contrast, our eagerness to make contact with British and American investigators had been "dammed up." Contacts became gradually possible through occasional visits of individuals, and at a number of international conferences.

At the occasion of my attending the fisheries conference in London in 1946 I went to Cambridge to meet Dr. Thorpe for the first time; two years later I had the opportunity to visit workers on behavior in Oxford (Lack, Chitty) and Cambridge (Lissmann, Kennedy, Gray) more extensively. However, the first official international event I attended, one that was to have important consequences for the development of ethology, was the international symposium held by the Society of Experimental biology in Cambridge (U.K.) in 1949 on "Physiological Mechanisms in Animal Behaviour" (S.E.B., 1950, Baerends, 1950). Thorpe was one of the organizers; in addition, he arranged a workshop with the purpose of standardizing the main terminology and defining the concepts used in behavior studies. Physiologists, ethologists, and learning psychologists of six different countries had been brought together and attempts were made to evoke discussion on some important theoretical controversies. One of these concerned Lorenz's concept of "reaktionspezifischer Erregungstoff," which he tended to consider as a different biochemical compound for each activity, decomposed when the activity was performed and built up in the intervals. This proposition was strongly opposed by most of the physiologists.

Another dealt with the problem of to what extent elementary behavior patterns, such as those for locomotion, were peripherally and/or centrally controlled. Gray and his coworkers strongly tended to see these movements as reflexes; in contrast, Von Holst was of the opinion that their initiation and patterning was primarily due to central activity. Due to illness, Von Holst was not able to attend. Instead he had sent a letter (in German) explaining his point of view. The letter arrived shortly before Von Holst had been expected to give his paper. Lissmann (whose native language is German) offered to translate it offhand. However, the aggressive and assertative tone of the letter made his task increasingly embarrassing the further he went on. The reaction of the meeting remained restricted to accepting the chairman's proposal to send Von Holst wishes for a speedy recovery.

To the Dutch ethologists in particular this letter came as a shock. We had great respect for Von Holst's work, admired its originality, and considered it of much importance for building a bridge between ethology and physiology. In our opinion it was insufficiently recognized in the Anglo-Saxon world, perhaps because most of it had been published in German and because unorthodox methods were often applied. However, the tone of the letter was of a type to which we had become allergic between 1940 and 1945. We felt thoroughly disappointed.

With considerable effort Niko Tinbergen managed to explain our reaction to Otto Koehler, who knew Von Holst well personally. It took them most of the train and boat voyage back to the Continent, but when they parted at the Hook of Holland Koehler had decided to go and see Von Holst in Wilhelmshaven to talk to him. Von Holst reacted by inviting a small international group of behavior workers to his place in 1950. This meeting was such a success that it was decided to repeat such informal meetings regularly. It became very clear that Von Holst's letter had given an erroneous impression of his personality. He proved to be charming and open to ideas and criticism from other people. During the time he was working in Wilhelmshaven (125 km from Groningen), we repeatedly discussed each other's work. I learned to like him very much and it became clear to me that he did not really feel so sure of himself as one would have expected from his great achievements.

Because this Wilhelmshaven meeting had been such a success, it was agreed that from then on the leaders of ethological centers, with their more advanced pupils, and some established ethologists working in different places independently, would come together every two years by invitation. The first of these meetings, initially called the first International Ethological Conference, was held in 1952 in Buldern, near Münster, where a German baron had given Lorenz's group facilities in his rural estate. In these conferences great value was placed on keeping the group small and informal in order to promote a free flow of thoughts. Great effort was made to overcome the language barriers. Summarized translations of German talks were given offhand in English, and vice versa, section by section. Lorenz, Kramer, and Tinbergen were particularly good at acting as interpreters; unfortunately we were not able to serve the small number of French-speaking participants equally well. This practice was certainly time-consuming but had the advantage that, when the speaker had finally finished his paper, his message was properly understood by everyone. Since most of the talks considered conceptual issues—new ones, amendments, or criticisms—this was of great importance. This way of becoming familiar with what an author really meant was far superior to just writing and reading.

In this stage, a year later, European ethology was forcefully struck by a severe and not very conciliatory criticism, originating from Schneirla and his coworkers at the American Museum of Natural History in New York and worded by Danny Lehrman (1953). The attack was directed primarily at several of the Lorenzian concepts that, because of the heuristic value they had proved to possess, had become so dear to us.

In the first place, Lehrman objected to the emphasis Lorenz had laid on the "innateness" of species-specific behavior, and expressed the view that "any instinct theory which regards 'instinct' as immanent, preformed, inherited, or based on specific neural structures is bound to divert the investigation of behavior development from fundamental analysis and the study of developmental problems."

Further Lehrman strongly objected to several concepts, to the casual way in which concepts were transferred from one level of organization to another, and to the fact that behavioral phenomena had been related to physiological ones. The aggressive tone of Lehrman's criticism undoubtedly promoted much attention; on the other hand, it also retarded the process of acceptance and incorporation, particular with German ethologists. Fortunately, an earlier draft that had been even harsher had been considerably softened through the influence of some American collegues, among whom was Dr. Frank Beach, who sympathized with both sides.

I think that the criticism had been strongly inspired by affective feelings against any kind of preformation in behavior, to a large extent due to indignation and alarm about the abusive and criminal applications of such ideas in some political ideologies. However, Lorenz's emphasis on "innateness" was also strongly affectively motivated; viz. by a disgust, resulting from his extensive knowledge of the consistent and adaptive differences in behavior between individuals of different species, also, in "naive" animals, for extreme Watsonian behaviorism and the generalization of conclusions based on learning experiments with a few kinds of laboratory animals, such as rats, to other species.

Jan van Iersel and I were the first European ethologists who met Lehrman personally after his paper was published. This happened in 1954 in Montreal at a symposium on the analysis of instinctive behavior, organized by Frank Beach during the program of the 14th Psychological Congress and in which Lehrman,

Verplanck, and Eckhard Hess (who was strongly supporting the Lorenzian ideas in North America) also participated. Lehrman was waiting for us when we arrived at the registration desk. We immediately started discussing the controversies and went on with it during the rest of the day.

It did not take us long to find out that there were no differences between us with respect to the problems we were interested in. The passing of some North American birds, such as nighthawks that Jan and I had not seen before, helped to reveal that Danny also was an enthusiastic field ornithologist and that he fully shared our interest in the species-specificity of behavior. It further became clear to the three of us that a considerable part of Danny's aversion to various ethological concepts was due to misunderstandings. Frequently he had read more into the descriptions of concepts than had actually been intended, since he was looking at them from a background different than the one from which they had arisen. It was a surprise to Danny that we at least considered the concepts only as temporary constructs that should be tested, amended, changed, or replaced as soon as they were no longer sufficiently supported by the available evidence. At the end of the day I do not think that any major essential difference remained between us, and we certainly had become very good friends.

The Montreal congress gave me an excellent introduction to American comparative psychology. So did a workshop with ethologists, psychologists, and physiologists in Boston a month later, in which, among others, Aronson, Beach, Bullock, Estes, Griffin, Harlow, Hinde, Van Iersel, Pribram and Verplanck participated. Even more valuable for me was a workshop Beach organized in 1957 in the Center for Advanced Study of Behavioral Sciences at Stanford, California. For six weeks, Beach, Harlow, Hebb, Hess, Van Iersel, Lehrman, Rosenblatt, Tinbergen, Vowles, and myself discussed our work and exchanged opinions and ideas in an open, informal atmosphere.

As a result of these contacts, North American workers on animal behavior began to come to the Ethological Conferences. This happened first in 1955 when the conference was held in Groningen and Frank Beach, Demorest Davenport, Eckhard Hess, and Danny Lehrman participated. Their numbers increased in following conferences. Whereas initially our contacts had mostly been with comparative psychologists, now increasing numbers of North American zoologists became interested in European ethology. And although controversies with respect to the nature/nurture problem remained, they gradually lost their relationship with particular nationalities.

The work of the French entomologists had been an inspiration for my *Ammophila* study. My first opportunity to meet with contemporary French students of insect behavior was when their nestor, Prof. P. P. Grassé, invited me to attend an international conference on the "Structure and Physiology of Animal Societies" in Paris in 1950 (C.N.R.S., 1952). It was the beginning of lasting good contacts with several French ethologists (e.g., Chauvin, Le Masne, Richard).

Moreover, at this meeting, I met for the first time Allee and Schneirla.

In 1954 Grassé organized an even more important symposium ("Instinct and Intelligence in Animal and Men") in which the work on behavior was well represented by some twenty scientists from six different countries (Autuori et al.,

1956). It was at this conference, just a few weeks after our discussion in Montreal, that Lorenz and Lehrman first met. My visit to the U.S. unfortunately prevented me from attending this second Paris conference. My best contacts with French ethologists have been with Gaston Richard and his active group at the University of Rennes. This group was also the first in France to receive, in 1969, an International Ethological Conference.

In the following decades, the international contacts became even more intense. However, the character of the meetings gradually changed. Initially particular "hot issues" concerning the theoretical basis of ethology, such as whether the state of "the" IRM is contant or variable, whether fixed action patterns are peripherally *or* centrally controlled, or whether a particular activity was either innate *or* learned, were dominating the discussions. Later the presentation of single studies gained in relative importance on the programs, and in evaluating their results, a wider scale of alternative interpretations was considered.

With the growth of the number of ethologists and its extension over more countries, the Ethology Conferences had also to become larger. Reluctantly we realized that they needed a more formal organization. This was established in 1967, and until 1971, I was entrusted with the task of ensuring its continuation and acting as its first secretary-general. This organization has recently come to the conclusion that the number of ethologists has become so great that the conferences can no longer be kept closed without being unfair to many persons eligible for attendance. Future conferences will therefore be open.

In a discipline such as ethology, the wording of concepts and ideas plays an important role. Misunderstandings easily occur, in particular between persons with different native languages and different scientific and/or cultural backgrounds. Since outside the borders of our small country, the Dutch language is of very little use, Dutch scientists are forced to express themselves in languages of other nations. However, against this handicap stands the advantage that through learning other languages one also gets insight into the appertaining literature and cultural backgrounds. For practical reasons it can be applauded that in so many meetings and journals one leading language nowadays has become dominant. With it, however, goes the danger that delicate differences in thinking and cultural background are no longer expressed and/or understood. The promotion of intensive personal contacts between workers in different countries is the best way to fight such dangers.

For this and various other reasons I am grateful that a considerable number of foreign guests from various countries could spend periods, varying from a few months to a couple of years, to carry out research in our laboratory. I hope that the contributions they made to our knowledge and skill have been duly reciprocated by the experience they obtained from staying with us.

The promotion of ethology in different countries as well as the importance to this new disciplines of having a periodical of its own, run by editors with knowledge of the discipline, were arguments for Niko Tinbergen to found the international journal *Behaviour* in 1948. Since the journal is published in Holland, Niko asked me to take over his job as executive editor when he left for Oxford in 1949. The editorial work and my contacts with our coeditors and authors have been of great value to me. For years my most frequent contacts were with our coeditor Otto Koehler, who very dedicatedly checked German texts written by non-German authors and in this way greatly contributed to my insight both into the German language and into problems of ethology.

Growth of the Institute

After 1946 the number of biology students rapidly increased. It soon became clear that their training required more teaching personnel and more laboratory facilities than had been originally envisaged. However, the trustees of the university kept their promise; the burden of carrying the growing teaching load was not lightened at the cost of our research. On the contrary, our staff was increased and the research benefitted too; in addition, more students could be influenced by it. As a consequence, different working groups formed around our senior staff members, each concentrating on their own area of study, all of which, however, concerned some aspects of behavior. In spite of the obvious advantages in investigating different aspects coherently in the same species, I have never tried to exert pressure to achieve this. I consider it a better policy to respect, when possible, rational as well as irrational preferences for particular species and problems. Such preferences have often turned out to facilitate the work. Moreover, a diversity of species kept in and around the laboratory stimulates comparison, which is of great heuristic value.

One of these groups I wish to dwell on a bit further because of the influence it had on my thoughts. It grew out of my initial plan to make a comparative study of the display behavior of gallinaceous birds. With the grant obtained for this project, I hired Jaap Kruijt, who as a student from the University of Utrecht had earlier assisted me in the egg recognition work. Our plan was to compare homologous signal activities in different species of pheasants and fowl to obtain insight into the causation and possible functions of interspecies differences. A purpose of this study was to test and possibly expand the "conflict hypothesis," developed by Tinbergen and his coworkers (Baerends, 1975a) to explain the derivation of the form of communication activities.

In the course of the first year, however, Jaap felt increasingly uneasy about the way in which statements about homology of behavior patterns had so far been substantiated (Baerends, 1958). He leaned toward the view that for a judgment on the homology of behavior patterns, knowledge about their causation is required. Since a causal study could not be undertaken in several species simultaneously, he chose for one representative of the family, the red jungle fowl, wild ancestor of our domestic fowl.

With his attempts to trace the causation of different communicative patterns, however, the conviction grew in Jaap that for an understanding of their causation, knowledge of the ontogeny of the patterns was essential. We agreed that the emphasis of the study should therefore be switched to that aspect. The result was a description of the consecutive stages in the development of the communicative behavior of cockerels, combined with experimental attempts to interfere with normal development through depriving individuals of specific experiences for varying periods. On the one hand, this approach was inspired by Lorenz's findings and statements about imprinting, but on the other hand, Kruijt's attitude was strongly influenced by Lehrman's criticism of Lorenz's dichotomous treatment of the innate/learned issue. In fact, Kruijt put the emphasis on a different aspect than Lorenz did. Whereas the latter was primarily concerned with showing *that* genetic differences are at the basis of species-specific behavior, Kruijt wanted to study *how* the ontogeny of such behavior is actually controlled by genes. Kruijt has emphasized that for answers to this problem one should not merely deprive an animal of the adequate object normally releasing the response concerned (Kaspar-Hauser experiment), but also test the influence of various other experiences to which the animal is exposed in the developmental phase of its life.

Kruijt found that cocks that had missed the opportunity to interact agonistically in early life were later unable to use their display patterns appropriately in social and sexual contacts with suitable partners. He suggests that early experience in the use of behavior systems, singly or in simple combinations, is essential for an animal to control them harmoniously in more complicated situations later in life (Kruijt, 1964).

My close contact with Jaap Kruijt has been of great importance in making me realize that the structure of the behavioral organization of a species is the product of interaction between genes and experience, controlled by developmental programs, often superimposed on each other. As a consequence, the ontogenetical aspect also entered our program of work on the behavioral organization, although the analysis of the ontogenetical processes remained primarily Jaap's domain in our laboratory. One of my Ph.D. students, N. Williams, worked on the ontogeny of behavior patterns in a cichlid fish. My wife—after having successfully raised four daughters—embarked on a study of the ontogeny of predatory and social behavior in domestic kittens (Baerends–Van Roon & Baerends, 1979), and in our present research on communication behavior in gulls and terns, in which Kruijt is actively participating, ontogeny is an important component.

Administrative Tasks

In 1947 it became clear that we were indeed going to have a new laboratory; the planning for it had to be started. Because of the character of the research we expected to do, much attention was paid to designing facilities for keeping animals under conditions that reasonably approached natural ones. The building was to include a large aquarium house and pens of various sizes. It was to be situated on the grounds of a large botanical garden that since the twenties had been developed in Haren, a suburb of Groningen. This garden was designed for the study and demonstration of different types of vegetation; it forms a very atttractive environment for our laboratory, and is suited for small-scale fieldwork, with the laboratory facilities near at hand. The grounds give us plenty of opportunity for keeping animals in relatively large enclosures. Building was started in 1950 and finished in 1954. In 1967 the facilities for keeping and observing animals were greatly expanded and in the early seventies a new building was erected for accommodating our subdepartments for behavioral physiology and functional morphology, which had in the meantime outgrown their original housing. In the same period new laboratories had been built for the other biological departments (botany, microbiology, genetics, and also experimental psychology) so that a Biological Center was created at Haren. In addition to the buildings at Haren, our laboratory obtained field facilities on the nearby island of Schiermonnikoog and in a moorland reservation area 30 km southwest of the laboratory.

My share in the preparation of these developments took a considerable amount of my attention and time, of course at the cost of my own research. In addition I became increasingly involved in the organization of biological work in general in our country. For my own experience the most important task in this context was to follow, as a deputy of the Dutch Zoological Society, the planning for turning the "Zoological Station" in Den Helder, in close cooperation with the director, Dr. Jan Verwey, into the Netherlands Institute for Sea Research. I also worked with Dr. Verwey to obtain a limnological institute in our country. The reason why such activities tended to accumulate is that because of experience gained and connections made with authorities in one function, one becomes suitable and desirable for other similar functions; this is a logic that is difficult to dispute, in particular when important projects with which one sympathizes, are at stake. Since in this period our own scientific staff was still increasing considerably, I felt that our laboratory was duly compensated. Nevertheless, in 1964, after having also finished four years as dean of the Faculty of Sciences at our university, I was glad to be able to escape from administrative duties, thanks to an invitation to spend a year at the Center for the Advanced Study in the Behavioral Sciences at Stanford. This year allowed me to make a start with working out the data collected on incubation behavior in the herring gull. Besides, the regular contacts with behavior workers from many other disciplines has been of very great value when later I became involved in human ethology.

Tropical Experience

When I returned from Stanford, the Institute for Sea Research kept making appeals to me. In 1956–66 I even combined my function in Groningen with a part-time directorship of that institute, to bridge a period needed for finding a satisfactory solution regarding its directorate after Verwey had retired. Then, from 1968 to 1974, my interest in marine biology was utilized when I was asked to direct a project to study the ecological consequences of the regular discharge of waste water from sugar, starch, and cellulose factories near Groningen into the Waddenzee.

In 1967, however, an entirely new aspect was added to my engagements. It started when Dr. John Owen, the director of Tanzanian National Parks, acting on a suggestion by Niko Tinbergen, asked me to visit the Serengeti Research Institute (S.R.I.) to give advice in matters of behavioral ecology regarding the studies carried out in support of the management of the parks. To be directly confronted with numerous species of larger mammals and the tremendous possibilities for studying them in the savannah areas was fascinating. I easily became committed to it, particularly when I was also asked by our Dutch science foundation to act to help promote ecological field research of promising young scientists in East Africa. Between 1967 and 1975 I visited this area for several weeks each year. I became a member of the Scientific Council of S.R.I. and functioned as its chairman from 1972 until the board of the institute was fully Africanized in 1975. The institute had an interdisciplinary character; landscape morphologists, soil scientists, botanists, zoologists, and anthropologists were working together, just as in our research of the sea. One of my coworkers, Dirk Kreulen, studied the migration of wildebeeste in the Serengeti area.

Experience with a quite different tropical area came in 1968 when I visited, with a small group of collegues, our former colonies in the Caribbean area, the Netherlands Antilles and Surinam. The promotion of research in tropical areas was again one of the purposes of this visit. I especially committed myself to developing plans for the survival of the Caribbean Marine Biological Institute on Curaçao, for which I conceived a project for coral reef research. My experiences in Africa and Central America have made me an advocate of the combination of nature conservation with tourism—the latter to make the former pay for the host countries involved—but guided and controlled on the basis of research in ecology and behavior, not only of animals but also of humans.

Symbiosis of Arts and Science

A new important, exciting and rewarding experience began when, in 1969, I was visited by the Dutch cinematographer Bert Haanstra, internationally renowned for his documentary films, and his friend and scriptwriter Anton Koolhaas. Both had become captivated by ethology through the popular writings of Lorenz, Ardrey, Eibl-Eibesfeldt, Morris, and Wickler. They very much wanted to make a ninety-minute 35 mm movie that would, besides entertaining the general public, give it an idea of the importance of ethological research on animals for the understanding of human behavior. They asked me to act as scientific adviser for this film.

On the one hand, their request very much appealed to me. In my opinion there was an urgent need for a responsible popularization of ethology and I was fully aware of the great power of cinematography as a medium to communicate. I personally enjoyed filming as a tool in our research. Moreover I knew and admired Haanstra's work and felt that he was looking at animals and man in the same way as we did.

On the other hand, the many hostile reactions to statements in the popular books, and, generally, the aversion of a great many people to serious comparison of humans with animals, had made me aware of the great difficulties of bringing our message home. Would it be possible for an artist and a scientist to cooperate in a product that would satisfy the demands of both? Haanstra gave me the right of veto with respect to the scientific contents and he has always abided by it.

My first task was to draft, within the framework designed by Haanstra and Koolhaas, a program of the issues to be treated. For each issue, examples had to be found as well as suitable locations where scenes could be obtained within a reasonable time period. Here my experience in traveling in the tropics and elsewhere, and my connections with collegues in many places, proved to be very useful.

The film team worked as devotedly as any ethologist I have ever seen at work. In the assemblage phase, where Haanstra of course had the initiative and I was-supposed to criticize, I learned a great deal about the art of presentation to a public that I have found of great use for my teaching work.

The demonstration of links between animal and human behavior proved to be the most difficult part of the film. Actual research in this field was still in an early phase and it was often closely interwoven with the more conventional approaches of psychology. Making it obvious to a wide range of people in which way an ethological approach could contribute asked for a considerable amount of clearing up in my own mind!

The movie, *Ape and Super Ape,* came out in 1973, in the same year the Nobel prize for physiology and medicine was awarded to Von Frisch, Lorenz, and Tinbergen. It has been shown in many countries of the world with great success; it even was nominated for an Oscar. Unfortunately the U.S. distributor who had bought the rights for the whole Western hemisphere, felt that he had to cut out everything dealing with violence, sex, and evolution in order to make the movie suitable for the "American-family," and he replaced these parts with scenes from other nature or pseudonature movies. To the present day, the unmutilated version has been shown only to limited audiences in the American continents.

In our country the film greatly increased interest in ethology and its possible application for understanding and solving problems of human behavior. However, the film also stimulated the resistance of people who rejected a biological approach. I considered it part of my commitment to the film to enter into discussion with both groups.

Farewell to Administrative Duties

Around 1970 the universities in this country became involved in the growing tendency to democratize the administration of any kind of institution. A new system, enforced by law, implied that decisions on all matters—on domestic affairs, educational regulations, appointments, and the planning and execution of research—be laid in the hands of elected councils operating at different levels of integration and all consisting of scientists, technicians, and students. Every person had an equal vote; politics entered university business. Although I had always tried to follow the principle that decisions should not be taken without carefully consulting those involved, I considered this new system unworkable. Running an institute in accordance with these rules would, in my opinion, no longer be compatible with research. Since I was willing to take on the burden of administrative duties only if I expected them to serve the research aims I had in mind—through studies carried out by myself or other people—I decided to look for other positions.

A very attractive opportunity turned up when, in 1972, as a consequence of Konrad Lorenz's retirement, the Max Planck Society in West Germany, invited me to move to the Max Planck Institute für Verhaltensphysiologie at Seewiesen. However, in an advanced phase of the negotiations, when I was almost prepared to leave, our Ministry of Education and Sciences came with a counter proposal, the result of an attempt, by our Royal Academy, my colleagues in Groningen, and the governing board of our university, to keep me in the Netherlands. The offer amounted to relieving me of all administrative duties by creating a new personal chair chiefly for research, a research staff of four persons, my own budget, and space in the zoological laboratory. It was a difficult decision. I had certainly been looking forward to working in Seewiesen. However, this would still have implied a considerable amount of administrative work, and building a laboratory again. I finally felt that staying in our laboratory at Haren, with all the pleasant and useful contacts it was providing, would be more favorable for proceeding with the scientific aims I was pursuing, than moving to another place, no matter how attractive and honorable this might be. At the end of 1973 my new position was realized.

The task I set for myself in this final period of my scientific career was twofold. First, I wanted to promote critical constructive research based on the theoretical framework Lorenz had postulated, but directed at critically testing it and correcting or replacing it where necessary. It was clear that many of the younger ethologists did not recognize its value or understand the purpose for which it had been conceived. Moreover, they were discouraged by the tediousness of the research required and since sociobiology had become fashionable, they were also more interested in function than in causation. Second, I wanted to give support to the growing number of attempts in our country to apply ethology in other disciplines.

In my privileged position I began giving the highest priority to finishing my analysis of the large amount of data collected in our experiments on the responses of herring gulls to egg dummies of different appearance, undertaken about twenty years earlier with the purpose of initially investigating the Lorenzian concept of innate releasing mechanism. Work on these data had been greatly delayed because of competition with administrative duties. In particular I was interested in the relationship between this mechanism, which was thought to be responsible for stimulus selection and sensitivity for supernormal stimuli, and the mechanisms allowing the same animal to respond to constellations of details of a situation. The behavioral parameters noted down during the tests indicated that the relative importance of each of the two mechanisms for determining a response depends on the motivational state of the animal, that is, the balance between its tendencies to incubate and to leave the nest. Birds showing signs of being scared rely much more strongly on the mechanism sensitive for details. A bird's responsiveness for these details was likely to be acquired through experience with its own eggs. In contrast, the high responsiveness with incubation behavior to a few key stimuli, found similarly in various species of gulls resulted from a maturation process; it was considered to cause the strong reaction to supernormality. I came to the conclusion that the IRM enables the bird to extend its knowledge about a vitally important component of its environment through learning from experience with it. The information encoded in the IRM and the acquired information were found to work in combination. Reactions to supernormality occur when the IRM dominates the engram, for instance in rapid responses or confusing experimental situations (Baerends & Drent 1982).

The impact on information processing of the motivational state of the bird was recognized as a consequence of the hierarchical structure of the behavioral organization. Further investigation of the validity of this concept forms an important part of the program of my present working group. It is again carried out with gulls and cichlid fish as subjects and deals with the study of social behavior in these animals. It merges with a third issue, testing the validity of the conflict hypothesis for explaining not only the evolution but also the present causation of communication behavior (Baerends, 1975a).

This field of research has advanced only little beyond the stage it had reached in the early fifties. Young investigators at that time seem to have become discouraged in tackling these problems because of the methodological difficulties of studying them with a satisfactory degree of sophistication. I am of the opinion that the strong development of research techniques in the last decade will make it possible to overcome these obstacles. Thanks to audiovisual equipment, interactions between animals can now be recorded on tape and film, and repeatedly observed and analyzed. In this way no aspects of a situation need to be missed, in strong contrast with the earlier technical restrictions of pencil and notebook. The data resulting from such work are certainly huge, but can be mastered nowadays with the help of computers.

Simultaneous registration of visual and acoustic activities in an exact and objective way is also likely to be of great importance in these investigations. The more recent approach, developed in sociobiology and inspired by game theory, of testing hypotheses about the causation of communicatory activities by the likelihood that they could really be functional and thus could have evolved in the postulated way, deserves a proper place in our approach.

The appeals that I receive for ethological contributions to the study of human behavior chiefly come from people engaged in investigating and combating human activities that are considered socially undesirable and for which no obvious rationality seems to exist. Phenomena such as individual educational problems, maltreatment of children, vandalism, persistent and violent criminal offenses, group discrimination, severe nostalgia, depression, neurotic behavior, etc. come into this category. They all seem to concern problems of people who are no longer able to respond in a functional way to situations in which they find themselves. We can experimentally evoke such behavior in animals by placing them in an environment that does not correspond to their species-specific niche and deprive them of feedback stimuli that would result from their activities under natural conditions. In such cases, the behavioral organization of the animal is not suited to cope with the situation. Although many people seem to find it an unattractive idea that humans also would possess a species-specific behavioral organization (of which each individual has a variant) primarily based on genes, it is difficult to see how a proper functioning of the morphology and physiology—for which the species-specific genetic basis is generally accepted—would be possible without a corresponding genome-based behavioral control. Consequently the most important and characteristic contribution ethology can make in the human context is to emphasize the importance of acquiring knowledge about the structure of our own behavioral organization and to contribute to the improvement and development of methods for the study of it (Baerends, 1975b).

As important as the structure itself is the way it develops in the individual. In behavior, an important way through which genes express themselves is by facilitating (programming) learning processes. In the ontogeny of the behavioral organization of our own species, this method has been extensively applied and in combination with cultural transmission of learned behavior, especially through verbal and written instructions, it accounts for the success of our species. It enabled us to accommodate to natural environments very different from the one in which we originally evolved, and also to create artificial environments. Thus this capacity must also enable us to produce environments for which our behavioral organization is not suited when our knowledge about the latter lags behind. Biological thinking should help to prevent a situation in which the unavoidable corrections will be executed by evolutionary means.

In my own work I always felt attracted to the analysis of the complicated behavioral interactions observed, in order to derive hypotheses about the underlying causal mechanisms from which they result. I also recommend this approach for the study of the organization of human behavior. As a rule, hypotheses derived should be experimentally tested. For ethical reasons this will usually be unacceptable in case of humans; experimental testing will—via a stage of comparative research—have to come from experiments with animals (but also not without obeying ethical rules).

References

Autuori, M. et al. 1956. L'instinct dans le comportement des animaux et de l'homme. Coll. Int. Fondation Singer-Polignac. Paris: Masson, 769 pp.

Baerends, G. P. 1941a. On the life-history of *Ammophila campestris* Jur. *Proc. Ned. Akad. Wet.* 44:483–88.

———. 1941b. Fortpflanzungsverhalten und Orientierung der Grabwespe *Ammophila campestris* Jur. *Tijdschr. v. Entomol.* 84:68–275.

———. 1947. De rationele exploitatie van den zeevischstand, in het bijzonder van den vischstand van de Noordzee. *Min. Landb. Viss. & Voedselv. Meded.* 36:1–99. Engl. transl. The rational exploitation of the sea fisheries with particular reference to the fish stock of the North Sea. U. S. A. Dept. of Int. Fish Wildl. Serv., *Spec. Sci. Rep. Fish.* 13 (1950): 1–102.

———. 1950. Specialization in organs and movements with a releasing function. In *Symp. S. E. B.* 4:337–60.

———. 1958. Comparative methods and the concept of homology in the study of behaviour. *Arch. neerl. de Zool.* 13 (suppl. 1):401–17.

————. 1971. The ethological analysis of fish behavior. In *Fish Physiology,* vol. 7 ed. W. S. Hoar and D. J. Randall. London, New York: Academic Press.

————. 1975a. An evaluation of the conflict hypothesis as an explanatory principle for the evolution of displays. In *Function And Evolution of Behaviour,* ed. G. P. Baerends, C. G. Beer, and A. Manning. Oxford: Clarendon Press.

————. 1975b. The functional organization of behaviour. *Anim. Behav.* 24:726–38.

Baerends, G. P., and Baerends-Van Roon, J. M. 1950a. Embryological and ecological investigations on the development of the egg of *Ammophila campestris* Jur. *Tijdschr. Entomol.* 92:53–112.

————. 1950b. An introduction to the study of ethology of cichlid fishes. *Behaviour, Suppl.* 1:1–242.

————. 1960. Ueber die Schnappbewegung des Fischreihers, *Ardea cinerea* L. *Ardea* 48:137–50.

Baerends, G. P., Baggerman, B., Heikens, H. S., and Mook, J. H. 1956. Observations on the behavior of the black tern, *Chlidonias n. niger* (L.) in the breeding area. *Ardea* 44:1–71.

Baerends, G. P. Brouwer, R. and Waterbolk, H. T. 1955. Ethological studies on *Lebistes reticulatus* (Peters). I. An analysis of the male courtship pattern. *Behaviour* 8:249–334.

Baerends, G. P. and Cingel, N. van der. 1962. On the phylogenetic origin of the snap display in the common heron (*Ardea cinerea* L.) *Symp. Zool. Soc. Lond.* 8:7–24.

Baerends, G. P., and Drent, R. H., eds. 1970. The herring gull and its egg. Part I. *Behaviour (Suppl. 17):* 1–312.

Baerends, G. P. and Drent, R. H. (eds.) 1982. The herring gull and its egg. Part II. *Behaviour* 82:1–415.

Baerends-Van Roon, J. M., and Baerends, G. P. 1979. The morphogenesis of the behaviour of the domestic cat. *Verh. Kon. Ned. Akad. Wet., Afd. Natuurk.,* 2:72. Amsterdam, Oxford, New York: North-Holland Publ. Co. 116 pp.

Baggerman, B. 1957. An experimental study of the timing of breeding and migration in the three-spined stickleback. *Arch. neerl. de Zool.* 12:1–213.

Beusekom, G. van. 1948. Some experiments on the optical orientation in *Philanthus triangulum* Fabr. *Behaviour* 1:195–225.

Colloques Internationaux du Centre National de La Recherce Scientifique. 1952. Structure et Physiologie des societes animales. Coll. Int. C.N.R. S., Paris, 1950. Paris: Masson. 353 pp.

Hogan-Warburg, A. J. 1966. Social behaviour of the ruff *Philomachus pugnax* (L.) *Ardea* 54:109–229.

Kortlandt, A. 1940a. Eine Uebersicht der angeborenen Verhaltensweisen des Mittel-Europaischen Kormorans (*Phalacrocorax carbo sinensis* Shaw & Nodd), ihre Funktion, ontogenetische Entwicklung und phylogenetische Herkunft. *Arch. neerl. de Zool.* 4:401–42.

Kortlandt, A. 1940b. Wechselwirkungen zwischen Instinkten. *Arch. neerl. de Zool.* 4:442–50.

Kruijt, J. P. 1964. Ontogeny of social behaviour in Burmese red junglefowl. *Behaviour* (Suppl. 12): 1–201.

Lehrman, D. S. 1953. A critique of Konrad Lorenz's theory of instinctive behavior. *Q. Rev. Biol.* 28:337–63.

Lorenz, K. 1935. Der Kumpan in der Umwelt des Vogels. *J. Ornithol.* 83:137–215, 289–413.

Postuma, K. H. 1974. The nucleus of the herring otolith as a racial character. *J. Cons. Int. Explor. Mer.* 35:121–29.

Postuma, K. H. and Zijlstra, J. J. 1958. On the distinction between herring races in the autumn and winter spawning herring of the North Sea and English Channel by means of the otoliths and an application of this method in tracing the offspring of the races along the continental coast of the North Sea. *Rapp. & Proc. Verb. Cons. Int. Explor. Mer.* 143 (2): 130–33.

Rhijn, J. G. van. 1973. Behaviour dimorphism in male ruff *(Philomachus pugnax)*. *Behaviour* 47:153–229.

Society for Experimental Biology. 1950. Physiological mechanisms in animal behaviour. Symp. Soc. Exp. Biol. IV. Cambridge Univ. Press. 482 pp.

Tinbergen, L. 1960. The dynamics of insect and bird populations in pine woods. *Arch. neerl. de Zool.* 13:269–472.

Tinbergen, N. 1932. Ueber die Orientierung des Bienenwolfes (*Philanthus triangulum* Fabr.). *Z. vergl. Physiol.* 16:305–35.

Tinbergen, N. and Kruyt, W. 1938. Ueber die Orientierung des Bienenwolfes (*Philanthus triangulum* Fabr.). III. Die Bevorzugung bestimmter Wegmarken. *Z. Vergl. Physiol.* 25:292–334.

Verwey, J. 1930. Die Paarungsbiologie des Fischreihers. Zool. Jahrb. Abt. *Alg. Zool. u. Physiol.* 48:1–120.

Wiepkema, P. R. 1961. An ethological analysis of the reproductive behaviour of the bitterling (*Rhodeux amarus* Bloch). *Arch. neerl. de Zool.* 14:103–99.

Wolvekamp, H. P., Baerends, G. P., Kok, B., and Mommaerts, W. F. H. M. 1942. O_2 and CO_2 binding properties of the blood of the cuttle fish (*Sepia officinalis* L.) and the common squid (*Loligo vulgaris* Lam.,) *Arch. neerl. Physiol.* 26:203–11.

Zijlstra, J. J. 1969. On the "racial" structure of North Sea autumn-spawning herring. *J. Cons. Int. Explor. Mer.* 31:197–206.

Vincent G. Dethier (b. 20 February 1915). Photograph by Fabian Bachrach

2

Curiosity, Milieu, and Era

Vincent G. Dethier

Many years after I had become an established scientist (meaning that I could successfully earn my living at that profession) I, together with some colleagues, was asked to take a test devised by some psychologists who were intent on discovering what made a scientist and particularly what determined the kind of scientist he became. One element of the exercise was a Rorschach test. Some of the ink blots were very interesting; they resembled fish and butterflies, and landscapes; but mostly they looked like ink blots. I never did learn whether scientists were made of "snippets and snails and puppy dogs' tails," or "sugar and spice and everything nice."

In retrospect, the origin and development of my scientific career seems to have followed no set plan nor to have been directed toward any consciously formulated goals. Yet I can now discern a causal chain of events and forks in the path I trod. At the forks, however, decisions seem to have been made on the basis of immediate, perceived outcomes rather than upon postulated ultimate consequences or set goals. This bent of mind might explain why I could, for example, never excel at chess; I seldom saw beyond the first, or at most, the second move. Paradoxically my experimental strategies did not embrace these tactics at all.

The raw ingredients from which my career developed were an insatiable childish curiosity that I have not to this day outgrown, an urge to collect that I have almost brought under control, and an abiding sensitivity to the beauty of nature. The starting point, as for most of us, was my parents; the selective pressures were the times and environment of my childhood and adolescence.

On the paternal side I came from a family of professional musicians. My grandfather, Émile, trained at the Royal Conservatory of Liège, received first prize in fugue, and spent his life as composer, organist, and Maître de Chapelle of the Grand-Seminaire de Liège. He received the Order of the Crown from King Leopold and was made Chevalier by Pope Leo XIII. My father, Jean, also graduated from the Royal Conservatory of Liège, won second prize in piano at age eighteen, and first prize at age nineteen. Shortly thereafter he emigrated to America where he eventually became Director of Music in the Norwood, Massachusetts public school system and organist and choir director at St. Catherine's Church.

My maternal grandfather, Thomas Patrick Lally, was a Boston fireman and skilled cabinetmaker. My mother, Marguerite Frances Lally, traced her lineage back to the Irish royal clan O'Mullaly. Before her marriage she taught primary school in the slums of Boston.

Thus both of my parents were teachers. More importantly they both had an unusually fine appreciation of education and scholarship. Both loved music and nature, and my mother possessed the talents and sensitivity of an artist. This was the family background. Not one iota of science existed anywhere in the lineage. The great gifts that my parents did proffer were a liking for learning, an appreciation of the great and lasting things of the world, mental discipline, and a sense of responsibility—to one's self and to others. We four children grew up in a home environment of constantly effervescing intellectual ferment.

This background together with the environmental circumstances of my early childhood certainly fostered and enhanced a predilection for natural history. Possibly the first reading material made available to me also had some directive influence. At age five I received two books for Christmas, Thornton W. Burgess' *Reddy Fox* and Hugh Lofting's *The Story of Doctor Doolittle*. Although these two books are still in print, they and their like are not held in high repute by educators, editors of juvenile books, child psychologists, and a goodly number of librarians. They are anthropomorphic, saccharine, trivial, and full of cloying sentimentality! Worst of all, they contain no "messages," do not relate to any of the problems of society, and are complete fantasy! But in time and place of my childhood, there was no need to read books of social relevancy; we lived it. We were children of an immigrant family, a minority, not well-off, and Roman Catholic in the Boston of "George Apley." These books played a crucial part in my leaning toward natural history. They synergized certain special environmental influences. It all began with butterflies.

Butterflies are difficult to find in a city—unless the city is blessed with a few oases of greenery strategically situated so that any wayfarer, whether it be bird, insect, or seed, drifting in from the surrounding countryside can come to rest. Then, having discovered his green pasture the visitor is unlikely to depart unless he is like the monarch butterfly or summer birds upon whom the shortening days of fall bring an irresistable urge to migrate. So the urban green patches become for many animals attractive traps surrounded by inhospitable grays of concrete, macadam, and dingy buildings.

The Boston of my youth had such spots—not the Common, the Public Gardens, and the Fenway; they were in the neighborhood of the city where the Boston Brahmins lived. There were three green oases in our immediate neighborhood. One was a triangular vacant lot wherein grew a profusion of weeds: buttercups, daisies, dandelions, plantain, sheep sorrel, dock, burdock, beggars'-lice, clover, bachelor's buttons, tansey, butter-and-egg plants, shepherd's purse, goldenrod, milkweed, and untold species of grasses. At one end stood a single large weeping willow.

The second oasis was a small oval, euphemistically called a park, at the top of our dead-end street. The oval was common ground. In those days it was ringed with hedges of barberry, lilac, wisteria, and privet and boasted also a crab apple, a ginkgo, and some maples. This together with the vacant lot and a few green backyards attracted and held butterflies.

My first acquaintance with a live butterfly resulted entirely from the initiative of the butterfly. I had wandered up to the oval late one hot, humid, summer day. The long slanting rays of the sun illuminated my white shirt. Suddenly something rocketed across the street, made a few zigzags, and landed on my shirt just above the pocket. I stood stock still and slowly lowered my head to see what it was. There with its wings slowly expanding clung a brown butterfly with a red band extending down each wing. This red admiral was the first live butterfly I had ever seen at close range, and I was fascinated.

Two more formal situations now built upon the groundwork of family and childhood experience. These were the art school of the Boston Museum of Fine Arts and the Children's Museum, then located in Jamaica Plain. When I was ten years old, my mother enrolled me in art classes. There I received training in pencil drawing, watercolors, and charcoal. Nature again asserted itself, because animal drawings of Pisanello, Japanese bird, fish, and flower prints, Egyptian scarabs, sacred ibis, and cats, and hunting scenes on Greek vases appealed to me most. This training in art later stood me in good stead when the time came to learn scientific illustrating.

The Children's Museum was a small but well-stocked museum of the times. What made it special were the clubs—bird club, rock club, stamp club, etc., the illustrated lectures, and the field trips. Another feature, absolutely unique, was the monthly magazine *Our Hobbies*. It was produced entirely by children. We designed it, wrote all of the articles, stories, poems, and notices, did the editorial work, proofed the galleys, pasted the dummy, solicited the advertisements, and, in short, did everything but the printing. Under enlightened supervision it became a satisfactorily successful magazine. From this experience I acquired many of the skills associated with preparing manuscripts and illustrations for publication. The importance of the museum and its staff of dedicated women to my scientific development cannot be overestimated.

In the meantime I was progressing through a parochial school that instilled in me disciplined study habits and basic skills, without, as is so often alleged of such schools, stifling my creativity, warping my judgment, or impairing my ability to think for myself. Equipped with the basic skills, I continued in the local public high school to acquire a solid old-fashioned classical education: Latin, French, literature, art, composition, history, chemistry, physics, and mathematics. I took no course in biology. The courses that served me best in later years were Latin, French, and English composition.

At the Children's Museum I came to the attention of Professor Joseph Bequaert of Harvard. Until that moment the possibility of college had never entered into discussion at home. Bequaert persuaded my parents to have me apply to Harvard—a wild dream considering our background. Nevertheless, that became the goal.

The original idea behind acquisition of a college education was to prepare myself for a teaching career in secondary schools, and there were three areas of concentration open: music, art, and science. Both my father and I agreed that my talent in music was a small one, and it is an indication of his sagacity that he did not push me in that direction. As between art and science there was general agreement that art might be a useful adjunct to science whereas the reverse did not hold true. Accordingly I entered college with a declared major in biology.

Harvard nearly nipped that career in the bud then and there. In addition to English composition, European history, philosophy, and German, that first year I studied zoology and botany. These last two were what nearly drove me from science. In zoology Professor George Parker, resplendent in neatly trimmed gray beard, presented each week two lectures broadly covering all aspects of zoology except classification. That subject warranted one special lecture each week by Dr. John Welsh. The content of that lecture was dull and obfuscated by yards of Latin names alien even to my Latin-tuned ear (I had learned ecclesiastic as well as classic Latin). Parker, on the other hand, had more intrinsically interesting information to dispense. He was also a master of expository style embellished with dry wit. It is of interest to note parenthetically that although Parker was the foremost authority on chemical senses at the time and author of *Smell, Taste and Allied Senses in the Vertebrates,* this had no role in directing me into that specialty. I discovered Parker much later.

Zoology laboratory began, as it probably has ever since the microscope was invented, with our inspection of the letter *e*. We then progressed through the animal kingdom, amoeba, paramecium, hydra, earthworm, and so on up to frog. We slaved away with prepared slides and pickled specimens in the dark nether regions of that magnificent mausoleum, the Museum of Comparative Zoology. Nothing could have been more remote from the biological world of my experience.

The companion course in botany was taught by Professor Ralph Wetmore, one of the kindest, most approachable gentlemen I have ever known but one of the least inspiring lecturers. Again the classification discouraged me, and trying to comprehend the tissue structure of the three planes in which a tree trunk could be cut approached philosophy in its elusiveness as far as I was concerned. Plant alternation of generations was equally stultifying. Laboratory exercises followed the plant kingdom from algae to angiosperms.

The features of the introductory courses that almost induced me to abandon science were: the inexhaustible vocabulary of technical terms that riddle biology, the general dullness of lectures, the welter of details to the exclusion, or subordination, of principles and generalizations, the interminable Latin nomenclature, and the preoccupation with dead, mounted or pickled specimens. To one whose prior experience had been with living organisms and who was attracted to nature by the *behavior* of organisms and the beauty and mystery of the living world, these were bitter pills to swallow.

The original impetus for going to college had been to obtain certification for public school teaching. The Department of Education at Harvard to which I had gone for advice was less than helpful. It was suggested to me, however, that I continue on to graduate study, which would then prepare me to teach in a college. Despite my disillusionment with undergraduate courses, I majored in biology and then entered graduate school. At that point the whole world of biology opened up to me.

Despite the Depression, and despite being a day student (and hence persona non grata) and having to work my way through college and university, the decade of the thirties represented vintage years. Being a graduate student at Harvard in the years preceding World War II was an unforgettable experience. That era represented a benchmark in the evolution of biology at Harvard. Many of the great figures of earlier times were still active, a living heritage of the post-Louis Agassiz period, that is, the Eliot period. They included Edward Mark (anatomy), William Morton Wheeler (entomology), Edward Jeffrey (plant morphology), Nathan Banks (curator of insects, M.C.A.), William Castle (genetics), Hubert Lyman Clark (marine invertebrates), and Merritt Fernald (botany).

The formal courses were essentially grist for our mill; the real education devolved from close informal association with an enormously rich variety of personages, the great and the near great, from fellow graduate students, from research, and from places and organizations. The latter included the Museum of Comparative Zoology and its staff, the Farlow Herbarium, the Arnold Arboretum and its staff, the Cambridge Entomological Club, and the scientific fraternity Gamma Alpha. Other sources of inspiration included seminars and those parts of Harvard not located in Cambridge, namely, the Harvard Forest at Petersham, Massachusetts, and the Atkins Institution, Soledad, at Cienfuegas, Cuba. There was also Woods Hole. This rich diversity, affording as it did opportunities for close personal association with persons of many specialities, produced a mix, an atmosphere, that being built on a sound, broad undergraduate base, enabled a graduate student to specialize to his heart's content while inevitably being exposed to beneficent contamination from multiple intellectual fields. Here was a curious intellectual ecology, specialization within diversity and interaction from all directions.

Ever since the butterfly had landed on my shirt, I had continued to study butterflies, to collect them, and to rear them. This last endeavor worked fine when I knew which plant was the preferred food. I was impressed by the fact that some caterpillars actually starved to death rather than eat other than their favorite plant. I wondered why. This puzzle prompted in turn another question: how do they recognize food plants? Beginning in the early thirties, I designed crude behavioral experiments, including elimination of various putative sense organs that might influence feeding behavior. In the process I came to the conclusion that caterpillars possessed well-developed senses of taste and smell and that these were vital to plant selection. Although this conclusion seems trivial now, it must be remembered that at the time the possession and discreteness of such organs in insects were matters of controversy. Also at this time I attempted to discover what kinds of compounds constituted adequate stimuli. Having learned about plant essential oils in Oakes Ames's course in economic botany, I even attempted to steam-distill some of the essences of milkweed, carrot, and oak. The resulting effluvia and stinks bore no resemblance whatsoever to the natural aromas of these plants.

Nothing might ever have come from these experiments had it not been for Harvard's tutorial system and my particular tutor. Obligatory tutorial sessions were patterned after the English university system, wherein the tutor acted in loco parentis. In the biology department, each undergraduate met with his tutor once a week. Mine was Dr. T. J. B. Stier, a physiologist who investigated yeast cytochrome. Each week I was expected to read a book selected from a list. Among others I remember Zinnser's *Rats, Lice, and History*, Valery-Radot's *Life of Pasteur*, de Kruif's *Microbehunters*, Smith's *Kamongo*, Walter B. Cannon's *Wisdom of the Body*, Nordenskiöld's *The History of Biology*, and Sinclair Lewis's *Arrowsmith*.

For the first time I developed an idea of what research was all about, what constituted a control experiment, how to analyze data, and so forth. For the first time I realized that teaching in a college or university could be combined with research. I discussed my experiments with Stier, and to my great astonishment he suggested that I prepare the results for publication. He not only provided the initiative, he began my education in the art of scientific writing and preparation of manuscripts for press. This paper was published in the *Biological Bulletin* in 1937.

When I began graduate school, I was fully committed to the study of host-plant selection by lepidopterous larvae. The obvious person to direct my research was Professor C. T. Brues. He had pioneered in this field and had resurrected the now classical paper of the Dutch botanist Verschaffelt.

It is important to remember that there was then no gas chromatography, no mass spectroscopy, no electronmicroscopy, and to all intent and purposes no electrophysiology. Hence, separation and identification of secondary plant substances was all but impossible in the average biology department. Examination of the fine structure of sense organs was limited to light microscopy, and the identification of sense organs was possible only by behavioral analysis before and after extirpation of parts. Furthermore, insect physiology as a distinct discipline did not exist. Wigglesworth's *Principles of Insect Physiology* did not appear until 1939 (a slim volume, *Insect Physiology*, was published in 1934).

Biologists who could be characterized as insect physiologists could be numbered on ten fingers. In America there were: A. Glenn Richards, J. F. Yeager, John Buck, William Trager, N. E. McIndoo, D. E. Minnich, J. H. Bodine, F. Crescitelli, T. J. Jahn, and D. Ludwig. Dietrich Bodenstein and Gottfried Fraenkel had just arrived from Germany and Kenneth D. Roeder only slightly earlier from England. There was nobody teaching insect physiology at Harvard, and we graduate students essentially taught each other.

For my part I decided that the only way to approach the problem of host-plant selection was through a thorough study of the chemical senses of caterpillars. This launched me into the general area of chemoreception.

In 1937 there was only scanty evidence that insects possessed discrete gustatory and olfactory senses; no known insect chemoreceptor had been functionally identified. As I began my behavioral and morphological studies, it soon became clear that the technical limitations were severe. Through the literature I was aware that Adrian had recorded electrophysiologically from sense organs. I was also aware from Woods Hole contacts that C. Ladd Prosser, then at Clark University in Worcester, Massachusetts, was employing electrophysiological methods to study the nervous system of earthworms. Roeder, at that period, was not yet completely immersed in electrophysiology. I took my problem to Prosser, and one summer, through his great generosity and kindness, I enjoyed the use of his laboratory and equipment.

I need not dwell on the primitive nature of the equipment. Needless to say, it was unequal to the task of detecting action potentials in the very fine nerves of insect chemoreceptors, and, indeed, fifteen years were to pass before these receptors were to give up their secrets. In the meantime I spent my summers in Cuba, where I augmented my knowledge of tropical plants and their insects, at Woods Hole where I was exposed to more marine biology, and at the Franklin, New Hampshire laboratory of Harvard physicist Professor George W. Pierce, where I assisted in his pioneering research on the songs of insects. At Franklin I came to know Walter B. Cannon, Robert Yerkes, and Hubert Lyman Clark on an informal basis. These acquaintanships were enriching experiences for a young graduate student.

For the first six months following the award of my doctorate in 1939, I was unemployed. In the middle of the fall term I fell into a position at John Carroll University in Cleveland, Ohio where during a period of three years, the first two as an instructor, I taught introductory biology each year, an evening course in comparative anatomy each year, and genetics, embryology, parasitology, and histology the rest of the time. All, I might add, were with laboratory. There were no student laboratory assistants.

In my spare time I completed for publication a paper on food-plant selection and its evolution in caterpillars of swallowtail butterflies. The field experience in Cuba had provided many insights into the problem, as had also the numerous botany courses I had taken earlier. I presented this paper at a meeting of the Columbus (Ohio) Entomological Society to which I was a complete stranger from a small Jesuit university. I mention this incident only because in terms of later events it turned out to be a turning point in my career.

Lacking the facilities to pursue research on the chemical senses, I decided to round out my knowledge of the total sensory complement of caterpillars by investigating their photoreceptors. To this end I taught myself a smattering of optics, built an optical bench and lens system to work with lenses as small as those of caterpillar ocelli, and completed for publication two studies on the visual system of the woolly bear caterpillar *Isia isabella*. The optical object on which I focused in these studies of dioptrics was the letter *e* (could I have been imprinted on this letter in Zoology 1?) in a newspaper headline that proclaimed "Moscow Next Goal of Nazis." In the spring of 1942, I began a four-year stint with the Army Air Corps and spent half of that time in the African-Middle East theater of operations.

In wartime there are many hours of waiting. The boredom of war is one feature that is seldom noted in commentaries. I managed to put much of this time to good use by working on a duplicate manuscript of the book on insects and chemicals that I had begun in graduate school. Parts of this manuscript were typed on a "liberated" Italian typewriter that I carried with me on long routine flights. The only other aspect of the war that had any direct bearing on my scientific career was a chance meeting with Joseph Bequaert in Liberia. I even found time to go with him officially on a one-week trip into the Liberian bush.

When rotated stateside at long last, I was detached from the Air Corps to be assigned as liaison officer to the Office of the Chief of Chemical Warfare Service in Washington. This was the first assignment where my specific scientific training had any relevance. Three features of this assignment are of import: I had to visit periodically the laboratories at Edgewood Arsenal in Maryland where an insect physiology section under Leigh Chadwick's direction was studying the effects of certain chemicals, including anticholinesterases, on insects; I had to visit various university campuses where work on insect physiology was being conducted under contract with the Chemical Corps; together with the Chief, I represented the Chemical Corps in high-level scientific meetings in Washington. As subsequent events proved, these were very valuable experiences.

Few contemporary young scientists realize the invaluable contributions that the armed services made to insect physiology, and to basic research in general, during the late years of the war, but more especially during the immediate postwar period. There was no National Science Foundation then, and there were no National Institutes of Health research grants. The Chemical Corps, and to a lesser extent the Quartermaster Corps and the Corps of Engineers, had contracts with leading insect physiologists of the day among them Roeder at Tufts and Richards at Minnesota. It was through this liason assignment that I became au courant with the progress of insect physiology since I had left graduate school and saw that it was developing into a cohesive and definitive field. I also saw during my visits to Roeder's laboratory, the great strides that he had made in electrophysiology. Kenneth Roeder was to play a crucial role later in the study of the chemical senses. If I were asked to designate the leading insect physiologist in America, I would choose Kenneth Roeder.

Attendance at high-level meetings injected me into the postwar Washington scene and led, as the years passed, to membership in the National Research Council Committee on Sanitary Engineering, N.S.F. Panel for Molecular Biology, N.I.H. Tropical Medicine and Parasitology Study Section, Office of the Surgeon General's Medical Research and Development Committee, National Research Council Advisory Board on Quartermaster Research and Development, Executive Committee of the American Institute of Biological Science Curriculum Study, N.S.F. Panel for Psychology, Armed Forces Epidemiological Boards, and Commission on Undergraduate Education in the Biological Sciences. The broadening of horizons that these associations provided for me proved of incalculable value in my development as a scientist.

When I first returned to mufti, two significant events occurred immediately. First, Ohio State University needed a professor of zoology and entomology. Alvah Peterson remembered my presentation at the Columbus Entomological Society and offered me a full professorship. Several months intervened between the time of my discharge from active duty and the beginning of the academic year. For the interval I was offered employment as a research scientist at the Army Chemical Center at Edgewood, Maryland. This position provided me with an opportunity to work in what was probably the leading laboratory of insect physiology, where Dietrich Bodenstein also worked and Leigh Chadwick was Chief. I could choose, within limits, my own program of research.

After an absence of four years from science, I began to think once again of the chemical senses of insects. By now I had realized that, given the technology of the day, caterpillars were not the experimental animals of choice. What I required was an insect that could be cultured in the laboratory in large numbers through the year and whose behavioral responses to chemicals, especially sapid substances, could be studied without the nutritional state being altered in the process. At this point I remembered Minnich's studies of 1921 onward, in which he demonstrated that butterflies and flies possessed chemoreceptors on the legs and that when tarsi were placed in contact with sugars, the retractable proboscis was extended. Following Minnich, Hubert Frings, then at Edgewood, had extended investigations from carbohydrates to mono- and divalent salts with the horsefly *Tabanus sulcifrons* as experimental animal. With these facts in mind I sought a large, placid, rugged, prolific fly that would fare well under laboratory conditions. The choice was the black blowfly, *Phormia regina*.

Initially I had two goals; one to try to relate the structure of carbohydrate molecules to their stimulating effectiveness; the other to find a molecular basis of rejection. Studies with caterpillars had shown that some sugars were more effective than others. With honeybees, von Frisch had failed to discover any orderly relation between structure and stimulating effectiveness. Students of vertebrate gustation had fared no better. At Edgewood, Charles Hasset and I had demonstrated that there was no orderly relation between behavioral thresholds to sugars and their nutritive values.

On the rejection side, Frings had found a correlation between ionic mobilities of salts and behavioral threshold. I sought comparable correlations in the case of organic molecules the structures of which could be altered in a progressive fashion and for which data on chemical properties were known. Aliphatic alcohols provided the homologous series I desired because each succeeding member differed by one CH_2 group. Furthermore, the position, species, and number of substituted polar groups could be changed.

Thus at Edgewood I began an attempt to relate structure to stimulation with the goal of understanding the transduction mechanism of chemoreceptors. There were two formidable obstacles lying in the path of this realization. The first was that the criterion of stimulating effectiveness of salts and aliphatic organic compounds had to be failure to respond behaviorally to a standard sugar solution with which the test compounds were mixed. The second was that the test compounds had to be water soluble to some extent. The results obtained raised two questions: how could compounds that were infinitely soluble in water and those that were more oil than water soluble both act effectively? How did any of the molecules reach the dendrite?

It was at that time an axiom that insect cuticle constituted a complete unbroken covering. We were forced, therefore, to postulate a two-phase cuticular membrane at receptor tips. It is interesting to recall that N. E. McIndoo, a United States Department of Agriculture entomologist, had the correct answer in 1914 for the wrong reasons. He argued that small circular and oval sense organs scattered over the body were olfactory pores, open to the air. His ideas were never accepted. We now know that the organs in question are mechanoreceptors, that they lack pores. McIndoo was correct, however, in insisting that chemoreceptors had to have pores. Eleanor Slifer proved the correctness of this view years later, and electronmicroscopy has since provided beautiful pictures of the openings, which, incidentally, McIndoo could never have seen because they are below the resolution of the light microscope.

The second question referred to above was resolved with the advent of electrophysiology, which proved that salts acted on a salt receptor while alcohols acted by inhibiting the sugar receptor. Behavioral rejection was not a unitary phenomenon!

When I took up residence at Ohio State I extended behavioral studies of rejection until nearly four hundred compounds had been tested. The relationships that were revealed paralleled almost exactly data reported in the literature on anesthetics, various narcotics, hemolysis of red blood cells, and other phenomena. The same thermodynamic principles applied to all. I should have realized then the dual nature of rejection. I note wryly in concluding the account of this era that laboratory safety standards were informal to say the least. Little did I realize the hazards attending experimentation with some of the compounds, especially the brominated ones.

Kind though my colleagues at Columbus were, I felt that I was working in isolation; consequently, when the opportunity to move to Johns Hopkins University presented itself, I made a decision that seemed idiotic to my friends but in the final analysis proved correct: I reisgned a full professorship with tenure to accept an associate professorship without tenure.

The period at Hopkins (1947–58) proved to be one of the most productive, educational, and adventuresome of my career. With renewed vigor, I continued attempts to fathom the nature of chemoreception, but the goal seemed as elusive as ever. Furthermore, it was drawing me away from biology and behavior and deeper into physical chemistry. One final attempt to understand tranducing mechanisms is worth recording.

The Army Chemical Corps, which had awarded me a research contract, was interested in developing more effective repellents and fly baits. About this time Clyde Barnhardt and Leigh Chadwick discovered that bait that had been visited by flies became more attractive than unattended bait. They ascribed the enhancement to something that they labeled "fly factor." Seizing this idea, I began to investigate the effect of flies' visits on the attractiveness of pure sucrose. Thinking that perhaps flies regurgitated or defecated on the sucrose I tested the effect of visits by flies each of which had the proboscis and anus plugged. These flies still produced the effect. Could the tarsi be producing an enzyme that hydrolized sucrose? I washed the legs of flies (plugged fore and aft) and then allowed the tarsi to hang in solutions of sucrose and other sugars. A Benedict test revealed that di- and trisaccharides with alpha-glucoside linkages had been hydrolized. Clearly the tarsi produced alpha-glucosidases. My immediate thought was that this was the receptor enzyme. Upon further reflection, since certain hexoses and pentoses also stimulated effectively, I concluded that glucosidaes were not the long-sought-for enzymes.

Later Kai Hansen in Germany picked up the original data, discovered that the strength of enzyme could be correlated with the number of chemoreceptive hairs on each segment (these had not yet been discovered in the period of which I write), and characterized some of its properties. The Japanese, under Morita's aegis, began the exquisite analyses and characterization of the enzyme (more than one, it developed) from single hairs.

A few years ago Hansen asked me why I had never followed up the original discovery, and I told him about my doubt about it being the critical receptor enzyme. It is now doubted more widely that this is in fact the long-sought-after enzyme. The search continues.

As my initial research interest had been in behavior, and I had ventured into chemoreception only because I could not conceive of understanding feeding behavior without understanding chemoreception, I decided to turn my attention to other matters. Despite all the behavioral and chemical work to date, the identity of the receptors remained hidden. There was a hint in the literature as to what to seek. In 1926 Minnich, in a short abstract, reported that he had elicited proboscis extension in flies by stimulating with sugar a single long recurved hair on a fly's labellum. In 1933 Eltringham in England suggested that certain multi-innervated, thin-walled hairs on the tarsi of the butterfly *Pyrameis* might be the receptors mediating proboscis extension.

With these leads, a graduate student in embryology, Casmir Grabowski, and I set out around 1950 to locate the tarsal receptors of *Phormia*. Grabowski made complete maps of the distribution of hairs and together we began stimulating each hair individually with sugar. It soon became apparent that a particular type of hair mediated the response. Under the microscope, this hair appeared to have two tones of color, a light side and a dark side. Methylene blue staining revealed that each two-toned hair was innervated by three bipolar neurons, one of which terminated at the base of the socket. A few more years passed before we established the correct number of neurons as five. For a while, interpretations of physiological studies were constrained by the miscount.

By now I had measured the tarsal acceptance and rejection thresholds for over four hundred compounds. In the process, data had been accumulated on 17,000 individual flies. It was natural to search the literature for comparable data relating to vertebrates. The difficulty with making meaningful comparisons was aggravated by the habit of each investigator of using his own unit of concentration. As a result of discussions with several colleagues, Carl Pfaffmann, P. T. Young, Curt Richter, Eliot Stellar, and I set in motion plans to convene a symposium on chemoreception, possibly the first of its kind, at Hopkins. Although this 1951 symposium spread its net wide, the five of us addressed our joint need for uniformity and precision in the preparation and specifications of solutions (we favored molarity). Of more particular significance was the drawing power of the symposium in bringing together the principal American workers from backgrounds as diverse as industry, food science, the military, academic psychology, and biology. Bonds were established that endured over many decades.

A sad postscript is that no further gatherings followed. Chemoreception was an orphan science then. It was not until eleven years later that chemoreception again held center stage at the First International Symposium in Olfaction and Taste held in Stockholm. The list of participants was long and impressive, including as it did two of the most eminent figures in the field, Lord Adrian and Yngve Zotterman.

Interspersed in the Hopkins years were three trips abroad, all of which contributed to my research and continuing education. Each owed its inception to antecedent events that were almost chance. The book *Chemical Insect Attractants and Repellents* had been published by Blakiston in 1947. Not long afterwards, I received an invitation from Jan de Wilde to present a paper at a symposium on insect-plant relations that he was organizing for the IXth International Congress of Entomology in Amsterdam in 1951. As I learned many years later, Wigglesworth had seen this book and asked de Wilde who this fellow Dethier was. The invitation had followed.

I dug into my pockets for the fare and flew to Amsterdam. There I discovered that an active international (mostly European) group interested in insect–host-plant relations had evolved since the end of the war. It included most of the active workers who were to become pioneers in this rapidly developing field: Jan de Wilde, John Kennedy, Gottfried Fraenkel, Pierre Grison, Hille Ris Lambers, Reginald H. Painter, and Chun-Teh Chin.

I had by no means forgotten herbivorous insects, and this meeting rekindled my interest. It was during this meeting, on a side trip to Groningen with Jan de Wilde, that I met Luke Tinbergen who showed me a paper he had written (and I had overlooked) in 1939 in which he had described the identical hairs in *Calliphora* that we had found in *Phormia*. Apropos of this, on a subsequent trip to London in 1954, I visited George Varley in Oxford and was shown Eltringham's original slides of butterfly tarsi. He, Tinbergen, and I were all talking about the same kind of hairs.

Another direct outcome of publication of the book was that Ernst Mayr, on the strength of some tentative ideas that I had expressed in a chapter on evolution of feeding preferences, asked me to develop the ideas for publication in the journal *Evolution*. This paper was subsequently translated into Japanese. Thus, while working with flies, I was maintaining my ties with herbivorous insects.

I had hardly returned from the Netherlands when Joseph Bequaert dropped in for a visit at Hopkins. As we talked about flies and recalled our wartime meeting in the Liberian bush, he suggested that it might be to my advantage to look in on tsetse flies in the Belgian Congo. The Dark Continent was still enshrouded in romance and mystery. Although my experiences there had been those of wartime, I, who so much enjoyed snowy New England winters, felt, nevertheless, the pull of Africa. I had picked up a working knowledge of Arabic and Swahili and knew my way, so to speak, around Africa. I had, since the first visit, developed an interest in African history and had collected a nearly complete library on the subject, from early Arabic sources through the period of European exploration. A second visit during the period of growing anticolonialism and active Mau Mau revolution fit with these interests. Scientifically it would enable me to collect more field data regarding insect-plant relationships while at the same time casting a new perspective on the chemoreceptors of flies.

The 1951 trip to Amsterdam led directly to a year in London in 1954 in the department of Patrick Buxton at the London School of Hygiene and Tropical

Medicine. I had met Professor Buxton twice in my life and only briefly on each occasion. In a bus in Amsterdam he had turned briefly in his seat ahead of me and announced apparently apropos of nothing, "Young man, those sparrows in the street are really finches." A year later, when I was passing through London on my way home from the Congo, I stopped long enough to attend a cocktail party at the home of a friend outside of London. Buxton was there. The total conversation on his part was "How d'y do" on meeting and "Good afternoon" on leaving. Two years later he invited me to spend a year in his laboratory as a senior Fulbright research scholar. That released me from a heavy teaching load at Hopkins and provided a hiatus in which to fill in experimental gaps.

As the situation stood in the early fifties, insect contact chemoreceptors had finally been identified, a start had been made on the relevant neuroanatomy, and many of the properties of the receptors had been inferred from behavioral studies. Another attempt at electrophysiological recording was now made. Having discovered putative chemoreceptors on the ovipositors of parasitic ichneumonid wasps, I reasoned that the long (250 mm.) sensory nerve might yield to the electrode. With this thought in mind I journeyed to Kenneth Roeder's laboratory at Tufts College to make the attempt. We failed.

The breakthrough finally came in 1955 when Edward S. Hodgson, having received his doctorate, left my laboratory for a year of postdoctoral study at Tufts. In collaboration with Roeder and Lettvin, he succeeded in recording from neurons of the fly labellar chemosensory hairs by picking up backfiring of the axonic potentials through or around the dendrite. This feat was accomplished by placing a pipette electrode over the tip of the hair. It acted as electrode and as a source of stimulation. At about the same time (1957), Morita and his associates independently developed the same technique. They then proceeded to refine the technique until ultimately they were able to record with electrodes inserted through the side wall of the hair.

With an early crude electronmicroscope we attempted to elucidate the plan of innervation of the hairs and came to the conclusion that there were five neurons. While these studies were in progress, Jean Adams at Rutgers completed a dissertation that showed that the hairs of the stable fly, *Stomoxys calcitrans,* also had five neurons each. This was confirmed for *Phormia.* Later Joseph Larsen, doing postdoctoral research in my laboratory, clarified most of the mystery still shrouding the relations of these cells to the hair structure and accessory cells.

These electrophysiological and neuroanatomical studies confirmed the broad behavioral conclusions, namely, that there was a sugar receptor and a salt receptor. Within a very short time, Myron Wolbarsht, then my graduate student, recorded potentials from the mechanoreceptor and the water receptor. Not long afterwards a second salt receptor was found, bringing the complement of receptors in the hair to five.

The creation of the new biophysics department at Hopkins during the fifties and the arrival of Detlev Bronk, Frank Brink, Martin Larabee, Philip Davies, Theodore MacNichols, and Keffer Hartline provided a technical and intellectually stimulating atmosphere that firmly launched the blowfly studies on an electrophysiological course that not only continues to this day but finally enabled me to attack some of the problems of receptor specificity and such aspects of peripheral integration as synergism and competitive inhibition. Electrophysiological studies also cleared the way for studies of the neural control of feeding behavior.

Until now behavioral studies had exploited only the proboscis response. With this relatively simple reflex it had been possible to separate receptors behaviorally, and to study the time course of adaptation, spatial and temporal summation, relations between threshold and chemical structure, and changes in threshold sensitivity with state of deprivation. It was even possible to derive postulates concerning neural circuitry beyond the level of first order neurons. Now, however, more extensive behavioral analyses became necessary.

To this end I was particularly desirous of developing a method for presenting flies with a free-choice situation in which solutions could be paired and the volume of each imbibed could be measured. A psychology student, Marion Rhoades, collaborated in this effort. Together we put to use a two-bottle choice apparatus designed by him, whereby it was possible to study stimulating effectiveness of solutions as a function of nutritive value and of deprivation. The apparatus also provided a more sensitive measure of threshold and a means of investigating discrimination ($\Delta I/I$).

The original apparatus was undeniably crude; nevertheless, it yielded results that stood up well over time as successive generations of more sophsticated "two-bottle" apparatus were developed. Even the prototype was automated so that it registered number and duration of drinks. On one occasion Curt Richter dropped by and suggested that I study intake and choice over a prolonged period of time in order to understand the relation between sensory behavior and nutritional balance (as he had done in his classical rat studies). With other agenda ahead of this suggestion it was several years before I got around to it.

Two significant vistas opened up as a result of this new technique. First, we were able to study quantitatively the effects of mixing different sugars, a procedure previously employed by Kunze (1927) and von Frisch (1935) to study the acceptability of sugars that by themselves had only low stimulating power. Not only did we find that all sugars are not strictly additive, we found instances of specific inhibition and synergism. Ten years later, Elizabeth Omand, then a graduate student, and I confirmed these findings electrophysiologically.

My original behavioral studies on sugars plus my data of 1955 were reanalyzed by David Evans, also a graduate student. From his analyses he developed the idea that the sugar receptor (neuron) actually possessed at the molecular level two sites of action, a glucose site and a fructose site. Suddenly an answer to the difficulty that von Frisch in 1935 and I in 1955 had experienced in relating stimulating effectiveness to structure emerged. A whole new molecular approach to transduction dawned and led to the subsequent extensive biochemical and electrophysiological studies that have become the hallmark of Morita and his associates.

While all these new approaches were expanding, continuing behavioral explorations opened still another door. Until now my attention had been restricted largely to events occurring in the peripheral nervous system; however, the astounding change in behavioral acceptance thresholds with food deprivation (reported first in the 1920s by Minnich and subsequently by many other investigators) presented a challenge. Electrophysiological evidence proved that whatever the change that was taking place, it was not at the receptor level. Where then?

By analogy with human experience (I have always felt that a certain amount of subjectivity and tightly reined anthropomorphism has considerable heuristic value) I began to suspect such factors as blood sugar level and gut mobility or fullness. Thus began a long sustained attack on the insides of the fly. Techniques were developed for transfusing blood, loading the gut, feeding rectally, ligating and removing various regions of the gut, making parabiotic flies, and analyzing blood sugar levels. During the last-mentioned endeavor, Evans identified the blood sugar as trehalose just before Wyatt's paper on trehalose in insects appeared. Other metabolic studies were being pursued by Dr. Anne Hudson who had come to us from London. Dr. Lindsay Barton Browne from Australia had also joined the group.

Having eliminated blood sugar, hormones, crop, and midgut as direct factors terminating ingestion I was left with the foregut. I had been unsuccessful in my attempts to cannulate it. During a discussion of this difficulty with Dietrich Bodenstein, who periodically came in from Edgewood to savor the atmosphere of the groves of academe, he mentioned that he experienced no difficulty in removing the corpus allatum of *Drosophila* so it ought to be possible to get at the foregut in huge *Phormia*. Then it occurred to me that it might be possible to denervate the foregut by transecting the recurrent nerve. Together we perfected this operation and after the first ones, were able to gaze in awe at the bursting hyperphagic flies that we had produced.

We postulated that the recurrent nerve acted as a feedback system that monitored the transfer of ingested fluid from the storage crop to the mainline midgut. In our view it provided central neural inhibition to the stimulating input from peripheral chemoreceptors. Thus "hunger" and satiation could be satisfactorily explained in terms of balance between peripheral chemosensory input and inhibitory mechanosensory input from the foregut. No central mechanism other than the integration of these two inputs was required to explain the behavior (later a second inhibitory input from abdominal nerves was found).

It was during the progress of this work that, together with Kenneth Roeder, I attended the 1954 Ethological Conference in Freiburg, Germany. This was my first contact with ethologists and with Niko Tinbergen, Konrad Lorenz, and E. von Holst. These were exciting times because Daniel Lehrman had just written his celebrated critique of ethology. Presentation of the blowfly data caused an excitement that puzzled me. I did not fully appreciate its impact until I looked at it in terms of contemporary views of consummatory behavior in which repeated performance of a motor act was presumed to lower "something" that shut off the behavior. The blowfly continued to suck and never shut off.

The period at Hopkins illustrates the tremendous impact that milieu and epoch can have on research. In the years immediately following World War II and during the fifties there was an aura of excitement and optimism. It might have been a normal reaction to the end of the war or a measure of our youthful exuberance. I doubt that it owed much to the latter because the enthusiasm was not confined to our generation.

The biology department was oriented toward classical embryology. Under the firm hand of B. H. Willier it was the training ground for a whole generation of embryologists. It seemed, however, to represent the winding down of the era of Spemann, Hamburger, Weiss, and others, while the neural sciences seemed to be in the ascendancy. This might not be a fair perception and could represent my personal bias; nevertheless, there was a large number of neurally oriented people on the Homewood Campus and in the Medical School.

There were two quite different informal organizations that contributed to the vitality of the neural network. One, irreverently known as the Knownothings, in contradistinction to the Greybeards (officially the Biology Dinner Club), consisted of active research workers below the rank of full professor. Any member so unfortunate as to be promoted to the rank of full professor was summarily dismissed (although a few exceptions were eventually made). Among the members were David Bodian, Vernon Mountcastle, Stephen Kuffler, Eliot Stellar, James Sprague, Louis Flexner, Frank Brink, and Martin Larabee.

Once a month the Knownothings met in Welch Library (Medical School) under the Sargent portrait *The Four Doctors,* also known as "The Four Horsemen" (Welch, Halsted, Kelly, and Osler), and quaffed cocktails from an assortment of medical and pharmaceutical glassware. Everyone then repaired to Hausner's restaurant where we consumed a monumental dinner under the gaze of marble busts of the twelve caesars. After dinner we returned to the Medical School's anatomy amphitheater where one of our number presented the latest results of his current research. No more critical audience could be found anywhere; critical interruptions, banter, and exclamations of incredulity punctured the talk. Being able to speak to that group fitted one for speaking under any circumstances.

In stark contrast was the Biology Dinner Club, whose members foregathered for a sedate dinner at the Faculty Club on the Homewood Campus. Following dinner the elders retired to an upstairs meeting room painted in Williamsburg green and furnished for gentlemanly comfort. There the members sank into overstuffed chairs for the presentation of the evening. Among the notables in this group were B. H. Willier, Curt Richter, and Mansfield Clark.

This was an exciting decade, but like plant succession and evolution and life in general it had to run its course. By the end of the fifties, change was already in the air at Hopkins. The establishment of the McCollum-Pratt Institute for the study of trace elements and molecular biology instituted a marked change in the flavor of the biology department. Under William D. McElroy's directorship, the institute gradually overshadowed the department. This was the era of molecular biology. As these events were developing, Bronk moved to Rockefeller taking Brink and Hartline with him. Stellar left the psychology department to go to the University of Pennsylvania. Across town at the Medical School, Flexner, Sprague, and others also emigrated to Philadelphia. At one time there were so many Hopkins men at Penn that it was often referred to as the Little Johns. In 1958 I too left Hopkins for Pennsylvania.

While at Hopkins I had established contact with physiological psychologists who were housed in the same building as Biology, and with others during the 1951 conference on chemoreception. My acquaintance with Curt Richter and Eliot Stellar had been stimulated by a shared interest in feeding behavior. I now had entrée to two opposing camps, physiological psychology and ethology. The experience, plus that already acquired in neurophysiology, provided me with a breadth of view that greatly enhanced my appreciation of the multiple approaches to animal behavior. No curriculum or formal training could ever had had this educational value.

At the University of Pennsylvania I found the same aura of excitement and enthusiasm that had characterized Hopkins. Louis Flexner had established in the medical school the Institute of Neurological Sciences, the first, I believe, in the country. Operating out of the Department of Anatomy, its members included Flexner, Stellar, Sprague, Chambers, Brobeck, Koelle, Erulkar. This unique organization became an intellectual center for neurally oriented scientists. Under its auspices there was held twice a month a "Feeding Seminar," appropriately at lunch time. Here one particular facet of behavior knit together a group of workers whose organizational allegiance was to one of a diversity of departments: biology, psychology, anatomy, veterinary medicine, pharmacology, surgery, psychiatry, and physiology. At this time the psychology department was rising to eminence under the leadership of Robert Bush. Among its members were Maurice Vitales, Francis Erwin, Leo and Dorothy Hurvich, Henry Gleitman, Duncan Luce, Bert Rosner, Richard Solomon, Philip Teitelbaum, and Paul Rozin. I held a joint appointment in biology and psychology.

The period of the sixties witnessed a development and expansion of the research begun at Hopkins. Electrophysiological analyses of the chemoreceptive system of *Phormia* continued. Frank Hanson was exploring many organic compounds for his doctoral dissertation. Additionally, he and I together finally overcame the formidable problems of recording from oral gustatory receptors (the interpseudotracheal papillae) of *Phormia*. That venture is a saga in itself. It began as a weekend task and concluded three years later. I remember with amusement the exclamations of von Bekesy who had visited us and was watching us turning a fly's mouthparts inside out to get at the oral papillae. This work completed the identification and survey of the fly's chemosensory system.

Other graduate students were also extending our knowledge of the fly. Mary Caroline McCutchan probed the tarsal receptors electrophysiologically, Elizabeth Omand employed the same approach to refine our understanding of inhibition and synergism by sugars. Dr. Marie Wilczek from Poland worked out a map of the distribution of labellar chemosensory hairs (now the standard guide) similar to the one Grabowski and I had prepared for the tarsi. She also mapped the complete peripheral neuroanatomy. This information was important in confirming that the axons of the receptor cells did indeed proceed directly to the central nervous system without synapsing, a fact crucial to correct interpretations of behavioral and electrophysiological studies. Joseph Larsen, who had come up from Baltimore for postdoctoral studies, began a detailed histological investigation of the brain. Katarina Tomljenovic arrived from Yugoslavia to begin studies on behavioral responses to protein (until now attention had been restricted almost exclusively to carbohydrates and salts).

As our understanding grew, we suddenly found ourselves in a situation where we probably knew more about the chemical senses and feeding behavior of this animal than existed at that time for any other animal, with the possible exception of man. I felt reasonably certain that all the chemoreceptors, olfactory as well as gustatory, had been located. The peripheral neuroanatomy was well known. The physiology of the gustatory receptors was detailed, and the behavioral parameters had been extensively, though not exhaustively, explored. I might add parenthetically that although there was continuing experimentation on olfaction, I placed emphasis on the gustatory system. Dietrich Schneider and his group in Germany had the olfactory system well in hand.

Obviously, if it was the phenomenon of chemoreception that was to be understood, one had to venture beyond the bounds of the blowfly. A detailed acquaintance with vertebrate systems was essential. Carl Pfaffmann and his associates were the forerunners in work in vertebrate gustation. As we were doing with the fly, Pfaffmann was combining behavioral and electrophysiological studies. He and I remained in contact ever since 1951. Another link was established by way of Lloyd Beidler, then a graduate student in the biophysics department at Hopkins where he was beginning the work that led ultimately to his theory of mechanism of action of rat salt receptors. I was doing unit recording on the fly and was unable to record from whole nerve; he was restricted then to studying integrated responses from whole nerve, the rat chorda tympani. In later years he commented on the fact that I had progressed from behavior to electrophysiology and he from electrophysiology to behavior.

The wisdom of not being restricted conceptually to one's own animal was demonstrated to me time and time again. Despite the early evolutionary split in the pathways leading to insects and mammals, the student of one has much to learn from the student of the other. As I became better acquainted with the rat, hamster, and squirrel monkey I understood the fly better.

The same alchemy occurred with respect to feeding behavior, as studied by physiological psychologists. As my knowledge of the neural mechanisms underlying feeding behavior of the fly increased, the fly in my mind became the white rat of the invertebrates. Rat psychologists like Eliot Stellar and his students were interested in such matters as the regulation of ingestion, the machinery of hunger and satiation, and motivational systems related to feeding—the same kinds of problems for which we sought solutions in the fly.

It is noteworthy in reviewing some of the history of these efforts that students of the rat, stimulated by Brobeck's and Anand's experiments on hypothalamic lesions, attacked the problem of feeding from the central nervous system outward. Stellar, Teitlebaum, Epstein, and others at Penn and elsewhere also began here but moved slowly toward the periphery in search of input mechanisms. We, with the fly, had started at the periphery and were slowly attempting to work our way centrally. Although I had actually made a brain atlas of the fly with the help of Marie Wilczek, and a stereotactic instrument for stimulating and making lesions, minimal success attended these efforts. The obstacle was not small size; it was neuroarchitecture. With modern techniques it is surely possible to stimulate and lesion the brain in a meaningful way. We never got closer than the stomatogastric system, the analogue of the vertebrate sympathetic systems.

Without by any means abandoning the chemoreceptive system as such, I began to explore other facets of behavior during this period because behavior was my prime interest as it had always been. Bodenstein and I discovered one feedback loop regulating food intake. The loop that connected the recurrent nerve mechanosystem with ultimate metabolic requirements remained unknown. Gelperin joined my graduate student group at Penn and began a long series of studies of factors controlling crop emptying. Together we were able to complete the story begun at Hopkins.

Ingestion, however, is only one aspect of feeding behavior; locomotion is another. George Green arrived from Wellington's laboratory in Canada as a graduate student to investigate this component. Employing miniature actographs, parabiosis, and other techniques developed earlier, he was able to tease out some of the mechanisms relating feeding and locomotion and to show that whereas ingestion was influenced by neural factors, the proximate influences in locomotion were hormonal.

Until this point all our studies on feeding behavior had been restricted to ingestion of carbohydrates because these were the principal energy sources for blowflies, no other nutrients being required since all other necessary items had been bequeathed by the larvae. It had been known, however, that some female insects needed a proteinaceous meal in order to bring eggs to maturity. With this thought in mind I adapted the two-bottle preference test to measure the comparative intake of proteinaceous and carbohydrate solutions by females over their lifetimes. The cyclic relations revealed were correlated with egg development and oviposition and stimulated a long series of behavioral, electrophysiological, and histological studies designed to elucidate the underlying mechanism. The fact that, unknown to us, Strangways-Dixon in England was following the same lines is still another illustration of how the state of knowledge at a given time causes ideas to crystallize simultaneously and independently in widely separated places. Other individuals in our laboratory who were involved in the protein studies were Katarina Tomljenovic, Dr. Blanka Bennettova-Rezabova, William Belzer, and Nancy Rachman. Our interest in this phenomenon was reinforced by the thought that it might be relevant to the general problem of specific hungers as studied in vertebrates.

Another aspect of feeding that commanded attention was that of plasticity. To date we had been able to explain feeding behavior in terms of rigidly wired-in systems, that is, reflexes and fixed action patterns. Feeding seemed to owe little to learning. Eliot Stellar and I had tried strenuously in the fifties to demonstrate classical conditioning in *Phormia*. All attempts had failed. At Penn I suggested the problem to Margaret Nelson, a graduate student in psychology. In the meantime Richard Solomon, Lucille Turner, and I were investigating a phenomenon that we termed central excitatory state, meaning nothing more or less than that stimulation of sugar receptors established in the central nervous system a perseverating state (site and nature unknown) that increased the behavioral sensitivity of the fly to a stimulus (water) that had previously been an ineffective stimulus. These discoveries cleared the way for a clean study of conditioning, and Nelson finally succeeded in demonstrating its existence with respect to feeding.

At no time during these years had I abandoned my original interests in insect-plant relations. During summer vacations in Maine, I put flies out of my mind and interspersed leisurely pursuits with field studies. These dealt principally with populations and resource utilization. They involved following successive generations of the butterfly *Melitaea (Closyne) harrissi* in the same fields for a period of ten years. The final paper reporting on this study was written in collaboration with the population biologist Robert MacArthur while we both were still at Penn. It was important in that it dealt with a fundamental ecological question that had seldom been investigated. It demonstrated experimentally that an area does in fact have a "carrying capacity" for a species of animal.

Attention to insect/plant relations increased in intensity when first Louis Schoonhoven from the Netherlands and then Tibor Jermy from Hungary came for a year of postdoctoral research. At long last it was possible to embark on a serious electrophysiological analysis of caterpiller chemoreceptors. All the techniques had been perfected on the blowfly and all the lessons learned were immediately applicable. In 1966, at the invitation of Jan de Wilde and with the assistance of a Guggenhein Memorial Fellowship, I took a sabbatical leave to expand at the Landbouwhogeschool at Wageningen the caterpillar studies begun in Philadelphia. I had been preceded in the electrophysiological phases of this work by Ishikawa and Morita; nevertheless, we were able to work out in detail the action spectra of the gustatory receptors of four species of caterpillars and relate the findings to host-plant discrimination. The results laid the groundwork for a long series of studies that eventually led to the threshold of decoding the neural message transmitted by the chemoreceptors.

The visit by Tibor Jermy prompted me to resurrect other plans that had lain dormant for many years, namely, an investigation of the role of experience in food-plant choice. Learning seemed unimportant to *Phormia;* perhaps caterpillars with their more discriminating feeding were different. Jermy, Hanson, and I developed a technique for measuring preference in a multiple-choice test. We then showed that preference could be reversed by feeding larvae exclusively on a less-preferred plant. This phenomenon we termed *induction* (meaning nothing more than that a new preference had been established) because we did not, in our ignorance of the facts, wish to equate it with any particular category of associative learning.

In any case, the demonstration raised fundamental issues and at the same time provided a valuable tool for further studies of feeding behavior. The investigation continues in the laboratories of Schoonhoven in the Netherlands, Erich Staedler in Switzlerland, and Hanson in Maryland. A fundamental question is: does the change in preference involve alteration of receptor response or a change in the central nervous system interpretation of an unaltered sensory message? In the work mentioned above, evidence favoring a sensory change has been found. My own inclination, unsupported by any firm evidence, favors a central change.

Thus the decade at the University of Pennsylvania saw a culmination of many of the studies begun at Hopkins, studies that have been greatly expanded by Morita and his associates in Japan, Reese in England, and den Otter and van der Starre in the Netherlands. That decade also witnessed a migration from (but not abandonment of) the periphery to matters of central integration.

In a different but equal way the University of Pennsylvania offered as stimulating and productive milieu as had the Johns Hopkins University. Again it was evolution in the biology department with consequent defections that changed the milieu. The psychology department urged me to join them as a full member, but I was not, after all, a bona fide psychologist. Instead of accepting their gracious invitation, I moved on to an endowed chair at Princeton, where MacArthur had already emigrated.

The ambience at Princeton differed from everything I had previously experienced. It presented a cloistered atmosphere. It was a superlatively intellectual oasis lying between New Brunswick and Trenton. A friend who had spent a sabbatical there captured, I believe, the essence of the school in the remark, "Princeton University is a superb college."

Three of my students, Margaret Nelson, William Belzer, and Nancy Rachman, and one postdoctoral fellow, Blanka Bennettova-Rezabova followed me to Princeton. There experiments on the role of proteinaceous material and on learning in the blowfly continued. Other blowfly studies were also extended. To date most of my electrophysiological analyses had dealt with sensory matters, and motor events had been investigated only behaviorally. The brain was still terra incognita.

I had planned to monitor activity in the motor nerves serving the extensor muscles of the proboscis, but that study had just been concluded by Getting. Two new graduate students joined the group and were persuaded to complete the story of the proboscis motor system. Gerald Pollack worked out the sensory input/motor output relations of the spreading of labellar lobes and David Folk accomplished the same for the muscular pump effecting sucking. Thus, after fifteen years of work, we had constructed as complete a picture of feeding behavior as had been made for any animal heretofore. The major missing links were the integration centers in the central nervous system. With modern techniques there is no reason why this gap cannot be filled in soon.

At Princeton caterpillar studies continued apace. With an undergraduate, Jeffrey Kuch, I began an extensive electrophysiological survey of gustatory responses to a selection of compounds normally occurring in plants (carbohydrates, amino acids, glycosides, alkaloids, tannins, etc.). Schoonhoven returned for a second visit to the States and joined me in extending the study to olfactory and thermal receptors.

Quite apart from the significance of these studies in providing an understanding of insect-plant relations, there was an open invitation to attack the problem of sensory coding. I was back where I had started in 1937, still wondering how caterpillars discriminated among different plants. In the field of vertebrate perception, a controversy had been going on for years with respect to how the taste system coded information about the four primary tastes and whether in fact there were four primary tastes. The two extreme views were designated as the "labeled line hypothesis" and the "across-fiber pattern hypothesis." Pfaffmann originally proposed the latter for taste but later veered back to the former, while Erickson espoused the latter. Both of these views bore on the question of how caterpillars discriminated among plants. When I first looked into this question, the view, stemming from Verschaffelt's work in 1910, was that certain specific compounds in plants acting as token stimuli were detected by highly specific receptors and so identified the plant. This involved a labeled line system. Much of the early work on *Phormia* tended to support the labeled line hypothesis; there was a specific receptor for suger and one for salt. The caterpillar experiments at Princeton pointed to more broadly tuned receptors and a complex multineuronal response to plant saps.

Since the *Phormia* system was reasonably well understood, I reasoned that it could in one sense serve as a control system. Accordingly I began stimulating *Phormia* with exotic compounds, alkaloids, glycosides, etc. At the same time we recorded responses to complex stimuli (stale beer, rotting fish, vintage wines, etc.). Dr. Louis Mukwaya arrived from Uganda during this period and joined in the study.

The results of the combined work on caterpillars and *Phormia* bore on the specific question of host-plant discrimination and on the more fundamental question of gustatory coding. I advanced evidence for a hypothesis that discrimination of plants depended upon an assessment of their chemical gestalt and that the gustatory system, while indeed sensitive to specific token stimuli, relayed a much more detailed picture of the plant to the central nervous system. On the more general issue, I was able to add a different facet to the vertebrate controversy and adduce evidence in invertebrates of the importance both of labeled lines and of across-fiber patterning.

I had come almost full circle on caterpillars, but the blowfly had become such a useful tool in the intervening years that I kept studying it. My research was now candidly following two paths simultaneously.

Having now taught Introductory Biology (and a potpourri of other courses) for thirty-seven years, with a four-year hiatus during World War II, and having had my turn at the chairmanship at Princeton, I decided upon an early retirement to our home in Maine. Here, so I dreamed, I would pursue my avocation of writing fiction. In this pursuit I had been moderately successful. I had published two novels, a book of children's stories, and nine short stories. One of the latter was anthologized in the *Best of the Kenyon Review* and another in the *Best American Short Stories of 1981*. But it was not to be. Children still in college and the national economy dictated otherwise. I accepted a research professorship at the University of Massachusetts where for the first time I was essentially a totally free spirit.

Now I tackled the question of gustatory coding with a vengeance. Three postdoctoral fellows joined the laboratory: Elizabeth Bowdan of Hull, England, Mary Behan from Dublin, and Roberto Crnjar from Italy. Martha Yost, who had been my first research assistant at Hopkins, now lived in Amherst and rejoined me.

In thinking about the problem of coding I concluded that however useful the blowfly had proven, it had too many sensory neurons and much too limited a behavioral repertoire. What was required was an animal with few sensory neurons and a rich repertoire of behavior. Caterpillars fulfilled these requirements. Behavioral choice tests showed that tobacco hornworm larvae could discriminate among tobacco, tomato, and Jerusalem cherry and that preferences could be induced by experience and reversed. Furthermore, the behavior could be executed when the entire chemosensory system except eight gustatory cells had been surgically removed. This meant that somewhere in the electrophysiological response from the receptors lay the information identifying the plant. Examining the results of 900 experiments over a five-year period, Crnjar and I narrowed the candidate codes to four. The most promising at this writing is an ensemble code, the across-fiber pattern.

Although the studies of behavior that I have pursued over the years have touched many areas, I believe that there has been a philosophical consistency throughout. My abiding interest has been to satisfy a curiosity about the nature of animal behavior, especially causal aspects. In the pursuit I have attempted to dismantle the animal experimentally and reassemble it conceptually. Partly because I worked with insects, which have a reputation (not totally deserved) of being "instinctive" animals and partly because my work began in the era of reflex physiology and stretched into the era of cybernetics, I tended to emphasize input/ output relations. It did not take long to discover, however, that chains of reflexes were weak servants. The fact that input/output systems (e.g., the proboscis response) were sensitive to various physiological variables, together with the case that Kenneth Roeder made for the role of spontaneous neural activity in behavior, detoured me from a dead-end reflex path. What remained, however, was the proverbial black box. Ethologists and psychologists alike were puzzling over it. I was not content to accept ethological action-specific potentials as an answer. On the other hand, psychologists, who were not hunting Lorenzian hydraulic pumps inside the animal, were merely substituting drives and motivation. All these attempts to understand what was transpiring within the animal disturbed me because I felt that they went either too far or not far enough. I also felt that concepts like drive and motivation were being reified.

My thoughts at the time were gathered together in an address to the Eastern Psychological Association and enlarged at the 1966 Nebraska Symposium on Motivation, where I presented three versions of feeding behavior in the blowfly as told by: a naturalist, a physiologist, and a psychologist. I concluded that application of the concept of motivation added nothing to our understanding of the fly's behavior. If that was so, did it add to our understanding of feeding behavior of the rat, and, if so, what were the fundamental differences between the two animals?

I still had a lot to learn, and over the years I tried to educate myself more fully on these matters and refine my thoughts on animal behavior. These attempts brought me a succession of invitations to present these views. From each presentation I came away with a clearer perception of my position. My contact with my psychologist friends became ever closer the more we disagreed. Numerous fruitful exchanges occurred with Philip Teitelbaum, Eliot Stellar, Alan Epstein, and Randy Gallistel. At this point we have agreed to disagree.

Among the questions raised by many discussions was that of the extent to which there might be universal laws of behavior. To what extent are animals essentially different and how does this relate to the evolution of behavior? Evolution, theories of levels of behavior, and emergent mechanisms raise some very sticky questions indeed.

Having the background of an evolutionary biologist and believing firmly in the neural basis of behavior, I am disturbed by the segregation that persists between many students of invertebrate and vertebrate behavior. In my view it has placed unnecessary obstacles in the way of understanding how animals do what they do. Indeed, each of us has his experimental animal of choice; however, to remain in ignorance of what is being learned about other animals, and of other approaches and perspectives, is to miss opportunities to probe fully the depth and breadth of animal behavior.

Select Bibliography

Dethier, V. G. 1937. Gustation and olfaction in lepidopterous larvae. *Biol. Bull.* 72:7–23.

———. 1941a. Chemical factors determining the choice of food plants by *Papilio* larvae. *Amer. Nat.* 75:61–73.

———. 1941b. The function of the antennal receptors in lepidopterous larvae. *Biol. Bull.* 80:403–14.

———. 1942. The dioptric apparatus of lateral ocilli. I. The corneal lens. *J. cell. comp. Physiol.* 19:301–13.

———. 1943. The dioptric apparatus of lateral ocelli. II. Visual capacities of the ocellus. *J. cell. comp. Physiol.* 22:115–26.

———. 1947. *Chemical Insect Attractants and Repellents.* Philadelphia: Blakiston. Pp. 1–289.

———. 1951. The limiting mechanism in tarsal chemoreception. *J. gen. Physiol.* 35:55–65.

———. 1953. Host plant perception in phytophagous insects. *Trans. IXth Internat. Congr. Entom., Amsterdam* 2:81–88.

———. 1954a. Evolution of feeding preferences in phytophagous insects. *Evolution* 8:33–54.

———. 1954b. Sugar preference-aversion functions for the blowfly. *J. exp. Zool.* 126:177–204.

———. 1955a. Mode of action of sugar-baited fly traps. *J. Econ. Ent.* 48:235–39.

———. 1955b. The physiology and histology of the contact chemoreceptors of the blowfly. *Quart. Rev. Biol.* 30:348–71.

———. 1956. Some factors controlling the ingestion of carbohydrates by the blowfly. *Biol. Bull.* 111:204–22.

———. 1957. Communication by insects: physiology of dancing. *Science* 125:331–36.

———. 1959a. Egg-laying habits of Lepidoptera in relation to available food. *Canad. Ent.* 41:554–61.

———. 1959b. Food-plant distribution and density and larval dispersal as factors affecting insect populations. *Canad. Ent.* 41:581–96.

———. 1961. Behavioral aspects of protein ingestion by the blowfly, *Phormia regina* Meigen. *Biol. Bull.* 121:456–70.

————. 1962. *To Know a Fly*. San Francisco: Holden-Day. Pp. 1–119.

————. 1963. *The Physiology of Insect Senses*. London: Methuen. Pp. 1–266.

————. 1964. Microscopic brains. *Science* 143:1138–45.

————. 1966. Insects and the concept of motivation. In *Nebr. Symp. Motivation*, 1966. Pp. 105–36.

————. 1970. Chemical interactions between plants and insects. In *Chemical Ecology*, ed. E. Sondheim and J. B. Simeone, pp. 83–102. New York: Academic Press.

————. 1973. Electrophysiological studies of gustation in lepidopterous larvae. *J. comp. Physiol.* 82:102–134.

————. 1976. *The Hungry Fly*. Cambridge, Mass.: Harvard Univ. Press. Pp. 1–489.

————. 1977. Gustatory sensing of complex stimuli by insects. In *Olfaction and Taste VI*, ed J. LeMagnen and P. MacLeod, pp 323–31. Paris.

————. 1980. Food-aversion learning in two polyphagous caterpillars, *Diacrisia virginica* and *Estigmene congrua*. *Physiol. Ent.* 5:321–25.

————. 1981. Fly, rat, and man: the continuing quest for an understanding of behavior. *Proc. Amer. Philos. Soc.* 125:460–66.

Dethier, V. G., and Bodenstein, D. 1958. Hunger in the blowfly. *Z. Tierpsychol.* 15:129–40.

Dethier, V. G., and Chadwick, L. E. 1947. Rejection thresholds of the blowfly for a series of aliphatic alcohols. *J. gen. Physiol.* 30:247–53.

Dethier, V. G., and Crnjar, R. M. 1982. Candidate codes in the gustatory system of caterpillars. *J. gen. Physiol.* 79:549–69.

Dethier, V. G., and Gelperin, A. 1967. Hyperphagia in the blowfly. *J. exp. Biol.* 47:191–200.

Dethier, V. G., and Hanson, F. E. 1965. Taste papillae of the blowfly. *J. cell. comp. Physiol.* 65:93–100.

Dethier, V. G., and MacArthur, R. H. 1964. A field's capacity to support a butterfly population. *Nature* 201:729.

Dethier, V. G., Solomon, R. L., and Turner, L. H. 1965. Sensory input and central excitation and inhibition in the blowfly. *J. comp. Psychol.* 60:303–13.

Evans, D. R., and Dethier, V. G. 1957. The regulation of taste thresholds for sugars in the blowfly. *J. Insect Physiol.* 1:3–17.

Grabowski, C. T., and Dethier, V. G. 1954. The structure of the tarsal chemoreceptors of the blowfly, *Phormia regina* Meigen. *J. Morph.* 94:1–19.

Jermy, T., Hanson, F. E., and Dethier, V. G. 1968. Induction of specific food preferences in lepidopterous larvae. *Ent. exp. and Appl.* 11:211–30.

Omand, E., and Dethier, V. G. 1969. An electrophysiological analysis of the action of carbohydrates on the sugar receptor of the blowfly. *Proc. Nat. Acad. Sci., U.S.A.* 62:136–43.

Schoonhoven, L. M., and Dethier, V. G. 1966. Sensory aspects of hostplant discrimination by lepidopterous larvae. *Arch. neerl. Zool.* 16:497–530.

Wolbarsht, M. L., and Dethier, V. G. 1958. Electrical activity in the chemoreceptors of the blowfly. *J. gen. Physiol.* 42:393–412.

Irenäus Eibl-Eibesfeldt (b. 15 June 1928). Photograph about 1975 by H. Ahlborn

3

"Fishy, Fishy, Fishy"
Autobiographical Sketches

Irenäus Eibl-Eibesfeldt

I have a vivid memory of a glass jar with goldfish. It stood on a small table covered with a white tablecloth in our house chapel. I must have been very excited, as my mother told me later that I did not stop stuttering "fishy, fishy, fishy," which caused my father to remark in mock alarm: "Jessas, der Bua wird deppert [By jove, the boy is nuts]". But this is just a childhood tale—the pictorial engram, however, is reality and remains engraved in my memory. It happened on the occasion of my third birthday.

There exists furthermore a mosaic of pictorial memories in my mind—like short film scenes—of anthills in our garden with ants transporting caterpillars, of spiders that I fed with flies, and of toads swimming in a nearby pond that I caught and released in our garden. All this, of course, reflects nothing but the curiosity of a small boy growing up in a little village near Vienna. The only fact of interest is that this fascination did not fade away with age, but deepened with time. I assume that this must in part be credited to the understanding of my father who, as a botanist, tolerated, if not enjoyed, my childhood interests. He taught in Klosterneuburg near Vienna at the "Höhere Bundesanstalt für Wein-, Obst- und Gartenbau," a special school for horticulture. He loved mountaineering and by the age of nine, I often accompanied him on long excursions, such as one to the Lungau in the Tauern in the heart of the Alps in the summer of 1937. I saw my first marmots, learned about plants, fished for fresh-water lampreys in the ice cold waters of the Weissbriach, and collected blackberries and mushrooms. I had time to look, since my father liked to draw. He always carried a sketchbook with him and whenever he saw a picturesque farm house, an old church, or a weathered larch tree, he sat down and drew with skillful strokes. I often wonder what difference it would have made, had he just photographed the scene and walked on after a few seconds.

In every person's life, key experiences can be traced that in a decisive way direct the growth of one's individual interests, and one's outlook on the world. The first experience that I would consider to be such occurred during this trip. We made an excursion to the "Zinkwand" and got trapped well above the tree line on a steep slope strewn with large boulders. There was nothing in sight but the bare rocks around us. We sat in silence, wrapped in our cloaks, with the water dripping from our felt hats and watched the moving fog lifting its veil now and then only to enveil us seconds later again, so that I could only see the silhouette of my father.

After a while, in silence, he began to scrape upon the surface of one of the seemingly lifeless rocks with his pocket knife and the rock turned bright green. My father explained how even in this hostile environment, bitten by frost and baked by an unfiltered sun, life managed to hold out. Here, where no other plants grew, lichens still made a living by undermining the surface of a rock, leaving the outer layer intact, thus producing a micro-greenhouse, with small openings for the exchange of water and atmosphere. And with these endolithic lichens, a chain of life was founded with small snails and minute mites as consumers. I was impressed and ever since my eyes have been open for details—a rock with a few colorful patches of lichens competing for space or a patch of moss growing on a tree have caught my attention ever since. One year later I got my first microscope and little aquaria and jars filled my room, each populated with a rich variety of protozoa thriving on decaying matter. Soon I knew them by name and habit.

I was born in Vienna on June 15, 1928 and given the name Irenäus. The name is derived from the Greek word *eirene,* which means "peace," and can be translated as "man of peace." The name was intentionally chosen by my parents who wished in vain for peace for the generations to come.

My mother Maria, née von Hauninger, was a delicate blond art historian by training and painter and poet by passion. I loved her dearly, but the relationship was not without conflict. Her father was a "Feldmarschalleutnant" of the Austro-Hungarian army. I do not remember much about him, except that he played the violin with great virtuosity.

My father was also of delicate build, blue-eyed and dark-haired. He was very affectionate with us and more outgoing. In the evening he used to tell me and my younger sister bedtime stories.

My early childhood days in Kierling remain strong in my memory. I was very happy then and I still feel attached to Kierling, even though I go there only on rare occasions. The place has changed, of course, but the herbs on the side of the paths remain the same, as do the hills and forests. Kierling is a typical, small, lower-Austrian village in the midst of rolling hills covered with beech forests, meadows full of flowers, and patches of farmed land.

Our house was called the "Theresienschlössl." It was built in the eighteenth century to provide a resting station for official hunting guests. The spacious building still exists and is under protection. For us the building provided a rich environment. when it was raining, We spent the day playing in the attic, jumping from the beams down into the soft hay. There were also old storerooms, where I sat surrounded by spider webs with a beam of sunlight from the small skylight penetrating the dark interior, the little speckles of dust dancing in the beam. Here I sat and read old journals from my father's childhood, sagas, well-illustrated travel adventures, and old postcards. I spent hours among the crates, where all these treasures were hidden and I had my secret hiding places, where I stored away my favorite belongings.

Another exciting place was the pitch-dark cellar, which went deep into the hill and had side chambers. It had been previously used as a wine cellar, as well as a refuge. Here I played caveman, as a book, *The Höhlenkinder* by Sonnleitner, had caught my imagination. I made stone implements and read what I could about prehistory. Once when I was hunting crows with my bow and stone-tipped arrow, I crawled over a freshly plowed field and found finely worked flint scrapers. After this, I started to search actively in the fields around Vienna, which abounded with archaeological remains such as scrapers and fragments of old pottery from Roman and pre-Roman times.

Our garden had three separate parts, the flower garden, the vegetable garden, and the orchard with old walnut, cherry, pear, and apple trees on the hillside behind our house. I built little tree houses in those large trees, and I dug burrows in the ground, sometimes in cooperation with our neighbor's son. I had a band of friends, with whom I roamed the countryside, climbing other peoples' cherry trees even though our own garden was full of them. I had some fights with other children, particularly in spring, when I fought bitter battles to defend toads against the cruelty of some boys.

In March 1938, airplanes dropped propaganda leaflets for the "Anschluss" and Kierling throbbed with excitement. Cheering crowds chanting slogans milled through the main street of the village. At one evening rally they carried burning torches, which they threw in the creek. For us boys it was all very exciting and I still have a vivid memory of the cheering crowd. We collected the torches and illuminated our caveman's cellar.

In 1939 we moved to Vienna. Even though we were close to the country and even though my room had a special door leading directly to the garden, my movements were greatly restricted. I turned my room into a zoological garden, in which a hedgehog roamed freely. Between the inner and outer window, I kept white mice. I also had several leaky aquaria, in which I kept communities of plants and invertebrates just as I found them in the ponds and creeks of the Viennese forest.

In the spring I studied the courtship of newts, and later the development of their larvae. I kept tadpoles through metamorphoses, raised a nestling crow, and made plaster nests in which to study red forest ants. It was clear to me that I would study zoology and I dreamed of traveling to faraway places.

I started to read zoological texts when I was ten. Floerike and Bölsche in particular had a decisive influence on me as did the journal *Kosmos* with its accompanying little booklets. Whenever I got hold of a travel account of some explorer, I devoured it. When I was thirteen I had already read Hesse Doflein's *Tierbau und Tierleben* and on my fourteenth birthday I got Doflein-Reichenow's *Lehrbuch der Protozoenkunde*.

Zoology absorbed me completely and I neglected other subjects. I played hooky in order to roam in the forests, fish in the ponds, or watch insects in flowering meadows. "Der Bub lernt keine Wörter [The boy doesn't learn vocabulary]," my teacher in Latin used to complain. I performed "poorly," to put it mildly, in all subjects except biology and English. Were it not for the understanding of my teachers at that time, I would have failed. When my father died in March 1941, my performance worsened. My mother, my teacher, and my two aunts decided to engage in a rescue operation: no animal pets anymore, and, from then on, boarding school, with strict schedules of Latin, physics, history, mathematics, geography. I had nothing else to do but to learn, and my school performance improved dramatically. Still, I felt depressed and the relationship with my mother became difficult. I was left on my own while my mother escaped into her world. Looking back now, I realize she was very unhappy and anxious because many of her friends had disappeared for political reasons that, for fear of endangering me, she could not speak to me about. To try to communicate, she wrote some poems for me. On the second anniversary of my father's death, she wrote me a very long letter about herself, her life, and her family background, tracing her roots for generations.

In January 1944, my whole class got drafted for military service; I was fifteen and a half. After four weeks of military training, we were sent to serve in an antiaircraft battery at Breitenlee, north of Vienna. I ran the 10.5 cm gun together with two schoolmates, some regular soldiers, and Russian volunteers. My task was to control the vertical movements of the gun. We were a strange mixture, united in despair, and were kept going by the belief that it was up to us youths to save our city Vienna from being turned into rubble. During this time, we still received education from our teachers, who came to our battery. I often wonder how they felt, when they saw their pupils in uniform, running to the guns when the alarm bell sounded, serving combat for an hour or two and then returning to the bench.

Many hours were spent simply waiting, ready at the gun. I read a lot of paperback classics—Shakespeare, Grillparzer, Molière, Kleist, Ibsen, Goethe—whatever I could get hold of. We were largely left in isolation. Our battery was located in the midst of the fields and only every second or third week were we given a weekend off to visit our relatives. We discussed the events of the war among ourselves, but certainly were not well informed. In the second half of 1944, Vienna was bombed and we were increasingly involved in combat, our battery downing seventeen planes. They were targets, and, for us, the representation of the evil, since they were bombing our people. But once, when we captured an American who had parachuted when his plane was shot down, we felt curiosity and not hatred. We simply did not associate the bombers with people. He, by the way, looked similarly puzzled, towering over us school kids by a foot or so.

In the summer of 1944, my cousin Manfred paid a farewell visit to me. He was a very friendly person and had just been ordained a Catholic priest. He had been drafted to serve as a priest in the war. He tried to cheer me up, speaking about God's will, but I could not accept his views at that time. We argued until the tears ran down his chubby face. In parting he made the sign of the cross on my forehead and kissed me goodbye, which touched me very much. At least he had convinced me that there was still some good in man. He lived up to his name, Manfred, which was given to him by his parents in the hope that he would serve peace *(Manfred—der zum Frieden Mahnende)*. He was later killed by partisans, who did not care whether their prisoners were carrying arms or a cross.

From January to March 1945, we received special military training for the front within the "Reichsarbeitsdienst." Before I was transferred to another unit, I got civilian status for fourteen days, beginning March 27 and during the first days of April, the Russian conquest of Vienna ended the war for me.

In my days of newly won civilian status, I decided to leave the city to go to western upper Austria where my mother and sister had been evacuated. I do not remember now whether it was the third or fourth of April when I left my aunt's flat in Vienna. The streets were empty, except for occasional soldiers and patrols checking papers. Mine were in order. When I was asked by a patrol where I was going, I said, "to Döbling," where I was in fact registered and which lay en route. The center of the city was burning; barricades against tanks were blocking the streets; bombers, this time Russian, flew overhead; and our artillery was kept busy.

I arrived safely at the beginning of the "Döblinger Hauptstrasse," where the Stadtbahn Viaducts cross the street. Before I entered the main road, on the corner before Döblinger Hauptstrasse, I asked soldiers waiting at their machine gun, whether the Russians had already reached Döbling or if they knew where they were. They had no idea. I wished them good luck, rounded the corner, and there they were, right where the drugstore was and still is. There was a tank, a horse cart with a dead horse and a few Russian soldiers in uniform and with their typical helmets, and a civilian. The dreaded Russians!

They shouted and beckoned me to come. With my heart in my throat, I obeyed; I had a pistol in my pocket which I handed to them, before they could ask for it. "Du Soldat!" was the immediate accusation. "Nix Soldat," was my reply. We repeated this dialogue several times, the Russian with a rough barking voice and I desperately timid. The civilian fortunately spoke German, as he was a foreign worker, like many others from all over Europe who had been drafted into the labor force. I explained to him that I brought this gun only in case of emergency, since I wanted to go to my sister and that my official status was one of a civilian.

They chatted back and forth and finally in my despair, I pulled out the official document that declared me a civilian, even though I feared that the insignia of the Third Reich would infuriate them. But a document is a document. The civilian read slowly "Reichsarbeitsdienstentlassungsschein," then that I indeed lived in Döbling, that I had even been born there. I gave a nod of reassurance; they handed the document back to me and motioned that I could proceed. I could not really believe that they would let me go. The street is a rather straight alley, and as I walked slowly away, I expected every second for an endless 200 meters to feel the impact of a bullet. But they showed mercy on a sixteen-year-old boy.

The months to follow fueled my thoughts about the extremes of human behavior and brought a new awareness. We learned about the atrocities and murders in the concentration camps, about the systematic mass killing of Jews, about the brutality of our occupation troops in Russia, and of the millions of Russian soldiers who were left starving by intentional neglect. My beliefs and values were quite thoroughly shattered and a deep resentment against those who had abused their power to misguide a whole nation grew.

The information about the atrocities committed in the name of our people brought about a deep crisis in identity. It was so hard to believe; even now it haunts my thoughts. It took me years to accept my roots again. Basic values like loyalty could not be trusted. Later, they called my generation "the angry young people." We lived in shame, and many of us radically broke with tradition. It was no longer easy to identify with our people, history, and culture.

Nevertheless, life went on and conditions slowly improved. Electricity came back, at first for only a few hours a day. The streets were cleared of rubble, work for which we had to volunteer during the first weeks after occupation. Law was reestablished and some streetcars resumed their routes. Vienna was coming back to life again.

Because the education we received during our military service had been sort of a crash program, without the usual holidays, I had officially finished seven years of gymnasium, and the note of "maturation" was entered in my certificate. This allowed me to enroll in the University of Vienna, which opened in May. Bombs had destroyed many parts of the building, and between lectures we had to clear up the rubble.

We were half-starved at that time and often desperate; not at all subdued, however, but in an elation brought about by despair. We worked hard and with new enthusiasm. There was no fear of bombs anymore, but the misery of those of our kin who had been expelled by our neighboring country was depressing. There were signs of improvement in other sectors of life, however, and there was always hope for a better world.

Of my teachers, Ludwig von Bertalanffy and Wilhelm von Marinelli influenced me greatly. At that time, there were not many students, and our professors addressed us with the respectful and friendly term "Herr Kollege." Bertalanffy invited those who were interested into his home to discuss mathematics for the natural sciences. Marinelli had a small zoological *Arbeitsgemeinschaft*. It was there that I met Otto Koenig, a vigorous man in his early thirties. He had located a number of abandoned army barracks in the Viennese forest opposite the Castle Wilhelminenberg and decided that we should occupy them and turn the site, including a lovely pond, into a biological field station for behavioral research. Along with several others I joined his group, and I got my first lessons in practical ethology. Otto Koenig had large posters printed: "Biologische Station Wilhelminenberg"—"Eintritt verboten!" and placed these property markers all around the occupied site. This completely illegal venture worked. We lived there, invited the press, and were acknowledged. The station still exists.

We were seven students, old, close friends from boarding school: Wolfgang Schleidt, Ilse Gilles, Heinz Prechtl, Edmund Frühmann, Eberhard Trumler, Kurt Gratzl and me. There were six barracks, which we had to clean. We each had rooms of our own, and with boards, which we collected wherever we could find them, we made our beds and furniture. By painting with a dark stain, we covered up all irregularities so that our work looked professional. We were very proud of our skills.

The station soon was crawling with tame animals. Koenig raised herons, which we brought from Lake Neusiedl in unforgettable excursions. They flew around freely. I began by observing the breeding behavior of the common toad and through translocation experiments I learned about their migration to ponds (Eibl-Eibesfeldt, 1950a). In addition, I raised a baby badger, the most wonderful pet I ever had. From him, I learned much of relevance to my future work in human behavior. Once weaned, the badger lived under my barrack. For months he remained strongly attached to me and every evening he came to visit. He would open the door, enter the room, and begin to be a nuisance. He would tip over a can with water, then proceed to open my bureau and pull out my clothing. When I tried to chase him away with "Pfui Dachsi," I only got a blank stare. Whereas a dog would learn easily to adapt to commands, this badger proved to have a more stubborn character. I became aware that in order to obey and submit, an animal must be programmed for it, as is the case with dogs, which are adapted to living in groups with dominance hierarchies. Badgers are not so adapted, and therefore lack the capacity to submit to authority. When I finally gave him a slap to prevent any further destruction, he would growl and attack.

Otherwise he was a playful, friendly creature and his inventiveness during play caught my attention. One day, by accident, he discovered somersaults and practiced them until he was able to somersault to perfection down a hillside. Once we went down an icy road to town. In a playful invitation to a chase, he ran in front of me, but when he tried to brake for a sudden halt, he started skidding. Immediately he forgot me and the chasing game and experimented with skidding. He was freely inventing new motor coordinations as well as new applications of skills, upon which he had hit by chance (Eibl-Eibesfeldt, 1950b).

Another peculiarity about his play struck me: he was combining motor patterns from different functional systems, which normally are mutually exclusive. Patterns belonging to the repertories of fighting, hunting, and sex were combined in play. Furthermore he freely exchanged behavior from playful fight and attack with flight behavior, autochthonous motivation evidently being absent. During playful fighting, the badger showed no signs of fearful arousal, nor did he act, during such play, as if he were motivated to fight. Among other things, he showed strong inhibitions against biting and never broke my skin during these play fights.

This led me to postulate that by a special mechanism of detachment, movement patterns of the lower level in the hierarchical organization, as described by Tinbergen's model, were activated independently of the mechanisms that activate them under normal circumstances. Thus it is not the whole "instinct" that is activated during play. Play then lacks higher motivational arousal and this allows an animal to dispose freely of the repertory of motor patterns available to it and to experiment with its skills in ever-new combinations. Play is thus experimentation with one's own abilities and with objects in the environment. It was selected as a means of active information acquisition. Ever since my experience with this badger, the subject of ontogeny of behavior has been of central interest to me.

In winter, our little barracks were invaded by mice. First I started to trap them, but then became tired of finding their little corpses in traps and so I decided to let them be and study them. This started my comparative work in the social behavior of rodents. Starting with the concept of the ethogram, I went into the study of mice and other rodents with particular concern for the ontogeny of their behavior

(Eibl-Eibesfeldt, 1950c, 1951a,b; 1952a,b, 1953a). In addition, I continued my studies of anurans, keeping a wide variety of frogs in terraria. Later I added salamanders and reptiles.

In December 1946, we published the first issue of a journal named *Umwelt*. Koenig was the chief editor and each one of us was responsible for one section. I signed up for the reptiles and amphibians. We had to write many of the contributions ourselves, and thus acquired writing skills. The journal was appreciated by many, but times were still too tough for such an enterprise and it lasted only two years.

In 1947, I got my first glimpse of an intact world. Otto Koenig had to make a shipment of herons (white herons and purple herons) to Heini Hediger who was the head of the zoo in Basel. They had to be fed during transport and Koenig chose me to accompany them. It was my first visit to a foreign country and I decided to take advantage of the entry permit and make a small side trip. What a contrast Switzerland was to our beaten country! After a wonderful day with Hediger, I stood at the Basel train station waiting for the connection to Lucerne. As I was standing there, somewhat forlorn, I saw a lovely young lady whispering to what seemed to be her mother. Then she went to a machine, put in a coin and pulled out a large Suchard chocolate bar and handed it to me. I was taken by her charm and accepted it. This event touched me deeply and I have always associated it with the Swiss. So far nothing has changed this friendly image and I do not think that anything ever could. The fact that such a minor event involving just a small token led to a strong attachment made me contemplate the roots of friendly behaviors in man. Under the given situation, with general hostility still reigning in the world, the incident meant, of course, much more than it would today.

In 1947, I met my future wife Lorle Siegel, who was also studying zoology at the University of Vienna. From then on we raised pets together. My badger resented her for a while and bit her in her calves, half playfully, half maliciously, providing me with repeated opportunities to rescue her.

Winters were rough in 1946 and 1947, cold and with little food. We chopped trees illegally to feed our little stoves. Lorle exerted a cultivating influence on me. We went to operas and concerts at the time when Wilhelm Furtwängler and Herbert von Karajan were conducting the Vienna Philharmonic Orchestra in turn in the Wiener Musikverein. Vienna was split into two parties, those who hailed Furtwängler and who found Karajan to be too much of a showman, and the others, who considered Furtwängler to be too stiff. There was no solution except to enjoy them both, which we thoroughly did. A standing place at that time cost only a few shillings or about twenty-five cents.

I, in turn, read Lorenz's papers to Lorle. I had acquired some reprints of his papers and they fascinated us. In this sense, we were his pupils before he returned in 1948 after four years as a prisoner of war in Russia. He returned carrying a manuscript and a cage that he had made of wire, in which he kept a hand-reared starling. From this manuscript, which was written on paper torn off concrete sacks, he gave his first lectures to our small group at Wilhelminenberg. I had already read most of his publications, but to listen to him was enthralling.

Shortly after his return, Konrad Lorenz suggested to Lorle and me that we be on "du" terms with him, which in German signifies the establishment of a close personal relationship. I have looked upon him as a fatherly friend ever since. Lorle's family had had a longer relationship with Lorenz, since his father, Adolf Lorenz, had been a tutor in the home of her grandmother's family in his student days. Furthermore, Konrad Lorenz's grandfather, Zacharias Lecher, was one of the founders of the Free Press *(Freie Presse)* in Vienna, where Lorle's great-grandfather was chief editor. For two generations, both families shared a trait—a loud voice—and their explanation was that they had to outdo the loud printing press.

Lorle and I were married in February 1950. She was occupied with her thesis, while I had finished my exams and was teaching at the "Piaristen Gymnasium," a school with a long tradition. In the summer months, we often visited Konrad in Altenberg, spending the day on an island on the other side of the Danube basking in the sun and chatting.

We had left Wilhelminenberg and moved into a flat in the city. We had little space at that time, but still kept animals. Wild rats, which I caught in the sewer systems of Vienna, were our favorite and I tried in vain to breed them—the wild ones simply stayed wild. We were successful only when I found some nestlings, which Lorle nursed with a bottle. In addition, we had in our apartment two hand-raised squirrels, which we kept in a cage, although I often let them run free. In spring 1951 they started to chase each other. The male was courting. He was extremely aggressive during this time, attacked us, and, in displacement activity, furiously gnawed on our furniture. It still shows the traces. But they finally bred, and it was the first time that red squirrels were bred in captivity.

In the meantime, Lorenz had received an invitation from the Max Planck Society to build, jointly with Erich von Holst, an institute within the Society. I, along with Wolfgang Schleidt, was invited to join him as a research assistant. Our squirrel had just given birth and so we moved, with squirrels, rats, and mice to Buldern in Westphalia. Our address was Schloss Buldern in Westfalen. It sounded nice, Castle Buldern, but we were not living *in,* but *at* the castle for the first weeks, all crammed into a small water mill that belonged to the castle. Lorle, our animals, and I lived in one room within Konrad Lorenz's flat. The squirrels ran free in order to prevent any frustrations that would have endangered the life of their young, which soon made their first excursions and were lovely pets. Soon the male, "Fritzi," was chasing the female, "Mucki," again. It was in this setting that Otto Hahn, the president of the Max Planck Society paid his first visit to us. He appeared to be impressed.

After a couple of weeks we moved out of Lorenz's mill into the castle's bowling alley. We had a nice quadrangular room with, at one end, a miniature bathroom with no bathtub or shower, and a narrow alley, of which a few meters belonged to us. Here we lived for years together with our animals, including a tame galago, who marked his path by urinating in his palms; rubbing the urine into the soles of his feet, and "tap, tap" leaving traces wherever he walked (Eibl-Eibesfeldt, 1953b). We soon discovered his traces but it took time until we realized how they had been made.

Our zoo was gradually expanded by the addition of polecats and small weasels. Our institute was a big family. Lorenz's mill was open to everyone, and we usually went there for coffee after lunch or in the evening to discuss a wide variety of issues in an intimate and relaxed atmosphere. Lorenz spent the day with his geese in a part of the castle's park, which was rented by the Max Planck Society, where a stream had been damned up to form a shallow lake. Here we had a little platform and a wooden cabin. On sunny days we all went there to chat and bask.

Lorenz's fame attracted many visitors, such as Nikolaas Tinbergen, Gerard Baerends, Eckhard Hess, and Jocelyn Crane. We also had students from other countries who stayed with us for months or even years. Mike Cullen and Uli Weidmann were among them. We were happy, enthusiastic, and full of hope for a better world.

My interests went in two directions. First was the study of ontogeny with particular emphasis on the question how innate and individually acquired components of behavior interact in mammalian behavior (Eibl-Eibesfeldt, 1955f, 1956a, b). I used the study of the normal ontogeny coupled with deprivation experiments to tackle these problems. At the same time, I was interested in animal communication and such questions as: how do signals evolve, how are they understood, and in which way do they function to control social interaction. Here I used a typically "Lorenzian" comparative approach, but credit must also be given to my teacher Wilhelm von Marinelli, who trained us thoroughly in comparative morphology. I had learned from him the concepts of homology and analogy and the significance of both in the comparative approach. It was Lorenz who added the perspective that motor patterns could be treated like morphological structures.

In autumn 1951, Heinz Sielmann approached me. He had heard about my work with frogs and rodents and he asked me whether I would cooperate with him in the production of educational films (Eibl-Eibesfeldt, 1954a,b,c; 1955g). In the spring we filmed frogs in ponds and aquaria watching patiently for hours in order to catch the moment of spawning.

In the summer, we made a film on the European hamster. We constructed an artificial burrow and succeeded in filming the entire reproductive cycle from birth to propagation, including their secret subterranean life. In the following year, we made a film on polecats. Since this time Heinz and I have been close friends. I learned a lot about filming, including patience and self-control. Animals, after all, tend to do the most important things either when the film needs to be changed, when there is a blockage in the camera, or when you take off a minute to stretch. Since that time, I have worked in close cooperation with the Institut für den Wissenschaftlichen Film in Göttingen.

In 1952 we were honored to be the hosts of the first International Ethological Conference. Niko Tinbergen, Gerard Baerends, Otto Koehler, Erich von Holst, Lars Hartmann, Jan van Iersel, and many of the younger people who met at this conference have stayed in contact since. We were a small group, particularly since behaviorism was still very much in vogue in the United States. But slowly European ethology began to make its inroads. The first response came in the form of an attack by Danny Lehrman, whose 1953 review article started off a lively discussion and a personal friendship between the opponents, Lorenz and Lehrman. Both discovered during their first personal encounter that, at least outwardly, they shared one physical trait—both were overweight. Later they found that they had much more in common. I was inspired by Lehrman's publication and decided to concentrate on systematic deprivation experiments with rats, squirrels, and small carnivores. I had already begun along this line with my study on the development of nut-opening techniques in squirrels.

Two events marked 1953—the birth of our son Bernolf and receipt of an invitation to join a yearlong excursion to the Galápagos Islands. The events took place in the reverse order. In June, Lorenz came to me as I was feeding my polecats and said: "Renki, would you like to join Hans Hass for a year on a diving expedition to the Carribean Sea and the Galápagos Islands? I have already asked Lorle and she said it was alright with her." I was in a quandary. On the one hand, we were expecting our baby in October, and on the other, it was too good an opportunity to miss. I knew nothing about diving, but was sure that I would be able to learn. I had read William Beebe's *Galapagos: World's End,* which had struck me as outstanding because of how open his eyes were to "hidden wonders." His lively description of the islands had caught my imagination. Now I could actually see those islands that had inspired Darwin.

What if Lorle had said no? And what if Konrad had not agreed? There were many things to be done at the Institute. I will always be grateful for their decisions. During this trip, as well as later during my long absence, my wife ran my office, took good care of all my numerous animals, and ran experiments. She was a trained ethologist and had gotten her doctoral degree with Lorenz.

My first meeting with Hans Hass was awkward. He had expected a settled, distinguished scientist, and here came a young man who was little more than a boy—I looked much younger than I was, and at congresses, people often asked me, after I had introduced myself, whether my father was also here. And so there I stood in front of Hans Hass, who was still looking for the expected distinguished scientist. I introduced myself to Hass who took one look at me and said that he was not yet sure whether there was really a place for another scientist on board his research ship, but that I should nevertheless come home with him. There I met Lotte Hass and we three spent the evening, chatting and listening to music. Before I departed, Hass had decided that there was a free place on his ship!

On the twenty-third of August, our ship *Xarifa,* a three-masted sailing schooner, of 350 tons, sailed down the Elbe and our great adventure began. First of all, I became seasick, but by the time we reached the Azores, I had overcome this handicap and thoroughly enjoyed the sailing, even in rough sea. It would take me hundreds of pages to describe the numerous impressive events of this trip. From Emmo Ankel, who joined us for the first couple of weeks, I learned about the invertebrates of the surface of the ocean. We fished Janthina, Velella, and Glaucus and other pelagic animals. At the Azores, we went with Portuguese whalers and observed their sperm whale hunts. Hass filmed sperm whales under water for the first time, as well as the white-tipped sharks, that arrived in schools when the blood of a harpooned whale tainted the water. I made observations on schooling behavior. On our way to the Canary Islands, another storm caught us and in the midst of the turmoil, we received a message by amateur radio: "To Dr. Eibl: It's a son!" We celebrated the event.

Whenever we came to an island, I took off on an excursion and soaked up new impressions, particularly on Tenerife, which, at that time, had not yet been discovered by tourists. Many of the invertebrates and xerophytic flora were related to similar ones in Africa. I had read Alexander von Humboldt's description of his ascent of Pico de Teide and climbed a considerable stretch to learn about volcanism.

We first reached the New World at Santa Lucia and I went off in search of frogs, spending half of the first night in the bush. Los Roques and Bonaire were our next stations. For the first time I dived with tanks. We had breathing equipment that was light and comfortable, but at first I had difficulties getting down from the surface. Hass finally pulled me down, grasping one of my flippers, and I swallowed lots of seawater, as the mouthpiece did not close well. I struggled to come up, but Hass pulled me down until I managed to fight my way up. He explained that it was only my imagination, but agreed to switch equipment. Soon he was swallowing water, while I, waiting for him to come, sat peaceful and content on the ocean floor in the midst of coral gardens teeming with fish. The teeming life filled me with awe. Those dozens of colorful fish! Would I ever be able to know them by name? Would I ever find any structure, any way to get a hold on a problem? I was confused, but only until a strange event caught my attention. Along came a fat grouper, paddling along with slow movements of his fins. He came to a halt above a coral block just in front of me and to my astonishment, two little blue fish with dark lateral stripes along their sides swam up to the fish and started to search around his head. Upon this the grouper opened his mouth and lifted one operculum. One small fish slipped under it, searching for something between the gills, while the other entered its gaping mouth! Why would a small fish ever enter the mouth of a predator! The grouper closed his mouth, but, to my surprise, not fully. Upon this sign, the little fish left the mouth cavity of the grouper, which only then closed his mouth fully and started to breath. He then shook his head and the two fish returned to the coral block from which they had come. The grouper swam away, only to be followed by a damselfish, which took his place. Again, the two blue fish came and inspected the damselfish. I had happened upon an interesting phenomenon. I soon found out that I had discovered a clear case of mutualism. The small fish were cleaners removing ectoparasites from the body surface, gills, and mouth cavity of other fish. The cleaners and their hosts communicated by simple signals. I coined the terms *cleaner* and *cleaning symbiosis* and studied the phenomenon in detail (Eibl-Eibesfeldt, 1955e).

We were a fine team. Hans and Lotte were both experienced divers. Lotte, in addition, knew when a party was needed to raise our spirits. Her charm was an important asset on this trip. Another excellent diver and fine person was Jimmy Hodges, a lieutenant commander in the British Navy, who was hired as an underwater cameraman. Heino Sommer was our ship doctor and radio operator, Kurt Hirschel technician, and Georg Scheer and I were the two hired zoologists.

In December we sailed on. We spent Christmas in Panama where we undertook long excursions to the Chilibri bat caves and forests. For several nights, I followed the course of creeks, in order not to lose my orientation, caught helmet basilisks *(Basiliscus vittatus)* and watched armadillos in the moonlight. Then on we went to the Galápagos Islands, which we reached on the fourth of January 1954. The next day I moved ashore onto the tiny islet that William Beebe had named Osborn. I was fascinated by the sea lions and studied their harem life and parental behavior both on land and under water. Their territorial behavior in water struck me as peculiar, as did the paternal behavior of the bulls. Whenever a calf ventured into the deeper water, a bull cut it off and edged it back to the shore again to prevent it from falling prey to the numerous sharks (Eibl-Eibesfeldt, 1955b,c,d). We toured the archipelago for two months, diving and observing.

I studied the ritualized fighting of the marine iguanas (Eibl-Eibesfeldt, 1955a) on Narborough and was enchanted by these islands, which had so many faces and whose tame animals allowed close approach. But I was also appalled when I saw the damage that had been inflicted on the flora and fauna of some islands by fishermen and settlers. I realized that a biological station should be established in order to observe the development and promote conservation measures.

On our return trip, we passed the Cocos Islands, where we inspected a wreck in Wafer Bay and where I saw my first school of large hammerhead sharks. Back in the Carribean, we passed the San Blas Islands and paid a short visit to the Cuna Indians. A month of "heavy diving" near Bonaire followed. Hans Hass had committed himself to produce a theater film and we had to perform both under and above water. The weather was not good. When clouds shielded the sun, we could not film, since we used color films that, at that time, were not very light-sensitive. We would sit waiting in the reef, until our boatman gave us the sign to proceed by lowering one of our oxygen bottles on a rope. Then there would be sun for a few minutes. We were freezing, as diving suits were uncommon at that time.

It was hard work. Once the script required that Dr. Scheer and I place a large mirror in the reef, so that the fish could see their own images. Under normal conditions mirrors trigger territorial aggression against the "intruder." We had to place it neatly, look into the mirror, make some important remark, and leave the scene. So there we came with the large mirror, each of us holding one end and drifting by. We placed it neatly and Hans grunted "Mmmh, Mmmh." From behind his camera he signaled to us again. Perhaps one of us had obscured the view. We picked up the mirror again, but this time a cloud passed in the middle of the scene. "Stop!" And now again! Another "Mmmh, Mmmh, Mmmh." Hans motioned that we should surface. Then he explained that we should put the mirror down, look in it for a few minutes, and act as if we were engaged in dialogue. He added, "please act naturally [*ganz natürlich*]." "Ganz natürlich! freilich!" Easy to say, but it is hard to act naturally when you are already freezing and in a rebellious mood. Again we drifted in with the mirror and again and again. It was an agony. The end came when I stumbled during the process of placing the mirror and broke it into three parts that collapsed in slow motion, or at least slowly enough for us to see Hans's expression of despair. We chuckled until the water rose in our masks.

It was tiring, but rewarding work. Later we were the first to use underwater spotlights to bring to life all the colors of the reef. In caves and under the coral heads, a multicolored community of sponges, ascidians, and soft corals shone red, violet, yellow, or green, whenever they were illuminated with the spotlight. We learned what patience and persistence meant. And Lotte Hass, ever enthusiastic, devoted, and most of the time in high spirits, set a good example. She also had a very civilizing effect on us, who, after months at sea, were inclined to be rough-mannered.

We lost our best diver, Henry James Hodges, during those weeks while we were taping the sounds of harpooned fish to find out what attracts sharks over distances. We never found out exactly what happened to him, but during this work, he might have followed a fish too far down, using oxygen equipment that was later found to be dangerous. The last time we saw him, he was acting normally. He left our group to surface without apparent signs of distress. We found him dead, ten minutes later, not far from where we had worked in the reef. He and I had shared a room, so I deeply felt the emptiness caused by the loss of a friend.

The working rooms of our ship had turned into a small zoological garden in the meantime. I had taken lizards and iguanas from Panama and frogs from Carribean islands. These had to be fed. By radio, I had asked my wife Lorle to send meal-worms to feed my animals. One day Lotte Hass came to me complaining, "Can you imagine, there are worms in the letter I just got. Unglaublich!" I could imagine it, but I kept silent. I received an empty little parcel with a tiny hole in it.

At the end of May, our expedition ended. Lorle met me in Geneva and we returned to Buldern in Westphalia. I held my little son for the first time. He resented me at first, as he had reached the age at which every healthy child discriminates and shows fear of strangers. Back in the Institute, I was very busy, eager to publish my observations. In addition, I felt that something needed to be done to stop the deterioration of habitats and animal life on the Galápagos. I wrote a report to UNESCO and IUCN, proposing, among other things, the establishment of a biological station. The following two years were happy ones, since we lived at the Institute and work and family life were one. I had time to play with my little son; we went on many excursions and he could be with us at the little house by the goose pond. In April 1955 our daughter Roswitha was born, a healthy, cute baby girl.

My son Bernolf was then one year and seven months of age. He looked at his sister very thoroughly and carefully and then classified her as a "wau wau." Evidently he recognized her as something like a little puppy, a mammal. But when Lorle brought Roswitha to her breast, Bernolf thought that this was too much and tried to pull the baby away, but we did not let him. He turned abruptly, took my finger and said, "Let's go."

Our financial situation, which had started on a very narrow base, improved during this period. In order to arouse public interest in the Galápagos, I gave many lectures, traveling on my motor bike all over Germany. I also had frequent correspondence with IUCN and many other colleagues who were interested in the Galápagos. My proposal had awakened interest and in 1957, I was asked to head a survey on the Galápagos Islands whose goals were to provide information on the present state of affairs, to work out proposals for conservation measures, and to propose a suitable location for a biological station. I was accompanied by Robert

Bowman, an ornithologist from San Francisco and by the two *Life* reporters Rudi Freund and Alfred Eisenstaedt. From mid-July to the beginning of November we traveled in the islands, climbing volcanoes in search of tortoises and land iguanas. My attachment to the islands grew. In addition, the trip was a success. I wrote my UNESCO mission report (Eibl-Eibesfeldt, 1959b), and in 1959, the Charles Darwin Foundation was established in Brussels and the green light was given for the establishment of the proposed Charles Darwin Station, which grew and is now flourishing. I am still a member of the executive committee.

In the same year, I worked on a number of projects with Hans Hass, with whom I was by then a very close friend. He had entrusted me with the scientific director-ship of his International Institute for Submarine Research, which organized the Xarifa expeditions. We prepared for another diving expedition to the Indian Ocean in the following year.

I returned from the Galápagos on the fifteenth of November and had to depart on the twenty-eighth of November in order to catch the *Xarifa* in Aden. Roswitha had just become familiar with me again and Bernolf followed me around as if he knew that I was about to leave. It was hard to depart and when I returned in November 1958, Roswitha asked, "Is that daddy?"

We were six scientists this time: Hans Hass, Sebastian Gerlach, Wolfgang Klausewitz, Georg Scheer, Ludwig Franzisket, and myself. On the twenty-first of December we reached the Maldives. The next day we dived in the Addu Atoll. We dived for at least two hours every day, sometimes more. By then, I was an experienced diver and I knew what to look for. Again I studied the cleaning symbioses and discovered a mimic fish tht in coloration and behavior imitated the "cleaner-wrasse" in order to get close to the host (Eibl-Eibesfeldt, 1959a). But instead of cleaning, he attacked him and bit chunks from the soft parts of his fins and skin. I also made extensive collections, gathering data on ecology and behav-ior at the same time. Together with my friend Wolfgang Klausewitz, I described a number of species, including garden eels, which at that time were little-known (Eibl-Eibesfeldt & Klausewitz, 1959). I made the first photographs of the amazing fields of eels. Studying the adaptive radiations of fish was like taking a course in phylogeny.

We slowly traveled the islands, from south to north. In March we spent exciting days at Goha Faro where we decided to feed sharks in order to study their behavior toward repellents (Eibl-Eibesfeldt & Hass, 1959). We accidently hit upon a good method to do so. While exploring a wreck, I hit a coral block, which tumbled down the outer reef slope. This triggered the response of two sharks, which shot up from the deep, evidently prepared to hunt. This gave me an idea. I gave Hans a signal, shot a grouper, and quickly cut him off the gun. I had just finished when two sharks came. The first picked up the grouper and the second followed him, shaking his head in excitement. Hans, who had anticipated what I would do, filmed the whole event. We repeated the procedure and found that it worked. One could feed sharks under water, and if one took care, it was not very risky. The only tricky moment was cutting the bleeding fish loose, while the sharks were milling around. Once the sharks were accustomed to the procedure, we began to try our shark repellent and found that it did not work!

After four months of diving, we left the Maldives and continued via Ceylon to the Nicobar Islands. Exciting months of diving followed. This time, I concentrated on the anemone fishes, in order to find out what protected them against the sting of underwater nettles (Eibl-Eibesfeldt, 1960b). We made contact with the practically unknown ethnic group of the Shompen. On Kondul, I collected scaredevils of the animistic Nicobarese and my latent interest in humans was fueled. All of the waters that we had entered were virgin lands for divers, with fish life abounding. Unaware of any danger from man, they came close to inspect us intruders and often milled around us, forming a tightly packed wall of fish bodies with hundreds of inquisitive eyes. The islands of Malaya were our last exploring grounds. Our collections were rich and our notebooks full, so in November 1958, I returned home again (Eibl-Eibesfeldt, 1964). In June of the previous year, just prior to my long trip, our Institute had moved to Seewiesen in Bavaria. We had spacious new buildings and Konrad Lorenz and Erich von Holst each headed one department of the Max Planck Institute für Verhaltensphysiologie. Our group was still small enough to enjoy a familiar intimacy and the intellectual atmosphere was challenging and inspiring. We had many discussions and I continued with my deprivation experiments and also worked on my data from the last expedition.

In 1959, I made a trip to South America and Bermuda, where I did some diving again, studying shrimp associated with anemones (Eibl-Eibesfeldt & Holthuis, 1964). The following year, Heinz Sielmann and I went to the Galápagos Islands, this time to film the behavior of numerous animals on the islands. Among other things, we filmed the ritualized combat of the male marine iguanas and the egg-guarding behavior of females, which until then was unknown. We also filmed the tool-using behavior of the woodpecker finch, and the underwater foraging of the marine iguanas. From the material that we filmed on this trip, we made a well-received film for the public, published fifteen films with Encyclopaedia Cinematographica in Göttingen, and provided ourselves with new scientific material.

For me the trip was a reunion *(Wiedersehen)* with an area to which I had become strongly attached (Eibl-Eibesfeldt, 1960a). Ever since I first stumbled over the rugged boulders, I have felt drawn to this archipelago, and the smell of the Galápagos herbs evokes in me a feeling of belonging. Despite my other current research commitments, I take advantage of every opportunity that arises to visit the islands, as a member of the executive committee of the Charles Darwin Foundation, as consultant for the Charles Darwin Research Station, or on personal visits.

In 1975, I made a trip to the islands with Lorle and in spring 1982 with my daughter Roswitha. We cruised the archipelago on board the *Beagle III* together with Heide and Friedemann Köster and Gail and Gary Robinson. We dived at the exposed cliffs of Gardener near Floreana, and at Roca Redonda, Cousins Rocks, and Gordon Rocks, all places with sheer precipices, strong currents, and brilliantly clear water teaming with sea lions and fish life, including hammerhead sharks. Roswitha performed bravely, even though it was her first diving experience in the sea! In fact, I was more nervous than she. I was happy to get the opportunity to collect a previously unknown species of garden eel (Eibl-Eibesfeldt and Köster, 1983).

In 1961 I paid my first long visit to the United States. Eckhard Hess had invited me to lecture for a term at the Department of Psychology of the University of Chicago. It was a challenging experience to fit into the American educational system and the "American way of life." Erich Klinghammer and his wife were superb hosts throughout my stay.

After my term was over, I traveled west, first to Hawaii on a charter flight to meet Ernie Reese from the Marine Biological Station on Coconut Island. We established a lasting friendship and see each other often in Hawaii. On the trip back, I saw California, and took the Greyhound bus from San Francisco to Chicago, stopping at places of interest. During this visit to the United States, I developed a liking for the country and have visited it thirteen times since, including another term of teaching at the Department of Child Development at the University of Minnesota in Minneapolis. There, Maria and Bill Charlesworth took good care of me and at the end of my term, Lorle was able to join me and we spent a couple of weeks in the coral reefs of Curacao with Bill.

In the early sixties, my interest shifted gradually from animal behavior to human behavior. In 1963, I completed my "Habilitation," a procedure that is needed in Germany before one is allowed to teach as a professor at the university. I worked at the University of Munich in zoology and, at the same time finished my work with mammals. I began to investigate the behavior of children born deaf and blind and, in the same year, I started to discuss with Hans Hass the feasibility of documenting human behavior objectively. Our interests coincided, since Hans wanted to make a television series on "Man," so he had asked me to assist him as an ethological advisor. In principle, we knew what we needed: unstaged documents of human behavior or humans behaving naturally. After numerous unsuccessful trials, Hans found the solution. In front of his camera, he attached a device that looked like a lens, but that had a prism inside and an opening on the side. With this "mirror lens" it was possible to film at an angle without pointing the camera directly at people. Prior to this, we had tried in vain to film mothers interacting "naturally" with their children, people greeting each other, people engaged in conversations, and so forth (Eibl-Eibesfelt & Hass, 1966, 1967a). However, whenever we pointed the camera at the subject, even from a distance, they changed their behavior significantly.

Now that we had the key to the documentation of unstaged social interactions in our hands, we experimented with the method on two trips. The first, in 1964, took us to East Africa and the second, in 1965, around the world. Both were exciting and most enjoyable. We toured East Africa in a Land Rover, camping in a little tent and filming along the way whenever an opportunity arose. In Nairobi, at the market and in the streets, we filmed both from the top of the car and on the sidewalks—and it worked.

In addition to documenting unstaged behavior as it occurred, we also experimented with provoking certain responses, such as a startle-reaction, yes and no responses, and refusal and acceptance. To get a startle reaction, we had a jack-in-the-box in a peanut can. It was my task to offer it to people while Hans filmed their reactions. I had to overcome strong inhibitions in order to try this on strangers, but to my relief, practically everyone took it as a joke.

We also experimented with filming at different speeds in order to learn what speed best fit our purpose. We varied the speed from one to seven frames per second and filmed a variety of activities, some of which we provoked. For instance, in Nairobi Hans posted himself with his camera at a window on the third floor of a hotel, while I deposited a ten-shilling bill on the sidewalk. When the bill just lay there, it was usually picked up by a passerby within a short time. But when I drew a white circle on the ground around the bill, nobody dared to touch it. People usually just congregated around the bill in the circle and discussed it. The only exception was a white man who saw it and pocketed it. He was followed by a crowd, arguing and protesting.

We visited Jane and Hugo van Lawick-Godall during this trip. They received us very warmly, but since they did not know how their chimpanzees would respond to strangers, they put us behind a blind. We were going to peek through a curtain, when suddenly a hairy arm came from the side and pulled the curtain away. There we sat, completely exposed; the chimps were friendly—to the relief of all of us. One, in fact, soon started jostling me into friendly rough-and-tumble play. Later I groomed him. It was the only time I have had such close contact with wild chimpanzees and the visit remains vivid in my memory.

Making a wide arc, we drove via Fort Portal to Uganda. Close to the Sudanese border, we turned east, heading for the land of the Karamajong. The area in the north of Kenya and Uganda was actually not open, since it was not under government control, but this we did not know at that time.

On the afternoon of August 26 we came upon some Karamajong Kraal. It was a breathtaking encounter. Tall men, stark naked, except for pieces of leather draped like capes over the shoulders of some, armed with spears and with strange-looking headdresses surrounded curiously. Some of the young men wore red wigs, tightly fitted to their scalps, that looked as if they were composed of strings and a mixture of cow dung and clay. The women, bare breasted, wore great numbers of beads and aprons of cowhide.

The response of the people was basically friendly and curious, but there was also underlying tension on both sides. We bought some of their beautifully carved headrests and distributed candy first to the women and children and then to the men, who finally also wanted some. We gave the headman a knife and then started to film. Everything went well. Late in the afternoon, the young warriors started to dance, which, as we later learned, is typical for the nilotohamitic people. To the chanting and clapping, one male engaged in series of high jumps. I approached with my taperecorder, nodded, and said "hello." He made no response but just kept on jumping, not recognizing our presence but singing a song that sounded like bulls roaring. We retired to our tent, listening with some discomfort to the song. It became dark, and as the night proceeded, bursts of shrill yelling sounds accompanied the deep voices of the singers. We became anxious and left the tent to hide in the rocks, but nothing happened. We finally went back to our tents to sleep.

The next morning, our hosts came to our camp. We played the tape of the dance to them and they enjoyed it enormously. They had just performed one of their traditional dances; that was all. We treated some of the sick, filmed the bloodletting of cows and a number of other customs, and pushed on to the country of the Turkana, Samburu and the El Molo.

We felt experienced by this time and were a bit too self-confident. Our wives, Lorle and Lotte, had arrived in the meantime, and we returned to Nairobi to pick them up and show them what we had accomplished. Driving without much hesitation to the next Massai Kraal, Lorle, Lotte, and I distributed small presents, while Hans set up the tripod and camera on the roof of our Land Rover. We could feel that the situation was tense and finally were made to understand that the dry bush over which we had parked our car was their "door" used to close the entrance to their kraal at night. Then a young man who spoke English told us that he resented our filming and that we should stop. Hans tried to argue, upon which the man grabbed one leg of the tripod. Without thinking, I interfered and grasped his arm. He was much taller than I and with a glare turned toward me, jumped a few paces back, drew his Kiri and threw it at me. It hit my leg and broke into pieces. This infuriated the young man even more and he drew his sword—I still remember the awful sound, "swish"! Immediately the others interfered, shielding me from him with crossed spears and calming him down. We realized then that he had been drinking, but that we had been tactless in our eagerness to show Lorle and Lotte what we had accomplished. Clearly, although we had mastered the mechanics of the camera, we had still much to learn about people.

And so we began our study of human behavior in a round-the-world trip, starting with the carnival in Rio in 1965. From there we went to Cuzco and other places in Peru and Mexico. In California we filmed in Disneyland, continued to Hawaii, Japan, Hong King, and finally to Bali, where we worked for several weeks.

In the years to follow, I elaborated on this technique and developed a program of cross-cultural documentation of human behavior (Eibl-Eibesfeldt, 1967, 1970b, 1972a, 1973, 1975a, 1976). In 1967, I was in New Guinea among the Kukukuku and Woitapmin; in 1969, I went to visit the Yanomami on the upper Orinoco; and in 1970, I started documentation among the !Ko Bushmen of the central Kalahari. In the meantime, I had gotten the professorship at the University of Munich and I was given the opportunity by the Max Planck Society to build up a "Arbeitsgruppe für Humanethologie" in response to the proposal that I had made. This research group, which I still head, became an independent unit of the Max Planck Society in 1975.

It is at this point that I will end my autobiographical sketches, although I feel that it is a new beginning rather than an end. The lack of a time perspective on the last ten years means that my mind is so full of names and events that it is not possible at this time to single out the most important ones. So, I will close with a short description of my life and work today. Perhaps in twenty years, I will be in a better position to say which of these were significant or formative in my second career.

For field research, I have decided to focus on a longitudinal study of a number of cultures that present different socioeconomic strategies of survival and that could also be in part serve as models for different "stages" of cultural evolution.

These cultures include the Bushmen of the Kalahari (G/wi, !Ko, !Kung) who are hunters and gatherers; the Yanomami (Waika) of the upper Orinoco who are incipient horticulturists, and the Eipo in West New Guinea (Irian Jaya), neolithic horticulturists, with whom we work in an interdisciplinary enterprise initiated by the Berlin Museum of Ethnology.

The Eipo were contacted by our group in 1974 and at that time were a completely intact neolithic people. Previously I had worked with the Daribi and Biami in Papua-New Guinea. As representatives for pastoralists in 1973 I chose the Himba, a Herero-speaking people of the Kaokoland (South West Africa) who still live entirely according to their traditional customs. In addition, since 1965 we have been working with rice-farming Balinese. In 1982, we incorporated the Trobriand Islands into our longitudinal research program.

In all of these groups, research and documentation by film and tape recording will be continued over years in repeated visits by a team of scientists from my group, as well as by guests and foreign scientists attached to our Institute.

In addition to those longitudinal studies, I have taken advantage of every chance that arises to visit people of other cultures, such as the Tasaday in the Philippines, the Pintubi, Walbiri, and Gidjingali in Australia, and many others more, in order to increase the breadth of our sample. In those cultures, I concentrate on specific behaviors, such as mother-child interactions, or certain reactions that I provoke, such as a startled or coy reaction.

The film documentation rests mainly on my shoulders, but I derive much pleasure from observing people for many hours a day through my camera. I start in the morning and finish at dusk. The fascination of the work helps me overcome the discomforts—the insects, the climate, and separation from family. It is an exciting life, full of adventure. During each trip, I acquired new insights into human behavior and fortunately so, because I need this knowledge to live all alone in a Yanomami village, for example, far from any outposts of our civilization. But I like people and this helps me to adapt. Having one's hammock under the communal roof of the Shabono, the hammocks of other Yanomami mothers, fathers, children within arm's reach, is a heart-warming experience.

In all, I have spent close to 9,000 days in the field, approximately half of these on zoological expeditions. To document human behavior I have spent 4,480 days with people of different cultures. By now, we have approximately 180 km of film in our archives, mainly of unstaged social interactions and rituals. From these, we can demonstrate the universality of a number of motor patterns and other behavioral traits. However, our most interesting finding concerns the existence of universal strategies of social interactions.

In order to achieve a particular social goal, say to ward off an aggressive act, to put oneself into the focus of attention, to provoke an attack, to get something from another person, to evoke pity and so forth, people have a limited set of effective strategies at their disposal. These strategies are universal and children employ them in nonverbal behavior in nearly identical ways in all cultures studied so far. Adults verbalize most of their interactions, but in doing so, they follow the same rules that structure the nonverbal interactions. Thus we can say that verbal and nonverbal social interactions are governed by the same rules and that within this system of rules, a variety of actions as well as verbal utterances can be substituted for each other as functional equivalents.

Our investigation of this universal grammar of social interaction that structures verbal and nonverbal behavior alike will be my main area of investigation for the years to come. I will continue the cross-cultural work in coordination with other members of my institute, ethnologists, linguists, and anthropologists, concentrating more on verbal behavior than we have in the past. In addition, to study the ontogeny of these strategies and gain further understanding of our own culture, we have an ongoing project to study the behavior of children in our local kindergarten. After the manuscript went to press I published a book on human ethology summarizing the concepts and our work in particular (Eibl-Eibesfeldt, 1984).

I am now fifty-four years old and Lorle and I celebrated the thirty-third anniversary of our marriage in February 1982. Our children are grown-up and married. My daughter Roswitha is a psychologist and has one son and my son Bernolf is a medical doctor and has two daughters. My wife has started her own professional career as a family therapist and consultant. I am very lucky to have my children living nearby. Roswitha's ten-year-old son Fabian and Bernolf's daughters Anna, five and Mara, two and my son- and daughter-in-law (Johannes and Monika) are often with us and this compensates for the time when I am absent in the field.

Being away in the field for several months a year certainly imposed a strain on me. The happier memories are of those trips on which I had my wife and members of my family with me. Lorle accompanied me on several occasions, some already mentioned, including trips to the Himba and Balinese. When my son Bernolf was seventeen, he accompanied me to the Yanamami and on another occasion to Central Australia and New Guinea. Only recently, Roswitha was able to join me. Her obligations as a young mother had not permitted an earlier trip.

My warm personal relationship with Konrad and Gretl Lorenz persists and I visit them several times a year. I am also in contact with my old friends Hans and Lotte Hass and we still dive together on holiday trips to the Great Barrier Reef, the Seychelles and Amirants, and last year to the Maldives. Hans is coeditor of our film archives and one of many strong links to Vienna, which remains my hometown.

My work rhythm varies. For several months a year I work with one or more of the groups mentioned. The rest of the year I work on my films and write, often retiring to the Tyrolise mountains or to the Cinqueterra in Italy, where there is no telephone. Even though my interest has shifted from animal to human behavior, the fish that impressed me so much at the age of three still exert their fascination upon me.

References

Eibl-Eibesfeldt, I. 1950a. Ein Beitrag zur Paarungsbiologie der Erdkröte (*Bufo bufo* L.). *Behaviour* 2:217–36.

———. 1950b. Über die Jugendentwicklung des Verhaltens eines männlichen Dachses (*Meles meles* L.) unter bes. Berücksichtigung des Spieles. *Z. Tierpsychol* 7:327–55.

————. 1950c. Beiträge zur Biologie der Haus-und Ährenmaus nebst einigen Beobachtungen an anderen Nagern. *Z. Tierpsychol.* 7:558–87.

————. 1951a. Beobachtungen zur Fortpflanzungsbiologie und Jugendentwicklung des Eichhörnchens (*Sciurus vulgaris* L.). *Z. Tierpsychol.* 8:370–400.

————. 1951b. Gefangenschaftsbeobachtungen an der perischen Wüstenmaus (*Meriones persicus* Blanf.): Ein Beitrag zur vergleichenden Ethologie der Nager. *Z. Tierpsychol.* 8:400–2.

————. 1952a. Ethologische Unterschiede zwischen Hausratte und Wanderratte. *Verh. Dt. Zool. Ges.*, Freiburg, 169–80.

————. 1952b. Beobachtungen an einer in Freiheit gehaltenen weiblichen Biberratte (*Myocastor coypus*). *D. Zool. Garten (NF)* 19:277–83.

————. 1953a. Zur Ethologie des Hamsters. *Cricetus cricetus* L. *Z. Tierpsychol.* 10:204–54.

————. 1953b. Eine besondere Form des Duftmarkierens beim Riesengalago. *Galago crassicaudatus. Säugetierkundl. Mitt.* 1:171–73.

————. 1954a. Paarungsbiologie der Anuren (Grasfrosch, Erdkröte, Laubfrosch, Wasserfrosch). Wiss. Film C 628, Inst. Wiss. Film, Göttingen.

————. 1954b. Die Iltiskoppel. Film F 417, Inst. f. Film u. Bild, München.

————. 1954c. Im Hamsterrevier. Film F 401, Inst. f. Film u. Bild, Büchen.

————. 1955a. Der Kommentkampf der Meereschse. *Amblyrhynchus eristatus* Bell, nebst einigen Notizen zur Biologie dieser Art. *Z. Tierpsychol.* 12:49–62.

————. 1955b. Einige Bemerkungen über den Galapagos-Seelöwen, *Zalophus wollebacki* Sivertsen. *Säugetierkundl. Mitt.* 3:101–3.

————. 1955c. Beobachtungen über territoriales Verhalten und Brutpflege des Galápagos-Seelöwen. *Z. Säugetierkunde.* 20:75–77.

————. 1955d. Ethologische Studien am Galápagos-Seelöwen. *Zalophus wollebacki* Sivertsen. *Z. Tierpsychol.* 12:286–303.

————. 1955e. Über Symbiosen, Parasitimus und andere besondere zwischenartliche Bezienhungen tropischer Meeresfische. *Z. Tierpsychol.* 12:205–19.

————. 1955f. Angeborenes und Erworbenes im Nestbauverhalten der Wanderratte. *Die Naturwiss.* 42:633–34.

————. 1955g. Wiss. Filme d. Encly. Cinemat. d. Inst. Wiss. Film. Gottingen, C 697 Biologie des Iltisses, C 698 Sexualverhalten und Eiablage beim Alpenmolch.

————. 1956a. Angeborenes und Erworbenes in der Technik des Beutetötens (Versuche am Iltis). *Z. Säugetierkunde* 21:135–37.

————. 1956b. Über die ontogenetische Entwicklung der Technik des Nüsseöffnens vom Eichhörnchen. *Z. Säugetierkunde,* 21:132–34.

————. 1959a. Der Fisch Aspidontus taeniatus als Nachahmer des Putzers *Labroides dimidiatus. Z. Tierpsychol.* 16:19–25.

————. 1959b. Survey of the Galápagos Islands. *UNESCO Mission Reports* 8. 33pp.

————. 1960a. *Galápagos. Die Arche Noah im Pazifik.* München: Piper.

————. 1960b. Beobachtungen und Versuche an Anemonenfischen der Malediven und Nikobaren. *Z. Tierpsychol.* 17:1–10.

————. 1961. The fighting behavior of animals. *Sci. Amer.* 205:112–21.

————. 1964. *Im Reich der tausend Atolle.* München: Piper.

———. 1967. *Grundriß der vergleichenden Verhaltensforschung.* München: Piper.

———. 1970a. *Ethology, Biology of Behavior.* 2nd ed. 1975. New York: Holt, Rinehart & Wilson.

———. 1970b. *Liebe und Hass. Zur Naturgeschichte elementarer Verhaltensweisen.* München: Piper.

———. 1971. Transcultural patterns of ritualized contact behavior. In *Behavior and Environment. The Use of Space by Animals and Men,* ed. A. H. Esser, pp. 238–46. New York-London: Plenum Press.

———. 1972a. *Die !Ko-Buschmanngesellschaft. Aggressionskontrolle und Gruppenbindung. Monographien zur Humanethologie I.* München: Piper.

———. 1972b. Similarities and differences between cultures in expressive moments. In *Non-Verbal Communication,* ed. R. A. Hinde, pp. 297–314. London: Cambridge Univ. Press.

———. 1973. *Der vorprogrammierte Mensch. Das Ererbte als bestimmender Faktor im menschlichen Verhalten.* Wien-Zurich-München: Molden.

———. 1975a. *Krieg und Frieden aus der Sicht der Verhaltensforschung.* München: Piper.

———. 1975b. Aggression in the !Ko-Bushmen. In *Psychological Anthropology,* ed. R. T. Williams, pp. 317–31. The Hague: Mouton Publishers.

———. 1976. *Menschenforschung auf neuen Wegen. Die naturwissenschaftliche Betrachtung kultureller Verhaltensweisen.* Wien: Molden.

———. 1977. Evolution of destructive aggression. *Aggressive Behavior* 3: 127–44.

———. 1979. Konrad Lorenz. In *International Encyclopedia of the Social Sciences,* 18: 455–63.

———. 1984. *Die Biologie menschlichen Verhaltens. Grundriss der Humanethologie.* München: Piper Verlag.

Eibl-Eibesfeldt, I., and Hass, H. 1959. Erfahrungen mit Haien. *Z. Tierpsychol.* 16: 739–46.

———. 1966. Zum Projekt einer ethologisch orientierten Untersuchung menschlichen Verhaltens.

———. 1967a. Neue Wege der Humanethologie. *Homo* 18: 13–23.

———. 1967b. Film studies in human ethology. *Current Anthrop.* 8: 477–79.

Eibl-Eibesfeldt, I., and Holthuis, L. B. 1964. A new species of Periclimenes from Bermuda (Crustacea, Decapoda, Palaeimonidae). *Senck, biol.* 45: 185–92.

Eibl-Eibesfeldt, I., and Klausewitz, W. 1959. Neue Röhrenaale von den Malediven und Nikobaren. *Senck biol.* 40: 135–53.

Eibl-Eibesfeldt, I. und Köster, F. 1983. *Taenioconger Klausewitzi, A New Garden-Eel from the Galapagos. Noticias de Galapagos,* 38: 26–27.

Eibl-Eibesfeldt, I., and Kramer, S. 1958. Ethology, the comparative study of animal behavior. *Quart. Rev. Biol.* 33: 181–211.

Eibl-Eibesfeldt, I., and Pitcairn, T. K. 1976. Concerning the evolution of nonverbal communication in man. In *Communicative Behavior and Evolution,* ed. E. C. Simmel and M. Hahn, pp. 81–113. New York: Academic Press.

John Langworthy Fuller (b. 22 July 1910)

4

Of Dogs, Mice, People, and Me

John Langworthy Fuller

Origins, Youth, and Education

My father, John Harold Fuller, was a direct descendant of Samuel Fuller, "the Pilgrim Doctor" of the Mayflower. My mother, Joyce Langworthy, traced her ancestry to Andrew Langworthy, who had immigrated to what is now Rhode Island about 1642. The genetic and environmental factors that have shaped the New England character were certainly present at my birth in Brandon, Vermont, July 22, 1910. Father was the high school principal and a Democrat at a time when in Vermont Democrats were viewed as misguided or mentally unbalanced. Nevertheless he was highly respected in the community. Mother was a beautiful woman who devoted her life to the welfare of her husband and four children, of whom I was the oldest.

I have no recollection of Brandon nor of a time when sister Mary and brother Sam were not part of the household. From postmarks on old letters it appears that we moved in 1913 to Hardwick, Vermont where brother Robert was born in 1916. To ease my mother's burdens I was sometimes sent to my maternal grandparents for the summer. Grandpa and Grandma had charge of a large dormitory at the University of Vermont, and I was a favorite of the teachers who lived there during the summer sessions. Grandma took me to visit her friends and to my first circus. Grandpa brought me to his cottage on Lake Champlain where we fished for perch and hornpout.

When the younger children had matured a bit, our whole family traveled by train to the lake for some of the happiest summers ever. Father was knowledgeable about flowers, trees, birds, and animals. He showed me marine fossils in rocky outcrops and explained their significance. Together we searched for Indian arrowheads at a sandy beach near the mouth of Lewis Creek.

In 1920 we moved a few miles east to Lancaster, New Hampshire, on the Connecticut river. Here my father bought a house on the edge of town overlooking the broad river valley. Six years later he accepted an invitation to become superintendent of schools in a district of six townships just south of the Presidential Range. We settled in North Conway, which from the start I liked better than Lancaster. I made friends quickly, played on the football and basketball teams, and was elected class president. Some of my happiest recollections are of exploring the White Mountains and trout fishing in the Saco River and its tributaries.

Occasionally I hunted, more for the pleasure of tramping through the woods than for killing game. In fact I cannot remember firing a single shot at any animal.

High School and College

I was a good student and graduated when still sixteen as valedictorian of my class at Kennett High School in Conway. I had begun to notice that girls existed and were often attractive, but I felt uneasy in their presence and, to avoid making a fool of myself, avoided emotional commitment. My plans for a career were still vague. I liked the sciences, but had little knowledge of what scientists did. In a rural area there were few professionals to serve as models beyond physician, dentist, lawyer, preacher, and teacher. I did know that I wanted a college education and applied to Yale, where my father had graduated in 1898. The entrance exams were scheduled for the week after commencement. Unfortunately, three hours after giving my valedictory speech I was in the hospital being treated for streptococcal septicemia, a life-threatening disease in those preantibiotic days. Yale would not admit me without the entrance exams despite my excellent record. I could have taken the tests in the fall at New Haven, but the uncertainty about qualifying for a scholarship was disquieting. Without it the financial burden would have been intolerable. Thus I accepted an offer from Bates College in Lewiston, Maine. Perhaps Yale would have given me a better education in the sciences, but I have no regrets about my choice of Bates. I won letters in track and football (twice state championships in the latter), was active on the college newspaper, literary magazine, and outing club. As business manager of the class yearbook I sold enough advertising to make a small profit. The previous manager had been unable to pay the printer. My classmates voted me the "most versatile" individual. Only in a small, liberal arts college would I have had the opportunity to take part in such diverse activities.

My academic major was biology. At Bates this was zoology, mostly anatomy and taxonomy, with two terms of botany. Hours were spent drawing the vascular system of the cat, the appendages of the crayfish, etc. Most of the male majors went on to medical school; the women seldom applied for what was considered to be a masculine profession. Professors Pomeroy and Sawyer were not distinguished scholars, but they were good teachers and took a personal interest in their students. My courses in the social sciences were unneccessarily boring except for one in systematic psychology that was really an introduction to the philosophy of science. When possible I elected courses in literature. My grades were good, but low marks on some freshman language courses disqualified me for Phi Beta Kappa. Later I was elected as a graduate member.

Graduate School

Inevitably senior year came round and it was necessary to plan for the future. I had thoughts of medicine, but nearing graduation I realized that I was more interested in keeping people healthy than in treating their illnesses. Thus, when

Professor Clair Turner of the Massachusetts Institute of Technology and a Bates alumnus invited me to apply for their public health program, I accepted. And so in the fall of 1931 a country boy just turned twenty-one came to the city and enrolled in one of the most prestigious academic institutions in America. One of the first things I discovered was that to be awarded any degree at MIT it was necessary to pass three courses in calculus. I had none. It was embarrassing to be enrolled in a freshman course, and even more so to barely pass the first examination. Then one morning just as I awoke there was a sudden flood of understanding of what the symbols meant. I passed the first course with a top grade and met the remainder of the requirement by self-study and examinations. There have been other sudden insights in my life, but somehow this one impresses me most.

MIT biology in the early thirties emphasized practical applications to public health and food technology. Its present eminence in neuroscience, biomedicine, and moleular biology was years ahead. I took courses in bacteriology, physiology, biochemistry, and public health with a minor in physical chemistry. My assistantships in general biology and physiology paid $700 in 1934, a raise of $100 from the previous year. During one term I taught general biology to a group of foreign students working for a master's degree in public-health engineering.

Although the instruction at MIT was of high quality, I was not attracted to Professor Turner's research nor to that of other senior professors. I might have changed institutions, but I had become interested in Turner's research assistant, Ruth Parsons. It started when a fellow student invited Ruth's roommate to dinner, and asked if I would make a foursome. On a bright moonlight evening, Ruth and I stood on the deck of the Winthrop Ferry crossing Boston harbor, and fell in love on the spot. A year-and-a-half later we were married in spite of the fact that Ruth was forced as a result to resign her position in the public schools of Malden. In retrospect I wonder why we did not simply conceal our marriage, or live together without benefit of legalization. It seems strange that we and others would tolerate a sexist regulation that deprived a woman of her employment because of marriage, while exacting no penalty for a man. Somehow we had been shaped by our culture to accede to regulations no matter how pernicious and unfair they might be. At seventy-one I am less respectful of authority than I was at twenty-two. Although many abuses have been corrected, every social advance has been opposed by conservative elements in our country.

My dissertation was unusual for an MIT biologist. I enjoyed my assistantship with Charles H. Blake, who taught the basic biology courses. He was only a few years older than I and was particularly interested in terrestrial isopods. I decided to work with these animals and eventually produced a dissertation entitled "A Comparison of the Physiology, Ecology and Distribution of Some New England Wood Lice." Its production involved both field and laboratory research plus an intensive search of the literature on the distribution of the fifteen species native to New England. It was a sound piece of work but has never been published. Why not? I do not really know. The years immediately following receipt of the doctorate were marked by frequent moves and the birth of our first child. Looking it over forty-seven years later I find it still interesting and regret that I did not split it into several papers of manageable size.

Transition Years

After finishing my Ph.D. requirements in 1935, I was invited to remain at Tech with a miniscule increase in salary, which seemed to be a small reward for four years of intense labor. I had been granted a three-month fellowship from the Woods Hole Oceanographic Institution to work with George Clarke on the nutrient cycle of *Calanus finmarchicus,* the small copepod that is an essential link in the food chain that maintains the North Atlantic fishery. Ruth and I enjoyed that summer at Woods Hole, making new friends, exploring Cape Cod, and attending lectures at the Marine Laboratory. My research went well and the results were published in the Biological Bulletin. A second experiment conducted in 1936 demonstrated that *C. finmarchicus* was primarily an automatic filter feeder, and that its rate of feeding did not vary with the concentration of diatoms in its watery environment. I was well on the way to becoming a physiological ecologist, or perhaps an ecological physiologist, and would have been happy to continue working at the WHOI.

During the summer of 1935 I was informed of a one-year opening at Sarah Lawrence, a new, unconventional women's college in Bronxville, New York. I was offered the position at a salary good for the times, and Ruth and I decided to take a chance on finding another position at the end of my appointment. The nation was in the depths of the Great Depression and academic positions were scarce, but we were confident that something would turn up. In many ways it was a good year. We explored New York City and its environs, and became well acquainted with my uncle, Raymond Fuller, a sociologist who was one of the first to study the hereditary aspects of the major psychoses. This was my first contact with the genetics of behavior. Sarah Lawrence students were a mix of unusually bright and creative young women, and others whose major qualification was the ability of their families to pay the high tuition. The most important event of the year was that Ruth became pregnant and on July 22, 1936 (my twenty-sixth birthday) she presented me with a daughter, Mary Jean.

The only problem with our having a baby was that I did not have a full-time job for the upcoming academic year. I had applied for a postdoctoral fellowship in limnology to supplement my experience in marine biology, but it was not funded. A half-time position for one year was available at Clark University and in September the three Fullers settled down on the middle floor of a three-decker in Worcester. Apart from poverty it was not too bad a year. I assisted Ladd Prosser in his studies of synaptic transmission in crayfish ganglia, and observed Hudson Hoagland's pioneer research in electroencephalography. I enjoyed sharing responsibility for Mary Jean, and it was a splendid way to learn about behavioral development, which was later to become a major research interest. At the end of the academic year I had a summer job with the New Hampshire Fish and Game Department surveying and testing the water quality of lakes and streams. We moved in with my parents in North Conway, where Mary Jean was the center of attention. Our only problem was supporting ourselves when the survey was suspended at the end of August.

Unexpectedly a letter from President Atwood of Clark arrived informing me that the faculty member for whom I had substituted was not returning, and offering me a second year of full-time service at a salary of $1600. Since I had received $900 for half-time, I was unwilling to accept his terms and requested that I be paid at least $1800. Belatedly Atwood agreed, giving me the pleasure of turning him down. A good friend from MIT days had alerted me to an opening in zoology at the University of Maine where he was employed. Ruth and I drove to Orono. I had a good conversation with Joseph Murray of the zoology department, and in September the three Fullers settled down for the fifth time since we had left Tech. We could not foresee in 1937 that World War II was imminent and that the America in which we had grown up would never be the same.

The Orono Years: 1937–47

Orono is a university town situated between the industrial and commercial cities of Bangor and Old Town. The social environment was more open and friendly than at Sarah Lawrence and Clark and we made lifelong friends during our ten years of residence. This favorable feature was countered by a harsh climate, remoteness from other institutions, and inadequate resources for research. Agriculture and engineering were better supported than liberal arts, although efforts were being made to strengthen the latter. Graduate programs were limited appropriately to the Master's level. My major teaching responsibilities were in physiology and general zoology.

For several summers I assisted my colleague, Gerry Cooper, in his role as biologist for the Maine Fish and Game Commission. We coauthored several reports on the hydrography and biology of Maine lakes. I was becoming a limnologist through the back door. The alternation between classroom and the field was very pleasant, and when Gerry left Maine to take a position in his native Michigan I was tempted to accept an offer to replace him. However, I realized that I needed more academic training in fisheries biology to become proficient, and this would have been financially difficult, particularly with a second daughter, Sarah Ann.

By this time World War II had disrupted regular university programs. English professors were converted to teachers of mathematics, and I became the coordinator of a new nursing program that involved the university with the three largest hospitals in Maine. It was an interesting assignment. The hospitals were cooperative, and the well-motivated students did not seem to mind having a male adviser with no credentials in the art of nursing. In fact, the first graduating class conspired with my wife to give a party in my honor, and insisted that I be included in their class picture.

In odd moments I continued marine research in collaboration with the WHOI. My assignment was to determine the rate at which objects submerged in the sea would be colonized by sessile plants and animals. The information was needed for scheduling the maintenance of underwater detection devices along the New England coast. When the war ended I was invited to spend a year at Woods Hole collating my data with those of others for a comprehensive report. I declined because I felt that educating the great influx of veterans had a high priority and prepared a report on my own that was published in *The Biological Bulletin*. It was my last contribution to marine biology.

Teaching physiology was enjoyable, but the paucity of resources at Orono severely limited research. Gradually I began to realize that my interest in the organism as a whole was greater than in cell and organ physiology. Thus when two seniors requested an honors seminar on animal behavior I agreed to sponsor them. Our mainstay was Maier and Schneirla's *Principles of Animal Psychology*. I remember ordering a book entitled *The Behavior of Organisms* by an unknown named B. F. Skinner, and was disappointed when I found that it dealt with a few white rats pressing bars. It was, however, a good introduction to behaviorism, which had been neglected in my college psychology courses.

By 1946 Joe Murray had become dean of Arts and Sciences and my fishing companion, Ben Speicher, had been appointed chairman of the zoology department. I had advanced to associate professor with tenure, and had acquired enough money to purchase a spacious house conveniently located to schools, shopping, and work. It seemed likely that the Fullers would remain in Orono until my retirement. Then, early in the year I was invited with other zoologists to lunch at Dean Murray's home with Paul Scott, a new member of the staff at the Roscoe B. Jackson Memorial Laboratory in Bar Harbor. Paul told the assembled group about the new Rockefeller Foundation-supported grant in the genetics of behavior. The idea for the projected ten-year program had originated with Alan Gregg of the Foundation and C. C. Little, director of the Laboratory. Scott mentioned that a few summer research fellowships were available and I applied. In retrospect this luncheon was the most significant event in my scientific career.

In early June the Fullers moved into a cottage near the shore of Frenchman's Bay about two miles from the restored millionaire's show farm that was now the Hamilton Station of the Jackson Laboratory, and housed the Behavior Laboratory. The summer of 1946 was not remarkable for its research results. I spent much of my time assembling apparatus to measure physiological reactions to varied forms of stimulation in loosely restrained dogs. In this I was aided by William T. James, who had been active in C. B. Stockard's earlier dog project linking genetics and endocrines with body type and behavior. Calvin Hall, a pioneer in applying genetics to psychology (and later well-known for dream research) was discovering that strains of mice differed tremendously in susceptibility to audiogenic seizures. Benson Ginsburg and Elizabeth Beeman (like Scott, former students of W. C. Allee at Chicago) were involved respectively with genetic variation in neurotransmitters and aggressive behavior among mouse strains. The daily intellectual interactions at lunchtime were more exciting than anything I had experienced before, even at Woods Hole. This ferment of minds persisted throughout the existence of the Division of Behavior Studies.

The climax of the summer was a conference on "Genetics and Social Behavior" chaired by Robert M. Yerkes. Animal behavior was represented by Theodore C. Schneirla, Clarence R. Carpenter, and Frank Beach. Among the contingent from psychology were Gardner and Lois Murphy, Neal Miller (later a member of Jackson's Board of Scientific Overseers), and O. Hobart Mowrer. I returned to Orono full of ideas for next summer's research.

Early in 1947 I received a telephone call from Scott asking if I would be interested in joining the Jackson staff. My salary would come from the Rockefeller grant that was guaranteed for nine more years. A few days later, after discussing the move with Ruth, I accepted. There have never been any regrets. We sold our Orono house and moved to Bar Harbor in June. In 1947 the town depended economically on the "summer people" (to be distinguished from tourists) whose estates either occupied the shore front or were perched on a hillside overlooking dazzling vistas of forest, rocks, and sea. The center of town was dominated by exclusive shops. All this changed in October when a forest fire destroyed many of the homes, acres of forest, and the main building of the Jackson Laboratory. The center of town was largely spared but the economy of the town and the nature of the Jackson Laboratory were drastically changed. Few of the mansions were rebuilt; many of those that had survived were demolished, motels proliferated, and, surprisingly, the Jackson Laboratory, rising like a phoenix from the ashes, grew to the point that it is now the major employer on Mt. Desert Island, with outstanding research facilities and staff. Hamilton Station was spared the holocaust, and my research continued with minor inconvenience. Bar Harbor was our home for twenty-three stimulating years. We are still in love with the island and will be buried there in a quiet spot encircled by low granite hills.

Research Activities: Behavior Genetics and Developmental Psychobiology

Up to this point my narrative has proceded chronologically. This section is organized around the major themes that guided my research at Jackson Laboratory (1947–70) and at the State University of New York at Binghamton (1970–77). The members of the 1946 conference made many detailed recommendations to the Division of Behavior Studies. Briefly summarized, the division was advised to concentrate on laboratory rather than fieldwork in animal behavior, and on larger animals that could not be housed easily in most university facilities. Priority should be given to genetic influences on behavior, and the research programs of the staff should be augmented by encouraging visiting investigators and advanced students. This good advice was followed.

Genetics and Social Behavior of Dogs

A major factor in the establishment of the dog behavior program had been Alan Gregg's interest in the nature-nurture issue. He believed that most social scientists had given insufficient attention to biological factors affecting behavior, and that they had insufficient understanding of genetics. The choice of dogs as subjects for a genetic investigation was based on their interesting social behavior and the availability of numerous breeds. No other species is as phenotypically variable as *Canis familiaris*. It has been hypothesized that the social organization of wolves, and presumably that of the ancestors of modern dogs, is an adaptation to group hunting and scavenging. The same mode of life probably characterized early *Homo sapiens*. Such a common ecological background may explain the fact that dogs fit better than any other species into a human society. Whatever the validity of this line of reasoning, the choice of species and breeds had been made before I moved to Bar Harbor, and I was able immediately to begin work in the laboratory.

My first major task was to collaborate with Paul Scott in developing standard test procedures that would be appropriate from birth to one year of age. This was a major task involving inventiveness with constraints imposed by space, time, and personnel. Once committed to a standard sequence of testing we were bound to it for approximately ten years of data collection. The final schedule was a full one. When a litter of pups was born, I knew exactly where I would be for each day of the next twelve months. During these years of data collection, Paul and I had to plan carefully for vacations, which were usually short. We were fortunate in the young men and women who were research assistants. In the early days several of them had been students of mine in Orono. Testing was not, however, turned over completely to our helpers. Paul and I not only directed the research; we tested dogs, inspected them individually on a regular basis, and analyzed our data on old-fashioned mechanical calculators. How wonderful it would have been to have a modern computer!

Our informal name for the project was "The School for Dogs." Many of our training procedures and tests were formalized versions of standard techniques used to make a puppy become a well-mannered member of a human household. Other tests involved problem solving, development of social behavior, and physiological reactions to a variety of physical and social stimuli. From the genetic point of view, there were two major experiments. The first was a comparison of the performances of five breeds in our test battery. These breeds, Basenji, Beagle, Cocker Spaniel, Shetland Sheep Dog, and Wire-haired Fox Terrier represented a variety of "personalities," presumably the result of selection. Although physically distinctive, all five were similar in size and free of major anomalies such as the achondroplasia seen in Dachshunds.

The second experiment was a standard Mendelian cross between Basenji and Cocker, chosen because these breeds were the most divergent among the five. These experiments are described in *Genetics and the Social Behavior of the Dog* (Scott & Fuller, 1965). The book was well reviewed by scientists, and to our surprise by serious breeders and trainers of dogs. The Dog Writers of America awarded Paul and me a certificate for "the best dog book of 1965." Few scientific books win such recognition. Sales in a paperback edition still continue and we suspect that most are to "dog people."

It is impossible to summarize the monograph in a few words. In addition to reporting our own research, we reviewed the history of the domestic dog, compared the social behavior of dogs and wolves, expounded the theory of critical periods, and pointed out parallels in the evolution of dogs and humans. A concluding chapter, "Towards a Science of Social Genetics," was an early essay in sociobiology. It differs markedly from today's sociobiology with its "selfish genes" and emphasis on population rather than developmental genetics. We emphasized the interaction of genotype and environment in the ontogeny of social behavior in contemporary individuals. The evolutionary and the developmental approaches to sociobiology have tended to remain separate. Granted that they do deal with different issues and different data, it would be helpful to work toward a synthesis.

The Kaspar Hauser Experiments. In the late 1950s I began a new series of experiments on the effects of early experiential deprivation on dogs. In part the research was a follow-up of experiments by Thompson and Heron at McGill and by Alan Fisher who had done his Ph.D. research at Jackson. In part the ideas were an offshoot of an experiment on the delayed behavioral effects of administration of psychoactive drugs in immature individuals. Collaborating in the study were Lincoln Clark, a psychiatrist, and Marcus Waller, a postdoctoral fellow. Some psychiatrists were promoting the use of drugs like chlorpromazine (CPZ) for reducing hyperactivity in young children, but there were concerns about possible harmful long-term effects. Our first experiment involved two types of rearing (permissive and disciplined) and administration of CPZ or a placebo. To observe possible age differences in sensitivity, some individuals were shifted from one treatment to another midway through the study. The results were clear. Although both punishment and CPZ depressed social interaction and play in the arena, they had no permanent effects. Pups reared by the disciplined procedure were as strongly attracted to humans when called as were the permissive groups. Being shocked for touching one toy did not inhibit play with an unwired toy. The implications of these findings for child rearing are interesting.

Another interesting observation in the CPZ experiment was unrelated to its original purpose. To obtain strict control of the experience of our subjects, we had held them in closed, individual cages from which they were removed for only a few minutes each day for the arena tests. Differences between these pups with limited opportunity for play and social interaction and standard-reared pups were undetectable by observation. I wondered if there were a critical dosage of experience that was necessary to permit full development of social and manipulative behavior. To test this idea, groups of pups were reared from three to fifteen weeks of age in well-ventilated and lighted cages. A one-way window permitted us to observe the physical condition of our subjects. Except for a tendency to overweight, they thrived like standard-reared animals. Pups were removed and placed in the arena for approximately 10 minutes for 0, 1, 2, or 4 times per week. As expected, the 0-emergence group was retarded in social interaction. We called them the K-pups after Kaspar Hauser, the boy who surfaced in Germany during the early nineteenth century after solitary confinement for more than a decade. Pups removed for 10 minutes 2 or 4 times per week were essentially like standard-reared pups in social and manipulative behavior. At maturity they were readily trained and handled, and performed well on discrimination tests. Females bred and reared young normally.

In another study I compared K-pups with pet-reared (P-pups) on a reversal learning task. The P-pups with twenty 10-minute sessions in the arena might be described as rehabilitated. K-pups made more errors than P-pups but the difference was significant only on the first reversal. The best performer turned out to be a K-pup, but so were the two poorest. I concluded that the effects of early deprivation might not be as severe as some have claimed. A genetic influence on the effects of experiential deprivation was demonstrated in a comparison of Beagles and Wire-haired Terriers. In postisolation tests, K-terriers were significantly more active and K-beagles less active than their pet-reared siblings. This disordinal genotype-environment interaction was not found for social contacts. Qualitatively the effect of experiential deprivation was to widen the behavioral differences between the two breeds. Crudely stated, in the absence of a stimulating environment, gene differences were expressed more clearly.

The Emergence-Stress Hypothesis. Perhaps the most interesting feature of the isolation experiments was the efficacy of very small doses of experience (less than 1 percent of standard rearing) in promoting development of typical social and manipulative behavior. The same doses of experience delayed twelve weeks were less effective. One explanation is that there are critical or sensitive periods during which developing pups are open to the stimulus input and respond appropriately. Sensitive is probably a better choice than critical because K-pups did respond to a degree. Critical period theories imply that the capacity to learn certain skills is temporary and this explains the poor prognosis for learning skills at a later age.

Clark and I proposed an alternative explanation for the isolation syndrome. The brief exposures to the arena at early ages did not correspond with critical periods for learning social and manipulative skills. Instead they provided an opportunity to habituate to complex arrays of stimuli while perceptual and emotional mechanisms were only partially developed. Delaying the same experiences to a later age when these mechanisms had matured autonomously produced a massive fear reaction that interfered with learning. This hypothesis predicts that lessening the emotional impact of sudden emergence from isolation would prevent the isolation syndrome.

We tested the idea by administering CPZ to some of our K-pups immediately prior to their first visit to the arena. The effectiveness of this procedure is most dramatically shown in a film comparing two pairs of terriers on their first day of emergence. Kaspar and Karla (nontreated controls) struggled to escape from the arms of their handler, stared vacuously at each other when they were placed together, and paid no attention to toys. Theron and Thea (CPZ therapy) were at ease with the experimenter, followed him closely about the arena, played vigorously with a ball, and interacted socially with each other. Kaspar and Karla behaved like autistic children; Theron and Thea like ordinary puppies. The drug was withdrawn after two days and the typical behavior patterns were unchanged.

This experiment is possibly the most significant I have performed, but it has not been widely cited nor followed up by others. Those who have seen the film are amazed, but they have not been motivated to continue this line of research. The fault is probably my own in not following up with a series of experiments. There were extenuating circumstances. I was now associate director of the Laboratory with half of my time assigned to administration. My office and mouse lab had been moved to the main building, an inconvenient location for overseeing the dogs. Paul Scott had moved to Bowling Green State University, and the resources of the Laboratory were diverted from the dog project to mice. This was a rational choice, but it presaged the decline of Jackson's behavior program. I considered an attractive offer to work with dogs at a medical school on the Pacific coast, but decided to remain in the East. I still believe that a dog center similar to the primate centers would serve a useful purpose for many types of behavioral research.

Guide Dogs for the Blind. Before turning to behavioral work with mice, it is appropriate to mention a spin-off from the school for dogs, a cooperative project with Guide Dogs for the Blind in San Rafael, California. Clarence J. Pfaffenberger, one of Guide Dog's trustees, had visited Bar Harbor, observed our work, and concluded that research should be done on the breeding and rearing of guide dogs. Many of the animals donated by the public were unsuitable for training. Pffaf obtained a grant for this purpose and I, along with Paul Scott and Benson Ginsburg, made regular trips to California over a six-year period (1961–67). My part involved statistical analysis of records and field observations of the dog-human partnership. I concluded that the social bond between these partners was as important as the visual information transmitted from dog to person. This is my only published work on human behavior. The small venture into applied research was rewarding, and a much-appreciated bonus was the opportunity to explore San Francisco and environs with Pffaf as a guide.

Mice and Behavior Genetics

Although dogs were my major research interest from 1947 to the mid-sixties I was involved with mice as soon as I arrived in Bar Harbor. It is unavoidable at the Jackson Laboratory. Mice are limited behaviorally but they are genetic gold mines. Also they are cheaper and breed and mature more rapidly; publishable results come more quickly. I actually grew fond of them as I learned more about them.

Audiogenic Seizures. Audiogenic seizures (AS) are known in rats, rabbits, and mice. Exposure to a loud, high-frequency sound leads to wild running, and often to clonic or tonic convulsions. They seem to have no adaptive value; tonic seizures are usually fatal. For this reason the genetics of AS has less ethological significance than the genetics of aggression and other social behaviors. Its advantages to the geneticist are stereotypy, ease of observation, and high variability among strains.

Following Hall's report of a striking difference in AS susceptibility in C57BL and DBA mice, a number of investigators became interested in the mode of inheritance (Hall made this discovery at Bar Harbor during my first summer as a visiting investigator). Witt and Hall offered a simple, one-locus, Mendelian model with susceptibility dominant. In my laboratory we were unable to corroborate their results, and postulated a polygenic model with a threshold for the manifestation of AS (Fuller, Easler, & Smith, 1950). Several other hypotheses have been proposed, but I have never lost confidence in my original threshold theory. In fact the 1950 paper may be the first application to animal behavior of Sewall Wright's original work on threshold systems and structural anomalies. It was a pleasure when Seyfried and associates (1980) using recombinant inbred lines demonstrated that the AS susceptibility genes were distributed over several chromosomes. How welcome it would have been to have these lines available in 1950.

Although never my chief research activity, AS has fascinated me for years, and

I have authored and coauthored about fifteen research reports and three reviews. Early studies dealt with the time course of the seizures and the "forbidden latencies." In 1964 Sally Huff and I described sensitization of resistant mice by exposure to intense sound. We failed, however, to recognize the potential significance of the phenomenon, and it remained for Kenneth Henry, then a graduate student, to point these out. Following his 1967 paper, Robert Collins and I measured the rate at which sensitization developed, and the duration of susceptibility. We also demonstrated unilateral priming by blocking the acoustic meatus of one ear during exposure to sound. This indicated that there were two independent foci for priming, although the seizures themselves were bilateral.

Research on AS continued after I moved to the State University of New York. In a synthesized random-bred heterogeneous stock of mice I found that susceptibility to spontaneous seizures and to priming were inherited independently. Later C. S. Chen, visiting from Australia, and I demonstrated that the two forms of susceptibility could be separated by selection. The significance of this research, with the exception of the Binghamton studies, is considered in *Genetics of Audiogenic Seizures in Mice: A Parable for Psychiatrists* (Fuller & Collins, 1970). We pointed out that there are interesting parallels between the genetics of audiogenic seizures and the genetics of some psychiatric syndromes. The postulated similarities between the murine and human syndromes are based on levels of explanation rather than on common chemical, psychological, or genetic mechanisms. These ideas are pertinent to other instances in which animals serve as models for the understanding of humans.

Genetics and Regulatory Behavior. Eating, drinking, activity level, and shelter building are examples of behavior that helps to maintain a relatively constant internal environment. Genetic factors affecting the regulation of eating have been one of my long-term interests. A number of mutations in mice result in hyperphagia and obesity. These include obese *(ob)*, diabetes *(db)* and viable yellow *(A^{vy})*. These mutants have played an important part in research on neural and endocrinological factors in hyperphagia. My interest has been primarily in behavioral responses to sensory quality and nutritive density of their diet.

In the first of these studies I was assisted by a student from the summer research program. We found that obese mice were less able than their normal sibs to adjust food intake appropriately when their diet was: (a) diluted with an inert filler, (b) adulterated with quinine, and (c) made more calorific by adding fat (Fuller & Jacoby, 1955). A later experiment compared three mutant types on a variety of tests including age changes in food intake, response to changes in palatability, and the effects of food deprivation (Fuller, 1972). I concluded that a distinction could be made between problems created by inappropriate information input to the brain, and those resulting from differences in central processing of the information. Research in this area was continued at Jackson Laboratory by Richard Sprott who was joined later by Israel Ramirez, one of my doctoral students at Binghamton. A brief review of the current status of genetics and food intake regulation has just been published (Fuller, 1981).

My research on the genetics of saccharin preference was an offshoot of the work on regulation of food intake. Again a major collaborator was a member of the college student research program. We found that food-deprived mice reduced intake of water, but increased intake of a weak saccharin solution. Furthermore saccharin preference was much stronger in C57BL mice than in DBA/2 (Fuller & Cooper, 1967). Later I crossed these strains in a standard Mendelian design and showed that a single-locus hypothesis with the high-preference allele dominant fitted the data. Later experiments using other inbred strains and a heterogeneous strain proved that other loci also contribute in saccharin preference (Ramirez & Fuller, 1976). Results with a single pair of strains cannot be uncritically extended to the species as a whole. Ramirez showed that the heritability of fluid consumption was higher than would be expected for a trait subjected to natural selection, suggesting that the strain differences are the result of random fixation of alleles during the inbreeding process.

Alcohol Preference. In 1961 McClearn and Rodgers published a paper on the heritability of alcohol preference in inbred mice. Their measure was the proportion of 10 percent ethanol consumed when an animal had access also to water. I felt that this ratio was misleading for a genetic analysis because of its statistical properties. The variance of probabilities near 0.5 is much greater than that for 0.01 or 0.99, and this could lead to ambiguities in heterogeneous groups such as back-crosses and F_2s. To eliminate this problem I invented a multiple-choice technique, with each subject having access to six randomly positioned drinking tubes containing from 0.5 to 16 percent ethanol. By a somewhat complicated process I calculated an "alcohol score" for each individual. It was an elegant treatment but has never caught on as a standard technique. It probably is too time consuming and unnecessary for most experiments. My estimates of heritability from a half-diallel cross were very close to those of McClearn & Rodgers (Fuller, 1964). While the research was in progress, an invitation to prepare an exhibit for the Seattle World Fair was received. Ruth and I with our younger daughter attended the exhibit and were pleased to see our mice a center of attention, and drinking or abstaining from alcohol as reliably as they had done in the laboratory.

Research on strain differences in alcohol consumption continued after I moved to Binghamton. The subjects were two strains selected for sensitivity to injected alcohol by McClearn, and generously shared with me. Short sleep (SS) mice have a brief loss of the righting reflex following a standard dose of alcohol; long-sleep (LS) mice are immobile much longer. To minimize the effects of taste our subjects were presented with a choice between an extremely attractive sugar-saccharin solution and the same mixture spiked with alcohol. SS mice ingested a higher proportion of the alcoholic mix regardless of prior experience; LS mice preferred the unspiked beverage, but accepted the alcoholic mix if the only alternative was water. (Church, Fuller & Dann, 1979). A follow-up experiment tested the ability of these two strains to meter their alcohol intake when the strength of the spiked option was varied. As expected SS mice were heavier drinkers; both strains compensated for increased alcohol concentration. A surprising result was a sex difference in the accuracy of regulation. Females regulated so accurately that their daily alcohol intake was about the same with a 16 percent or a 2 percent choice.

Males were less successful. With the same eightfold increase in alcohol concentration their intake doubled (Fuller, 1980). I would have liked to investigate this gender difference more thoroughly, but retirement came too soon.

Psychopharmacogenetics. My research on alcohol preference in mice and the effects of CPZ on behavioral development in dogs led to a more general interest in psychopharmacogenetics. The area is important to medicine for the behavioral effects of a drug differ among individuals. Animal experiments comparing several strains provide useful models for analyzing idiosyncratic responses to psychoactive agents. An example is a comparison of two drugs, CPZ and chlordiazepoxide (CDE) on the learning of active and passive avoidance in a shuttle box (Fuller, 1970). Mouse strains that learned the active task readily performed poorly on the passive task and vice versa. I concluded that success on these tasks was not dependent on a generalized learning ability, but on inherited tendencies to run or to freeze in response to pain. The effects of CPZ varied with the type of training. It decreased activity in subjects trained to avoid actively and increased activity in those trained to avoid passively. In this experiment there were no significant differences among strains in the action of drugs.

A second study dealt with drugs and audiogenic seizures. Two laboratories, one of them mine, had obtained conflicting results on the efficacy of a drug as a blocker of the priming effect. The two differences in procedure were the strains used, and unspecified variations in handling the subjects. Repeating the procedure with the two strains of mice in the same laboratory showed that the disagreement was a function of the choice of subjects (Siporin & Fuller, 1976). This point may apply to natural populations, particularly when there is a degree of genetic isolation. My ideas on this subject were expressed in an invited lecture at the International Ethological Conference in Vancouver, and are about to appear in a multiauthored book of which I am coeditor.

Selection for Brain Size. Selection for behavior (e.g., open field activity, aggression) has often been successful, and many investigators have sought to identify the biological correlates of the process. In the late 1960s I thought that it would be interesting to reverse this design, select for a physical characteristic, and look for changes in behavior. Originally I planned to select for a biochemical character, and spent eight months in Alfred Pope's neurochemistry laboratory at McLean Hospital. Formally I was a visiting fellow in the Harvard Medical School. However, the first attempts to select for a neurochemical character were unsuccessful, and my laboratory was not well equipped for this type of research. Instead I chose the ratio of brain weight to body weight, based on the premise that the species we consider to be most intelligent have high *brain weight/body weight* ratios. The first steps were made in collaboration with two Jackson colleagues, Richard Wimer and Thomas Roderick, but for logistic reasons they and I decided to conduct separate selections.

Selection was successful and high, intermediate, and low brain weight lines were established. Variations in the rate of development of sensory capacities, reflexes, and motor skills were found, but there was no simple relationship with brain size (Fuller & Geils, 1973; Fuller & Herman, 1974). However, an analysis of the relation of brain size to several learning measures in the heterogeneous stock from which the selected lines were derived yielded modest but significant correlations (Jensen & Fuller, 1978). For the amount of effort that went into the brain-weight selection the results were somewhat disappointing. There were tantalizing indications of correlations between brain size and learning, and between brain size and susceptibility to audiogenic seizures, but no clear cause-effect relations appeared. I am sure that they can be found and that this approach to brain-behavior relations is worth continuing. The brain-weight lines are still being maintained by Martin Hahn at William Paterson State College. Martin collaborated with two of my graduate students at Binghamton in organizing a symposium on "Development and Evolution of Brain Size." The proceedings have been published (Hahn, Jensen, & Dudek, 1979). The book contains contributions from many investigators with diverse approaches to brain-behavior relationships. I was extremely pleased to open my copy and read the dedication: "To John L. Fuller, our mentor."

Remembrance of Times Past

In bringing this section on research activity to a close I must acknowledge strange feelings in rereading many of my older papers. Yes, I recalled the experiments but the details of procedure and the discussions might have been the work of another person. The redeeming feature of this lapse of memory was that I liked what this alter ego had written several decades ago. He was intelligible, not too long-winded, and I agreed with most of his conclusions. With the wisdom of age I could suggest improvements in some of the experiments, but they were not bad for their time.

Exposition and Interpretation

Whatever note may be taken of me in a future history of animal behavior will probably refer to the publication of *Behavior Genetics* (Fuller & Thompson, 1960). This book gave a name to a body of research whose literature had been widely scattered, and made that literature available to the scientific community. My collaboration with Bob Thompson began when he was a visiting investigator at the Jackson Laboratory in the early 1950s. Our ideas on the genetics and the development of behavior were similar, and we enjoyed each other's company. In 1955–56 I received a Guggenheim Fellowship to work on the book at Yale and at the University of California at Berkeley. At Yale I was sponsored by Frank Beach, became well acquainted with Neal Miller, and participated in a neuropsychology seminar conducted by Karl Pribram. Most of my time was spent in Yale's excellent library.

At the end of the first semester, Ruth and I drove to Berkeley (our first cross-country expedition) where I was provided space by Curt Stern, eminent in the genetics of *Drosophila* and a magnificent teacher of human genetics. What was probably the world's first organized seminar on behavior genetics took place that spring in Berkeley. Robert Tryon, famed for the selection of rats for rapid and slow maze learning was a major figure. Geneticists Mike Lerner and Ed Dempster often attended. Participants Jerry Hirsch, Tom Roderick, and Ruth Guttman later made important contributions to the field. Mark Rosenzweig, who has made great contributions to another of my interests, behavioral development, was another regular. Altogether it was an exciting and profitable sabbatical year.

In June our daughters joined us in Berkeley and the four of us returned east with stops at Yellowstone National Park and other scenic areas. After conveying the two girls to summer jobs, Ruth and I sailed from Montreal to Liverpool on the Empress of Britain. After a few days in London we rented a car and visited animal behavior laboratories in Cambridge, Oxford, and Edinburgh. From Britain we set off on a grand tour of Scandinavia, Germany, Switzerland, France, and the Netherlands. En route we attended the first Congress of Human Genetics in Copenhagen, my first international meeting. In Germany we visited Buldern, where Konrad Lorenz did much of his early research. He was elsewhere but the Eibl-Eibesfeldts were splendid hosts and showed us the charms of that portion of Westphalia. Switzerland and France were all pleasure; mountains, historical buildings, museums, and a delightful cuisine.

Returning to Bar Harbor, the writing of *Behavior Genetics* began, but it had to compete for time with the analysis of the data from the Rockefeller study. The developmental studies were in full swing, supported by grants from the Ford Foundation and the National Institute of Mental Health. Thus, even with Thompson's collaboration, *Behavior Genetics* needed four years to appear in print. It was well reviewed and sold well considering its specialized nature and the small number of courses in the subject. After more than two decades it is clearly outdated, but I still pick it up and am surprised at how few of its generalizing statements need to be changed. Perhaps this indicates my failure to keep up with the times, but I do not think so. There is little in contemporary writings on nature and nurture that contradicts the ideas put forward in *Behavior Genetics*. One of my pleasant memories is hearing Irving Gottesman in his 1977 presidential address to the Behavior Genetic Association advise beginning students of the subject to read Fuller & Thompson. Another compliment was paid by Hans van Abeelen who dedicated *The Genetics of Behavior*, a collection of chapters by British and continental researchers, to "John L. Fuller, explorer and founder."

In 1978 Bob and I completed *Foundations of Behavior Genetics*, a more massive and technical work than its predecessor. The literature of behavior has grown tremendously even since *Foundations*, and its scope has widened to the point that no one person (or even two) can be conversant with all of it. Nevertheless, there are now several good introductions to behavior genetics, each with its own emphasis. I doubt that a second edition of *Foundations* will materialize, but if it does it will be without Bob Thompson who died in 1979 at the age of fifty-six.

In 1980 I was invited by the Canadian Psychological Association to present a lecture in honor of Bob. For me it was an emotional experience. One usually memorializes a senior figure who has retired with honor; Bob was much younger than I. In my talk (Fuller, 1982), I tried to describe my thoughts on the relations between genetics and behavioral science. I hope that Bob would have agreed with my views that it is a happy marriage, but one with some tensions.

Lectures, Chapters, and Symposia

Once one has gained a reputation for expertise in an area of general interest, requests to speak, contribute chapters, participate in symposia, and review books arrive. I have generally accepted invitations if they do not conflict with other obligations. A number of my lectures exist in typescript or outline, but I have no complete record of them. This is no great loss; a listing would be tedious to prepare and even more tedious to read. I have presented talks in many American colleges and universities, and a few in Canada and Europe. These occasions always gave me pleasure, especially when the program included an opportunity to meet with students and faculty for an informal interchange of ideas. Other audiences have included teachers, secondary schools, and church groups. It is important to communicate with people who are not scientists, but who are curious about what we do and what our work means to society.

My invited chapters have ranged from basic treatments of the entire field of behavior genetics for textbooks to reviews of such specialized areas as psychopharmacology, audiogenic seizures, the regulation of food consumption in animals, and genetic factors in deviant behavior of animals. Some chapters are an effort to bridge the gap between scientific and humanistic approaches to human nature (Fuller, 1968, 1978). In retirement I continue to write essays on the relations between genetics and the behavioral sciences, psychology and ethology. A recent chapter contrasts the approach to social behavior of people who call themselves *sociobiologists* and those labeled *behavior geneticist.* Both use genetics, but in different ways. Sociobiologists emphasize changes in gene frequency as the moving force in the evolution of species-typical sociality. Behavior geneticists are interested in the genetic basis of individual variability, and in the interaction of genotype and environment during development. I find sociobiological hypotheses stimulating and heuristic, but am concerned with the difficulties of testing them empirically. My possibly biased view is that a developmental-genetic approach to sociality will be more productive.

Symposia are a form of continuing education. Some take place at society meetings, feature three to five speakers during a two-hour session, and sometimes leave a few minutes at the end for comments from the audience. Others feature a larger number of speakers, last several days, and allow time for the participants to talk with each other over drinks. These are true symposia in the etymological sense of drinking together. Again it would be tedious to list these, but a few merit attention. In 1954 I organized what was probably the first symposium on genetics and behavior presented to the American Psychological Association. Thompson, Hirsch, and Raymond B. Cattell were the speakers. I thought we had done a good job of demonstrating the importance of genetics to psychology until a noted behaviorist rose and said, "It seems to me that all we have heard is that there is another variable to consider besides schedules of reinforcement." Another first was the organization of the first behavioral session to be included in the International Congress of Genetics. Gerald McClearn, Aubrey Manning, Ernst Caspari, and I spoke before a good audience. This time J.B.S. Haldane congratulated the speakers. Since that time a session of behavioral papers has been a standard feature of the Congress.

A couple of recent symposia are worth mentioning. In 1977 I contributed to a series of talks on "Sociobiology and Human Nature" at San Francisco State University. It was my first exposure to the disruptive tactics of the organized left wing. My own talk was permitted to proceed in orderly fashion, but others were not as fortunate. The use of such tactics turns me against the views of the demonstrators regardless of any validity they may have. Despite the unpleasantness, I enjoyed the exchange of ideas and have become somewhat of an armchair sociobiologist.

Another memorable symposium on "Theoretical Advances in Behavior Genetics" was held in Banff, Alberta, in 1978. The site was magnificent. The speakers and invited listeners lived together for a week so that there was opportunity to converse outside of the formal sessions. The conference, sponsored by NATO, was the idea of Joseph Royce who had gathered the data for his doctoral dissertation at the School for Dogs twenty-five years earlier. Two years earlier I had visited him at Edmonton where he heads the Center for Advanced Study in Theoretical Psychology, and we had discussed the composition of the group so as to present a variety of views. A special personal pleasure was the attendance of eleven individuals who had been associated with me at Jackson Laboratory or Binghamton. The resulting collection of papers and comments provide a good summary of contemporary ideas of genetics and behavior. Unfortunately its price is very high.

We are a social species. Those of us with kindred interests have a propensity to form organized groups with a hierarchical structure. A good feature of our scientific culture is that a person can belong simultaneously to many groups. This may be the most important difference between us and other mammals. During my scientific career I joined fifteen scientific societies. As my interests changed I dropped some memberships, but still belong to eleven. Fortunately, most of them give a break to retired members by reducing or eliminating annual dues.

I have a special interest in the three societies of which I was a founding member. The Animal Behavior Society originated from the fusion in 1957 of the Section of Animal Behavior and Sociobiology in the Ecological Society of America with the British Association for the Study of Animal Behaviour. Paul Scott was active in the negotiations that led to the transatlantic merger. At the birth of ABS I worried that there might be too few workers in animal behavior to support an independent society. Obviously I was wrong. Later my election as a fellow of the society pleased me greatly.

The International Society for Developmental Psychobiology was organized in 1966. During a three-year period I served as its treasurer and member of the executive committee. As a candidate for president I lost to Austin Riesen, which was not in the least embarrassing. During my active years as an officer, meetings were held just prior to the large neuroscience meetings. ISDP is now testing the response to individual meetings. This is a good idea as satellites tend to lose their individuality in a crowd.

In 1970 I participated in the founding of the Behavior Genetic Association, and I was honored to serve as president in 1973. My major accomplishments were related to obtaining tax-exempt status for the society, and conducting negotiations with the publisher of our journal. At the annual meeting in 1974 at Minneapolis I spoke on "Reflections on Zero Heritability," a critique of extreme environmentalism in the behavioral sciences. In my opinion, denial of a biological basis for individual differences is not only bad science, but potentially damaging to individuals and to society. The presidential address has not been published but with a little updating perhaps it should be.

The American Psychological Association elected me a fellow in the Division of Physiological and Comparative Psychology in 1969. I have long served as a consulting editor of their journal. I also served a term on APA's Committee on Animal Research and Experimentation, becoming chairman in 1977. In that year we revised the policy statement on humane treatment of research animals. Two kinds of issues must be dealt with. Standards for routine care must be established. It is relatively easy to identify violations, and corrections benefit the researcher as well as his charges. A second type of problem arises when experimental animals are subjected to pain or impaired surgically. Complete agreement on whether the value to society of such an experiment is commensurate with the cost to the animal is unlikely. There are obvious cultural biases in making such judgments. Humans extend more sympathy to dogs, cats, and horses than to chickens, rats, and mice. We also differ in our evaluation of knowledge that does not promise immediate personal benefits. The present tensions between animal experimenters and advocates for animal rights will continue, and scientists will need to educate both the public and themselves.

One other societal activity has just begun. Just as I started to write this autobiography I was elected president of the Society for the Study of Social Biology for a three-year term. I appreciate the confidence of the electorate in assuming that I will still be able to function at age seventy-four. The society is devoted to furthering the discussion, advancement, and dissemination of knowledge of the biological and sociocultural forces that affect the structure and composition of human populations. Few topics are of greater importance on this crowded planet, where medical technology and rapid environmental changes must inevitably have effects on our biology. We must avoid the error of zoomorphism, directly applying our knowledge of animal behavior and animal population regulation to ourselves, but I am convinced that inputs from ethologists, animal ecologists, and animal geneticists will be helpful in considering human problems. Perhaps I can demonstrate this during the next three years.

The Educator

Although the move from Orono to Bar Harbor meant that my major responsibility was research I did not cease being an educator. I have already mentioned the summer programs for college and precollege students and their contributions to my research. In addition there were usually a few Ph.D. candidates from universities doing dissertation research in the Division of Behavior Studies. We also accepted postdoctoral fellows for one- or two-year terms. Four who worked with me, Mark Waller, David Blizard, Richard Sprott, and Roger Ward, have made substantial contributions to animal behavior.

In 1958 Earl Green, director of the laboratory, asked me to accept the new position of Assistant Director for Training. I was responsible for training grant applications, recruitment of summer help, and, with the aid of committee, selection of students. In the precollege program we would have 500 applications for 20 places. Ninety percent of the applicants were qualified, and choices were difficult to make. Not all of these scientifically bent young people continued in the field, but many are now prominent researchers and two have shared a Nobel prize. In the 1960s there was little difficulty in obtaining training grants from federal agencies. Now in the eighties our political leaders are sharply curtailing support. Are we really that much poorer, or have our values changed? I believe it is the latter.

A very different type of program was the two-week summer course in Medical Genetics directed to established researchers, clinicians, and educators. The idea was conceived over lunch with Victor McKusick of the Johns Hopkins School of Medicine. In 1960 the first two-week session was held at the elegant Oakes estate on the shore of Frenchman's Bay with thirty students and a star-studded faculty. As codirector for nine years, I was responsible for local arrangements and shared the recruitment of faculty with McKusick. And of course I managed to put in a few words on genetics and behavior as applied to medicine. The intensity of the course was frightening, but the students loved it. We outgrew the intimate Oakes Center and transferred the operation to a public school auditorium. The course was not all work. Many participants brought their families to enjoy the beauty of Mount Desert Island, and all had a grand time at the annual picnic which was sometimes held at Fuller's 150-year-old farm house on the Breakneck Road.

Indian Interlude

My most exotic educational experience was a visit to India in 1968 as a participant in a National Science Foundation program to improve science education in that country. Originally scheduled to spend six weeks in the spring at Osmania University in Hyderabad I was informed that because of student riots the session would be shifted to the summer. Since I had conflicting responsibilities at Jackson Laboratory I regretfully withdrew. A few days later a telephone caller asked if I would accept an appointment as a special lecturer visiting several universities for a few days each. This created a problem. Ruth had planned to accompany me to Hyderabad, where we would have our own apartment. The new assignment involved a great deal of travel, which would be tiring and expensive. We decided that I should go alone. It was a sound decision. The tropical heat was oppressive and we could not have found suitable housing.

My assignments took me to New Delhi, Ahmedabad, Hyderabad, Gorahkpur, Trivandrum, and Chandigar; from the southern tip of the continent to the foothills of the Himalayas. My Indian hosts were most courteous; several entertained me with other Americans in their homes, which ranged from bungalows that might be found in an American suburb to an elegant, walled enclave housing an extended family of forty with their servants. The faculty in these universities were well educated and welcomed our visits. Their task of raising the standards of science education was formidable. The newer universities had attractive campuses with spacious buildings, but equipment and library facilities were limited. Maintenance of laboratories was neglected. The result was that most of a student's education would depend on lectures and a single textbook. Many of the teachers who attended my lectures had trouble with the elementary mathematics necessary for understanding simple population genetics. I attribute this to the lack of a scientific tradition in the Indian educational system. It was a great learning experience for me, and I would like to return someday. However, I was happy to return to Maine just in time for the birth of my third grandchild.

Back to the University

Had anyone asked me on New Year's Day 1970, about my future plans I would have stated my intention of continuing in my present position until my scheduled retirement in 1975. Since 1963 I had divided my efforts equally between research and my duties as associate director. In addition to standing in for the director when necessary I had major responsibility for long-range planning. Inevitably there was competition between these two jobs for my time and energy, but I enjoyed them both. Nevertheless there were stirrings of discontent. My experimentation was in the hands of capable assistants but there were occasional failures of communication. Space for behavioral research at the main laboratory was of good quality, but crowded for three staff members with independent programs. The dog colony had been discontinued and the space allocated to other functions. I agreed with the decision in principle, but I missed the opportunity to work with these intelligent and friendly animals.

Out of the blue, Andrew Strouthes, a former research assistant at Jackson, telephoned and asked if I would be interested in joining the psychology department at the State University of New York at Binghamton. Ruth and I visited the campus, an offer was made, and we had to make a decision. After twenty-three years of residence we had a strong affection for Bar Harbor and the Jackson Laboratory. Ruth was active in community organizations and was a volunteer in the schools. I had served a term on the School Committee and chaired the planning board during the preparation and submission of a zoning ordinance. Perhaps the high point of my tenure was standing before the Bar Harbor special town meeting, explaining the ordinance, replying to opponents, and winding up with a large favorable majority. On an earlier occasion, by a bit of impassioned oratory, I had been successful in countering an attempt by a conservative cabal to take over our middle-of-the-road town government. Perhaps my career should have been politics, but the closest I came was the chairmanship of the local Democratic committee in a predominantly Republican town.

To leave all this and our 150-year-old home on five acres with a trout brook was a difficult choice. We had put much thought and energy into restoring and improving the property, but recognized that after retirement it would be burdensome and costly to maintain. We liked what we had seen of Binghamton, and I liked the feature of not having to write grant applications for my basic salary. Another advantage was the mandatory retirement age of seventy, which would give me

flexibility in planning. Somewhat in a spirit of adventure we accepted the Binghamton offer.

SUNY at Binghamton was metamorphosing from an elite liberal-arts college to
a research university with technical schools. Growth was in the air and a new
science center was under construction. As usual, completion was far behind
schedule. The building won an architectural award for design, but in the animal
quarters it had serious flaws that were expensive to correct. Nevertheless it was
an improvement and there were funds for new equipment. A major problem was
dealing with the Albany bureaucracy who dragged out the requisition process
while inflation continuously gnawed at the value of our allocations. Jackson Laboratory had been much more efficient than New York State.

I enjoyed the return to the classroom and the opportunity for the first time to
teach courses in behavior genetics and comparative psychology. As a professor I
read more broadly than when research was my major activity. Thus, in a seminar
on aggressive behavior, ideas from ethology, endocrinology, learning theory,
sociology, and anthropology were brought together and evaluated. My graduate
students chose their dissertation topics subject to the availability of facilities. In
the give and take of our discussions I learned from them as they were learning
from me.

Moving to a university did not bring an escape from committees. Jackson Laboratory had essentially one administrative committee, of which I was a member. It
handled budgets, appointments, promotions, planning, and personnel problems.
SUNY-Binghamton had numerous committees at the university, school, and department levels. My share of such duty was reasonable, but I am convinced that
some academics enjoy contention for its own sake and waste the time of others.
The big surprise was being "drafted" as chairman of the psychology department in
1974. As a tribal elder, close to retirement and without personal ambitions, I was
reasonably successful in keeping peace and guiding the growth of the department.
At any rate I survived without getting ulcers or losing all my faculty friends.

On balance, the move to Binghamton was a plus. My individual research productivity decreased, but there was compensation in the work of my graduate
students, who often gave me joint authorship of their papers on the basis of a few
procedural hints, and severe editing of proposals and reports. Library facilities
were better than at Bar Harbor, a great asset in writing *Foundations.* Another
benefit was frequent interactions with scholars in other disciplines. I represented
psychology on the steering committee of a program in philosophy and the social
sciences. Finally, Ruth and I enjoyed the cultural events, particularly the concerts, of this small city set in the attractive countryside of New York's southern
tier. We try to return once a year to renew our friendships.

Retirement

My decision to retire before seventy was based on an inner voice saying, "Now
is the time." We had purchased a house in York, a historic town on the Maine
coast sixty-five miles northeast of Boston. Our choice of York was based on its fit
with a long list of criteria I had composed several years earlier during a boring

plane ride. We discovered later that for Ruth moving to York was a return to the home of her ancestors. She is the eighth generation from John Parsons who settled in York about 1672 and was killed in an Indian raid in 1692 close to where we now live.

We moved our household goods in July 1976, and I spent the remainder of the year working on *Foundations*. In 1977 we returned to Binghamton where I taught two semesters. My original plan was to taper off by teaching one semester each year until 1980, but the 1977 experience convinced us that seasonal migration is for birds. Thus on January 1, 1978, I became Professor Emeritus.

Retirement has not brought an end to professional activities such as editing, consulting, reviewing grant applications, and writing. My study is more spacious than any I occupied as a hired hand, but it still overflows with reprints to be filed, letters to answer, notebooks with half-finished projects that compete for attention. Some contain data that have never been adequately analyzed. Perhaps I will get around to it someday. Whether I do or not, I expect to keep professionally active for a few more years.

Two features of retirement lead to changes in life-style. One is the absence of a schedule of activity imposed by one's employer. Paradoxically the freedom to allocate one's own time in retirement can be as stressful as having one's schedule dictated by academic schedules and application deadlines. A second factor is a gradual decline in energy. One's choices of activity are constrained by the weakness of the flesh. It is necessary to set priorities and allocate time and resources just as one did when working for a salary.

Continuation of some professional activities does not exclude enjoying other interests that may have been crowded out during the working years. Both Ruth and I find more satisfaction in creative activities than in being entertained. We enjoy occasional plays and some television, mostly PBS or athletic events. More frequently we entertain ourselves with Double-Crostics, which we justify by pointing to the growth of our vocabulary. Ruth is an accomplished needleworker and shares her interest with others. Next to my study, is the "maintenance center" with a small shop, a stationary exercise bicycle, and an illuminated stand for growing plants. Gardening has become a major interest and I am gradually learning how to manage a perennial border. Our major exercise is walking, often combined with bird-watching. When it is pleasanter to stay inside, I design and execute wall hangings or try to interpret the simpler piano pieces of Chopin, Mozart, and company. Our socializing is done mainly in small groups where the conversation is intelligent and differences in opinion are voiced with wit and good humor. My biased opinion is that people who retire to Maine are a select group with a good sense of values. As for travel, we do less than we did a few years ago. The costs in money and in energy dissuade us from visiting parts of the world that would certainly give us pleasure. For now we enjoy our memories and photographs collected in Europe, India, Canada, and our own country.

Looking Ahead

It is a new experience to write one's autobiography and to realize that the major part of one's life is over. The idea of death is not frightening. I only hope that it will not be a drawn-out painful affair. Although I have maintained contact with organized religion, I do not expect to be reincarnated or to be a spirit in Heaven or Hell. Ruth and I have prepared for our eventual demise by organizing our affairs so that they can be settled easily. Having done so we look forward to an unknown period of participation in mankind's quest for fulfillment. Love in the old is not the same as love in the young, but it is as real and is much more stable.

It may be a universal trait of the elderly to compare the past with the present and to speculate on the future. It is clear that I am better off materially than were my parents and grandparents, but I look back on my childhood and youth as being happier than they would be today. Our world was simpler and though growing up was stressful, we were better able to cope with its problems. It is almost embarrassing to confess to my own lack of participation in the cultural fads and changes of a half-century and I shall give no details. However, I have no sense that Ruth and I missed anything important. We have no difficulty in relating to persons of all ages who share our old-fashioned values. Our family bonds are strong and so are our friendships both old and new. Although the news is dominated by accounts of violence and deceit, our personal relations with people in many walks of life are characterized by good will and helpfulness. Tragedy, illness, and sorrow come to everyone and we have not been spared. But our joy in living has not been destroyed. We humans are capable of concern for others, the *agape* of the early Christian church.

It is this helping quality of humanity that may enable our descendants to cope with the diminution of natural resources, increased population, and the nuclear threat. Unfortunately I do not see this in the policies of the two most powerful nations of the present time. Perhaps scientists and artists will be the moving force in solving these planetary problems. More than any other human enterprise, science is cooperative on a worldwide basis. Artists are more individualistic, but their communication is less dependent upon a common language and level of education. I like to think that those of us who study the behavior of animals combine the qualities of the scientist and the artist, and that we can help in shaping the adaptation of our own species to a changing world.

References

Church, A. C., Fuller, J. L., and Dann, L. 1979. Alcohol intake in selected lines of mice: importance of sex and genotype. *J. Comp. Physiol. Psychol.* 93:242–46.

Fuller, J. L. 1937. Feeding rate of *Calanus finmarchicus* in relation to environmental conditions. *Biol. Bull.* 70:223–46.

———. 1946. Season of attachment and growth of sedentary marine organisms at Lamoine, Maine. *Ecology* 27:150–58.

———. 1954. *Nature and Nurture: A Modern Synthesis.* New York: Doubleday.

———. 1956. Photoperiodic control of estrus in the basenji. *J. Hered. 47:179–180.*

———. *1960. Behavior genetics. Ann. Rev. Psychol.* 11:41–70.

———. 1963. Effects of experiential deprivation upon behavior in animals. *Proc. 3rd Wor. Cong. Psychiat.* 3:223–27.

———. 1964. Measurement of alcohol preference in genetic experiments. *J. Comp. Physiol. Psychol.* 57:85–88.

————. 1967. Experiential deprivation and later behavior. *Science* 158:1645–52.

————. 1970. Strain differences in the effects of chlorpromazine and chlordiazepoxide upon active and passive avoidance in mice. *Psychopharmacologia* 16:261–71.

————. 1972. Genetic aspects of regulation of food intake. *Adv. Psychosom. Med.* 7:2–24.

————. 1974. Single-locus control of saccharin preference in mice. *J. Hered.* 65:33–36.

————. 1975. Independence of inherited susceptibility to spontaneous and primed audiogenic seizures. *Behav. Genet.* 5:1–8.

————. 1976. Genetics and communication. In *Communicative Behavior and Evolution*, ed. M. E. Hahn and E. C. Simmel, pp. 23–38. New York: Academic Press.

————. 1978. Genes, brains and behavior. In *Sociobiology and Human Nature*, ed. M. Gregory, A. Silvers, and D. Sutch, pp. 98–115. San Francisco: Jossey-Bass.

————. 1979a. Fuller BWS-lines: history and results. In *Development and Evolution of Brain Size*, Ed. M. E. Hahn, C. Jensen, and B. C. Dudek, pp. 187–204. New York: Academic Press.

————. 1979b. The taxonomy of psychophenes. In *Theoretical Advances in Behavior Genetics*, Ed. J. R. Royce and L. P. Mos, pp. 483–504. Alphen aan den Rijn: Sithoff & Noordhoff.

————. 1980. Regulation of alcohol intake in long- and short-sleep mice. In *Animal Models in Alcohol Research*, ed. K. Eriksson, J. Sinclair, and K. Kiianma, pp. 57–62. London: Academic Press.

————. 1981. Genetics of eating behavior in animals. In *Body Weight Regulatory System: Normal and Disturbed mechanisms*, ed. L. A. Cioffi, W. P. James, and T. B. Van Itallie, pp. 197–04. New York: Raven Press.

————. 1982. Psychology and genetics: a happy marriage? *Canad. Psychol.*, 23:11–21.

Fuller, J. L., and Clark, L. C. 1966. Genetic and treatment factors modifying the postisolation syndrome in dogs. *J. Comp. Physiol. Psychol.* 61:251–57.

Fuller, J. L., Clark, L. D., and Waller, M. B. 1960. Effects of chlorpromazine upon psychological development in the puppy. *Psychopharmacologia* 1:393–407.

Fuller, J. L., and Collins, R. L. 1970. Genetics of audiogenic seizures in mice: a parable for psychiatrists. *Semin. Psychiat.* 2:75–88.

Fuller, J. L., and Cooper, C. W. 1967. Saccharin reverses the effect of food deprivation upon fluid intake in mice. *Anim. Behav.* 15:403–08.

Fuller, J. L., Easler, C, and Banks, E. 1950. Formation of conditioned avoidance responses in young puppies. *Am. J. Physiol.* 160:462–66.

Fuller, J. L., Easler, C, and Smith, M. E. 1950. Inheritance of audiogenic seizure susceptibility in the mouse. *Genetics*, 35:622–32.

Fuller, J. L., and Geils, H. D. 1973. Behavioral development in mice selected for differences in brain weight. *Dev. Psychobiol.* 6:469–74.

Fuller, J. L., and Herman, B. H. Effect of genotype and practice upon behavioral development in mice. *Dev. Psychobiol.* 7:21–30.

Fuller, J. L., and Jacoby, G. A. 1955. Central and sensory control of food intake in genetically obese mice. *Am. J. Physiol.* 183:279–83.

Fuller, J. L., and Thompson, W. R. 1960. *Behavior Genetics.* New York: Wiley.

———. 1978. *Foundations of Behavior Genetics.* St. Louis: Mosby.

Fuller, J. L. & Waller, M. B. 1962. Is early experience different? In *Roots of Behavior,* Ed. E. L. Bliss, pp. 235–45. New York: Hoeber.

Hahn, M. E., Jensen, C., and Dudek, B. C. 1979. *Development and Evolution of Brain Size.* New York: Academic Press.

Huff, S. D., and Fuller, J. L. 1964. Audiogenic seizures, the dilute locus and phenylalanine hydroxylase in DBA/1 mice. *Science* 144:304–5.

Jensen, C., and Fuller, J. L. 1978. Learning performance varies with brain weight in heterogeneous mouse lines. *J. Comp. Physiol. Psychol* 92:830–36.

Ramirez, I., and Fuller, J. L. 1976. Genetic influence on water and sweetened water consumption in mice., *Physiol. Behav.* 16:163–68.

Scott, J. P., and Fuller, J. L. 1965. *Genetics and the Social Behavior of the Dog.* Chicago: Chicago Univ. Press.

Seyfried, T. N., Yu, R. K., and Glaser, G. H. 1980. Genetic analysis of audiogenic seizure susceptibility in C57BL/6J × DBA/2J recombinant inbred strains of mice. *Genetics* 94:701–18.

Siporin, S., and Fuller, J. L. 1976. Effect of amino-oxyacetic acid on audiogenic priming in C57BL/6J and SJL/J mice. *Biochem. Behav.* 5:269–72.

Donald R. Griffin (b. 3 August 1915). Photograph by Fabian Bachrach. Reprinted by permission from the Harry Frank Guggenheim Foundation

5

Recollections of an Experimental Naturalist

Donald R. Griffin

As I review my scientific activities, I realize how fortunate I have been in my family, friends, and general circumstances. Ever since boyhood, the people mentioned below have stimulated, advised, and guided me. And I have doubtless forgotten others who were also helpful in important ways. My earliest quasi-scientific recollections are from ages six to nine, when we lived in a somewhat rural area near Scarsdale, New York. The revolutionary era farm house that my parents rented was still surrounded by fields and second-growth woods, although now the area is built up and composed of suburban homes and a school. One vivid memory is of a farmer's boy carrying home a dead possum he had trapped.

My mother, Mary Whitney Redfield (1885–1968), read to her only child so much that my father was fearful that I would never learn to read for myself. One of my favorite books was the National Geographic *Mammals of North America* reproduced almost verbatim from two issues of the magazine published around 1918, with beautiful colored paintings by Louis Agassiz Fuertes. Not scientists or naturalists themselves, my parents knew people who were, and they did everything reasonably possible to encourage most of my enthusiasms. The one exception was a fascination with trapping, which they diplomatically discouraged. I remember seriously pondering my future somewhere around the age of nine. Recognizing that an earlier ambition to be a fireman was not quite what I really wanted, I was torn between the life of a trapper in the north woods and that of a sea captain.

My father Henry Farrand Griffin (1880–1954) had strong literary interests and was a scholarly amateur historian. When his father's financial affairs collapsed in 1907 he had followed his natural inclinations by becoming a newspaper reporter on the *New York Evening Sun*. He did well at this, but about the time I was born (August 3, 1915) he went into advertising to provide better for his wife and child. In the 1920s he established his own small but moderately successful agency; but he developed high blood pressure and was told he must lead a relatively quiet life if he expected to live more than a year or two. For the rest of his life he read widely, improved on standard translations of selected classics, wrote numerous unpublishable essays, and even sold two historical novels (H. F. Griffin, 1941, 1942). The timing of his semiretirement around 1928 was fortunate, and we moved to Barnstable, Massachusetts where my mother's family had for many years been summer visitors.

My maternal uncle, Alfred C. Redfield (1890–1983) was another important guiding influence on my scientific interests, although I saw him only occasionally until many years later. A great-grandson of William C. Redfield, who discovered the cyclonic nature of storms, Uncle Alfred was then a physiologist at the Harvard Medical School, especially interested in the respiratory pigments of invertebrate animals. He was a good amateur ornithologist, and had hunted waterfowl and shore birds in his youth. He spent the summers at the Marine Biological Laboratory in Woods Hole, and did his best in an unobtrusive way to stimulate my scientific interests.

Another encouraging influence was a series of monthly visits to the Boston Museum of Natural History. The trips to Boston were necessary because my teeth were badly misaligned and required several years of orthodontic attention. After I had become reasonably well acquainted with the public exhibits, the kind librarian (whose name unfortunately escapes me) opened up a whole new world of scientific books and journals. Two of the curators, Francis Harper and Clinton V. McCoy, also guided my clumsy efforts to become a mammalogist. For instance they introduced me to the pamphlets describing how to prepare study skins published by major museums. Partly because my aspirations to trap furbearers were not encouraged, I began to collect small mammals around 1930. Uncle Alfred also stimulated me to subscribe at age fifteen to the *Journal of Mammalogy*.

From kindergarten through fourth grade, I had attended a private school near Scarsdale named after Roger Ascham, the sixteenth-century scholar and tutor of Queen Elizabeth. But after our move to Barnstable in 1924, my schooling was extraordinarily irregular. I doubt whether it would now be accepted by any respectable college, and it worried my parents even then. One year in the Barnstable grammar school (1925–26) convinced them that I had learned nothing except how to play craps. But my most vivid memory from that year was one day when the forbidding white-haired principal suddenly let fly a violent diatribe against that hideous doctrine of evolution. The occasion was the death of Luther Burbank (1849–1926), the famous plant breeder. According to our Mr. Cornish, Burbank had been struck dead by the Lord because of his blasphemous advocacy of evolution. The harangue ended with a rhetorical "Do any of *you* believe your grandmother was a monkey?" My faint objections were drowned out by the chorus of horrified NOs.

From my next three years' schooling at Tabor Academy and Roger Ascham I recall mainly that I spent a great deal of time in study hall designing cages for a future fur farm. Much of my spontaneous reading was devoted to such learned periodicals as *Fur, Fish, Game* or the books published by Harding in Columbus, Ohio on how to hunt or trap various animals. This drove my father nearly to despair in the fear that my writing would be irreparably corrupted. After this second year at Tabor, my long-suffering parents decided that I might as well stay at home and study with tutors.

The principal tutor was my father, who felt quite able to instruct me in English, history, Latin, and French. His tutorial was almost continually punctuated with vivid comments on the stupidity of conventional education. His viewpoint was doubtless influenced by his own experiences with an English tutor during several summers that his family spent on the Isle of Wight while he was preparing for college. He had also traveled widely in Europe before college, and devoted his years at Yale (1899–1903) almost wholly to languages and literature. I am sure there were may gaps in my education corresponding to subjects that he disliked, but I am undyingly grateful for his methods of teaching composition. This was simply to have me write short essays about a wide variety of subjects, many of my own choosing. He then criticized my efforts, obliging me to rewrite endlessly until the final product met with his approval.

Because he despised German and mathematics (algebra to him was "a low form of cunning"), he hired a former high school teacher, Miss Lillian Decatur, who had been obliged to give up classroom teaching because she was hard of hearing. During her first visit to our house, Dad learned that she also knew Greek, and asked her help with a passage that was troubling him. She copied the sentence and promised to work out an accurate translation from her references at home. On her next visit she announced triumphantly "It's a dative of manner." But when asked what the passage meant, she could not improve on Dad's earlier guess. So my tutor was known ever after as "Miss Dative." Fortunately her partial deafness masked our occasional slips of the tongue during my many tutorial sessions over the next few years.

After two delightful years living at home, sailing in the summers, converting small mammals into study skins in the winter, and being tutored a couple of hours per day, my parents decided that I really should go to a proper school. Since Dad had attended Phillips Andover, I was vigorously tutored and managed to pass the examinations allowing me to enter tenth grade. I enjoyed the life of a boarding student at Andover, and after some initial academic disasters, managed to study hard enough to do reasonably well in algebra, French, and Latin. Although biology was not available to tenth-grade students, I also spent some time with the biology instructor (and track coach), Larry Shields, and worked actively at the student-operated bird-banding station. Unfortunately I had a long series of colds and finished both tenth and eleventh grades at home with more tutoring from Dad and Miss Dative. In those days one could take college board examinations in individual subjects at the end of each academic year, although this was already an obsolete procedure that was dropped shortly afterward.

When I passed the eleventh-grade college board examinations, my patient parents, with my enthusiastic concurrence, decided that for what would be equivalent to my senior year I could once again stay at home, enjoy sailing and mammal collecting, and receive vigorous tutoring in the hope of entering college in the fall of 1934. This program worked well enough that I was admitted to Harvard College in the class of 1938. We all felt rather pleased, but many years later I learned that in 1934 Harvard was admitting five out of six applicants whose fathers were willing to pay the full costs. At the time I learned this, while serving on some Harvard faculty committee, Harvard was admitting only one out of six applicants who did not request scholarship aid. At the time of admittance, my father was very far from affluent, and was essentially living on savings accumulated during his years as an advertising agent. A substantial fraction, perhaps as much as a quarter, of these savings was devoted to my Andover and Harvard tuition.

On turning sixteen and applying for a driver's license I discovered that I was quite myopic. Uncle Alfred suggested that this explained why I was more interested in nocturnal mammals than diurnal birds. For two or three years I banded birds in Barnstable, operating as substation of the Oliver L. Austin banding program in Eastham, which was at the time the largest in the United States measured in terms of the number of birds banded each year. This led to an ambition to band bats. I improvised crude methods of capturing *Myotis lucifugus* as they emerged from crevices in an abandoned frame house, and furtively placed standard bird bands on the legs of a few (Griffin, 1934). Having recaptured some the following summer, and doubtless with encouragement from the Austins, I ventured to apply to the U.S. Bureau of Biological Survey for permission to use bird bands on bats. The ensuing correspondence began with a very discouraging letter saying that bird bands had been tried on bats many years before and found to be totally useless, because the bats chewed bands so that the numbers could not be read. After persisting through further correspondence, I was finally told that the funds appropriated by Congress were meant for birds and not for bats. But the Bureau of Biological Survey relented to the extent of suggesting that I purchase from the National Band and Tag Company lots of bird bands that did not meet government specifications. This I was glad to do, and only a small fraction of the bands were actually defective.

During the summers of 1933 to 1938 I banded as many bats as I could catch at the two or three nursery colonies that I was able to locate on Cape Cod. Homing experiments showed that at least a few returned from distances up to 50 miles. I also enlisted college friends for bat-banding expeditions to caves in western New England. After I had reported the data from several thousand banded bats to the Bureau of Biological Survey they suddenly woke up to the fact that the rejected strings of bands duplicated numbers issued to bird banders. I received an outraged letter from a different member of the staff, who had to be reminded that this entire procedure was being followed at the suggestion of one of his colleagues. The upshot was that I was now issued regular bird bands. Somehow, between 1934 and 1937, the intent of Congress seemed to have changed concerning the use to which funds for the purchase of bands might be put. In fact, I do not believe that any recoveries ever led to ambiguity, since almost all of the recoveries of my banded bats were made by me or my friends.

My bat-banding efforts showed that *Myotis lucifugus* migrated from caves in the limestone belt of western New England to nursery colonies on Cape Cod (Griffin, 1940b, 1945). My sampling of intervening areas was nil, but later work by Davis and Hitchcock (1965) and by Gifford and Griffin (1960) showed that there was indeed a general southeasterly trend to the summer migration or dispersal of these bats. But probably the most surprising result of bat banding has been to show the great longevity that some bats attain, up to twenty years or more, (Hall, Cloutier, & Griffin, 1957; Griffiin & Hitchcock, 1965; Griffin, 1980).

Finally I should mention another sort of biological interest that grew directly out of my enthusiasm for sailing. In the 1930s, harbor seals *(Phoca vitulina)* were fairly common in Barnstable Harbor, and I had enjoyed many close encounters with them. Once a young harbor seal appeared at the bathing beach and remained for several days, allowing swimmers to play with it, and behaving very gently toward all of us. It was later shot for the five dollar bounty that the State of Massachusetts then paid for the noses of harbor seals. During the summer of 1934, shortly before entering Harvard College, I found several carcasses of seals that had been killed by two especially active bounty hunters. Since a seal usually sank after being shot, they carried in their outboard skiff a glass full of nitroglycerine. A cupful was poured into a smaller bottle equipped with a fuse. Then when a seal surfaced nearby, the fuse was lighted and the bottle tossed into the water. This apparently killed or stunned the seal with enough air in its lungs to prevent it from sinking. I was told later that the predictable accident ended this form of bounty hunting.

I decided it would be of great interest to preserve the stomachs and examine them to ascertain just what the seals were really eating. Very little information on this seemed to be available, and I approached Glover M. Allen, the curator of mammals at the Museum of Comparative Zoology, with the news that I would be delighted to bring in several large jars containing seal stomachs preserved in formaldehyde if he could help me identify the contents. He was kind enough to provide a table and some enamel trays into which I could decant the stomach contents. I had not known enough to inject the formaldehyde directly through the leathery walls of the seal stomachs, so that the preservation was less than perfect. The Curator of Fishes, whose name I have unfortunately forgotten, identified vertebral columns of alewives and one or two other common species of local fishes. Although certainly no earthshaking discovery, this warranted a brief note in the *Journal of Mammalogy* (Griffin, 1935). Many years later some young molecular biologist, who disliked my activities as chairman of the Harvard Biology Department, circulated reprint request cards in the names of distinguished molecular biologists asking for this paper. I did not catch on to the joke until I had dug out a few yellowing one-page reprints and despatched them to the doubtless astonished biochemists.

My undergraduate studies at Harvard concentrated heavily on biology and the related sciences. I was only a B student, and had difficulty with freshman calculus. The lack of any high school science did not seem a serious handicap except that it was quite late before I learned even the rudiments of chemistry and physics. I remember in particular my delight at being instructed, in a beginning physics course, during sophomore year, about that marvelous device the slide rule. Among my very few A's were Glover Allen's courses in mammals and birds. Unlike many such courses at American colleges, his perspective was truly worldwide; but the laboratory work consisted entirely of studying and drawing museum specimens.

John Welsh was then actively studying endogenous activity rhythms in various animals, principally invertebrates, although some of his students worked on laboratory rats. Since bats hibernate in deep caves where the environment is almost uniform over long periods, it was a natural for us to wonder how they coordinate their activities with the day-night cycle in the outside world. With his help and encouragement I maintained a few *Myotis lucifugus* in the most uniform available environment. We used a very crude form of activity recording device in which any crawling or jiggling of a small cage activated a kymograph pen. Smoked drum kymographs were standard equipment in physiological laboratories for recording details of muscle contraction. Our bats did have endogenous circadian activity rhythms under constant conditions. But sometimes there were also peaks of activity approximately twenty-four hours after each feeding session (Griffin & Welsh, 1937).

Harvard at that time still took its tutorial plan seriously, and honors candidates met at least two or three times a month with a tutor, read significant books or papers, and wrote tutorial papers. This was one of the rare times that Uncle Alfred explicitly exerted his influence. He had by that time moved from the Harvard Medical School to the biology department, and urged me to select as a tutor Jeffrey Wyman. Wyman's interests were in the physical chemistry of proteins, and he expected me to take physical chemistry and read what I now realize was relatively elementary material on the physical chemistry of biological systems. I particularly remember Langmuir's papers on surface tension. I dutifully worked hard at organic chemistry, physical chemistry, and even took a half-course in atomic physics with Oldenberg, which was slightly more than biologists were expected to attempt. I managed to get B's in the chemistry courses, but despite valiant efforts received only a C + in atomic physics, although I found it really exciting and tutored a floundering fellow biologist (who got a C −). What stands out principally in my recollection was the elementary development of the idea that one could account for the gas laws by simple "billard ball" assumptions about molecules. Nothing in my previous biology or chemistry had brought that home to me.

Physiology came to seem to me and most of my fellow students the substantial science of the future. Bat banding, mammalian systematics, or similar things were old-fashioned natural history, we were led to recognize—suitable for amateur dilettantes, perhaps, but not serious science suitable for one aspiring to a professional academic career. To be sure George Clarke represented the new science of ecology, and I enjoyed auditing his course, which emphasized oceanography and limnology. In my senior year, and simultaneous with physical chemistry and a specialized course in the physical chemistry of proteins with Wyman, I took the only course then offered at Harvard College in biochemistry. It was a half-course, without laboratory, given in the chemistry department, which I found fascinating though difficult. The final examination included a question about the intermediary metabolism of fats—beta oxidation and matters of that sort. I was sure I had that material down cold and wrote a lengthy essay. But when I returned to my lecture notes and textbook to see whether I had really got things right, I was horrified to find that my elaborate biochemical arguments had been figments of my imagination. To my astonishment I received an A. Either I must have guessed right about how biochemists would have handled a subject about which I was really quite ignorant, or else the reader did not devote very careful attention to my bluebook.

In retrospect I cannot really be sure how my interests in the biology of bats turned to their orientation in the dark. One could scarcely share totally dark caves with flying bats without being impressed by their agile flight and avoidance of obstacles. I had read elementary accounts of Spallanzani's experiments showing that blinded bats flew as well as ever. But I did not know about the paper by Hartridge (1920). Nor had I been at all well instructed in the physiology of audition. Hearing was then treated in a very cursory fashion in textbooks and courses in physiology, and I probably had picked up what little I knew from Boring's psychology course. Left to my own devices I doubt whether I would ever have learned about George Washington Pierce, a retired or nearly retired physics professor. But two people at Harvard told me I should visit him and see whether his apparatus for studying what we then called "supersonic" sounds might pick up something interesting from my bats. These two were James Fisk, later president of the Bell Telephone Laboratories, and Talbot Waterman, a slightly older student of biology subsequently a professor at Yale. It took a considerable effort during the winter of 1936–37 to bring myself to call on the distinguished physics professor, particularly in view of my C + from Oldenberg.

Once I made the effort, however, I found Pierce a jolly fellow who was delighted to find someone who knew one end of a bat from the other. He was already engaged in studying the high-frequency sounds of insects, and another fellow biology student, Vincent Dethier was serving as his research assistant to identify the grasshoppers whose sounds Pierce was studying at his summer place in New Hampshire. Pierce was enthusiastic about the possibility that bats might also be sources of supersonic sounds, and urged me to bring a cageful to his laboratory. Once he turned on his apparatus, its loudspeaker clicked and rattled delightfully whenever the bats were at all active. He explained that one had to be careful because many scratching sounds such as the claws of bats scraping on the wire mesh of a cage also contained strong components at frequencies above the range of human hearing. But by studying individual bats gently held in my hand and taking other precautions against the accidental generation of mechanical sounds, we soon convinced ourselves that bats were emitting definite sounds well above the range of human hearing.

As I have described in detail elsewhere (Griffin, 1958, 1980), when we tried to pick up sounds from flying bats we were only occasionally successful. Thus our joint paper in the *Journal of Mammalogy* (Pierce & Griffin, 1938) was appropriately cautious about assuming that these newly discovered supersonic sounds had anything to do with the orientation of flying bats. The situation was vastly clarified a year or so later when I returned to studying what we called the obstacle avoidance ability of flying bats in collaboration with Robert Galambos, as we have described elsewhere (Griffin & Galambos, 1941; Galambos, 1942; Grinnell, 1980).

During my senior year I decided to branch out and take up something new. I certainly wanted to continue as a graduate student in biology and looked forward hopefully to some sort of academic career. Bird navigation had begun to fascinate me, but wiser heads emphasized that if I really wanted to be a serious scientist I should put aside such childish interests and turn to some important subject such as physiology.

Resolution of this dilemma was greatly aided when I learned that Karl S. Lashley was coming to be a member of both psychology and biology departments. I had read the classic paper by Watson and Lashley (1915) on the homing of terns and sought him out to ask whether I could be a graduate student under his direction and study the homing of some locally available species. He was willing, and it was understood that I would be a graduate student in the Department of Biology, but do my research with him. He stipulated only that I take a couple of courses in psychology. I decided that Leach's petrels seemed a promising species, partly on the basis of enthusiastic conversations with a first-year graduate student in the department of geography, William Gross. Bill Gross had taken over from his father, A. O. Gross, the management of the Bowdoin College Research Station on Kent Island, near Grand Manan, New Brunswick. In addition to his graduate work in geography, he was busy planning a great expansion of the Kent Island Station for the summer of 1938. In his vision it was to be a serious rival to the Marine Biological Laboratory at Woods Hole. There would be not only research workers but courses for students of all levels from secondary school to graduate students. The fact that there was neither money nor facilities for anything remotely approaching so ambitious a program was merely an exciting challenge.

As a graduate student with a research project, namely the study of homing in petrels, I was able to negotiate a room-and-board charge for the summer of $40 instead of the standard fee of $200. This came out of a summer research budget of $200 provided by Lashley. Of course one should realize the extent of inflation since 1938; for instance a new Ford or Chevy cost about $700. I arranged to bring along my good friend, Douglas Robinson, and another Harvard student named Frederick Greeley. Infected by Bill's enthusiasm I agreed to give a few lectures in elementary genetics, based on the course in Mendelian genetics I was taking at the time.

Bill Gross persuaded companies that made staple foods, building materials, or engines for the station's boat, to donate them in return for the publicity that he promised from the "Bowdoin Outer Sea Islands Expedition of 1938." He was remarkably successful, securing a brand new Studebaker automobile engine, enough composition roofing for several new buildings, and a great deal of flour, uncolored margarine, and peanut butter. But on our arrival at Kent Island we all had to pitch in to complete the buildings, and food was often limited to whatever fish we could catch. Doug Robinson came to study the myoglobin of seals, equipped with a suitable high-power rifle, and sealburgers proved a welcome addition to the otherwise somewhat limited menu. One hopeful student said he knew how to bake rolls, and they were not altogether inedible.

Leach's petrels typically spend about four days with their eggs or young, while the mate is feeding at sea. Each experiment required catching twenty or thirty petrels, as far as possible taking birds only from burrows where previous recording of arrivals and departures showed that an adult had just returned, presumably well fed, from its long period of feeding at sea. They were then transported in covered cages to release points that ranged anywhere from a few miles away to several hundred miles along the coast or out to sea. To arrange releases far from land, birds were taken by the station's launch to a harbor on the western end of Nova Scotia, driven to Halifax, and placed in the custody of the deck officer of a freighter departing for the West Indies.

One experiment was to subject the petrels to rotation from time to time during their outward transportation in order to test whether or not their kinaesthetic sensory systems (presumably the inner ear labyrinth) might register the direction they were being carried away from home. This was accomplished with a windup phonograph on which each cage was placed for a couple of minutes. When we would stop somewhere in Nova Scotia to apply this procedure successively to several boxes, local residents were naturally curious and yet always politely accepted our explanations. About twenty years later Susan Billings of the anatomy department at the Harvard Medical School took up similar homing experiments with the petrels of Kent Island (Billings, 1968). Although rotation during transportation had been shown by me and several others to have no measurable effect, so that she was not doing anything of the sort, Nova Scotian villagers asked her, "Where is the phonograph?"

The key question that emerged at that point was whether birds displaced into unfamiliar territory could really determine the homeward direction and proceed directly or nearly directly back to their nests. While many, but by no means all, petrels returned from distances of as much as 470 miles, their speed of return was not especially impressive. They might have been well oriented but spent considerable time feeding or in other activities than flying straight home. I therefore decided that Leach's petrels were not the ideal species with which to measure accurately the nature of the homing ability (Griffin, 1940a).

Previous experiments in Germany had demonstrated that gulls and terns had a reasonably well-developed homing ability. In 1939 I therefore selected Penikese Island in Buzzards Bay as the most suitable nesting colony of herring gulls and common terns. Both species proved able to return reasonably well after being transported hundreds of miles along the coast or inland. I compared homing performance from coastal points southwest of Penikese in areas through which the birds probably passed during their migration, and points at similar distances northeast where it seemed unlikely they would have traveled previously. Another comparison was of homing from coastal and inland release points. Homing was rather good from all three directions, with a very slight superiority for the releases from points along the presumably familiar coast to the southwest. But again the results were consistent with the "nothing but" interpretation that when released in unfamiliar territory the birds either scattered more or less randomly or searched in a systematic pattern until they reached a familiar area that would presumably have been along the coast.

Somehow I had to observe the homing birds continuously during their homing flights. During the winter of 1940–41, I even tried radio tracking of a herring gull, but the smallest available transmitters were much too heavy (Griffin, 1963). How about direct observations from small airplanes? Alexander Forbes, the physiologist, yachtsman, pioneer aviator, and explorer of Labrador was the flier to whom I turned for advice. He was delighted to take me up in his single-engine plane to see how well gulls could be followed. Nothing could have been farther from any of my previous experience or aspirations than flying in light airplanes. But it was impossible to resist Forbes's infectious enthusiasm for such enterprises. In a typical early attempt at the Northampton Airport, we released a gull from the plane. "Wait 'til I pull the plane up in a stall, then you open the door Griffin, and throw the gull down as hard as you can, so it'll miss the horizontal stabilizer." Unfortunately the open door acted like right rudder and put the plane into a spin. I had no idea what was happening, and our altitude of perhaps 2000 feet was quite sufficient for Forbes to bring the plane out of the spin. But of course we had lost considerable altitude, and never saw that gull again. A simpler procedure was to have someone at the airport to release the gulls one at a time when Forbes signaled that we were ready by rocking the plane, and it was not too difficult to keep a gull in view while circling one or two thousand feet above it.

In 1941 I bought a six-year-old Rearwin tandem two-seater from the operators of the Falmouth Airport, and as soon as I obtained my private pilot's license I followed several gulls that certainly did not always head straight for Penikese Island. The results were again consistent with the idea that when released inland in what was presumably unfamiliar territory, herring gulls did not have clearly demonstrable ability to head directly toward home. The paper describing these observations was rejected by the *Auk* but published in *Bird Banding* (Griffin, 1943).

During the summer of 1939, on the recommendation of William J. Hamilton, the Cornell mammalogist, I was awarded a generous summer fellowship enabling me to work at the newly established E. N. Huyck Preserve southwest of Albany. I hastened to Rensselaerville with only the vaguest ideas of what I might study. Since there were large nursery colonies of *Myotis lucifugus* in several of the fine old houses of the village of Rensselaerville, I could not refrain from doing a little more banding, but then turned back to the problem of obstacle avoidance in bats. Although I had talked a great deal with my fellow graduate student Robert Galambos about bat hearing and the possibility that the newly discovered supersonic sounds might really be used to detect obstacles, we had decided that the high-frequency hearing of bats would be his thesis problem while I would concentrate on bird navigation.

But here I was at a biological field station with lots of bats available and no very satisfying problem on which to work. So I decided to repeat and extend the obstacle avoidance experiments that Hahn had carried out at Indiana University more than thirty years earlier (Hahn, 1908). Casting about the village of Rensselaerville I finally found one of the neighboring families who were willing to let me convert an 8-foot-square horse stall into a flight chamber. I divided it in half with a row of vertical wires and persuaded freshly caught *Myotis lucifugus* to fly back and forth between two 4 × 8 foot spaces. The agility with which some individuals dodged one-millimeter wires even when blindfolded was absolutely fascinating. I experimented with various types of earplugs and confirmed what Spallanzani and Jurine had reported long ago, and what Hahn had also found, even though his conclusion was that "the sixth sense of bats is located in the inner ear."

In Rensselaerville I did not have Pierce's apparatus for detecting high-frequency sounds, but the summer experiments motivated me to return to this problem on my return to Harvard in the autumn. Galambos was meanwhile recording cochlear microphonics from bats in the laboratory of Hallowell Davis at the Harvard Medical School and finding that their ears did indeed respond up to 90 or 100 kHz. I have already described elsewhere the development of our joint experiments (Griffin, 1958, 1980); and Galambos has also published some of his recollections of our collaborative experiments (in Grinnell, 1980). He was far more of a physiologist than I, and without his contributions I would very likely never have worked out as conclusive experiments. On the 'other hand I was the one who knew something about bats. Although our collaboration was intense, time consuming, and fruitful, we kept to our original plan concerning Ph.D. theses: Galambos's was on the hearing of bats and mine on the homing of birds.

Although my grades were inadequate for fellowship support, I was lucky enough to be nominated for a junior fellowship. These were magnificent, providing more money than anything else available to Harvard graduate students, providing it for three years, and having the stipulation that one *not* be a candidate for any degree. A junior fellow was free to take courses in any part of Harvard University but was not required to take examinations. When A. Lawrence Lowell, A. N. Whitehead, and L. J. Henderson set up the Society of Fellows in the early 1930s they felt that the traditional work toward the Ph.D. degree was wasteful and inappropriate for true scholars. They hoped that junior fellows would be so highly regarded that no one would expect them to bother with formal degrees; but most of us did so.

When a candidate was waiting to be interviewed by the six or seven senior fellows, his troubled spirits were soothed by a current junior fellow until his predecessor emerged pale and shaken from the inner sanctum. I vividly remember this interview, particularly because Mr. Lowell had a habit of frowning when I, and I presume other candidates, said something relatively inept. Then, as one was biting his tongue at having said anything so stupid, Lowell would thrust his hearing aid microphone across the table and say "Would you mind repeating that? I did not really understand you."

I was not elected during this year, but had the temerity to write to Henderson early the next year asking that I be considered once again. I do not know how often this happened, but I was interviewed a second time and was awarded a junior fellowship beginning in September 1940. I suppose I had become somewhat more confident, and also I could talk about my joint experiments with Galambos as well as about my plans for further studies of homing birds by following them from an airplane. In any event the Society of Fellows paid for my flying lessons and provided a small research budget that helped in the purchase of my Rearwin. Quite apart from the generous financial support, the Monday dinners with the senior fellows were a marvelous source of ideas and inspiration. Who could fail to profit from weekly dinners listening to Whitehead arguing with Henderson or with some of the older and less-timid junior fellows? We also met by ourselves for lunch on Tuesdays and these too were stimulating occasions.

After Pearl Harbor I was fortunate to be able to be employed in applied wartime research at Harvard in S. S. Stevens's Psychoacoustic Laboratory. Although very ignorant about almost everything that was going on, I found it an exhilarating experience to plunge directly into active work seeking to improve voice communications in the noisy environments of military aircraft and tanks. Tank crews or airmen often could not understand the orders they received by radio or even through intercommunication equipment within the same machine. This had tragic consequences when a Naval flyer could not understand the directions by which he was supposed to find his way back to an aircraft carrier.

The most severe problem with the equipment in current use by all the U.S. Forces was that it had a sharply peaked frequency response designed in World War I for receiving dot-dash telegraphy. When the gain was turned up in a noisy environment, the sharp peak at about 1 kHz became painfully loud or even damaging to the ears. But the U.S. Army and Navy would not consider changing equipment during wartime on the basis of such long-haired theorizing. Hard data were needed to demonstrate whether or not a broad-band system really would allow listeners to hear more in noise. So we conducted lengthy and laborious tests in a variety of noises and, sure enough, we proved to everyone's satisfaction that one could hear speech better in noise with a system having a roughly uniform frequency response. In retrospect it seems astonishing that all this effort was necessary to demonstrate the obvious. But Stevens was very effective in persuading colonels and commanders, as well as civilian bureaucrats, that even in the midst of frantic wartime conditions, our servicemen could fight more effectively, and many lives could be saved, by changing to new and better equipment for voice communication in noise.

By the end of 1942 I decided that as a physiologist I could be more useful in work that made more use of my biological background. So I managed to transfer to the Harvard Fatigue Laboratory, where applied physiologists were testing equipment and developing improved gear for aviators and soldiers forced to live and fight under adverse climatic conditions. One of my first tasks was to make sense of a series of experiments conducted in a low-pressure chamber where we would simulate the atmospheric conditions to which aviators were then subjected when flying at high altitudes (Griffin, Robinson, Belding, Darling, & Turrell, 1946).

We also tested cold-weather clothing and sleeping bags, and this led to one of our more hilarious wartime research projects. Late one afternoon when I was the only one in the laboratory, Bruce Dill, the former director of the laboratory and then a colonel in the Army Quartermaster Corps, which provided most of the laboratory's projects, telephoned to say that the Fatigue Laboratory would please conduct the following experiments starting at 0800 the following morning: A decision was to be made between one of three types of fly buttons to be installed on millions of pairs of winter Army trousers, but it was not known which of these would allow the most efficient operation of the buttons under conditions of cold stress. Furthermore the winter uniforms would include two or three types of gloves or mittens. We therefore ran carefully balanced experiments comparing the speed of unbuttoning and rebuttoning flies with and without various gloves or mittens under different conditions of cold and wind. I do not recall that we standardized other obvious physiological variables.

Although most of the work at the Fatigue Laboratory was of more practical importance and theoretical interest than the trouser button project, I was very happy when invited sometime in 1944 to join George Wald and Ruth Hubbard in a project sponsored by the U.S. Army Engineers in my old haunts, the Harvard Biological Laboratories. The problem was that night-vision devices had been developed by which invisible infrared light was converted into an image that allowed soldiers to see and shoot the enemy in total darkness. The only trouble was that, despite everything in the textbooks, the intended victims could see a dark red glow coming from the infrared searchlight. Did small leaks allow ordinary visible light to escape through the filters, or could the human eye really see intense sources of near infrared? The filters were still secret items, having recently been developed for this military use, as I recall by the Polaroid Corporation. Provided with sample filters (to be locked away in a safe at the end of every day's experiments), we were able to measure the human threshold into the near infrared by using a simple modification of the apparatus Wald had been using for years to study human dark adaptation.

The results showed that extrapolating the well-measured photopic and scotopic threshold curves for human vision could account for the visibility of the intense infrared searchlights. The only real surprise was that the photopic and scotopic curves cross again in the near infrared, so that at about 1000 nanometers, peripheral retina was more sensitive than the fovea as it is below about 650 nanometers. At these wavelengths one could feel the radiation as a slight warming of the skin at the visual threshold (Griffin, Wald, & Hubbard, 1947). By the time the infrared vision project was finished, the Navy asked Wald to study the practical problem that dark-adapted lookouts had difficulty focusing their high-quality binoculars. Simple experiments showed that many people change the focus of their eyes by as much as two diopters when fully dark-adapted (Wald & Griffin, 1947).

As war research ended in 1945, the senior fellows reappointed me beginning in January 1946. But by that time I had managed to obtain a real job at Cornell, to begin in July 1946. In many ways I regretted having less than the normal full three years as a junior fellow, but candidly I must admit that I learned more that was helpful in later scientific work through the wartime research projects.

My six months as a junior fellow in the first half of 1946 did represent a fine opportunity to apply to bats what I had learned about microphones and acoustic measurements. I was especially helped by Francis Wiener, who had worked throughout the war in L. L. Beranek's Physical Acoustics Laboratory. He lent me appropriate amplifiers and a 640AA condenser microphone, which at the time was by far the best available transducer to pick up airborne sounds above the frequency range of human hearing. Once I could return to bats, the obvious thing to do was to display on a cathode ray oscillograph those supersonic sounds that Pierce and I had discovered but had been unable to characterize except in a very general fashion. They turned out to be even briefer in duration than we had guessed, often only one or two milliseconds. They were not broad-band noise bursts but signals that swept progressively downward in frequency over approximately one octave. What I was able to learn during these first few months of 1946 began a new era in the acoustic analysis of the sounds used by bats to detect obstacles (Griffin, 1946, 1950, 1958).

Having been at Harvard ever since 1934, I felt strongly that I should vary my background and become better acquainted with the real, non-Harvard world. I had had some limited acquaintance with various zoologists at Cornell, including a fellow graduate student, William Wimsatt. He had come to Harvard Medical School to teach anatomy during the war and we had enjoyed a few trips to bat caves during that period. But he returned to Cornell somewhat before the end of the war to join the zoology department then being reorganized under the vigorous chairmanship of Howard Adelmann.

In those days young biologists were discouraged from actively seeking positions. One was supposed to wait for one's senior mentors to receive letters from other institutions asking for recommendations. Then one hoped to be invited for an interview. But I let Bill know that I would be delighted to become Cornell's comparative physiologist. Part of Adelmann's screening procedure was to ask me, and presumably other candidates, to prepare outlines of three courses in physiology which I would offer if appointed. He stipulated vertebrate physiology, invertebrate physiology, and general physiology. Since I had been brought up to believe that physiology was one subject and not readily divisible along phylogenetic lines, I managed to persuade him that there should be one course throughout the year in comparative physiology.

I started a small pigeon loft, but it never really flourished, primarily because of predation by rats. I started flying again, renting more modern planes than my 1935 Rearwin. I found I could follow flocks of pigeons, especially if they included some white or light colored birds. Although the resulting data were rather inconclusive, they did demonstrate the variety of flight paths flown by homing pigeons, including some cases in which they were apparently misled by topographical similarity between the Finger Lakes near Ithaca (Griffin, 1952a).

My principal effort in bird navigation was to activate my long and thoroughly planned homing experiments with the gannets of Bonaventure Island. Research support adequate for a trip to the Gaspé Peninsula, and rental of a suitable airplane was unheard of at that time, and I was almost discouraged from even trying. Then suddenly I learned from Laurence Irving and Pete Scholander at Swarthmore College that, mirabile dictu, the federal government was still supporting research projects, even though the war had ended. I and everyone I knew had assumed that this sort of federal largesse was a temporary wartime phenomenon. Certainly there had never been anything like it before the war. But Larry and Pete told me that there was now an agency called the Office of Naval Research that had established a biological research project at Point Barrow, Alaska, and that they would award research contracts similar to those with which I had been familiar during the war to university scientists who were willing to carry out research on the Arctic Coast of Alaska. What was even more astonishing, they did not really care what one studied, provided that one went to Point Barrow. It seemed that the Navy realized that academic scientists had made great contributions during the war, and now they wished to build up a body of scientists with Arctic experience in case they were needed in the future.

What I really wanted to do was to follow gannets over New Brunswick and Maine, but of course there were interesting birds nesting on the north coast of Alaska, including large white species such as snow geese and swans. So a proposal was worked up in which I maintained that I really wished to study the homing of birds in Arctic Alaska but that it was necessary first to develop and test the methods of airplane following, which could be done more economically with gannets in the northeastern United States and adjacent provinces of Canada. If I could do this in the summer of 1947, I would then go to Alaska in late summer, and make realistic plans and arrangements to do similar homing experiments there in the summer of 1948. The ONR did provide the necessary funds to rent a suitable plane for homing experiments with gannets.

A graduate student, R. J. Hock and his wife Ann helped me catch gannets nesting at the top of the cliffs where we could walk into the nesting area. We transported them by local boat and jeep to Caribou, Maine and repeated with minor improvements the sort of observation I had been making six years earlier with herring gulls in Massachusetts. Things worked rather better, primarily because of my greater flying experience and improved logistics. I managed to follow several gannets for considerable distances, but their flight paths deviated enormously from the direct route home. Nevertheless roughly two-thirds of them did get back to their nests, about the same proportion as with birds that we had not followed from an airplane (Griffin & Hock, 1949; also for a more popular account see Griffin, 1964).

These airplane observations, together with a thorough analysis of all the data then available concerning the homing ability of birds persuaded me that almost all homing could be accounted for without assuming that birds had any ability to choose the correct homeward direction when released in unfamiliar territory (Griffin, 1952b). This conclusion, which in retrospect seems so narrowly overconservative, was very much in keeping with the basic ideas on which I had been brought up in the biology and psychology departments at Harvard in the 1930s. Everything that animals did must be explained on the sort of very simple basis characteristic of Jacques Loeb's theories of tropisms or the somewhat more refined terminology developed by Fraenkel and Gunn (1961).

The simplest possible way to account for the homing of birds was to assume that in unfamiliar territory they would simply scatter at random or perhaps explore more or less systematically until they found landmarks that they had learned during previous travels. Another idea that seemed to have considerable merit was advanced by the English physicist Wilkinson (1952) who likened homing birds to molecules in a gas. His equations based on the random movements of molecules could be made to fit reasonably well with the then available data on homing performance. Yet when I reviewed all of the published evidence small nagging doubts remained. There were a few reports, especially of homing pigeons and the Manx shearwaters studied by Lack and Lockley that were hard to reconcile with theories of random scattering or systematic exploration, even though the great bulk of the available data could be accounted for in this fashion. But on balance "nothing but" explanations seemed in the early 1950s to be the most reasonable and appropriate.

It was only a few years later that Matthews in England and Kramer in Germany demonstrated that homing pigeons really do show far better than random homeward orientation when released in unfamiliar territory (reviewed by Matthews, 1968 and Kramer, 1961). My rather poorly trained pigeons in Ithaca did not do so, but Matthews's and Kramer's certainly did. This development had a sobering effect on my reductionistic thinking. At about the same time, an even more startling development shook up my whole scientific viewpoint in an even more drastic fashion.

In 1948 at an AAAS meeting in Washington I arranged a symposium on animal orientation at which Ernest Wolf startled everyone by summarizing Karl von Frisch's experiments on the dances of honeybees. I had heard a little about these dances that were alleged to convey information from one bee to another about the distance and direction to a food source, but was frankly incredulous, even though von Frisch's earlier work on color vision in bees and on hearing of fishes was well-known and highly regarded. Good God, if mere insects communicate abstract information about distance and direction, where does that leave Loebian tropisms? If bees do something like that, how can I be so sure that homing birds simply search for familiar landmarks?

I lost no time in studying von Frisch's papers and set up my own observation hive, which the Cornell apiculturalists were happy to stock with a functioning colony. Like many other initially skeptical biologists, I had to see for myself the striking correlation between the pattern of the waggle dances and the distance and direction of the food source. I even managed to work it into one of my comparative physiology courses as an "unknown" in which the students first watched the dances then estimated where the food source must be and saw for themselves that the bees really did seem to be communicating about directional and distance information.

I was so fascinated by this revolutionary discovery that I explored the possibility of inviting von Frisch to come to the United States for a series of lectures. The Rockefeller Foundation, which had been supporting him at Munich for many years before the war were happy to provide *part* of the costs, but they expected that universities where he visited would provide appropriate honoraria and contribute to his local travel expenses. I therefore launched into an extensive correspondence with biologists I knew or had heard of at major American universities. Everyone rallied around enthusiastically so that the problem was simply to arrange a reasonable sequence and schedule.

Von Frisch and his wife came in the early spring of 1949, starting from Graz in the Austrian province Styria, passing through Vienna, which was still occupied by the Russians, and on to Frankfurt. There they watched in amazement the almost continuous series of American planes taking off and landing in their shuttle flights to blockaded West Berlin. Ithaca was their first stop and everyone rallied around to.make them as comfortable as possible. The worst problem I recall was that like typical Europeans they wanted to walk through the countryside. But almost every motorist stopped to see what was the matter with these elderly people and where they wished to be driven. They found it a nuisance to explain to driver after driver that they really liked to walk.

After his lectures at Cornell, von Frisch went on tour, visiting most of the major American universities, including two on the Pacific Coast and, as far as I could tell, enjoyed all his visits. He certainly spoke warmly of them in his recollections (von Frisch, 1967). The von Frischs returned briefly to Ithaca in the late spring of 1949 although no more lectures were expected. By that time I had my observation hive functioning and von Frisch was able to give me a few pointers on how to conduct the type of experiment he had worked out. I recall that he was stung on the nose by one of my bees and only then confided in me that he was liable to severe reaction to bee stings. But he insisted that the only treatment he wanted was topical application of acetic acid, and, to our enormous relief, the incident led to nothing worse than a swollen nose.

One of the requirements of the lecture series was that the Cornell University Press should have the option of publishing the lectures. Von Frisch readily agreed, and within a few weeks I received a manuscript. It was the very same marvelous material he had presented in his lectures, but it was typed single spaced on thin paper with many strikeovers and a very few places where German word order prevailed over English usage. It required some effort to persuade the Cornell University Press that this messy manuscript was really worth publishing (von Frisch, 1950).

Von Frisch's discoveries that honeybees communicate symbolically and the demonstration by Matthews and Kramer that pigeons really can choose approximately the correct homeward direction when released in unfamiliar territory shook up my thinking about the capabilities of animals. Although I still considered myself primarily a physiologist and directed my efforts toward mechanistic explanations of animal behavior, I came to see that these mechanisms must be much more subtle and versatile than I had imagined.

By 1950 I had decided to concentrate for a while on further experiments devoted to echolocation in bats. It seemed clear by then that what was needed was some method to test just what bats could and could not detect and whether they could discriminate between different objects that returned echoes to their ears. I decided that the escape flights of a bat released in some sort of flight chamber would be the best approach. I constructed a plywood box with an entrance hole in the middle of one wall, so that when a bat was released and hastened to fly off it would have a choice of flying right or left toward two identical openings. One of these, however, would be impeded by obstacles of various sorts and I assumed that the bat would of course choose the unencumbered passageway.

With the help of an interested graduate student I began a series of quantitative experiments in the early summer. But the bats frustrated us at every turn. Either they developed position habits, always flying to the right or left regardless of the obstacles, or they flew toward the cluttered opening rather than the open one. We tried endless variations and permutations of this type of experiment, reconstructing the flight chamber several times to provide different lengths and widths of fight space; we tried both repeated releases of the same individual and releases of large numbers with each bat being exposed to the apparatus only once. But the resulting data were chaotic and told us absolutely nothing about what sorts of obstacles bats could or could not detect. This was the first summer that I had utilized my ONR contract funds almost entirely to study bat echolocation. Previously I had persuaded the ONR to support some mixture of bird navigation work and studies of the orientation of bats. Toward the end of the summer I was totally frustrated since we had accomplished absolutely nothing despite a vigorous effort and expenditure of what then seemed substantial funds.

In late August, in what I thought of as a desperate last-ditch effort to salvage something of at least minor interest from the summers' work, I decided to take my apparatus for recording the orientation sounds of bats to some outdoor situation where wild bats were pursuing flying insects. At that time, my method of recording bat sounds was to photograph oscilloscope traces with an ancient 35 mm movie camera that I had modified so that the film moved continuously rather than intermittently. With some effort, the necessary apparatus, including a small gasoline engine-driven generator, was organized into an ancient station wagon. After several exploratory sessions, I managed to get everything working at the times and places where, for fifteen or twenty minutes each evening, several *Eptesicus fuscus* hunted insects at altitudes of ten to fifty feet over a small pond near Ithaca. Since it was impossible to watch both the bats and the oscilloscope, I threw together a rough-and-ready audible detector by feeding the ultrasonic signals into the second detector stage of a battery-operated portable radio. This generated an audible click for each of the bat's orientation sounds and enabled me to monitor them by ear while watching the bat and aiming the microphone with its parabolic sound-collecting reflector. This long deep parabola was the only piece of apparatus left over from Pierce's supersonic detector of the 1930s.

The results were unexpectedly spectacular (Griffin, 1953, 1958). Although Galambos and I had learned a decade earlier that bats increase the repetition rate of their orientation sounds when faced with obstacles, the increases in pulse repetition rate during insect pursuit were very much greater. Neither I nor anyone else had previously suspected that bats might catch small flying insects by echolocation. We had always thought of echolocation as a method for detecting stationary obstacles and avoiding collisions or orienting the animal in darkness, but prior expectations were that small insects would not return strong enough echoes to be audible and that the rapid and intricate maneuvers must be guided by vision. It is difficult to realize three decades later how much of a change in viewpoint was necessitated by this evidence that bats used echolocation not only for locating and avoiding stationary obstacles but for their hunting of small rapidly moving insect prey. Echolocation of stationary objects had seemed remarkable enough, but our scientific imaginations had simply failed to consider, even speculatively, this other possibility with such far-reaching ramifications. Wholly conclusive resolution of this question had to wait for another decade when collaborative experiments with Frederic Webster in Cambridge showed that echolocation really was used at least by some insectivorous bats in the capture of small flying insects (Griffin, Webster, & Michael, 1960).

I have now brought this account up to the early 1950s, when I was in my mid-thirties. By this time whatever influences molded my scientific work and thinking had done their work. Having come to realize that bat echolocation was a highly versatile mode of perception, it was natural to inquire whether it might differ among the many groups of bats. Even some years later this notion was a difficult one to accept. For example, Georg von Bekesy, Nobel laureate and dean of investigators of the sense of hearing, told me it was a waste of time to examine other kinds of bats such as all those queer-looking ones from the tropics. A bat is a bat; those sounds are simply noise bursts, and nothing more is likely to emerge from further studies. Despite such authoritative discouragement, however, it did seem well worthwhile to undertake comparative studies, as F. P. Moehres in Germany had already started to do. The subsequent history of research on echolocation has amply justified my zoologist's faith.

Likewise it was only natural to continue attempts to discover the sensory basis for bird navigation. My extensive efforts to crack this nut have had only limited success, and indeed the nut remains nearly intact. In 1968 I returned energetically to a new attack on the problems of bird navigation by obtaining, with the aid of the National Science Foundation, a military surplus tracking radar with which to observe the flight paths of migrating birds. This amounted to trying to *be* a bat and apply echolocation to this supremely challenging scientific problem. The immediate reason for taking up radar ornithology was to determine more precisely than others had done whether migrants maintain well-oriented flight when "flying blind" inside of or between layers of opaque cloud. While most migrants avoid such conditions, many of those few that do fly blind seem reasonably well oriented (Griffin, 1973). Although the center of scientific interest has shifted to the possibility that birds sense and utilize the earth's magnetic field in their orientation, the marginal positive evidence leaves me unconvinced (Griffin, 1982).

As time passed I found myself more interested in behavior than in physiological mechanisms. I began to ask more and more what animals did with the physiological machinery that had been discovered by comparative physiologists. I had spent a sabbatical year (1960–61) at Cambridge University with W. H. Thorpe, because I had been impressed by his scholarly analyses of animal behavior. It was my first trip to Europe since I was a boy of thirteen, and I profited greatly from a series of visits to universities in England and on the continent. I even visited von Frisch at Brunnwinkl and had the great honor of lecturing on the echolocation of insects by flying bats in the famous lecture hall at Munich. With great effort, and much help from kind colleagues such as Marie-Claire Busnel in Paris and J. Schwartzkopff in Munich, I managed to lecture in French and German, but I never was able to handle questions or discussion.

After seven years at Cornell, I had been invited to return to the Harvard biology department, with the primary obligation of teaching Introductory Zoology. I approached this with great trepidation, and I am afraid did only moderately well at elementary teaching. But the stimulating environment of the Harvard Biological Laboratories and the superior facilities available at Harvard did permit better research work than I would have been able to accomplish at Cornell. I also managed to avoid the bitter political problems that wracked the Cornell zoology group during the late 1950s.

In 1965 the opportunity arose to move to New York and organize a research program in ethology. The original impetus for this came from discussions with Fairfield Osborn, the president of the New York Zoological Society, and it was he who persuaded Detlev Bronk, president of The Rockefeller University that this was a good enough idea to warrant generous support by both institutions. The resulting investigations of ethology and neurobiological mechanisms underlying animal behavior by Peter Marler, Fernando Nottebohm, and many of our younger colleagues have demonstrated the essential wisdom of Osborn and Bronk in making this whole development possible. In 1970 I was totally astonished by a telephone call from Frank Stubbs, one of four trustees of the Mary Flagler Cary Charitable Trust in Millbrook, New York asking whether we might like to establish a field research station there. The resulting Millbrook Field Station has certainly become an effective center for distinguished contributions to ethology.

Beginning in the mid-1970s I have undergone one further change in scientific outlook. Slowly over the years, I had become more and more dissatisfied with the reductionistic and behavioristic viewpoints of my colleagues in biology and psychology. In particular I had begun to doubt the wisdom of totally ignoring the possibility that animals might experience conscious thoughts and subjective feelings. This has led me to attempt to launch a subdiscipline of cognitive ethology (Griffin, 1981, 1984). My efforts to date have been largely verbal and except for attempts to reinterpret existing data, I feel that only limited progress has yet been achieved. I do have some optimistic confidence, however, that in due course ethologists and psychologists will begin to make headway in learning what nonhuman animals actually think and feel. I have often wondered in recent years why it took me so long to speak up on this subject. I believe the reason was my early indoctrination in the positivistic climate of science at Harvard and elsewhere in the 1930s. Many scientific developments and much shaking up of prior ideas were necessary before I was ready to think seriously about the thoughts and feelings of animals. Hindsight is always easy, and perhaps I am simply swimming with a changing tide in the history of ideas. But it does seem that my firsthand involvement in several surprising discoveries is what prepared me to shift my thinking into new and I hope fruitful channels.

References

Billings, S. M. 1968. Homing in Leach's petrel. *Auk* 85:36–53.

Davis, W. H., and Hitchcock, H. B. 1965. Biology and migration of the bat *Myotis lucifugus* in New England. *J. Mammal.* 46:296–313.

Fraenkel, G. S., and Gunn, D. L. 1961. *The Orientation of Animals*. 2nd Ed. New York: Dover.

Frisch, K. von 1950. *Bees, their Vision, Chemical Senses and Language*. Ithaca, N.Y.: Cornell University Press. (Revised edition, 1971).

———. 1967. *A Biologist Remembers*. Oxford: Pergamon. (Trans. L. Gombrich of *Errinnerungen eines Biologen*. Berlin: Springer, 1957.)

Galambos, R. 1942. The avoidance of obstacles by flying bats: Spallanzani's ideas (1794) and later theories. *Isis* 34:132–40.

Gifford, C. E., and Griffin, D. R. 1960. Notes on homing and migratory behavior of bats. *Ecology* 41:377–81.

Griffin, D. R. 1934. Marking bats. *J. Mammal.* 15:202–7.

———. 1935. Stomach contents of Atlantic Harbor Seals. *J. Mammal.* 17:65–66.

———. 1940a. Homing experiments with Leachs petrels. *Auk* 57:61–74.

———. 1940b. Migrations of New England bats. *Bull. Mus. Comp. Zool.* 86 (6).

————. 1943. Homing experiments with herring gulls and common terns. *Bird Banding* 14:7–33.

————. 1945. Travels of banded cave bats. *J. Mammal.* 26:15–23.

————. 1946. Supersonic cries of bats. *Nature* 158:46–48.

————. 1950. Measurements of the ultrasonic cries of bats. *J. Acoust. Soc. Amer.* 22:247–55.

————. 1952a. Airplane observations of homing pigeons. *Bull. Mus. Comp. Zool.* 107:411–40.

————. 1952b. Bird navigation. *Biol. Revs.* 27:359–93.

————. 1953. Bat sounds under natural conditions, with evidence for the echolocation of insect prey. *J. Exptl. Zool.* 123:435–66.

————. 1958. *Listening in the Dark.* New Haven, Conn.: Yale Univ.Press (Reprinted 1974 by Dover Publications, New York).

————. 1963. The potential for telemetry in studies of animal orientation. In *Bio-Telemetry,* ed. L. D. Slater. New York: Pergamon.

————. 1964. *Bird Migration.* Garden City, N.Y.: Doubleday (Reprinted 1974 by Dover Publications, New York.)

————. 1973. Oriented bird migration in or between opaque cloud layers. *Proc. Amer. Philos. Soc.* 117:117–41.

————. 1980. Early history of research on echolocation. In *Animal Sonar Systems,* eds. R.-G. Busnel and J. F. Fish. New York: Plenum.

————. 1981. *The Question of Animal Awareness.* 2nd ed. New York: The Rockefeller Univ. Press.

————. 1982. Ecology of migration: is magnetic orientation a reality? *Quart. Rev. Biol.* 57:293–95.

————. 1984. *Animal Thinking.* Cambridge, Mass.: Harvard Univ. Press.

Griffin, D. R., and Galambos, R. 1941. The sensory basis of obstacle avoidance by flying bats. *J. Exptl. Zool.* 86:481–505.

Griffin, D. R., and Hitchcock, H. B. 1965. Probable 24-year longevity records for *Myotis lucifugus. J. Mammal.* 46:332.

Griffin, D. R., and Hock, R. J. 1949. Airplane observations of homing birds. *Ecology* 30:176–98.

Griffin, D. R., Robinson, S., Belding, H. S., Darling, R. C., and Turrell, E. T. 1946. The effects of cold and rate of assent on aero-embolism. *J. Aviation Med.* 17:56–66.

Griffin, D. R., Wald, G., and Hubbard, R. 1947. The sensitivity of the human eye to infra-red radiation. *J. Optical Soc. Amer.* 37:546–54.

Griffin, D. R., and Welsh, J. H. 1937. Activity rhythms in bats under constant external conditions. *J. Mammal.* 18:337–42.

Griffin, D. R., Webster, F., and Michael, C. 1960. The echolocation of flying insects by bats. *Anim. Behav.* 8:141–54.

Griffin, H. F. 1941. *The White Cockade.* New York: Greystone Press.

————. 1943. *Paradise Street.* New York: Appleton-Century.

Grinnell, A. D. 1980. Dedication to *Animal Sonar Systems,* R. G. Busnel and J. F. Fish, eds., pp. xix–xxiv. New York: Plenum.

Hahn, W. L. 1908. Some habits and sensory adaptations of cave-inhabiting bats. *Biol. Bull.* 15:135–93.

Hall, J. S., Cloutier, R. J., and Griffin, D. R. 1957. Longevity records and tooth wear of bats. *J. Mammal.* 38:407–9.

Hartridge, H. 1920. The avoidance of objects by bats in their flight. *J. Physiol.* 54:54–57.

Kramer, G. 1961. Long-distance orientation. In *Biology and Comparative Physiology of Birds,* vol. 2, ed. A. J. Marshall, pp. 341–71. New York: Academic Press.

Matthews, G. V. T. 1968. *Bird Navigation.* 2nd ed. London: Cambridge University Press.

Pierce, G. W., and Griffin, D. R. 1938. Experimental determination of supersonic notes emitted by bats. *J. Mammal.* 19:454–55.

Wald, G., and Griffin, D. R. 1947. The change in refractive power of the human eye in dim and bright light. *J. Optical Soc. Amer.* 37:321–36.

Watson, J., and Lashley, K. S. 1915. Homing and related activities in birds. *Papers Tortugas Lab.* 7:1–104. (Publication No. 211, Carnegie Institution, Washington, D.C.).

Wilkinson, D. H. 1952. Randomness in bird "navigation." *J. Exptl. Biol.*29:532–60.

Heini Hediger (b. 30 November 1908). Photograph by Jorg Klages

6

A Lifelong Attempt to Understand Animals

Heini Hediger

For many families in Basel, Switzerland, where I was born on November 30, 1908, it was customary to send children off to the zoo during their free time. Even the youngest went, riding in baby carriages pushed by their nurses. During the course of these early visits, I was evidently imprinted on zoos. I wanted to be in contact with animals and, consequently, assembled an impressive menagerie composed of all sorts of plush animals, ranging from monkeys to elephants. At the age of six, the time at which I entered school, I began to surround myself with living animals, some of which I captured myself.

This private menagerie included sea anemones, fish, snakes, owls, a fox, an opposum, and a slow loris; its growth was negatively correlated with my success in school. I spent far too much time with my charges and the wild animals in a nearby forest, as well as in the wonderful swamps near Neudorf in neighboring France. It was a wonder that I managed to pass my matura in 1927. I never had to choose my career: I always wanted to study zoology and become a zoo director. I was especially fascinated by exotic animals and their habitats, particularly the orient. My father had lived in Algeria for a number of years as a young merchant. After returning to his hometown of Basel, he married a gardener's daughter and designed an Arabian room in their flat. The centerpiece of this room was a stuffed flamingo. My father told me many stories about Algeria that made a strong impression on me.

I was told that my maternal grandfather, the gardener, in addition to cultivating plants in his large greenhouse, also raised reptiles, such as blind-worms, snakes, and crocodiles. Later, these animals were donated to the Basel zoo, which was founded in 1874. It was no wonder that, as a boy, I sought ways to familiarize myself with the palm trees and animals of the Orient. It was like a sickness that could only be cured by a trip to this region. The first opportunity I had for such a trip came after my initial semester at the University of Basel. My brother, a merchant, had taken a position with a friend on a Swiss farm near Rabat, Morocco, and he encouraged me to visit him. This visit expanded to three months, and I used the time to study the animals, especially the reptiles, that lived on the farm and in its vicinity. This experience produced my first publication in a

---

Translated from German by Stefan Tigges, Emory University, Atlanta, Georgia, U.S.A.

scientific periodical, "Die Tierwelt auf einer marrokkanischen Farm [The Animal Life of a Morrocan Farm]." At the end of our stay in Rabat, our friend, Albrecht Manuel, the supervisor of the farm, my brother, and I set out on a long journey through the then Sultanate of Morocco. Our little car took us through the Middle Atlas mountains even over the High Atlas into the Sous Valley, and through many areas that were inaccessible to tourists, or had not even been taken by the French. Consequently, we depended on our tent and avoided battle zones. I collected many live reptiles and brought them back to Switzerland.

Besides the reptiles, I was particularly fascinated by the fiddler crabs *(Uca tangeri)* of the Oued Bou Regreg Delta near Rabat, an animal that I observed and photographed extensively. Here, for the first time, I measured flight distances and learned about territories and the ways in which they were marked. These observations were detailed in two articles, both published in 1933, and, of course, I am very proud that these humble early efforts were cited in Jocelyn Crane's monumental work on fiddler crabs.

This first crucial journey to Morocco was followed by several others, which also led to the discovery of new species and subspecies, and to scientific and journalistic publications. Through these articles, I tried to finance my private menagerie and my trips.

After this adventurous excursion, I continued my zoological studies in Basel. As secondary subjects, in which I was also tested, I chose botany, psychology, and ethnology. Unfortunately, I was never able to attend a lecture on animal psychology because this subject was not taught in Switzerland at that time. Of course, I tried to fill this void through readings in appropriate works. In 1929, the need for a dissertation theme had become pressing, and my boss, the respected Professor Adolf Portmann of the Zoological Institute of the University of Basel, suggested that I study yolk resorption in the crayfish *(Astacus fluviatilis)* egg. The microscope and laboratory work did not suit me, and I soon abandoned the project. A more appropriate theme for me was culled from the areas of herpetology and the zoogeography of Morocco. I had barely begun this project when the much-traveled ethnologist from the University of Basel, Professor Felix Speiser, began looking for an assistant to accompany him for a year and a half on an expedition to the South Seas. A trip to the tropics was precisely what I had dreamed of for so long. I was filled with enthusiasm and applied for the position immediately; luckily, I was accepted. As compensation for my work, I was allowed to collect zoological material for my dissertation.

First, I had to familiarize myself with the film to be used and several very cumbersome cameras. Photoplates up to 13 by 18 centimeters in size were used, all of which had to be developed while traveling. To manage this task, we brought along an entire photo-laboratory, a darkroom, thousands of plates, and kilometers of film, as well as antropological material, a complete carpentry workshop with an incredible number of nails and screws. These additional supplies were used to pack and prepare ethnographic material, from pottery to blowguns and canoes for the journey from the jungle back to Switzerland. Other essential supplies included various cases of zoological material, including conservation material, preparatory instruments, alcohol tanks, vials, and tags of all sizes. Naturally, we also brought along medical supplies, as well as kitchen and lighting appliances, cots, tables, and hunting and defensive weapons, including ammunition, clothes, etc.

In Australia, we added food supplies and articles for bartering. Our luggage comprised approximately fifty crates. At the time, 1929, travel by air was out of the question; there was no air link between Europe and Australia. The entire journey was made by ship. The newest and fastest ship, the 20,000-ton *Orontes,* made the journey from Toulon to Sydney in thirty-five days. In Sydney, we had to wait two weeks for a small, 300-ton steamer to take us to Rabaul, the capitol of New Britain and our operations headquarters. In contrast, modern field ethnologists climb into an airplane and are received at their destination by a staff of people who know the area and a vehicle suited to the terrain. The long ocean cruises were not without advantages, however. The time could be spent reading about the country, its people, and its animals and studying the language, in this case, "pidgin" English.

After about one and one-half years, I returned to Basel in the spring of 1931, safe and sound but heavily infested with malaria. I immediately resumed my studies and received my doctorate in 1932. My dissertation, "Beitrag zur Herpetologie und Zoogeographie Neu Britanniens und einiger umliegender Gebiete [Contributions to Herpetology and Zoo Geography in New Britain and Some Surrounding Areas], was hammered out in the Basel Museum of Natural History. The curator for zoology, the herpetologist Jean Roux, generously helped me with his advice and extensive library. To thank him, I named one of the lizards that I had discovered *Leiolepisma rouxi.* Professor Roux reciprocated by naming one of the snakes that I collected from the Solomon Islands and had left for his study, *Parapistocalamus hedigeri.* In this case, we were dealing not only with a new species, but with a new genus.

The Pacific reptiles that I had collected for my dissertation afforded me the chance to learn reptile systematics and international nomenclature rules so that I could recognize and name new species. I believe that it is valuable for every zoologist and ethologist to familiarize himself with the rules of systematics. In addition, my work with reptiles made me realize just how closely morphology and behavior are related. Therefore, I borrowed a saying from the German ecologist, Richard Hesse, as the motto for my dissertation, "We have enough preserved specimens in alcohol tanks. What we need are observations concerning the relationship between the animal and its environment."

In fact, this relationship had always interested me enormously. Consequently, I divided my dissertation into three parts: a systematic, an ecological, and a zoogeographical part. In the second and most general section, I included all I could concerning what fascinated me the most: the behavior of the animals that I had discussed in the systematic part. I had so much to say about flight behavior alone, however, that I decided to devote a separate article to this subject. Most of the animals that I had captured were caught by hand, a procedure which was almost always preceded by an attempt on the animal's part to flee. These flight reactions did not consist merely of running or flying away, but were species-specific, dictated by the animal's morphology, its environment, and the time of day. The flight tendency of free-living animals, as well as the ability of man to eliminate this characteristic, has occupied me throughout the rest of my life. The tameness of animals was the subject of a study that I published in 1935. This study, dedicated to J. von Uexküll, formed the basis for my later discussions of animal training, the theme of my habilitation work in 1938. Tameness is closely related not only to training, but also to assimilation, the process whereby different animals accept man and treat him as if he were a member of their own species. This phenomenon has captivated me ever since.

The third part of my dissertation, zoogeography, forced me to familiarize myself with this discipline's different schools of thought. In what ways were animals distributed over the earth's surface? An important role was certainly played by now-submerged primeval land bridges. Man himself has also played an important role, especially in dispersing reptiles, animals which I studied personally. Geckos, small snakes, and lizards could easily be carried among provisions such as wood, coconuts, roots, etc., that are taken from island to island by seafaring canoes. Tornadoes present another possible mode, carrying palm fronds and other plant materials for long distances. In this case, we have to keep in mind that not only adult animals could be transported by this means, but also eggs attached to plants. Floating islands, which occasionally menaced us in the mouth of New Guinea's Sepik River, could also provide a way for animals to disperse. These islands frequently drifted out to sea, some, perhaps, with a complement of animals on board. Later, I noticed this transportation possibility on a larger scale in the Congo. Huge chunks of forest often drifted by, forcing us to weigh anchor at night.

One zoogeographical question, however, has continued to intrigue me my entire life: why do many excellent flyers of the South Pacific Islands choose, with incredible stubbornness, to stay on certain islands even when there are other suitable islands within sight? We know that some birds, such as the collared turtle dove *(Streptopelia decaocto)* and the cattle egret *(Bubulcus ibis)*, animals that I later worked with in the zoo, range over large areas, but others still are extremely attached to their original habitat. Birds are not just little airplanes with a technically circumscribed flight radius, but living creatures that obviously form psychological attachments to certain places. This conviction never left me throughout my later zoogeographical studies: ecological conditions are not the only determining factors.

Naturally, the dissertation covered only a portion of the zoological yield. I had also collected insects, birds, and mammals for the Basel Museum of Natural History, and I had been able to visit all of Australia's zoos. Later, in 1964, I was called in as an expert by the government of New South Wales to reorganize one of these zoos, the Taronga Park Zoo in Sydney. On the islands, I kept many different animals, including flying foxes, and especially monitor lizards. These animals were kept in the small bungalows, usually used by district officers, that had been placed at our disposal. The aggressive lizards *(Veranus indicus)*, sometimes one meter long, were remarkable in that they soon became completely tame. At first, I chained them to a post beneath our bungalow. When I approached them, I was greeted with threats, lashing tails, and defecation. After just a few days, however, I could feed them crabs and mice by hand without any problems. After a few weeks time, one of these lizards had become as tame as a domestic cat. I soon learned that this monitor could climb amazingly well, run quickly, and swim both on top of and under the water. In the crystal-clear lagoon water, I discovered to my surprise that my monitor's tongue darted out even under water.

While preparing for our departure for another location on the south coast of New Britain, I took this tame monitor on board the schooner with me, but when it was time to put it in the alcohol tank, I was unable to perform my duty. Instead, I declared him officially taboo in the presence of the local chieftain and turned him loose.

Since that time, I have always been fascinated by monitors. In his comprehensive monitor monograph, the standard-setting German herpetologist Robert Mertens cited the observations that I had made as a young student. Monitors can rightly be considered one of the most intelligent reptiles and they are the only group shown to have play behavior.

Strangely enough, it was a monitor *(Varanus salvator)*, a 1.6-meter-long specimen, that gave me my only serious work-related injury in a career that spanned thirty-five years. At a press conference in which everything was going wrong, as is sometimes the case, I was bitten above the wrist through the ramus dorsalis of the nervus ulnaris, an injury that has since hampered my writing and perplexed graphologists.

Because I had to deal with the herpetology of the Pacific in my dissertation, I had also examined the reptiles that had been collected on the Admirality Islands by the ethnologist Alfred Bühler of the Basel Natural History Museum. At that time 1933, the Museum possessed one of the most important reptile collections in Europe.

The many lizards that I caught or tried to catch on the islands vividly demonstrated some of the laws of flight: the specificity of the flight reaction, the measurability of flight distances, etc., topics that I touched upon in my dissertation. Future voyages and zoo work convinced me that this flight tendency and its accompanying typical behavior was one of the original and most important behaviors in all animals. The fulfillment of nutritional and reproductive needs can be delayed, but immediate flight from predators that appear suddenly cannot be postponed. I tried to present this behavioral aspect in "Zur Biologie und Psychologie der Flucht bei Tieren [Biology and Psychology of Animal Flight Response]" in 1934. This work was surprisingly successful; it was especially well received by J. von Uexküll at the Institute for Environmental Studies in Hamburg, where I met both him and his successor, F. Brock. For my part, I was extremely impressed by von Uexküll's Umwelt-Lehre.

Consequently, I dedicated another paper following that on my flight work, a study of tameness, to von Uexküll. In a sense, these two subjects formed the basis for my later investigations concerning the relationship between animals and man, especially in the zoo. In addition, they led to the founding of zoo biology, a discipline that took up more and more of my time.

I closed out my herpetological activities, which had been permeated with observations of animal behavior from the start, with two articles on the herpetofauna of

Morocco in 1935 and 1937. The material for these studies came from expanded collecting trips taken with my friend, Albert Manuel, the farmer who had lived in Morocco since 1927, and with a colleague from Basel, Hans Ritter.

A skink, found in 1935 near Mogador, turned out to be a new subspecies, which I named *Chalcides ocellatus manueli*. This skink was later found again west of the High Atlas Mountains. At that time, I worked closely with the Viennese herpetologist, Franz Werner, who often collected in Morocco. He dedicated to me one of his studies of scorpions, animals that I had brought back from Morocco. Like all of my specimens, the insects found their way into the Basel Natural History Museum. Among them was a new genus of mantids, which Navas named *Hedigerella*. At that time, the discovery of new animals in little-known areas was almost unavoidable. My last herpetological publication appeared in 1936. Commissioned by CIBA, it was a richly illustrated brochure, intended primarily for practicing physicians, about the snakes of central Europe, and was published in German and French. The impetus for this assignment was provided by the tragic death in southern Switzerland of a vacationing foreign girl as the result of the bite of a venomous snake *(Vipera aspis)*. This incident caused quite a stir, and it turned out that the doctors, even those in the area where venomous snakes were known to live, were helpless. Antivenin was not available everywhere and our knowledge of the eight snake species of Switzerland was inadequate because at this time, "only" about every second year did a person die of snake bite. Thus my brochure successfully remedied the problem of venomous snake bites.

When I wrote the brochure, I was already a lecturer at the Zoological Institute of the University of Basel. Immediately after I received my doctorate in 1932, I became the second assistant, and soon after, first assistant to my respected teacher, supervisor, and friend, Adolf Portmann, who decisively influenced my career. My main interest remained the zoo, a direction in which I had been oriented since my childhood. Until 1929 though, there was only one zoo in Switzerland, in Basel, and it was not concerned with science at all; quite the contrary. Therefore, I continued to maintain my private menagerie, which included an opossum *(Didelphis)*, previously thought to be untamable and a representative of so-called marsupial stupidity. I soon discovered, however, that when they are approached with understanding, in other words, quietly and at dawn, they are very receptive. Thus, I was able—apparently for the first time ever—to tame and even train an opossum. For example, I could command it to wrap its prehensile tail around a stick, enabling me to carry the animal about. I detailed this success in 1934 in the journal *Der Zoologische Garten* [*The Zoological Garden*] and in *Umschau in Naturwissenschaft und Technik* [*Science and Technology Review*]. Here, I saw an important connection with understanding possibilities between man and animals, a theme I discussed extensively in 1967 and that continues to concern me, as for example in the present-day attempts to communicate with higher primates, a subject that I will discuss later. My time as an assistant at the University of Basel's Zoological Institute from 1932 to 1937 was actually a waiting period; I wanted to become a zoo director and I wished to obtain a position at a zoo as soon as possible, but the opportunities at that time were extremely limited. In Germany, no foreigners were being hired, for political reasons, and in Switzerland, there was only the zoo in Basel, whose director was an amateur ornithologist and was still firmly entrenched. The zoo in Zürich, however, opened in 1929 and three years later, was searching for a full-time director. Naturally, as a brand-new Ph.D., I applied for the job immediately. The choice came down to a native of Zürich and me. He had spent several years as a farmer in Sumatra. As an experienced planter and tiger and elephant hunter, he was understandably chosen over me, the greenhorn. This setback proved to be a stroke of good luck. If I had been accepted, my education would have suffered greatly. Twenty-five years later, in 1954, I was chosen as the successor to my onetime competitor, Felix Hofmann as director of the Zürich Zoo.

During my assistantship, at the University of Basel's Zoological Institute and at the museum, I made additional trips to Morocco and twice accompanied my supervisor, Professor Adolf Portmann, and a group of students to his beloved study retreats on the Mediterranean, Villefranche and Banjuls-sur-Mer. There, I learned something about marine biology.

In addition, I worked as a journalist and author. At the time, I began contributing to the CIBA periodical, founded in 1933, with whose editor, Dr. Karl Reucker, I developed a true friendship. This journal which appeared in several languages, gained international recognition not only for disseminating medical knowledge, but also for its cultural-historical contents. Reucker also originated the column, "Mixtum Compositum," in which small notices concerning science and medicine appeared and to which I made many zoological contributions.

In 1935 Dr. Reucker left an entire issue at my disposal. I explored the theme "Zähmung und Dressur wilder Tiere [Taming and Training Wild Animals]." Three years later, I had a similar opportunity and discussed "Wildtiere in Gefangenschaft [Wild Animals in Captivity]." Both works were closely related to my habilitation work and my later books. "Wild Animals in Captivity" was later expanded to become the basis of zoo biology and became known in the United States as the "Zoo Bible."

If my dissertation was born in the jungle, then my habilitation work came from the circus, a place abounding with man-animal contacts. This fascinating area was and still is underresearched. A new approach was pioneered in 1970 by Thomas A. Sebeok, who founded and intensively promoted zoo-semiotics, an area in which we have been working together for some time.

The previously mentioned habilitation work arose from the intensive study of various European circuses and an extended visit to the training school of Hagenbeck's Tierpark in Stellingen near Hamburg. The manuscript was so voluminous that it was never published in its entirety. Only an excerpt appeared in 1938 under the title "Ergebnisse tierpsychologischer Forschung im Zirkus [Results of Research in Animal Psychology in the Circus]." The positive way in which this unusual study was received indicates the openminded nature of my superior, Professor Adolf Portmann, and also of the entire faculty. One can hardly imagine it: circus at the university! At that time, this was unheard of.

During my assistantship, I was confronted by an incredible temptation: the founder of cultural morphology and of that discipline's institute in Frankfurt am Main, Geheimrat Professor Leo Frobenius, the famed Africa researcher, offered to take me along on one of his expeditions; in fact, the one from North Egypt through the fossil (former, old) Nile Valley to Abyssinia. He turned to me because of my work in the Sahara on my various Moroccan trips. How tempting this offer was! I discussed this situation with my supervisor, Professor Portmann, who needed my help as first assistant. In those days, people of this sort were few. The overabundance of young biologists occurred decades later. He offered me an academic career, which tempted me only in connection with the zoo. On the other hand, I knew that after an extended absence due to an expedition, I could expect difficulty in reestablishing myself upon my return. Thus, I decided to remain in Basel and wait for opportunities to work in a zoo.

In the meantime, my pay as an assistant was meager. However, in 1937, my herpetological advisor, Dr. Jean Roux, retired from his position as curator of the Zoological Division of the Natural Historical Museum of Basel. I applied for the job and was accepted with a general recommendation from my superiors. Of course, I was well acquainted with my predecessor's work through my study of the South Sea specimens and my Moroccan trips.

I was still connected with the Zoological Institute through my lectures. The Museum's vast wealth of zoological and ethnographic material and its huge library continued to impress me and the work I did with colleagues resulted in daily enriching experiences. I still think of my days at the Basel Museum with gratitude; I learned much there. However, something was missing; I was working exclusively with dead animals when I was oriented toward living animals.

For this reason, I jumped at the surprising opportunity that presented itself a year and a half after I began working at the Museum. The City Zoo of Dählhözli in Bern was looking for a new director. This zoo had opened in 1937, was still small, and was supervised by a veterinarian as a sort of second job. The personnel consisted of, in polite terms, a few whimsical characters, none of whom had any zoo experience. Insubordination had become rampant; consequently, a new director was needed urgently. In September 1937, I stepped into the new position. My wish had finally been granted. It was not a real zoo, but an animal park, primarily displaying European mammals. Some exotic specimens were found in the volière, aquarium, and terrarium.

The entire place was small enough to be supervised by me alone, and large enough to offer me a challenge. I enthusiastically plunged into my work. The attendants, however, threatened to throw me into the Aare River my first day on the job if I did not agree to their demands. This threat lent a powerful romantic spice which I welcomed, to my work. In addition, the threat indicated that zoos are not just zoological affairs, but contain countless human problems that, in fact, as I learned through personal experience, often outweigh other problems. In the small, young zoological garden, the elements of zoo biology were apparent. During my six years at Bern, my book, *Wildtiere in Gefangenschaft: Ein Grundriss der Tiergartenbiologie* [*Wild Animals in Captivity: An Outline of Zoo Biology*], took shape. This book was reprinted eight times and was translated into many languages.

Among the animals at Dählhölzli, I was especially fascinated by the moose *(Alces alces),* which were tricky to care for at that time. They were especially prone to parasitic infections. This situation intrigued me, and I was fortunate to work with a first-rate parasitologist, Dr. Hans A. Kreis, whom I met in Basel and who later worked for the Swiss Health Department at Bern. The long, intensive parasitological surveillance, especially of the young, was of crucial importance to the establishment of the first moose breeding colony in Switzerland.

The European hare *(Lepus europaeus)*, which was considered unbreedable in captivity, fascinated me as well. For me, the problem was twofold: parasitological and psychological. Because the coccidia-containing droppings are infectious after three days and every adult is a carrier, all of the 2 × 2 meter cages had to be thoroughly cleaned every other day, a procedure that tends to dangerously over-excite the hares. An effective treatment for coccidiosis did not exist at that time; therefore, we had to devise a cleaning system that would not cause too much excitement among the hares. For this reason, I invented the "mirror-symmetrical shift cage," which caused a considerable stir because it enabled people to routinely breed this popular game animal at a time when not even its gestation period was known. Information of this type is generally only obtainable through observation in the zoo.

The new shift cage, which also proved useful with other sensitive species, consisted simply of two identical 2 × 1 meter cages joined together with a movable wall between them. Because of the proverbial shyness (flight tendency) of the European hare, only one side of this cage was transparent, the side with wire mesh. The other three sides consisted of solid wood.

While one half of this double cage was occupied for 48 hours, the other half could be provisioned with fresh straw and food for the next two days without exciting the hares. After this was done, the attendant quietly lifted the dividing wall while remaining out of sight. The hares quickly became accustomed to this procedure and moved into the clean half of the cage. Young animals could be moved by hand from side to side.

Through this rather simple cleaning method, both the deadly coccidiosis infection and the fatal excitement could be avoided. Subsequently, the hare proved to be easily breedable, and astonishing discoveries followed. In the Bern Zoo, the first cages were located in such a way that I could see inside them from my office and home. Because I was interested in the length of the gestation period, I noted each copulation and observed each birth.

I noticed that the paired hares copulated soon after the birth of young and was reminded of the postpartum estrus, a common occurrence in mammals. Soon, I had to convince myself, however, that copulations occurred three days *before* birth: visibly pregnant hares were being inseminated. That struck me as being highly unusual, and my first thought was that I had made faulty observations. It was hardly possible, however, for newborn hares to have been overlooked during the cage cleanings which took place every other day. Was it possible that superfetation was occurring here? To determine whether this was true, I isolated pregnant hares that had been reinseminated three days before the expected birth. To my surprise, these hares, in the absence of a male, delivered two litters, one three days and the second forty-two days after the last copulation.

In 1941, I published a short communication on the breeding of these hares in captivity and I produced a more comprehensive description of my study in 1948. Through these efforts, I revealed that the gestation period was 42 days and that in many cases, superfetation occurred. In other words, two gestation periods overlap. This study generated much attention and controversy. Hunting organizations from all over Europe wished to learn more details about the breeding possibilities of the European hare. The most interest, however, was shown by physiologists and gynecologists the world over, some of whom greeted this discovery with interest while others still viewed my assertion with skepticism, arguing that what was accepted at the time about hormones precluded the possibility of superfetation. I was attacked especially vigorously by Professor H. Knaus of Vienna and his colleague, the Japanese scientist, Ogino, who had become known through their natural birth-control method. One could argue that sperm was being stored until the *first* birth in the genital tract of pregnant hares and that it only appeared as if superfetation were occurring. I had no real laboratory of my own during my entire career and, therefore, had to restrict myself to making simple observations. Therefore, I was very happy that a study group was formed at Bern that included the zoologist S. Bloch, the English biologist, H. G. Lloyd, the gynecologist, C. Müller, the anatomist, F. Strauss, and myself. The coordinated efforts of these specialists led to a joint publication in 1967, the summary of which I have included here:

> Superfetation is the fertilization of an already pregnant female which results in the simultaneous development in the genital tract of eggs stemming from different ovulatory cycles. The occurrence of this phenomenon has often been debated and discussed but has only now been proven beyond a reasonable doubt: a hare weighing 3,250 g carried an embryo in the right horn of its uterus. The embryo weighed approximately 107 g without the placenta. In addition, the right side of the fallopian tube contained at least one egg at the four cell stage with numerous spermatozoa embedded in the surrouding membrane. These sperm must have moved past the embryo to fertilize at least one egg of a different ovulatory cycle. We noted similar occurrences in other hares.

With this publication and the few years that I spent pursuing an expanded breeding program of these popular animals at the Basel Zoo, my studies on the European hare came to an end. This example proved to me that not only exotic animals can be puzzling, but native animals can also present surprises. In addition, I learned that the length of gestation periods and many other facts about a particular species' way of living can only be established through observations made in captivity. Naturally, the moose and hares were not the only species that inspired study and publication. Even a small zoo like Bern offered innumerable opportunities for observation.

Let me present one more example. Substituting for the wisent, which was impossible to acquire because of its rarity, the Dählhölzli Zoo maintained a fine herd of about a dozen American buffalo. Each animal had a distinctive personality and unique characteristics. One female was so exceptionally tame that, when she gave birth, she allowed me to stay right next to her. This situation enabled me, apparently for the first time, to record photographically each phase of a bison birth. The completely natural birthing process of these wild ungulates can give us several hints for the human-managed birthing behavior of domestic cattle.

The several births that I witnessed and recorded all took place while the animals were lying down. A surprising aspect of the birth was the speed with which the mother arose immediately following the expulsion of the calf. The mother then turned and proceeded to eat and drink the embryonic membranes, amniotic fluid and, eventually, the placenta itself with carnivorous greed.

This transformation of an otherwise exclusive herbivore into a greedy carnivore was most impressive. Because of the speed of this process, especially the immediate devouring of the membranes in which the calf is often still encased, suffocation of the calf is prevented. This instinct has been lost or diminished in several domesticated animals like bovines and equines. Because of this situation, man's help is often needed to prevent the suffocation of the young.

Often, however, man's interference can be detrimental. For example, ingestion of the placenta by the mother stimulates milk production. Removal of the placenta by man robs the mother of this particular stimulation.

A further interesting phenomenon observed during these births is the first wide opening of the calf's eyes. This behavior apparently serves to imprint the calf to its mother, that is, the nearest large moving object. Obviously, the observer must take pains not to be identified in similar fashion, although in the case of the mother's death, this situation may be desirable. In zoo biology and animal psychology, as well as human psychology, this imprinting behavior plays an important role.

Opportunities for observation and accumulation of practical as well as theoretical experiences were abundant at Dählhölzli. During my tenure there from 1938 to 1944, I realized my lifelong desire to be surrounded by living animals. Even at night, there were interesting contacts with the animals, such as births or other occurrences. On my weekly day off, I traveled to Basel to give my lectures at the University. Still, I found time to write some short articles, besides making a few journalistic contributions.

Otherwise, my activities were very restricted. A year after I had assumed the directorship, the Second World War broke out. All of the young animal attendants were drafted, leaving me with a few crotchety old men and some not-exactly-highly-qualified temporary help. That meant that I had to act as a roaming attendant, filling in at whichever department was shorthanded at the time. I drove a pony cart to get food, cared for the aquarium, cleaned the buffalo area, and had to feed the poisonous snakes; in brief, I became acquainted with all types of work in the zoo, an experience that I later came to appreciate.

During the unbelievably cold winter of 1928, I had had an accident in the recruit school and later, in the South Seas, I had caught malaria. Therefore I was declared exempt from the military service, but later I was pressed into an armed auxiliary service. The terrible war raging in all the countries around Switzerland and stern defense preparations precluded the possibility of expanding the zoo. There was, however, one small exception about which I reported in 1944: the centerpiece of our thirty-one-tank aquarium, a tank with a volume of 18,000 liters and a 6.5

meter-long piece of glass. The architect, who was ignorant of zoos and fish, intended that this gigantic tank be used to hold fish from the Aare River. The water for the tank was to be cold tap water. Clearly, fish cannot survive under these conditions; the situation improved following the installation of an air pump, a water filter, and a heater.

Another source of headaches were the four 1,700-liter tanks. The 15-millimeter-thick glass of these tanks were slanted to avoid annoying reflections. The alignment, however, presented only disadvantages: the curious distortion made some people nauseous and even ill; more dirt accumulated on this type of window; and, finally, the glass in these windows was under very high pressure. Before and during the period of my directorship, some of these windows gave way with a tremendous crack, but fortunately, always at night. It would have been tragic if a window had exploded while, say, a group of school children was visiting the aquarium.

I did experience such an explosion one morning just after the working day had begun but before the zoo had opened. If you have not witnessed such a break, you cannot imagine what it is like. Heavy, razor-sharp glass shards were spewn under tremendous pressure throughout the room. If someone were hit by one of these shards, the injuries would at least be grievous, if not fatal. This situation made me aware of the grave responsibilities associated with the structures and management of a zoo. Later, people tried to shake my conviction by foolishly asserting that the concept of responsibility was accepted by only stupid people.

During my entire zoo career, I was not just a careful director, but even an anxious one. This attitude has caused me to be criticized on many occasions, but I am rather pleased with these reproaches; it fills me with satisfaction that during the thirty-five years of my directorship at Bern, Basel, and Zürich, not a single coworker nor a single zoo visitor was the victim of a fatal accident. Not all of my colleagues can make the same claim.

I believe that every animal-related injury stems for a misunderstanding of the animal. People that understand animals well should have no accidents. I, myself, am a bit proud that, despite my lifelong association with animals from every step of the phylogenetic ladder, I have received only one significant injury, from the aforementioned monitor.

In the middle of the war, in spring 1944, I was appointed director of the Zoo in Basel, the place where I had become imprinted on zoos. An important event however, preceded this change. In 1942, I married Kathi Zurbuchen in Bern. She was a zoologist and even though she was from Bern, she had a close relationship with my ancestral home of Zug. She shared in the shaping of *Wildtiere in Gefangenschaft* [*Wild Animals in Captivity*] and, for two years, my official residence in the Bern Zoo. She still shares the sometimes difficult life of a zoo director and professor.

My appointment in Basel came during an obviously hard time for zoos, but it did give me the opportunity, when things returned to normal, to work with animals that were missing in Bern: elephants, giraffes, great cats, monkeys, apes, etc.; in short, animals that are part of a real zoo. The Basel Zoo was opened in 1874 and had had all of these types of animals, but lost many of them during World War II.

For example, when I took office in April 1944, there were no sea lions. Their area had been taken over by cormorants. There were also no anthropoid apes. Their cages were filled with owls, young bears, and hyenas. The giraffes had also died off. Their enclosure was occupied by a single eland antelope. The international animal trade had totally collapsed due to the war, and providing the remaining animals with meat and grain was extemely difficult. I felt sorry for the solitary antelope, so to give it some companionship, I introduced some Somali sheep to the enclosure. This act proved to be the biggest biological mistake of my zoo career. After a few weeks, the antelope became ill: the area surrounding its eyes became streaked with secretions. The apparently healthy sheep had infected the antelope with catarrhal fever. We had to shoot it. One of the attendants who was designated to dispose of the cadaver was a bit overeager. While grabbing the "dead" antelope, he received a kick that shattered his forearm. This was the only time in my thirty-five year career that a practically dead animal injured an attendant.

Physical expansion of the zoo was just as impossible during the war as was the acquisition of new animals. The needs of the military and the difficulty of procuring animal feed precluded the purchase of new animals. To offer the visitors something novel anyway, I hit upon new ideas. One of these was called "animals in hibernation." Native animals, such as marmots, hedgehogs, dormice, and bats, were displayed in cool, dimly lit rooms and were a great success. Another special attraction, which cost practically nothing and was called "food for a day," displayed to visitors the daily rations of animals ranging from robins to elephants.

In the face of the zoo's survival problems, the demands of its animals, and the draft that created a dearth of young attendants, there was little room for scientific activity. During the winter rounds that took place between 7:00 A.M. and 9:00 A.M., however, I studied the behavior of the free-living squirrels *(Sciurus vulgaris)* in the zoo park. I soon established that these indigenous rodents did not behave like the textbooks said they should. Instead of hibernating, these animals spent most of the day in their nests in the trees. Only early in the morning, before it becomes very light, did the squirrels descend from their nests to the ground to search of their hidden supplies of nuts. Most of these caches were found in the corner between the foot of tree trunks and the ground. Occasionally, these squirrels stubbornly, but unsuccessfully, tried to hide their supplies in the corners between the stone steps of our house. This substrate was simply too unyielding to provide the abundance of space supplied by the soft soil at the foot of the trees. This behavior made clear to me what was then known as "innate behavior," as described by Niko Tinbergen and Konrad Lorenz, with whom I had taken up friendly and fruitful correspondence.

During the long, snowy winter, I was constantly being reminded by the free-living squirrels that they did not hibernate. My early morning rounds revealed that these rodents, considered to be absolutely diurnal, were also active in winter at dawn, trying to get at their nut supplies. The European squirrel was, therefore, not a hibernating animal, as had been believed for many years.

The meager complement of animals at the Basel Zoo at the end of the war still gave me the chance to undertake a comprehensive study of urinating and defecating behaviors of various animals; the results of this more detailed and differentiating examination of localized and diffused defecation habits were published in 1944. Related to this study was my research on the animal's space-time perception. I had always been impressed by the incredibly conservative nature of animals, not only in the zoo, but also in the wild, as shown by my Moroccan and South Seas studies, as well as my observations in the native woods and meadows. Large-scale animal migrations, which appeared to be manifestations of freedom, began to fascinate me. I gradually came to the conclusion that animals in the wild were not actually "free," but bound by space and time, by sex and social status. I sought to correct the idea of the free-living animal.

The war not only prevented the importation of new animals but also the construction of new buildings. Therefore, using the modest materials at hand, I used my time to study the characteristics of deer antlers, to breed hyenas, and, in 1944, to write on "Biologische und psychologische Tiergartenprobleme [Problems in Animal Biology and Psychology of the Zoo].

At a meeting of the Society of Naturalists in Zürich concerning this subject, I proposed a new way of keeping antelopes. My method makes use of their flight response to maintain their health. By eliciting this reaction in a symbolic way their heart, lungs, and general musculature could be maintained.

The idea called for a number of enclosures of the usual size grouped around an oval track. Daily, during the cleaning of their cages, they were to be run around the track in an orderly fashion by one of the attendants. Regular exercise of this type would doubtless contribute to the well-being and fitness of various hooved animals. In general, the problem of zoo biology lies in finding ways to keep the animals sufficiently active. Unfortunately, I was not able to realize this idea due to the building and maintenance costs, even though such an enclosure would have been a first-rate attraction. Artifically-induced activity of apes, carnivores, and elephants was seen much more sympathically.

When the boundaries between countries reopened after the war, I was finally able to study the large animals of Africa in the wild, a "must" for every zoo director. With an eye to the rapidly increasingly tourist traffic and to correct the widespread, although totally untrue, impressions of the behavior of large and poisonous animals, I wrote a book called *Kleine Tropenzoologie [A Small Tropical Zoology]*. This book also served as a guide for my lectures at the Swiss Tropical Institute at Basel, where I was also active.

I was fortunate enough to meet the director of the Institute des Parcs Nationaux du Congo Belge, Professor Victor van Straelen. Through his innumerable investigative trips from the Institute in Brussels to the Belgian Congo, he became convinced that the management of the expanded national parks and projects, such as the state-run okapi capture and elephant domestication, should not be entrusted solely to administrative officials, hunters, and museum zoologists. Instead, an ethologist or an experienced animal psychologist should share in these duties. I was chosen for this position. A young Belgian zoology student, Jacques

Verschuren, was to accompany me. Even then, he was an enthusiastic protector of nature, and he later became general director of the national parks in the Belgian Congo. During the political tumult caused by the liberation of Zaire, Dr. Verschuren worked heroically to save its imperiled national parks.

On March 2, 1948, Jacques Verschuren and I flew to the Congo and spent half a year there. I believe that this expedition was the first with a goal that did not include taking samples or hunting. Rather, our task was purely to observe, to study animal psychology. Consequently, our report of 1951 was entitled "Observations sur la Psychologie animale dans les Parcs Nationaux du Congo Belge [Observations on the Psychology of Animals in the National Parks of the Belgian Congo]."

Our first task, even before we were to actually begin work, was to visit and observe the Group de Capture d'Okapis (GCO) in Bilota. This group had a monopoly on the okapi catch and was under the supervision of Colonel Offermann, chief of the Hunting and Fishing Services. The actual leader of the group was J. de Medina, the son of a Portuguese physician and a native of the Congo, who ideally combined the attributes needed for this responsibile and difficult work. He was in charge of twenty-five experienced natives who set up and maintained pitfalls in the middle of the rain forests. Each of these men was responsible for ten of these pitfalls, which were 2 meters long, 1.8 meters deep and 80 centimeters wide. These pitfalls were camouflaged and inspected twice a day. Whenever an okapi fell into a pit, an alarm was sounded in the camp. Next, in such a way as to avoid exciting the animal, a fence was constructed out of young tree trunks and erected around the pit. A second fence of about 10 meters in diameter was placed around this first fence. An earthen ramp was then constructed on one side of the pit to enable the animal to climb out. The first fence was then removed and the animal was left in this enclosure until a run, also consisting of young tree stems and extending to a truck, had been built. When this task was completed, which often took several weeks, the animal was allowed to run right up on the waiting truck, where it entered a cage constructed of the same material as the run. From the capture site, the okapi was transported to the base camp, a trip lasting only a few hours. Here, it could descend into a clean, rectangular enclosure, also built of young tree stems. In this fashion, these large, shy, jungle animals could be transported from the wild to captivity without undue excitement and without even being touched by man. This procedure was the most elegant, gentle method of capture that I have ever seen. The dart gun was unknown at this time.

Approximately one week after its arrival in this acclimation enclosure, the newly caught okapi let itself be touched and brushed over its entire body by a native attendant. In these enclosures, which were located in its native biotope, the okapi ate its natural food.

The tragedy began later, when the okapis were put into narrow transport boxes made of lumber. They were confined to these boxes during the trip to Stanleyville and the long journey to Europe. This situation was extremely traumatic for the animals: confined to narrow boxes, traveling by truck, rail, and ship, sometimes in rough seas, living in unclean wooden crates, which led to reinfections with all sorts of parasitic worms, made survival difficult. Several dozen okapis died during this voyage. Their corpses were simply thrown overbroad instead of being saved for museums or pathological studies.

After leaving the GCO, my companion and I traveled to Gangala-na-Bodio, at that time the site of a unique station for the domestication of elephants, about which I wrote an evaluation in 1954. This station was also the basis of the new Parc National de la Garamba that, at that time, only existed on maps.

Here, at the Dungu River, Commandant Marc Micha had prepared a simple round hut for each of us, which we occupied for two months. Our actual work began here: the study of the behavior of animals in a newly conceived national park. We were the guests of Commandant Micha, who placed at our disposal his wealth of knowledge and any help that we might need. Later we also stayed at the Parc National Albert and at the Parc National de la Kagera. From these bases, we undertook excusions into the countryside using a small Ford pickup truck.

Naturally, moving about on foot affords possibilities for observation that one does not have while driving, but the reverse is also true. Slowly driving through an area gives one a different perspective than walking. For example, one can drive through a greater area more often in a given amount of time. This gives the observer the chance to see individual animals in the same place at the same time, enabling him to form ideas about the structure of their territories, their fixed points and paths. At that time, little was known about mammalian territoriality.

We had complete freedom on our daily drives. We drove very slowly, stopped at any place for as long as we wanted, and searched areas on foot if it seemed useful. We were not required to study all the peculiarities of one species, a method that later necessarily became common practice. We were primarily interested in the naturally occurring behavior of these animals and the influence of beginning tourism on this behavior. The Parc National de la Garamba, where there was, as yet, no tourism, was perfect for this study.

A primitive path of only 25 kilometers led into the park. The regular trips we made on this path led us to conclude that use of this simple street did not frighten the animals. Instead, animals from turtles to giraffes chose to use this path rather than fighting their way through the undergrowth. Whenever possible, small animals used the paths made by man or by larger animals. This proved that animal paths were not species-specific; animals were not scared off by simple roads as long as the traffic did not exceed a certain limit. Heavily traveled roads, of course, do frighten animals. We compiled long lists of animals that used simple man-made paths.

These primitive roads, which usually consisted only of two grooves cut by jeeps, were widely used by a great number of animals. In other words, the road was absorbed, as it were, into the animal's territory. The roads built by man, however, differ significantly from the paths of animals; roads are straight while animal paths tend to be wavelike. Animal paths also have a specific width: mouse paths are 4 centimeters wide; zebras, 30 centimeters; hippopotami, 60 centimeters, etc. These paths are also constructed in a specific way.

The study of animal paths has interested me since my experience in the Congo. In my Congo report of 1951, I discussed this interesting territorial phenomenon and in 1967, I expanded on this theme in a book entitled *Die Strassen der Tiere* [*The Roads of Animals*] with many collaborators.

During our stay in the Congo, observation of a variety of animals confirmed the assertion made in 1942 that fixed points and connecting lines are the most important elements of a territory. Besides paths, we also studied fixed points, for example, homes, defecation spots, demarcation points, and, of course, flight distances, which we often measured and compiled in tables.

Of particular concern was the effect of man on the behavior of animal, especially the effects of tourism, which could be sensed, to a small extent, in the Parc National de la Kagera and in the Parc National Albert. The results of tourism were more strongly felt in Camp de la Rwindi, but even here, the intensity of human contact was not nearly at the level found today in East Africa.

One of the most striking consequences of the rise in tourism is the way in which animals have become accustomed to man's presence. This fact is manifested by a drop in the flight distance from 50 to 100 meters to almost nil. Photography, which used to require telephoto lenses, now is possible without special equipment because lions, leopards, elephants, and other animals can now be easily approached by car to within a few feet. Yesterday's national parks struggled to bring the tourists as close to the animals as possible, but today's problem is how to prevent the irreversible destruction of the animal world through tourism. Previously, it was customary to drag a piece of zebra behind a car in order to coax lions or other carnivores into camera range. The car, however, was also used to hunt animals. Therefore, automobiles were either strongly positive to an animal or strongly negative. My recommendation, though, was to make tourist vehicles in national parks neither positive nor negative, but absolutely neutral. A car full of tourists should be seen by the animals as a large, harmless, inedible, neutral animal that does not stimulate flight, but moves in a regular, predictable fashion. In short, the integration of tourists into national parks is best accomplished by restricting them to the inside of their cars and to designated paths and stopping points along the way. Cars must mimic the space-time system of larger animals. This basic procedure was successfully adopted by most national parks. In places where tourists were allowed to become positive influences, such as in the feeding of bears in North American parks, fatal accidents occurred. The importunate bears had to be killed and the feeding of the animals was subsequently forbidden.

During the Congo trip, we closely investigated some of the territories, specifically, their structure and the markings of animals in the area. One of these animals, the oribi *(Ourebia ourebia),* has a special marking organ, the antorbital gland, the function of which was first described in this African gazelle. Since then, this organ has been studied frequently in the oribi and in many other animals. We also analyzed the phenomenon of passing, that is, the tendency of many animals to cut across the path of fast-moving vehicles. A satisfactory explanation for this phenomenon, which is slowly disappearing due to the animals getting used to automobiles, has not been proposed.

I will not dwell on all of the results of the trip to the Congo; they are compiled in our report of 1951. The trip itself took place in 1948.

The Basel Zoo was approaching its seventy-fifth anniversary when I returned.

Our complement of animals had to be replenished, our buildings needed to be reorganized, and plans had to be made for coming improvements. One of the high points of this time was the arrival of an okapi that had been obtained in the Congo by my coworker, W. Wendnagel. Unfortunately, however, because of the unhealthy traveling conditions, it only lived for about two months. I tried to change this terrible okapi situation by documenting, with the help of numerous colleagues, my horrible experiences in this area. I hoped, thereby, to improve the prospects for successful okapi breeding. This goal, however, has not been fully realized to this day. Of course, biological factors are compounded by political problems.

Europe, and, in particular, Germany, had been in the forefront of developing zoos for many years, but the leading position was shifting to North America because money and space were more readily available there. This was especially true of the Bronx Zoo in New York. This situation made an extensive educational journey to the United States essential. The trip was made possible by the Basel Zoo in 1951. In two months, I visited more than thirty zoos and aquariums in the United Staes, both small and large, progressive and conservative. I crossed the continent by rail twice, gaining an impression of the countryside and its climate. The starting point and end point of the trip was New York. As it turned out, I probably saw more zoos there than did any of my American colleagues. Naturally, I did more than just look at the zoos; I also examined their structure, organization, finances, public relations, scientific activities, etc. Everywhere, I found friendly advice and help. I visited Buffalo, Chicago, San Francisco, San Diego, San Antonio, Miami, Washington, and many other cities. Before I returned to Switzerland, I visited the Bronx Zoo in New York, again where I met Fairfield Osborn, Lee Crandall, Lawrence Hagenbeck, Tee Van, Chris Coates, and had the opportunity to see the famous platypusarium, where, at that time, the only living platypuses outside Australia were displayed. Today, thirty-three years later, there are no living platypuses outside Australia. They were bred only once in captivity, in 1944, by the excellent Australian animal expert, David Fleay, whom I met in Brisbane.

While in New York, I visited with Charles Cordier and his wife, whom I had first encountered in 1948 south of Stanleyville in their camp just as Charles caught his first Congo peacock. In 1949, he was the first to import some of these extraordinary birds to the Bronx Zoo. He also imported an okapi and several other zoological treasures.

On my trip to the United States, I also had the chance to journey to the Yerkes Laboratories of Primate Biology in Orange Park, Florida, where I saw the famous chimpanzee Viki. Keith and Catharine Hayes had worked extremely hard to teach her human language. The possibilities for "Verstehens und Verständigungsmöglichkeiten zwischen Mensch und Tier [Agreement and Understanding between Men and Animals]" have interested me passionately from my childhood to the present. It was no small disappointment to learn that despite Herculean efforts, Viki had learned only three words: mama, papa, and cup.

Twenty years later, in Norman, Oklahoma, when W. B. Lemmon and Roger

Fouts worked with another chimpanzee, Washoe, they used the method invented by R. A. and B. T. Gardner. I visited them and was generously informed about the communication possibilities opened up by the use of ASL (American Sign Language). Unfortunately, I could not be persuaded that this was a valid approach to the study of communication. About ten years later (1980), I rushed to San Francisco, where Francine Patterson of Stanford University had recently made news with her communicating gorilla, Koko. I was impressed with the sureness with which the scientist handled this animal, allowing a stranger to play with her both inside and outside her trailer. Once again, however, I was not convinced that this experiment demonstrated actual communication by means of a formal language. I explained in my objections in a lecture entitled "The Clever Hans Phenomenon from an Animal Psychologist's Point of View" at a conference of the New York Academy of Sciences in 1981, organized by Thomas A. Sebeok and entitled "The Clever Hans Phenomenon: Communication with Horses, Whales, Apes, and People."

But let us return to Viki; specifically, to my stay at Yerkes, which was then located in Orange Park. There, I met Karl Lashley, who had just published his memorable work, "In Search of the Engram." This work and my conversation with Dr. Lashley deeply impressed me. I was especially surprised that there were no specialized brain cells designated for the storage of specific memories and that Lashley was not able to find the engram—and he admitted this latter puzzle freely and with humor.

During this American tour of 1951, I saw an oceanarium for the first time. I visited the marine studios in Marineland, Florida, that had opened in 1938 and can be called the prototypical oceanarium. The curator at that time was Forrest G. Wood, with whom I corresponded for many years. He was kind enough to show me everything that was of zoological or technical interest. This facility, originally built for the underwater scenes of movies, fascinated me greatly and I subsequently visited many oceanariums modeled after Marineland. During this time, man was first beginning to appreciate the extraordinary learning capabilities, specifically the trainability, of dolphins and other cetaceans. Because no one had any real experience with these animals, Adolf Frohn, whom I had known through his work in the circus with sea lions, came from Germany to Marineland. When I met him at work, he was still deeply impressed with the tamability and trainability of his new charges.

In 1952, I published my observations on this subject in the *Zeitschrift für Tierpsychologie*. In 1963, an article about training attempts on other whale species followed. This second article was dedicated to Konrad Lorenz on his sixtieth birthday. At that time, anatomy, psychology, and zoo biology applied to whales and the study of their brains, especially by G. Pilleri of the Gehirnanatomisches Institut der Psychiatrischen Universitätsklinik in Bern, Switzerland, with whom I had been in friendly contact, was making excellent progress.

After my return to Basel, I prepared a report detailing the richness of the impressions that I had gathered in the United States. I presented a special essay on "Seltene tropische Tiere und ihre Haltung in Zoologischen Gärten Nordamerikas [Rare Tropical Animals and Their Maintenance in North American Zoos] in *Acta Tropica,* the organ of the Swiss Tropical Institute in Basel, where I was active as a lecturer.

Naturally, I tried to apply to the Basel Zoo the many inspirations that I had gained through my recent trips. For example, in the Congo, I had noticed that the termite mounds, which often reached several meters in height, served men and animals in a variety of different ways. In my Congo report, I described these mounds extensively. I was especially impressed by the fact that many of the large animals of the African grasslands, such as the buffalo, zebra, antelope, and others, incorporated termite mounds into their space-time framework, and sought them out again and again in order to thoroughly scrub themselves. These termite mounds were important fixed points for the maintenance of the animals' hygiene. Accordingly, I placed an artificial mound made of cement inside the zebra pen at the Basel Zoo. After the zebras were released from their stall, they immediately proceeded to the apparently welcome fixed point of hygiene and scrubbed themselves so vigorously that the artificial mound tipped over. Henceforth, structures of this type were very solidly built. Since then, many zoos have successfully copied this idea.

I was also concerned that most zoo rhinoceroses had worn their horns, their distinguishing characteristic, down to stumps. In their natural biotopes, rhinoceroses keep their horns in good condition by rubbing them against trees, but in zoos, they rub them against concrete, causing them to wear away. To prevent this, I avoided using concrete in the rhinoceros enclosures. Instead, they were provided with tree trunks, which they rubbed so eagerly that the wood needed to be replaced constantly. The rhinoceroses also liked to chew on and eat meter-long chunks of wood.

I paid special attention to the improvement of nutrition, the floor covering, hygiene, and the management of space, and I also attempted to inform the public through more complete name tags and a scientific evaluation of the exhibited animals.

Soon after I assumed my position at the Basel Zoo, I introduced the so-called press cocktail, an arrangement that I later introduced in Zürich. The first of these took place in May 1944, and I continued them until I retired (1974). My successors in Basel and in Zürich have kept the press cocktail. The arrangement was simple enough. Every month throughout the year on an appointed day, representatives of the press, and later other media, were invited to come to the zoo at a specific time. During the first half hour, I, or infrequently a substitute, would show the journalists something of current interest, such as a newborn or newly arrived animal, a certain behavior, a new facility; in other words, a peek behind the scenes. During the second half hour, we sat in the zoo restaurant and, if necessary, discussed what we had seen. In the course of the years, we developed a friendly rapport. Shortly after these meetings, the journalists wrote instructive, often illustrated, reports for their newspapers and magazines. These articles piqued the reader's interest in the zoo, the number of visitors increased, and the zoo's income rose without an increase in expenditures. To my knowledge, through the thirty years that I conducted this press cocktail, not a single participant wrote anything negative about the zoo. For my part, I promised to keep no secrets and to be honest in all instances. This arrangement insured that even deaths and untoward incidents were always reported impartially.

My communications with the media also included lectures carried by radio, several of which appeared in book form, for example, *Exotische Freunde im Zoo* [*Exotic Friends in the Zoo*] and *Jagdzoologie für Nichtjäger* [*Zoology of Hunting for Nonhunters*]. These books went through several editions and were also translated.

In the summer semesters, I gave my lectures not in a university auditorium but in the zoo. During this time, the first doctoral dissertations appeared: R. Schenkel's "Ausdrucks-Studien an Wölfen [Studies of Expression in Wolves]," H. Bruhin's "Zur Biologie der Stirnaufsätze bei Huftieren [The Biology of the Horns and antlers of Ungulates]" and "Schwanzfunktionen bei Wirbeltieren [Functions of the Tail in Vertebrates]" by P. Bopp.

After the long, restrictive war years had passed, I undertook the reorganization of the Basel Zoo in all areas. I attempted to thoroughly and completely biologize the keeping of animals and to promote their scientific exploitation. I attempted to bring about this change through a series of scientific presentations, for example, "Beiträge zur Säugetier-Soziologie [Contributions to Mammalian Sociology] given in Paris in 1950, and "Observations on Reproductive Behavior in Zoo Animals" delivered in London in 1952.

Further works that were closely connected with my zoo activities are listed in my Bibliography. Many worthwhile inspirations stemmed not only from my zoo activities but through the friendly interactions with colleagues whom I met at conferences, and during private visits and with whom I corresponded through letters and the exchange of papers. These delightful relationships extended to numerous biologists and behaviorists who were not hampered by the administration of a zoo. These scientists included Bierens de Haan, Pierre Grassé, David Katz, Otto Koehler, Konrad Lorenz, Niko Tinbergen, Monika Holzapfel, and many others.

The year 1952 was a surprising one in many ways. On the occasion of the fiftieth anniversary of the Veterinary Medicine Faculty of the University of Zürich, I was awarded on May 24 an honorary doctorate in Veterinary Medicine. This degree was given "in recognition of his fundamental works in the area of behavior as well as his steadfast efforts to better understand the animal psyche and to place the care of wild and domesticated animals on a biological basis. Similar goals in veterinary medicine are supported by these endeavors."

This surprising and unexpected honor naturally made me sincerely happy. I received the degree on Saturday, and on the following Monday, the news of the award given to the Basel zoo director, was in the hometown papers. On Wednesday, May 28, 1952, I had an appointment with a delegation consisting of the president of the Zoo's Administrative Board, Dr. Rudolph Geigy, Dr. Adolf Portmann, president of the Union to Promote Zoological Gardens, and Ferdinand Kugler, the vice-president of the Administrative Board.

The first two of these men were both zoologists with whom I had been friends for many years. They had favorably reviewed several of my books, suggested my promotion to Special Professor, and appointed me in 1944 to the directorship of the Basel Zoo. Professor Portmann had been my supervisor twenty-five years earlier at the start of my studies and during my yearlong assistantship at his Institute. Eventually, we became friends.

Naturally, I assumed that the delegation had come to congratulate me, but I was badly mistaken. I was taken to task, especially by my two university colleagues. They accused me of stifling and sabotaging the scientific activities of the zoo. I could hardly believe my ears; this visit was the biggest disappointment of my life.

Of course, it took some time for me to recover from the shock and find an explanation for this change of opinion in my former friends. The leader undoubtedly was Geigy, to whom Portmann was curiously subordinated. I will not go into this matter more deeply here; I can only point to the biography of Portmann published in 1967 by Joachim Illies. It seems that I became too successful for these gentlemen, and Geigy, especially, thought it necessary to prove to me how much of a nobody I actually was. At that time, neither of them had been distinguished with an honorary doctorate.

In any case, this harrowing experience caused me to search for ways to escape this milieu. As it happened, my wife was not happy in Basel, she missed the mountains characteristic of her hometown of Interlaken. Fortunately, our son, who was born in 1947, was still too young to cause any problems relating to changing schools. While we were examining prospects in foreign countries, an opportunity soon surfaced close to home; the Zürich Zoo was looking for a new director. I informed the zoo of my interests, and was awarded a joint appointment as zoo director and professor at the University of Zürich, where I lectured on zoo biology and animal psychology until my retirement.

The Zürich Zoo had been founded in 1929 by amateurs and, like Bern Zoo, which had begun in a similar fashion, was run until 1932 by a part-time director. In 1932, the directorship had been made a full-time job and, as I mentioned above, I applied for it, but an older candidate who had twenty years of experience as a farmer in Sumatra was awarded the position. Hoffmann could legitimately call himself director of the Zoo, but the actual direction was controlled by a merchant named A. Kupper, who was in charge of purchasing the animals and used the zoo as a basis for his bustling trade.

For me, the trade in animals was a necessary evil, and I always resisted the treatment of animals as goods. Therefore, on January 1, 1954, after I assumed office, I officially prohibited any animal purchase that exceeded the actual needs of the zoo and took charge of the necessary acquisitions myself. This action, along with some other things, got me into hot water with my superiors. Even the zoo veterinarian belonged to the governing body of the Zoo Society of Zürich, a private organization to which the zoo belonged.

Trouble soon began between me and the zoo veterinarian and his wife. Their primary source of income was a small animal practice while the zoo's animals were being treated quasi secretly through an understanding between the veterinarian and the zoo attendants.

Various differences led to increasingly ugly relations between me and my superiors and finally, at their meeting on June 27, 1958, the administrative body of the society fired me. This action prompted a kind of revolution by the people of Zürich. A press campaign and even a torchlight procession were undertaken on my behalf. An official investigation into the case was demanded; it was presided over by a special judge and lasted about eight months.

Needless to say, the development of the zoo and my own scientific activities were not exactly enhanced through these lengthy hearings and confrontations with the zoo's administration and personnel. The result was, just as had happened in Bern, that I was vindicated and reinstated, while the administrative body was forced to resign. Only then could I begin the reconstruction of the zoo. But first, I had to confront financial problems that were eventually solved, as is customary in Switzerland, by a plebiscite. Fortunately, I had good friends in all circles who unselfishly helped me throughout my career. I am still grateful to these people. These examples show, and that is why I mention them, how outside circumstances hinder the efforts of the individual who is only concerned with his work, in this case, the betterment of the lot of zoo animals.

Today, zoos are more important than ever, not only as a last refuge for endangered species, but also as a substitute biotope for the city dweller who is becoming increasingly isolated from nature in a gray, noisy world filled with concrete, machines, and crowds of people.

I am convinced that zoos, even more today than before, can also perform an important service for the youth of big cities. These youths are often bereft of stimulation and become bored and indolent. An old maxim truthfully says that the devil finds work for idle hands. These days, that means drug abuse, drug dealing, and all kinds of criminal activities. If, as I have proposed over and over again, and with particular emphasis in 1965, it were possible to interest them in the inexhaustible and amazing variety of nature, especially the animal world with all its applications to environmental protection, crimes committed by the youth would drop considerably. Social duties, however, have been kept out of the domain of zoos for all too long, which is understandable since most zoos have enough problems keeping themselves financially afloat. But zoos, as I have repeatedly stated, are not only zoological concerns, but also human and humanitarian ones.

Many foreign zoos, especially those in America, were much more progressive in this area. During my twenty-year directorship, I tried in vain to at least appoint a zoo teacher to make the zoo more accessible to youths and to schools. Finally, my successor, Peter Weilenman, was able to hire a zoo teacher in 1981. In other zoos, in addition to the departments dealing with scientific research, there are specific departments concerned with teaching and education. During my tenure, it was all a one-man show, although I was occasionally helped by my assistant director, Dr. H. Graber, or my assistant lecturers, Dr. Chr. Schmidt and Dr. R. Keller. Still, I was able to raise the Zürich Zoo from the level of a mere animal show to that of a cultural institution like the theater or a museum. This change was of great financial significance; the zoo could now receive regular subsidies from the city and the canton.

Because of these achievements, at the end of my twenty years as director of the Zurich Zoo, the city awarded me the Culture Prize. I appreciated this honor very much, especially since it was presented by City President Dr. Sigmund Widmer, and was followed by a few kind words from Konrad Lorenz, who had come at the invitation of Dr. Widmer.

I think that I have adequately covered human problems and aspects. A discussion of this sort was necessary, however, to back up my thesis that zoos pose human as well as zoological problems. A zoo might, as contradictory as it may sound, present even more human problems, something that I have noticed throughout my life. The science of zoo biology, which I founded, involves not only animals, but humans as well, when they come into any sort of contact with a zoo. Let us leave this topic and return to the zoo animal, even though—in a way—the animals come last in the zoo, a situation that I detailed in my book, *Mensch und Tier im Zoo* [*Man and Animal in the Zoo*], in 1970.

It should also be mentioned that during my first twelve years at the Zürich Zoo and University, I also worked enthusiastically as a lecturer at cantonal teaching seminars. Here, I had the opportunity to familiarize prospective teachers with zoo biology and animal psychology, placing special emphasis on ways of integrating the zoo into a school biology course and how school children should behave in a zoo. Such a course comprised two semesters. The winter session took place in an auditorium, while the summer one was conducted in the zoo. Classes taught by teachers who had not taken this course were easily distinguished, when visiting the zoo, from classes taught by those who had: they were characterized by unruly and irresponsible behavior. After twelve years, I had to give up these additional lectures because I could no longer find the time for them. They were successfully continued a few more years by two short-term assistants of mine, Dr. H. Graber and my one-time coworker, Dr. E. Inhelder, and Dr. Robert Keller. These lectures were soon discontinued due to a lack of funds or, more precisely, a lack of official understanding.

The Zürich Zoo essentially remained a one-man enterprise, as was customary in Europe at this time. The director was often the only formally trained zoologist in the administration. Veterinarians were not usually full-time zoo employees, but, rather, treated zoo animals in addition to running a private practice. Sometimes, when needed, animals were treated on a contract basis. This custom continued during my twenty-year tenure as director.

Chronic money shortages used to be one of the main characteristics of zoos because they were mostly private or owned by cities, in contrast to botanical gardens which have, as a rule, been affiliated with universities. When I took office in Zürich in 1954, I was challenged by the aforementioned personnel problems and the financial problems of the zoo, which was totally independent from the university. An important time-consuming but largely successful activity consisted of soliciting contributions from wealthy, benevolent firms.

Because biologists engaged purely in science can hardly imagine it, I think that it should be emphasized within the framework of the autobiographies appearing in this volume how difficult it is to carry on research while constantly searching for funds and publicity.

Research, in most cases, came last and was often restricted to free time and vacations. In the Zürich Zoo, the cashier, who was seen as the most important person next to the manager of the restaurant, was required by my superiors to keep a record of the director's presence or absence. After the great trouble that I mentioned earlier, I was once required to detail the number of hours I planned to use for the preparation of my university lectures. These aggravations demanded that I be extremely idealistic but, of course, I had always wanted to be a zoo director.

It was clear to me that the Zürich Zoo needed to be thoroughly reorganized and considerably expanded to take a worthy and suitable place in the city of Zürich. In this respect, I was not short of ideas. To bring novelty and new interest to the zoo, I decided to convert the small bird house in the main building into an open flight room. The inspiration for this, the first open flight room in Europe, stemmed from my visit to the Philadelphia Zoo in 1951. Later, B. Grzimek in Frankfurt built a much larger, beautiful open flight room. This method of displaying birds rapidly gained popularity.

I had all the old cages and bars removed, thereby uncovering a gigantic mouse population. I had never seen such a large collection, even though all zoos suffer from mouse problems to some extent. Upon entering the birdhouse, a pungent mouse odor attacked one's nostrils. The numbers of mice were unbelievable; they occupied the space above the ceiling in the basement as well as the inside of the house. These mice had built intricate routes in each of the cages. At night, after the house had been locked, I observed an amazing flow of mice taking away giant portions of food the birds had not eaten.

The singular observations I was able to make on this large quasi free-living mouse population were described in 1955 in a publication honoring Dr. H. C. Erna Mohr, a mammal researcher from Hamburg. Here, I detailed the rarely described urinating posts in mouse territories.

The public was favorably impressed with the new barless flight room, despite its small size. I had always wanted to rid the zoo of bars, symbols of captivity. The first chimpanzee birth in Switzerland took place in Zürich amid a maze of iron. Therefore, I planned a barless anthropoid ape house but we lacked the funds for glass enclosures filled with tropical plants. So, despite the understanding help of architect M. E. Haefeli, the ape house, which opened in 1959, was, to my great regret, made into a house of bars. Still, I did manage to get glass placed on the visitors' side to prevent infection and the unauthorized feeding of the animals, and to guarantee proper air conditioning. I was even able to provide separate air conditioning for the gibbons and the great ape wing.

Air conditioners are known infection spreaders, not only in zoos but also in hotels where they are also superfluous and always great energy consumers. Monkeys should always be kept separate from great apes since there is a more rapid turnover among these smaller animals. The increased activity in the small animal wing, in turn, increases the risk of infection, despite all quarantine precautions. Certainly, this was the case in my experience.

In 1962, with the help of a gifted attendant F. Zweifel, I recorded some of my observations on primates in a publication honoring Adolph Schultz. We made clear that all behavior related to reproduction in chimpanzees was not instinctive, but had to be learned. These behaviors included copulation, proper relations with the newborn, treatment of the placenta and umbilical cord, the correct way to hold a suckling infant, and the proper method of placing the newborn on the mother's back. Normally, the young learn these behaviors by observing their parents or other adults. In the Basel Zoo, I encountered a pair of young, newly imported chimpanzees that lacked this knowledge. After puberty, the male and female masturbated independently, the female by introducing pieces of wood into her vagina. After the war, Japanese zoos had no chimpanzee matings as long as they imported only young, inexperienced chimps. Utilizing this information, cages in the new Zürich ape house were staggered, allowing the animals to be able to see one another, but also giving them the opportunity to avoid the gaze of their neighbors.

Chimpanzee mothers must also learn how to place an infant onto their backs after the infant has reached a certain age and can no longer simply cling to her stomach. The first chimpanzee born in the Züich Zoo, Miggel, was placed on his mother's back when he was one-half year old. Miggel, however, was placed in the wrong position: facing backwards. The infant sat in this position for months. I feared that the young ape would bump its head on a low bar, but I never saw this happen. Only when another mother who was living in the same cage placed her eight-month-old offspring in the correct position did Miggel's mother correct her mistake immediately and permanently, as I described in 1962.

We noticed that chimpanzee mothers actually present their newborns to attendants or familiar people. Another of these demonstrative gestures manifested itself when the mother wanted her newly self-reliant progeny to climb onto her back: she pointed to her back with her index finger.

A decisive, positive change after the tumult surrounding my dismissal and subsequent reinstatement in 1958 and 1959 was affected by the appointment of a new Zoo president, Dr. Hans Schinz, a world-renowned radiologist. Dr. Schinz, a member of the famous Schinz natural science dynasty, had an excellent understanding of zoo biology and the importance of the zoo for the city of Zürich. The friendly collaboration between us was the high point of my directorship in Zürich. Because he had prominent patients all over the world, Dr. Schinz was able to procure many exotic animals as gifts for the zoo. For example, we received a pair of pygmy hippopotami from President Tubman of Liberia.

The Schinz era was short; he died in 1966. Through noble actions and through his personal contacts, he helped the zoo get needed funds, or at least he made a significant start. At this time, I, with the help of architect R. Zürcher, built the Africa house, the only house with which I was finally able to realize some of my zoo biological ideas. That is, the building would have no cubicle rooms, no flat surfaces, no right angles, no stairs, no iron bars; in short, it would have no unnatural forms, nothing that did not occur in nature. In addition, the Africa house was to be filled with living plants. The basic theme of the house was the symbiosis between African birds and mammals, a familiar sight on an African safari. The two bird species I had in mind were the cattle egret *(Bubulcus ibis)* and the red-billed ox-pecker, or tick bird *(Buphagus erythrorhynchus)*. Naturally, this building could only be a prototype or model because there was neither space nor money for a larger facility.

The large animals that I wanted to keep in the Africa house included white and black rhinoceroses and a hippopotamus family. A smaller enclosure was intended for some shoebills. For me, the main attractions were the cattle egret, the red-billed ox-pecker, and the live plants. To insure the growth of these plants, the building needed and was given large windows and as a result I was accused of building a rhinoceros houses with church windows. Gum tree branches grew luxuriously where the white rhinoceroses were kept, but when these branches reached the levels of the black rhinoceroses and hippopotami, they were instantly devoured.

To enable the symbiosis to work, the partitioning walls were built low enough to allow birds to fly to all their symbionts and nest wherever they liked. Nesting boxes were placed for the red-billed ox-pecker, which had never before been bred in the zoo. Many young birds grew up in these boxes.

Naturally, there were risks inherent in this new display and I am uncomfortable with any risks, especially in a zoo. I did not even like small foul-ups because they stem from a misunderstanding of animal behavior, something that should never happen to an animal psychologist.

The symbiotic experiment was carefully researched beforehand. I had had my first contact with cattle egrets in 1927 in Morocco. During drives through the countryside, I often saw these birds riding on sheep, but on subsequent trips to Morocco, I was no longer able to observe this type of symbiosis. Roughly twenty years later, while at the Basel Zoo, I kept together with some cranes and other bird species, some cattle egrets and capybaras in a meadow with branches and a small hut. The cattle egrets were from Morocco and the capybaras came from Brazil. Therefore, these particular animals had never seen each other before. To my great surprise, the capybaras strongly attracted the cattle egrets: these birds readily rode on the capybaras, to which the capybaras did not object at all.

If cattle egrets could easily adapt to such unconventional symbionts as rodents, then I had no doubt that they would ride their natural symbionts, rhinoceroses and hippopotami, in the Africa house. I was unsure, though, about the red-billed ox-peckers because I had no experience with them. I had only seen one ox-pecker, and this was in the London Zoo. In its otherwise bare cage was mounted an artificial buffalo or rhinoceros back. I wanted to present these birds in a natural symbiosis with rhinoceroses, but I did not want to lose these precious animals, which had been raised by hand in Kenya. Accordingly, after the first ox-peckers arrived, I put them into a cage with pygmy goats. This cage was enclosed on all sides, on the visitor's side with glass. Even though the birds could sit on comfortable branches, they preferred to ride on the goats' backs. The goats, like the capybaras, did not seem to object.

After the Africa house was opened in 1965, the ox-peckers were first put into a traditional cage in the large room that they were to occupy. It was hoped that the birds would thereby familiarize themselves with the room and, especially, with the rhinoceroses. What occurred after we carefully opened the cage for the first time was one of the greatest experiences of my career: shuning the rich vegetation and comfortable perches available to them, the ox-peckers made a beeline for the rhinoceroses and settled on their backs, even though these hand-reared birds had never seen a rhinoceros before. This behavior reminded me of the concept of "angeboreness Schema" as described by Konrad Lorenz. Later, this term was replaced by others, such as IRM (innate releasing mechanism) or genetically fixed behavior, etc. I even see a certain, if distant, relationship with C. G. Jung's concept of "archetype." In all of these cases, we are dealing with programs that are somehow represented in the central nervous system. The "somehow" has remained a mystery to me to this day.

With this one example, I again want to show how seemingly banal, everyday observations made in the zoo can be applied to basic problems of behavioral research and biology. During my zoo career, there were many such instances, some of which I detailed after I retired in my book, *Tiere Verstehen* [*Understanding Animals*] (1980).

I think I need to clear up a few technical aspects concerning the Africa house, the first attempt to keep large African animals together with their bird symbionts. Many people who have heard of the arrangement cannot imagine why the birds did not try to escape when the rhinoceroses and hippopotami were let out into the open, which was the case every day except in the dead of winter. As I have already mentioned, the Africa house was built making use of new biological ideas: no cubicle areas, no flat surfaces, no right angles or rectangular doors. The Africa house's shape conformed, as much as possible to the natural surroundings and bodily shape of these giant animals: for example, the doors were oval. In Africa, I often noticed the paths cut by these animals through the dense bush or high grass were tunnellike. The cross-section of these tunnels was oval, so I chose this shape for the doors in the Africa house. Doors are unbiological; few doors are found in nature. Oval doors are, however, a bit more biological than the rectangular doors of men, which are forced into zoos because of convenience. Of course, the oval doors were not so narrow that the birds would bump into them; this was not the primary purpose of these doors. Rather, I wished to avoid the rectangular form. Even the doors that separated the display rooms from the sleeping areas were oval.

My strategy to keep the birds from escaping was different: while rhinoceroses and hippopotami are bound to the earth, birds live in three dimensions, in short, above the ground. There in the vegetation and later in their nests, they feel secure. When the attendants opened the door to the outside for the large animals, it created a certain, though slight, disturbance. What lay outside the doors was foreign and unfamiliar to the birds. This situation made the birds feel at home inside the house and they felt no drive to leave their familiar surroundings. We lost only one cattle egret during my time at the Zürich Zoo; somehow this bird was frightened and flew in the wrong direction, that is, out of the house. A few weeks later, we received reports that it had been found in Czechoslovakia. My successor, Peter Weilenman, told me about an ox-pecker that accidentally was carried out the door, clinging to the stomach of a rhinoceros. Fortunately, the bird was carried back into the Africa house, still clinging to the rhinoceros's stomach. Both bird species bred prolifically in the house, binding them even closer to their territory. My zoo policy was to biologize the keeping of wild animals in the zoo, in addition to biologizing the animal facilities, as I did in constructing the Africa house. Unfortunately, I had few opportunities to realize these ideas, since these facilities are very expensive.

I would like to mention another example of these biologizing efforts that has always fascinated me; the case of the European otter *(Lutra lutra)*. This animal had been inappropriately cared for by zoos; therefore, it was hardly ever bred and could be kept alive for only short periods of time, even though its tamability and liveliness made it an attractive feature of the zoo. Otters traditionally were kept singly in European zoos, mostly in a kind of cement bathtub attached to a small, wet indoor cage. The otter was long considered to be an aquatic animal. It does, in fact, move very elegantly in the water, but can more aptly be described as an amphibious mammal. These otters need a large piece of land covered with plants, places to sunbathe and, most importantly, areas of dryness. When the means to build such a facility became available in Zürich in 1970, I tried to put these insights to use and devise a breeding program. The facility was to have four divisions and, of course, no bars. Two of these divisions were separated from the public and consisted of a natural, as it were, densely overgrown bank area on a small pool where the animals could live totally undisturbed. If they needed it, they had access to the central part, where dry indoor cages, food, etc., were available. The other two divisions were symmetrical display enclosures with glass fronts, the bottoms of which were like aquariums, providing optimal viewing opportunities. Above the water's plane, one could clearly see the land, which was thickly covered with plants. The entrance had several narrow passages leading to the inner rooms and, finally, to the straw-covered sleeping area. The public was able to look into the sleeping area and could observe the otter's sleeping behavior, its sleeping positions, movements, etc. Double-paned windows shielded the animals from outside noise and the sleeping boxes were only dimly lit. Thus, the new facility allowed the public to observe the entire behavioral repertoire of the otters.

Earlier, the public regarded sleeping animals as boring and used all kinds of disturbances to scare them into activity. In the last few years, however, sleeping was discovered to be an interesting behavior with many fascinating aspects. By instructing zoo visitors through explanatory signs, publications, etc., they have come to appreciate the phenomenon of sleep. Many animals are rarely seen sleeping in zoos, for example, peacocks, giraffes, elephants, many antelopes, zebras, etc. The sleeping behavior of many species is still an enigma. Using the otter as an example, my publications called attention to that area of its territory, the particular fixed point where the vital behavior of sleep takes place.

The solid plate of glass completely separating the public from the river otters was necessary not only from an aesthetic standpoint, but also because of the bitter experience of my predecessor. Otters were regarded as dangerous to the fishing industry and until 1952, the Swiss government had a bounty on otters. Even the otters in the zoo were killed. Otters were placed under official protection only after I, with the support that I gained from the media through one of my press cocktails, called for an end to this horrible practice. The otters had practically become extinct but today efforts are underway to reintroduce these animals to Switzerland. With this example, I also hoped to illustrate the close relationship between the zoo and environmental protection.

Thanks to the initiative undertaken by Zoo president Rolf Balsiger, funds began to flow more readily for the Zürich Zoo. These funds enabled me to build an elephant house. The Indian elephants had been previously kept in a completely unsuitable room in the basement of the so-called main building, where breeding was impossible. One of the basic principles of zoo biology is to avoid the fruitless, solitary confinement of animals and, instead, strive to form breeding groups. At the time of my appointment to the directorship of the Zürich Zoo, not a single elephant had been born in a North American zoo. In European zoos, the statistics were also sad; for each elephant birth, there was one caretaker death. I reported these facts in 1963 in Professor Zimek's popular journal, *Das Tier* [*The Animal*].

Bull elephants, next to unwanted rats, are the most dangerous animals in the zoo. Therefore, I designed an absolutely safe elephant facility. In January 1969, I detailed the plans for this facility in three consecutive articles in the newspaper, *Neue Zürcher Zietung*. I hoped to call official and public attention to the need for a safe facility. This safety could only be guaranteed by building a double cage, similar to the one used for the European hares. Whenever the bull was in a dangerous state, which is not necessarily the same as the musth, he could be induced to move from one cage into the other from outside.

The city and the canton put two million franks at my disposal for this new fashioned elephant facility, with the condition that the new building, including planning expenses but adjusted to inflation, would not cost a single additional frank. This unfortunate clause led to massive deletions during the course of construction. This short-sighted and sad restriction first affected the facilities for the bull elephants. Elephants, of course, had to be shown in this house, but the safekeeping of bulls was made impossible. This example once again shows how closely related are zoo biology, finance, and politics, a situation that has haunted me all my life. Long after I retired in 1974, my successor, Peter Weilenman, managed to raise enough money through private contributions to build a facility for the bull elephant. So finally, in 1981, the basis for a safe breeding program was in place.

During my time at the Zürich Zoo, chronic money shortages made it impossible for me to realize many of my zoo biological ideas. Therefore, I turned to writing. In 1965, *Mensch und Tier im Zoo* [*Man and Animal in the Zoo*] appeared; the book was translated into English in 1970 and appeared as a paperback in 1974. In a sense, this work was a continuation and a completion of *Wildtiere in Gefangenschaft* [*Wild Animals in Captivity*] and, as Terry Maple informed me in 1981, helped inspire him to found the journal *Zoo Biology* in 1982.

In addition to writing, helping to plan or reorganize foreign zoos gave me an outlet for ideas that I could not realize in Switzerland. In 1956, I traveled to Brazil where my friend, Professor Mario Autuori, and I planned the São Paulo Zoo, which I revisited after its completion. Professor Autuori was a world-renowned entomologist who was primarily concerned with carpenter ants. We met in Paris in 1954 at a colloquium on instinct organized by Professor Pierre Grassé.

In 1969, the government of India commissioned me to plan a zoo for Simla in the Himalayas at an elevation of 2,400 meters. In 1968, using simple materials, I built a "zoo" at an elevation of 2900 meters in Piz Lagalp near Pontresina, Switzerland. This project fascinated me because it afforded me the opportunity to present animals native to our Alps in a different way. Ibex, chamois, marmots, snow hares and, eventually, eagles, snow hens and many other animals were displayed here. I restricted myself to these animals, which lived in areas above the timberline. Since there were no trees, I could not correctly call this facility a zoo or a park. Therefore, I coined a new word: *alpinarium.* It was designed to make the area attractive to tourists not only in winter, but in the summer as well, and also to allow the public to get a close look at animals native to this region. It was impossible to get this close to the animals in our Swiss national park. In the alpinarium, animals were kept in an enclosure with a natural floor and background and visitors watched the animals from a covered gallery. The only problem was caused by the massive amounts of snow that fell every winter. Of course, the animals could always be kept safely under the gallery, but this would prevent their being seen in their barless, open-air facilities during the winter. The specialized engineers who intimately knew this area and whom I had consulted, had assured me that even in winter the snow would not be much of a problem. The area, they claimed, could be reached by cable car and the snow could be removed by machines, electrically heated underground cables, flame throwers, or any other method. Unfortunately, these methods proved to be ineffective against 6 to 7-meter-high snowdrifts and the expense needed to keep the area clear of snow was so prohibitive that the alpinarium was given up after a few years.

These problems did not appear in 1966, when I was called by the government of New South Wales, Australia, to reorganize and restore Sydney's Taronga Park Zoo. This zoo had not changed at all since my first visit there in 1929 during my trip to Melanesia. It was high time to renew not only the outdated structures and technical installations, but also the frozen, archaic organization. The latter task was not easy since the president and director of the zoo, Sir Edward Halstrom, generously supported cancer research and other humanitarian works and had many friends. As self-proclaimed zoo director, however, he had also made numerous enemies who imperiously called for a thorough modernization of the zoo. Thus, I faced a ticklish situation. I even received threats from Sir Halstrom's fans, who told me that they would throw me into the Sydney Harbour Bay, where the biggest sharks were, if I did not leave immediately. However, I had been familiar with threats since my start in Bern in 1938. Threats had not kept me from my work either in Switzerland or in Australia. I have always openly expressed in a thoroughly undiplomatic fashion, both through writing and speaking, what I thought was right or wrong. I have never regretted it, even though it was not always beneficial to my career.

I always used my vacation time for such advisory trips to foreign countries. Although I was interested in the countryside and people, I undertook these trips primarily to learn about the animals of other countries. Often, zoo director conferences, ethological meetings, or lecture invitations provided me with travel opportunities. For example, I was invited to lecture at the Two-Stage Conference on Sex and Behavior organized by Frank Beach at the University of California, Berkeley, in 1961 and 1962. Here, I gave a talk on "Environmental Factors Influencing the Reproduction of Zoo Animals." I often remember such opportunities with deep gratitude. These meetings enabled me to exchange ideas with many longtime and new colleagues from both sides of the Atlantic.

This fondness also extends to a three-day conference entitled "Both Sides of the

Railings" which was organized by Robert I. Bowman in May 1968 in San Francisco. Here, I presented two lectures: "Approaches of a Zoo Director to the Study and Teaching of Animal Behavior" and "Animals as Social Partners of Man." On this trip, my wife and I visited various cities on the east and west coasts of the United States.

In 1974, while flying back from San Diego, where I received a gold medal from the local zoo society "for dedication and service to the cause of wildlife conservation," I stopped on the island of Haiti to visit our son, Peter, who was busy doing historical studies. I used this opportunity to observe New World geckos, anoles, tarantulas, and some bat species with which I had previously been unfamiliar.

In Zürich, my normal activities continued. I had made plans for a giraffe-okapi house, a sea lion facility, and many other things, but again, money was short. Even today, these plans remain unrealized. Restoration and repair costs simply consume too much money.

This situation made the abundance of new observations and insights presented by the animals all the more delightful. For example, because of their extreme tameness, Canadian beavers allowed me to observe their birthing behavior from a distance of less than one meter. The males seemed to act as midwives. For many years during the summer, I could show my students how the young were plucked from the water and carried back to the artificial beaver home by their parents or elder siblings on their forepaws, walking upright on their hind legs.

Strangely enough, the retirement age of the zoo was sixty-five, while that of the university was seventy. The last five years of my activity in the zoo consisted of lecturing. In other words, I was able to continue my zoological and animal psychological demonstrations in the zoo during the summer semesters. Because the demand for the course remained high and as it was impossible to demonstrate details of behavior to more than fifty students, I had to conduct two parallel classes in the summer and to restrict my lectures to large animals. For example, to my great sorrow, I had to totally forego the aquarium. Even during my last summer semester (1978), about one hundred students took my course, illustrating how interesting this theme was. Not only biologists took part, but also veterinarians, physicians, and even theologists and students of other disciplines when their stringent schedules allowed them to attend. My longtime lecture assistant and coworker, Dr. Robert Keller, successfully took over this much-loved course as my scientific successor until 1981. That year, to my and many others' great sorrow, he gave up the course as a result of personal problems, that are not uncommon in zoos.

Today, it is no longer possible to simply build a zoo and care for the animals; this is only half of what is needed. The other equally important half is a heavy commitment to education, teaching, and research. Even in the Zürich Zoo, the highly profitable restaurant was given priority over such simple educational facilities as a lecture hall.

After I retired as Zoo Director in 1974, I devoted more of my time to writing scientific and popular papers that I had had to delay for some time. I realize that most of my work is descriptive and that precisely this type of work has fallen out of favor in recent times. The deemphasis of descriptive work, however, may be regretted in the future.

Exact descriptions of species undoubtedly form the basis of all zoology. Similarly, exact description must be the basis of behavioral research. I see the literature concerning poisonous animals as proof of this assertion. Incredibly precise analyses of the poisons of various animals are accompanied by inexact classifications of the animal from which the poison was derived.

Another example of this type of thinking is behavioral analyses that are highly differentiated, computerized, statistically significant, delineated in exact curves, and computed to several decimal points. What is missing here is the delicate relationship between animal and observer, a theme that I have pursued since my youth because it relates directly to the little-researched possibilities of understanding between man and animal.

In Europe, this question, my original question, cannot be discussed without referring to a classic case that occurred in Berlin before the first world war: the Clever Hans case. Hans was a horse who could supposedly think like a human, speak, and do sums. The communication barrier between man and animals appeared to have been breached. Hans became a sensation in Europe but, unfortunately, in America not much attention was paid to this phenomenon. Otherwise, in my opinion, it would have been impossible that seventy years later, speaking apes were allegedly discovered in America. It was claimed in all seriousness that chimpanzees and gorillas were capable of conversing with humans. The accomplishments of Clever Hans were shown to be pseudoaccomplishments. It turned out that Hans relied on simple signals that had previously gone unnoticed. In various publications, I referred to this fact and recommended that experimental results be critically reviewed in light of the Clever Hans phenomenon and under strict laboratory conditions.

These thoughts brought me into contact with the linguist Thomas A. Sebeok of Indiana University. We met often in Zürich or Amsterdam and, in 1980, he organized a "Conference on the Clever Hans Phenomenon: Communication with Horses, Whales, Apes and People" at the New York Academy of Sciences. I delivered the introductory lecture, "The Clever Hans Phenomenon from an Animal Psychologist's Point of View."

I used this opportunity to travel to Atlanta and observe the famous chimpanzee, Lana, as well as the two younger chimps, Sherman and Austin, both of which had also been taught the artificial language, Yerkish. In 1979, I had already criticized this method of conversing with apes.

I believe that eventually an explanation for the extremely complex and, so far, underresearched problem of the relationships between man and animals will be obtained by means of signal study or semiotics, specifically, zoo-semiotics, an important and promising approach devised by Thomas A. Sebeok.

My childhood dream, my lifelong wish, would have been fulfilled if it had really been possible to converse with animals. I was passionately attracted to this goal, but I was disappointed again and again by the Clever Hans story, by Viki, Washoe, Lana and, finally, by Koko the gorilla, which I was allowed to see in its

home near San Francisco after my visit to Atlanta. First, however, I was required
to get a chest X ray and a checkup proving that I was healthy. These things were
demanded of me by telephone shortly before I was to leave Switzerland.

I was deeply impressed by the profound feeling of closeness between Penny
Patterson and Koko. This fragile woman so completely controlled this 90-kg
gorilla that even a stranger like myself could play with Koko both inside and
outside her living quarters without risk.

This incredible closeness, this singular rapport between Koko and her master,
Penny Patterson, reminds me of the special relationship and understanding between owner and dog. How many dog owners insist that their dogs "understand
every word"! In Koko's case, the attachment may be even stronger, but I do not
believe Koko's reactions follow rules of grammar or should be equated with
human speech.

From San Francisco, I returned home via Hong Kong, where we had a family
reunion. We had already had a family reunion there in 1977 after a visit to
Taiwan—and a subsequent one in 1983. I used this opportunity to study the
environmental problems of this rapidly growing city, and of its zoos.

After I retired, the Psychological Institute in Zürich provided me with an office,
for which I am very grateful. Here, I devoted myself to problems that had long
occupied me. I prepared a considerably expanded version of my "Ergebnisse
tierpsychologischer Forschung im Zirkus [Animal Psychology in the Zoo and in
the Circus], which appeared in 1979 in East Berlin. An informative foreword was
written by my friend, Heinrich Dathe, the director of the East Berlin Zoo and
editor of Der Zoologische Garten.

Finally, in 1976, I was able to commit to paper what I had previously only
described in lectures: "Proper Names in the Animal Kingdom". Even though
many authors writing about very diverse species from fish to monkeys had observed that the individuals of the social group, such as schools, swarms, and
herds, knew each other personally, it took surprisingly long before anyone dared
to call the signs, that make the recognition of an individual possible, names. I
pointed out that besides acoustic, there are also optic, olfactory, ultrasonic, and
other types of names. Man himself not only uses acoustics, but also optical means
of identifying himself, such as his face, signature, fingerprints, etc. By personal
names, I mean a sign or group of signs, specific to an individual that can be
perceived and can lead to identification of that individual.

I elaborated on the distinction between *nest* and *home,* two terms that have
been improperly used. The prototype of the nest is a bird's nest: solely a reposi-
tory for eggs and raising the young. It is never used as a place of refuge, which is
the function of the home.

Finally, I found the time after my retirement to formulate biological ideas that
had presented themselves during my decades of zoo work and my journeys into
the wilderness. I compiled these views in my book, *Tiere Verstehen*
[*Understanding Animals*], not because I was bored or for amusement, but because for me it was a necessity. I simply had to write it. Certain parts of this book,
however, were painful to write. In these sections, I disagreed with some of the

ideas of Konrad Lorenz and Niko Tinbergen, two men from whom I otherwise learned a great deal.

However, I must be true to myself. For example, I simply cannot accept the claim that "the two major constructors of speciation are mutation and selection." Too many of my zoo observations contradict this assertion. There must be other sources of speciation besides these two. My recent publication concerning the European cuckoo again reveals my distrust of the accepted evolution theory. Too often, hypotheses have been presented and accepted as facts and I have always attacked this tendency.

The book *Tiere Verstehen* [*Understanding Animals*] essentially marks the end of my career. After all, understanding animals was always the fondest wish of my childhood. Was I chasing an illusion? I do not believe so. Even though we have still not conversed with an animal, I believe we have moved much closer to an understanding of animals in the last seventy years.

A walk through the zoo convinces me that this belief is true: the prisonlike barred cages have disappeared. Animals are now in more natural surroundings, modeled on their native biotopes and territories. Efforts have been made to accommodate their social needs so that their lives are worth living.

In addition, words like *dull, destructive,* and *dangerous* are no longer applied to animals. Instead, we have made progress towards understanding their social, marking, and expressive behaviors. There are certainly fewer misunderstandings between humans and animals, and a greater appreciation of the animals' natural habitats. These facts make possible an imporved biological treatment of animals.

Now, more people than ever visit zoos. Through this man-animal contact, understanding of the need to protect endangered animals and their habitats has increased. The living animal is obviously the most important advertisement for the idea of environmental protection. Today, more and more people are becoming aware of and active in environmental protection; they have realized that animal and nature preservation is vital also to the survival of man himself.

References

Hediger, H. 1932. Zum Problem der "fliegenden" Schlangen. *Rev. Suisse Zool.* 39:239–46.

———. 1934a. Beitrag zur Herpetologie und Zoogeographie Neu Britanniens und einiger umliegender Gebiete. *Zool. Jahrb. Abt. Syst., Oekol. u. Geographie der Tiere.* 65:389–582. (Ph.D. diss.)

———. 1934b. Zur Biologie und Psychologie der Flucht bei Tieren. *Biol. Zentralbl.* 54:21–40.

———. 1935. Zähmung und Dressur wilder Tiere. *Cibz Zs. Basel* 3:27.

———. 1938. Ergenisse tierpsychologischer Forschung im Zirkus. *Die Naturwiss.* 26:242–52.

———. 1942. *Wildtiere in Gefangenschaft. Ein Grundriss der Tiergartenbiologie.* Basel: Benno Schwabe.

————. 1947. Ist das tierliche Bewusstsein enerofrschbar? *Behaviour* 1:130–37.

————. 1950. Das Okapi als ein Problem der Tiergartenbiologie. (Zusammen mit mehreren Autoren als Okapi-Sonderheft) *Acta Tropica. Basel.* 7:2.

————. 1951. *Observations sur la Psychologie Animale dans les Parcs Nationaux du Congo Belge.* Bruxelles: Institut des Parcs Nationaux du Congo Belge.

————. 1955. Mause im Zoo. *Der Zool. Garten (N. F.)* 22:76–85.

————. 1956. Instinkt und Territorium. In *L'Instinct dans le Comportement des Animaux et de l'Homme,* pp. 521–33. Paris: Masson. Fondation Singer-Polignac.

————. 1958a. Verhalten der Beuteltier (Marsupialia). *Handb. Zool. Berlin* 8:18. *Lieferg. 10:1–28.*

————. *1958b. Kleine Tropenzoologie. (Zweite neu bearbeitete Auflage).* Basel: Verlag fur Recht und Gesellschaft.

————. 1959. Die Angst Des Tieres. In *Die Angst. Studien aus dem C. G. Jung-Institut,* pp. 7–33. Zurich: Rascher Verlag.

————. 1962. Tierpsychologische Beobachtungen aus dem Terrarium des Zurcher Zoos. *Rev. Suisse Zool.* 69:317–24.

————. 1963. Weitere Dressurversuche mit Delphinen und anderen Walen. *Z. Tierpsychol.* 20:487–97.

————. 1964. *Wild Animals in Captivity, an Outline of the Biology of Zoological Gardens.* New York: Dover.

————. 1965a. Man as a social partner of animals and vice-versa. *Symp. Zool. Soc. London* 14:291–300.

————. 1965b. Environmental factors influencing the reproduction of zoo animals. In *Sex & Behavior,* ed. F. A. Beach, pp. 319–54. New York: John Wiley.

————. 1966. Report on Taronga Zoological Park, pp. 1–45. New Wales: Parliament of New South Wales. V.C.N. Blight, Government Printer.

————. 1969. Comparative observations on sleep. *Proc. Royal Soc. Med. London,* 62:153–56.

————. 1970a. Zum Fortpflanzungsverhalten des Kanadischen Bibers *(Castor fiber canadensis). Forma et Functio* 2:336–51.

————. 1970b. The development of the presentation and the viewing of animals in zoological gardens. In *Development and Evolution of Behavior. Essays in memory of T. C. Schneirla,* ed. L. R. Aronson, E. Tobach, D. S. Lehrman, and J. S. Rosenblatt, pp. 519–28. San Fraqncisco: Freeman & Co.

————. 1970c. *Man and Animal in the Zoo: Zoo Biology.* London: Routledge & Kegan Paul.

————. 1974a. Communication between man and animal. *Image Roche, Basel* 62:27–40.

————. 1974b. *Tiere sorgen vor.* Zurich: Manesse Verlag.

————. 1976. Proper names in the animal kingdom. *Experientia* 32:1357–64.

————. 1977a. Nest and home. *Folia Primatol.* 28:170–87.

————. 1977b. *Zoologische Gärten. Gestern—heute—morgen.* Bern/Stuttgart: Hallwag.

————. 1979. *Beobachtungen zur Tierpsychologie im Zoo und im Zirkus (Neue erweiterte Ausgabe).* Berlin: Henschelverlag.

————. 1980. *Tier verstehen. Erkenntnisse eines Tierpsychologen.* Munchen: Kindler.

————. 1981. The Clever Hans phenomenon from an animal psychologist's point of view. *Ann. N. Y. Acad.* 364:1–17.

Hediger, H., Bloch, S., Lloyd, H. G., Muller, C., and Strauss, F. 1967. Beobachtungen aur Superfetation beim Feldhasen *(Lepus europaeus).* *Zs. f. Jadgwiss.* 13:49–52.

Hediger, H. 1983. Natural sleep behaviour in vertebrates. *Functions of the nervous system,* pp. 105–30. Elsevier, Amsterdam/New York: Marcel Monnier Ed.

————. 1984. Geburtsort: Zoo. Silva Verlag Zürich.

Eckhard H. Hess (27 September 1916 – 21 February 1986). Photograph by Elihu Abbott, Jr.

7

The Wild-Goose Chase

Eckhard H. Hess

I have always wanted to be a scientist. Apparently I started early, for my first two years were spent in intensive examination of our family barnyard in East Prussia, Germany. I have no conscious recollection of it, but the smell of a barnyard, to this day, gives rise to pleasant, albeit muted "memories." My next years, up to the age of ten, were spent in a suburb of the industrial city of Bochum in the province of Westphalia. I was close enough to fields and ponds and forests to soon engage in that commonplace activity so usual in those who later become ethologists—I kept animals.

Konrad Lorenz once said that all of the ethologists he knew carried out this boyhood activity. I do not refer to the casual bringing home of some accidentally found frog or lizard, but rather the taking on of a responsibility for providing an environment suitable for that particular animal and seeing to it that normal activities could take place in such a period of confinement. I distinctly remember that if these conditions could not be met, my animals were returned to the wild. I mention this aspect of my early life because I feel it has something to do with the nature of the kind of research I have carried out over these years. It clearly supports the seemingly frivolous definition of "ethology" that was put forth by the late R. F. Ewer, "Ethologists are scientists who love the animals which they use in their research." The long and short of it is that I have designed my research to avoid cutting up the object of study. Clearly this has its advantages and disadvantages. It is, however, interesting to note that the two students of mine who are clearly identifiable as ethologists, both cared for animals in their youth, neither cuts up the animals they use in their research, and both continue their work in the analysis of the behavior of the complete organism in its environment. These two are Erich Klinghammer and Gordon Burghardt.

My family emigrated to the United States in 1927. I remember the new world it opened for me. Animals I had never seen before were everywhere in the forest and bush near the eastern Pennsylvania town that was to be my home for the next eight years. I soon knew every pond in the area that would provide frog and salamander eggs in the spring and daphnia to feed my fishes during most of the year. I began correspondence with a vivarium enthusiast in Germany, exchanged animals by mail, and many years later had the opportunity to see an aging spotted salamander that I had sent to him by boat mail seventeen years earlier.

Clearly there is a difference in the way that scientists deal with the organisms that make possible their scientific careers. One of my colleagues at the University of Chicago, David Bakan, asserted that the psychologists who cut up animals are very apt to have been raised in the country where they were exposed to the slaughter of food animals, and that those raised in the city ended up in clinical or other noninterventive sorts of psychology. Be that as it may, there is clearly a way for both the "cutter" and the "watcher" to make a contribution. There is no value judgment implied. Both have their place.

When I was thirteen, I made my first major scientific discovery, or so I thought. In observing ants, some of my favorite subjects, I used a mirror to illuminate them with sunlight as they moved in file along a partly shadowed pathway. To my astonishment, they reversed direction. When I removed this source of light they returned to their previous orientation. The reader will undoubtedly know that I had "discovered" the light compass orientation in ants, which had been described years previously. It was some time later that I realized this. Such "rediscoveries" are not exactly rare. In many cases they are published over and over as new findings. Subjective color (apparent color by intermittent black and white stimulation) was such a phenomenon. Over a period of about two hundred years it was independently found by more than a dozen scientists. Each time it was published as a new finding. Such illustrious names as Fechner and Benham are included in the list of "discoverers." Fortunately, just before I became another name in this long list I found the phenomenon in the literature. It has always been an object lesson for me to look thoroughly, and back more than the usual ten to twenty years when proposing a so-called "new" finding.

By the time I went to college, Blue Ridge College in Maryland under the difficulties imposed by the great depression, it seemed clear to me that my scientific interest could best be served in a "making a living" sense by entering a field like chemistry. So I became a chemistry major. Some time in my junior year I read a book, "The Animal Mind" by Margaret Floy Washburn. It changed my entire program of study. I thought, if all these things I have long been looking at are a part of a legitimate scientific field, then I want to be part of it. I changed my major to psychology and graduated with majors in psychology, chemistry, and literature.

It was also at college that I met my wife-to-be. Dorle was an art major. She too had been born in Germany and we found many interests in common. Although her loves were music, art, and interior design, she particularly shared with me a love of animals—that is birds and mammals. As will be seen later, this helped a great deal to make possible some "home experiments." We were married in 1942.

After graduating from Blue Ridge College in 1941 I worked for a couple of years for the Seagram-Calvert Distilling Company as an industrial psychologist. I ran the quality control laboratory, for the assessment of taste and odor of whiskey. It was a way of marking time prior to my induction, as an enemy alien, into the U.S. Armed Forces. When the time finally arrived, I was made a rifleman in the infantry.

After Cassino and Anzio I was back in the United States and then made plans for my university training. I entered Johns Hopkins in 1945 as the first student in their newly constituted psychology program. Clifford Morgan was chairman and most of the faculty appeared to be younger than I was. The time I spent at Johns Hopkins afforded me little pleasure, because I felt that time was slipping by. My interests were largely in visual perception and only W. C. H. Prentice of the faculty was interested in my ideas as to what might constitute an adequate research problem for a Ph.D. thesis. I remember his efforts on my behalf with gratitude. It was finally possible for me to carry out my thesis research with chicks, although exploratory work had to be carried out with rats—scarcely a visual animal. I went through my training quickly, receiving both a master's degree and a Ph.D. in under three years. By that time, however, I was thirty-one years old.

Now came the time for my first academic position. All my life I had wanted to be a scientist and a professor. Clifford Morgan called me into his office some months before graduation and gave me a list of six universities that seemed to provide good opportunities. All were top-notch positions. Nineteen forty-eight was a year in which there were more openings than there were qualified applicants, and I was asked which one I wanted to go to least and which I most wanted to try for. Without much hesitation I asked to try for Stanford. And I definitely stated that I did not want to go to the University of Chicago. However, it was Morgan's decision that I go to Chicago. Years later, in one of our conversations, I discovered that he thought I would do well there because there was nothing to do but work. Perhaps he was right.

Now that I was on the faculty at the university, I expected a magical change. There was none. Life was little different from graduate school, living was not elegant in the environment of the university, and there was indeed little to do but work.

Early on I met Roger Sperry and we formed a long-lasting friendship. We talked about innate visual orientation and how I might carry out some of the same sorts of experiments with chicks that he had so elegantly carried out by surgical techniques on fishes and amphibians. Roger offered to help me with the surgery, rotating the eyes of chicks in their sockets. I demurred because I did not want to cut the animals. Finally, on a Thanksgiving morning, I realized that I could accomplish my objective without surgery, and with the reversible use of hoods and prisms. The experiment worked beautifully and indicated the innate organization that allowed the young chick to know exactly where objects were in space and that no practice or learning was necessary for this knowledge. I remember basking in the praise of some of the senior scientists when I presented my findings at our annual psychological meetings.

But by far the most important impetus to the direction of my academic life came as the result of reading another publication. During the last year of my graduate studies at Johns Hopkins, I became acquainted with John Cushing, a biologist. Biology and psychology were at that time housed in one building. He gave me a translated copy of Konrad Lorenz's *Companion in the Bird's World*. The English translation was miserable. I obtained the German version and knew at once that the concept of "imprinting" that Lorenz formulated was a made-to-order area of research for me. Being trained as a physiological psychologist I, of course, scorned the naturalistic approaches to this phenomenon and considered ways that it could be studied under "controlled" laboratory conditions.

This work I carried out over a period of many years, first, with the collaboration of A. Ogden Ramsay, who proved to be a good friend, and then later at my own laboratory on the Eastern Shore of Maryland, where I finally turned to a more naturalistic approach to imprinting. In regard to the latter I am reminded of one of my first experiences in the academic world that also dealt directly with the future trend of animal behavior research and ethology. It was at a conference held many years ago at Penn State. In an impassioned and yet bitter statement, the late C. R. Carpenter addressed the problem of field studies and their acceptance by the scientific community in the "fifties." He said, in effect, "the worst laboratory study turned out by a student in a department will be more acceptable for a dissertation than the most elegantly carried-out field study." It was true, and I, along with many other young researchers at the time, must surely have had our enthusiasm for such endeavor dampened by this attitude.

How times have changed. The Schallers, Goodalls, and many of the European ethologists have certainly changed that picture. At any rate, progress in our imprinting research need not be outlined here. My Public Health Research grant to study imprinting spanned more than two decades. I believe our work served its purpose. It stimulated other research and God knows that I always gave my fellow scientists much to shoot at. So, as one reviewer of a book of mine said, "Hess is no stranger to controversy." This is true, but when I examine the reasons it becomes very clear to me that I find it incredibly boring to work at the details of established phenomena. It is just plain dull to work out a variation of a theme that has been put forth by more imaginative scientists. At any rate, most of my efforts have been directed toward breaking new ground, certainly for my own satisfaction more than for the academic recognition that may or may not follow publication of one's work. And that is not a "safe" procedure.

So much talent, energy, and taxpayer's money seem to be wasted on make-work, almost "public-works programs." Obviously whatever gets done must be published. Otherwise no further funds or academic advances will be forthcoming. This veritable torrent of papers, most of them not worth publishing, is possibly a product of the growing "publish-or perish" mentality. Never mind who is at fault, it exists, but it was not always so. At another time and place, certainly in psychology, this was not true. I remember that in my graduate student days and even beyond, Carl Pfaffmann came out with a paper every few years, but they were real contributions that formed the basis for much subsequent work in the area of the chemical senses.

The present strategy appears to be to publish much, get research funds of considerable magnitude quickly, and to attend as many conferences as the grants will support in travel funds. Somehow I often feel that these endeavors are more important to the investigators than the research itself.

All of this leads me to observe that many individuals appear to go into an academic endeavor without it being what they really want to do. One should ask oneself this simple question—"Is this what I would still do if I were independently wealthy and could do exactly what I wanted to do without the necessity of earning a living?" Of course there are those who would do exactly what they *are* doing. In my academic years I have known a number of persons who would fit this category. Probably the difference is that between a way of life and work and just a job.

There is another problem. It seems to me that I have always engaged in research activities because I have been curious about aspects of animal behavior in our world. This can have an unfortunate consequence. When I get the answer to a question, one that satisfies me, I am much more interested in going on to another problem than I am in doing the chore work of making the observations and doing the experiments necessary to writing a paper that can be submitted for publication. My filing cabinets bulge with outlines, partial drafts, even nearly complete research papers that somehow never get into that final form necessary for submission. I am not happy with this aspect of my personal way of doing things—there is after all a responsibility to publish one's work.

Similarly, there is a responsibility to publish good, popular accounts of one's findings and contributions so that the public, which supports our work, has some feedback and knowledge of what has been done. I think particularly of such publications as *Scientific American* and related magazines. Further, in this electronic age, a well-done television presentation of one's work, as in "Nova" is not to be treated with academic contempt. It is a responsibility and a contribution.

The fact that I have not published more of what I have done and found may be, I like to think, a form of self-censorship and an effort to cut down on the torrent of published material that often does not enlighten but only adds to confusion. This is probably a rationalization.

At any rate, in spite of the fact that I clearly did not want to go to the University of Chicago and join its faculty, I have spent more than half my life as a member of that institution. I did work in Maryland every spring on the problem of imprinting, and I went to several universities as a visiting professor: Berkeley, Swarthmore, and the University of Maryland. Somehow, although these schools had excellent faculty and facilities, I came to appreciate my own university even more. Certainly it is too late to change now.

I have referred earlier to the fact that I have often been embroiled in controversy—as a result of research findings that I have published. It reminds me of a story about Otto Koehler, the founder of the "Zeitschrift für Tierpsychologie" and a friend and publicist of ethology. When the late Koehler was a young member of the distinguished German scientific academy, a pompous, but important, scientist announced in somber tones, "The truth of today is the error of tomorrow!" Up jumps young Koehler. "False!" he cries, "The truth of today is the special case of tomorrow." And so it must be if the search for facts is carried out intelligently and honestly. Why should a fact turn out to be *wrong* if the work is well conceived and well carried out? I went through several periods in my academic career when there was much sound and fury. Most clearly I remember the early fuss about our work in the critical period in imprinting and the later work on the constriction of pupils as an indication of negative affect. Clearly both are examples of special cases when one knows much more, but neither of these research results, in their context, was "wrong."

It is probably not clear to many of my colleagues that I began my academic career, as far as college and the first year of graduate school are concerned, as a Watsonian behaviorist. There was something seductively simple about such an approach. Although I became disillusioned after seeing the failure of this approach in dealing with behavior, I probably retained some of the bias for several more years. Certainly it is true that my Ph.D. dissertation was clearly designed to answer the question, "How does the organism *learn* to use light and shade cues of depth?" Obviously I was taking for granted that surely such a subtle cue was learned. In fact, all my psychology texts called it a "learned" cue. Now, if I were to study the same problem again, I would ask the question another way. "Does the organism have to learn to use light and shade cues of depth and, if so, how does it learn?" There is a difference between these two ways of asking the question.

However, reading the already mentioned "Kumpan" paper of Konrad Lorenz, and then in 1951 "The Study of Instinct" by Niko Tinbergen, changed all that. Somehow I felt that I had come home to my childhood "scientific—investigation—play." Up to now, that has not changed. I seem to find an answer and then I try to find the question so that the answer makes more sense. Let me give an illustration.

When I imprinted my first Canada gosling I took it out for a short walk on the lawn. For some reason I stopped at some point and took one small backward step. Unfortunately the gosling's foot ended up under my heel. It had been following me too closely. Now it took off with loud screams and as it fled me there flashed through my mind the certainty of what every psychologist knows—it would from now on avoid me, the source of punishment. However, it did not work out that way. After a short excursion of about six or eight feet, the gosling ran back to me and then stayed so close that I had to be doubly careful not to step on it.

I related that story in my class that fall together with some tentative experimental approaches to the problem this whole episode suggested to me. It was an answer, and did not yet have a question. One of the students in the course was Joe Kovach. He came to me after class and suggested he might like to tackle the problem. He did, and his thesis on punishment and critical period was the result. Of course there have been other instances where logic dictates hypotheses that are then tested in the approved method. But I have had many instances where "keeping one's eyes open" resulted in meaningful progress. I have reported how my wife's observation led to my research in pupillometrics. It was her knowledge of portrait painting that made her realize that my pupils were inordinately large for a particular room illumination level. A quick informal "experiment," just to satisfy myself, the next day, and we were off and running to begin an area of research that keeps many investigators busy even after two decades.

I have spoken of the fact that my wife loves animals and, almost certainly, this made possible some work that I carried out at home. Our first human-imprinted Canada gosling was hatched in a makeshift incubator that was ordinarily used as a place for the rising of yeast dough. It also served to hatch starlings from the egg. This, in particular, was an object lesson for me. If I had been more conversant with the ornithological literature, I wouldn't have attempted this task. Such authorities as Heinroth flatly stated that certain birds, including the European starling, just could not be hand raised from the egg. The practice was to remove the young from the parent's nest after they had started their growth, but before some of the fear responses had matured. I was enthralled, however, with the idea of vocal imprinting, since the starling was a good mimic.

The long and short of it was that I raised starlings from the egg, and later took them from our country home in Maryland to the University of Chicago, where one of them lived to the age of seven years. The vocalization experiments were not a success. I reasoned that that starlings exposed to human speech played back on speeded up tape would easily be able to imitate the fast, high-pitched gibberish. Then, by recording the vocalization of the starling, we could later slow the tape and see whether there was indeed recognizable human speech. I remember using "Now is the time for all good men, etc." and although it seemed to be working to some degree, there was no way to objectively demonstrate this fact. A few years later I acquired a sonagraphic device that might have been helpful to tease out what was happening.

But the biggest upheaval in our household routine was undoubtedly caused by the imprinting of lambs. We imprinted a total of five over several years. I learned something from this effort but not enough to make me want to follow it through— at least not without a major farm or university agricultural facility. My wife and I shared the chores of midnight and four o'clock feedings and made motion-picture records of the behavior of the lambs to the human imprinters, especially during feeding time.

Blue jays and crows proved to be the most intelligent of the birds we hand raised. The crows, in particular, exhibited a curiosity and an ability to use tools that would have made a monkey out of many monkeys.

All of this material, along with studies on hoarding in the golden hamster, food preferences in *Drosophila,* and other projects carried out in either our country place in Maryland or the one on the dunes of Lake Michigan have certainly not seen the light of publication. Yet this work served a purpose. It was fun and it generated ideas I was to find useful in my more formal work on imprinting, both in our Maryland laboratory and at the University of Chicago.

Since I have mentioned something about my work with crows, I want to relate a personal experience that seems pertinent. It happened during an A.P.A. convention held in Chicago during the fifties. I had noticed that one of the downtown theaters had a variety program that included someone and his "pet crow." That I had to see. I determined the schedule and later arrived in the theater just as the performer was walking onstage with the crow on his wrist in the manner of a falcon. Since the performance was in progress I remained at the rear of the theater. That was my first mistake. The performer went through a series of actions, the crow placing a cigarette between the lips of the performer, making crowlike comments at one time and another, and so on. There was nothing that seemed impossible for a crow to do since I had had some experience with the ones I hand raised.

I decided that I had to know more about this particular animal and went backstage to see the performer, my second mistake. When he appeared, I introduced myself as a professor at the University of Chicago and a student of animal behavior who wanted to know more about the training and origin of the crow. He looked at me quizzically, and then, seeing that I was serious, launched into a story about a poor, orphaned crow, the victim of lightning striking his nesting tree and so on. To my request to see the crow he answered that the poor thing, exhausted from a rigorous performance schedule, was sleeping. I thanked him and left.

Later that evening I ran into a colleague with whom I shared an interest in close-up magic. We went to a hotel where one of the best magicians was performing and were duly entertained. The magician came to our table and we sat and chatted a bit. I mentioned that I had earlier been to the variety show and referred to the crow and his owner, assuming that since the magician was in show business he might know something more about the crow and his partner. "Oh," he said, "you mean the ventriloquist?" I was dumbfounded as it dawned on me that I had been beautifully suckered. The "crow" was actually a mechanical masterpiece, controlled by the fingers of the hand on which the ventriloquest's "dummy" rested. It was a chastening experience for an "expert" on animal behavior, and I learned something, not about crows, but rather about human behavior.

If the reader was wondering about the title of this chapter, the reason for it should become apparent now. There are two things that make it appropriate. First, I never have really had the opportunity to participate in the "expedition" type of behavior research that, as a youth, seemed a goal to be sought after. Except for a brief, two-week trip to Scammon's Lagoon in Baja, California to observe and study the gray whales, there have been no such adventures. Probably, for one reason or another, I may not have tried too hard to make them possible.

The other is the Canada goose story. For twenty-five years we have had a growing family population at my Lake Cove Laboratory. Beginning with a handful we now have close to twenty free-flying pairs of Canada goose nesting over the 250 acres of our marsh, along with another forty to fifty of their young. We have collected data on mating pairs for years and it is clear to me that we have no record of a pair ever forming between siblings born of the same pair. There are some ramifications, but there appears to be a good substantiation for an incest taboo. However, there's a rub. Many of the geese have lost their bands after some years. I was assured by the federal wildlife people, who banded more than half our goose population, that they were on the legs to stay—for years. We'll still try to pull together what we have, but there is always the nagging doubt that the two new pairs formed by young geese this past spring may indeed be brothers and sisters. Only one member of each pair had a leg band.

And of course there is always the final problem. I am less sure now of what imprinting is all about than I was thirty years ago when I first began. Certainly I am not helped to understand the process by perusing the many reports on laboratory approaches to imprinting that are published at a steady rate. Most of them are clearly not experiments that deal with the question of imprinting. A flashing light bulb does not really invite social attachment. There has been a degrading of the object of attachment and the experience allowed to the organism for that attachment to occur. Clearly there is no answer to the promising thoughts of yesterday when the hope was that here was the "tabula rasa" that would open the door to an understanding of learning, for here we had it in its simplest and most controllable form. It has been a wild-goose chase.

Some years ago I had lunch with my godson, who was then about six years old. Out of what seemed to be a clear, blue sky there suddenly came his question.

"How come you don't have to work, Dr. Hess?" he asked.

As I struggled to come up with some explanation that would account for the style in which we worked, as compared with those others he may have seen at more apparent jobs, he stopped my efforts. He answered his own question and I was sure he had that answer before he asked the question.

"Oh, that's right. You don't have to work because you're a 'fessor."

His grin told me to let it go at that, and, at that, there may be a grain of truth in his answer.

References

Burghardt, G. M., and Hess, E. H. 1968. Factors influencing the chemical release of prey attacks in newborn snakes. *J. Comp. Physiol. Psychol.* 66:289–95.

Hess, E. H. 1950. Development of the chick's responses to light and shade cues of depth. *J. Comp. Physiol. Psychol.* 43:112–22.

———. 1956. Space perception in the chick. *Sci. Amer.* 195 (10):71–80.

———. 1958. "Imprinting" in animals. *Sci. Amer.* 198 (3):81–90.

———. 1959. Imprinting. *Science* 130:133–41.

———. 1961. Shadows and depth perception. *Sci. Amer.* 204(3):138–48.

———. 1964. Imprinting in birds. *Science* 146:1128–39.

———. 1965. Attitude and pupil size. *Sci. Amer.* 212(4)(2): 46–54.

———. 1972. "Imprinting" in a natural laboratory. *Sci. Amer.* 227(2): 24–31.

———. 1973. *Imprinting: Early Experience and the Developmental Psychobiology of Attachment.* New York: Van Nostrand Reinhold.

———. 1975. *The Tell-Tale Eye: How Your Eyes Reveal Hidden Thoughts and Emotions.* New York: Van Nostrand Reinhold.

Hess, E. H., and Hess, D. B. 1969. Innate factors in imprinting. *Psychon. Sci.* 14:129–30.

Hess, E. H., Petrovich, S. B., and Goodwin, E. B. 1976. Induction of parental behavior in Japanese quail *(Coturnix coturnix japonica). J. Comp. Physiol. Psychol.* 90:244–51.

Klinghammer, E., and Hess, E. H. 1964. Imprinting in an altricial bird: the blond ring dove *(Streptopelia risoria). Science* 146:265–66.

Kovach, J. K., and Hess, E. H. 1963. Imprinting: effects of painful stimulation upon the following responses. *J. Comp. Physiol. Psychol.* 56:461–64.

Robert A. Hinde (b. 26 October 1923)

8

Ethology in Relation to Other Disciplines

Robert A. Hinde

Every contributor to this volume, looking round at the prospects available to graduate students today, must feel fortunate to have spent most of his active years in a period of expanding opportunities for scientists. But beyond that, in any era scientific careers depend on at least a moderate share of good fortune: I have had more than most.

To start near the beginning: of the four children in my family, I was the lucky lastborn. My father was a family doctor with some specialization in obstetrics and a mild interest in natural history. Most of my education was at Oundle, an English boarding school with an excellent tradition in natural history. I was fortunate enough to come under the influence of an ex-Indian Army major who had turned from shooting tigers to collecting butterflies when he was not teaching boys engineering; Ian Hepburn, a remarkable housemaster who taught me chemistry but who also and more importantly encouraged me in bird-watching and collecting beetles; Kenneth Fisher, a passionately ornithological headmaster who used often to take boys out bird-watching on Sundays; his son, James Fisher, a broadcaster and ornithologist; and Angus Fisher, James's wife, who came to teach in the first year of the war, continued to educate me by correspondence during the war, and became a close friend. So I was not short of figures with whom to identify.

In the war my elder brother died of wounds in an open boat in the Atlantic, and my parents had to suffer his loss, endure the years of blackout, rationing, and hard work, and see the fabric of the society they had loved gradually disintegrating. For my part, I was more fortunate. I trained as a pilot in Southern Rhodesia (now Zimbabwe) and as a navigator in South Africa, and flew flying boats (PBY Catalinas and Sunderlands) from a variety of bases from Scotland to the Maldive Islands. I was thus able to see a wide range of habitats from the northern islands through deserts to coral islands, and to feel that I was doing something useful without having to kill anyone.

I had failed in an attempt for a scholarship on leaving Oundle, but while I was in the RAFVR my bird-watching headmaster fixed me up with a close exhibition (i.e., a scholarship that could be awarded only to boys from certain schools) at St. John's College, Cambridge—such decadent nepotism is no longer possible. In 1945 I managed to persuade the Air Ministry that this was equivalent to a scholarship (which it wasn't) and entitled me to early demobilization. I thus got to Cambridge in January 1946 before the rush of ex-servicemen.

At Cambridge I read zoology, chemistry, and physiology for two years and then specialized in zoology. I owe special debts to Roland Winfield, an ex-R.A.F. doctor/pilot who taught me physiology superbly but was too colorful a character to be tolerated by Cambridge for long; Hugh Cott, whose studies of camouflage and of the Nile crocodile should have brought him more recognition than he received, and especially to William H. Thorpe. A special stroke of luck brought me into contact with the latter. I used to spend much of my time at the Cambridge sewage farm watching migrant birds, and was fortunate to locate nearby a breeding pair of moustached warblers *(Lusciniola melanopogon)* and their newly fledged young. Since this species had been recorded only twice in Britain before (and those records were probably fakes) this gave me a certain status as an ornithologist. I suppose that that was why David Lack offered me a job as research assistant/graduate student at Oxford when I graduated. He really wanted me to study the comparative feeding ecology of rooks and jackdaws, but generously allowed me to follow my own predilections and do a purely behavioral study of tits *(Parus* spp.). David Lack was no longer especially interested in behavior, but he taught me much about science, gave me a background interest in behavioral ecology, and made it difficult for me ever to think in other than individual selection terms—the sociobiological view that all ethologists were then group selectionists is nonsense.

During my second year with David Lack I had a special stroke of luck—the arrival of Niko Tinbergen in Oxford. For a while he had no graduate students of his own, and I was profoundly influenced by the times we spent together.

My first papers, based on undergraduate work, were on wader migration through the Cambridge sewage farm, and (with the encouragement of and in collaboration with James Fisher) the opening of milk bottles by birds. My doctoral thesis was a straight field behavior study with ecological overtones. I modeled it on David Lack's own study of the robin, Niko Tinbergen's study of the snow bunting, and Mrs. Nice's of the song sparrow. At that time, a fairly general field study was very good training for a career in behavior. It was intensely exciting; in any field study of individually known animals—I colored-ringed the birds—one becomes caught up in the vicissitudes of their lives. The territorial boundaries, the breeding success, the winter ranges of Red/Red Green, Red/Black and White and other favorites kept me fascinated for my two years in Oxford. More importantly, the field study provided me with problems to work on in the years ahead, and the study area was a most beautiful wood—carpeted with primroses in the spring, sumptuous with blackberries later in the year, and with countless lessons for me. Dawn and dusk I used to clock marsh tits, blue tits, great tits, nuthatches and great spotted woodpeckers in and out of their roosting holes—outside the subject of my thesis, but exemplifying important ecological principles concerned with body size and the use of daylight. David Lack imbued my behavioral observations with an ecological slant—and I profited too from the proximity of Charles Elton, Dennis Chitty, and Mick Southern at the Bureau of Animal Populations, next door to the Edward Grey Institute. But the most important influence at that time was Niko Tinbergen, who taught me how to analyze behavior—and it was his lessons especially that I carried with me to Madingley.

While I was involved in my D. Phil. research at Oxford, Bill Thorpe was trying to start an Ornithological Field Station at Madingley, near Cambridge. Fortunately for me, his two first choices for the post of curator—Konrad Lorenz and R. E. Moreau—both became established elsewhere, and I slipped into the niche. Thorpe was a marvelous mentor. It was only some years later, when Peter Marler came to Madingley, that I realized that Thorpe had wanted me to work on bird song. So far as I was concerned, he let me go my own way, and brought no pressures on me. Over the years he fought all my battles for me, so that I only once made a job application (fortunately for me unsuccessful) and have always been provided with the research facilities that I needed.

I started at Madingley in 1950. Bill Thorpe had begun the station with one laboratory assistant a few months earlier, but we had nothing more than a field surrounded by a wire-netting fence. Much of the first summer and autumn was spent in building cages, creosoting, and the like. As facilities became available, I started behavior studies of the same species that Bill Thorpe was beginning to use to study song learning—the chaffinch *(Fringilla coelebs)*. Though difficult to breed, it proved an excellent aviary bird for many purposes. One aim was to analyze threat and courtship in terms of conflicting tendencies—a problem I had earlier been interested in with tits. This, together with comparative studies of other finch species, kept me busy for much of several summers. During the winters, I studied another aspect of chaffinch behavior—the mobbing response to predators. Finding that this could be quantified readily, and that it waned with time, I started to study it in detail, with the hazy idea that I should be able to specify the detailed properties of the Lorenzian reservoir. As the work revealed complexity after complexity I became disillusioned with the reservoir model, and subsequently, on this and other grounds, came to reject energy models of motivation altogether.

Our first graduate student at Madingley was Peter Marler. He already had a Ph.D. in botany and a secure job, but felt impelled to follow his interests. While I was studying chaffinch courtship in aviaries, he carried out a field study of the same species. We had long and sometimes acrimonious disputes over such important issues as whether "lop-sided wings-drooped posture" was a suitable name for one of the chaffinch postures, and I profited much from his functional approach. He was followed by Richard Andrew, Pat Bateson, and many other students from whom I learned a great deal.

Soon after I arrived, the Cambridge professor of zoology ceased to be kindly disposed toward me; unfortunately, my job had no tenure. However, St. John's College was generous in electing me first to a research fellowship and later, while I still lacked a proper university post, making me College Steward—a task for which I, almost lacking in taste buds and a sense of smell, was extraordinarily ill-fitted. Later I served as a College tutor—"in loco parentis"—to undergraduates, which I found much more interesting.

In any case I did not worry too much about security, because I trusted Bill Thorpe to look after my interests and because in any case the early fifties were such a marvellous time to be an ethologist. The science was blossoming, and practically everything one touched was new. Ethological meetings were filled with intense discussion, and I was especially influenced by some Dutch colleagues. I remember particularly Jan van Iersel, sitting impassive through discussions and then quietly coming out with "There are three points I would like to make. . . ."— and one knew that each one would be spot on. I still have mementos of some of those meetings, proofs of one of Niko's articles with hastily sketched hierarchical schemes on the back.

Around that time, Bill Thorpe suggested that I should join him in a study of imprinting. It was not an especially good study, though it did demonstrate the lack of importance of tangible rewards, that imprinting was not irreversible, and that the sensitive period was not absolute. But it was important for me in initiating an interest in parent-offspring relationships that has influenced nearly all my work since. Here I had another stroke of luck. John Bowlby, a London psychoanalyst, was in the process of developing new views about child development. He ran a weekly seminar, and wanted an ethologist to take part. He invited Niko Tinbergen, who was too busy, and so on the basis of my experience in parent-offspring relations in birds, I went instead.

The seminar was an important experience. Bowlby, with an eclecticism truly remarkable in a psychoanalyst, gathered together a group that included a Hullian learning theorist, a Skinnerian learning theorist, a Freudian analyst, a Kleinian analyst, someone trained by Piaget, myself as an ethologist, several psychiatric social workers, and sometimes an anti-psychiatrist. All we had in common was an interest in parent-child relationships. I believe that such eclecticism is essential if the behavioral sciences are to make substantial progress. Subsequently Bowlby incorporated a substantial ethological element into his theorizing and also helped me to set up a rhesus monkey colony at Madingley with the specific purpose of using monkeys as models to assess the long-term effects of mother-child separations.

Another problem I tackled in the late fifties also stemmed in part from Bill Thorpe's work. In his *Learning and Instinct in Animals,* he discussed the extent to which nest building can be seen as a goal-directed activity. I determined to try to test some of his ideas experimentally. I made little progress toward that aim because the work became diverted into a study of the manner in which the different aspects of reproductive behavior are integrated, and involved sketching a picture of the interactions between behavior, external stimuli, and internal hormonal state. That canary work was started at a time when the U.S. Air Force was supporting a wide range of research, and I had no guilt about putting in an application for the study of nest-building "skills" (skill is what you fly a bomber with) and saying I had been a pilot to show that I knew its relevance. A few years later we were entering the space age and a second application with an emphasis on "sensory deprivation" (because we sometimes did not let the canaries manipulate nest material) was also successful. Much of this work was carried out in collaboration with Elizabeth Steel, a very conscientious colleague with whom I worked for many years. Perhaps only in retrospect did I realize how much I enjoyed doing defined experiments, at least some of which brought answers, in comparison with the more messy (but I believe more important) problems I am trying to tackle now.

The canary work cemented a close friendship with Danny Lehrman, who would surely have been a contributor to this volume had he lived. I first met in him in the early fifties, just before he wrote his famous "Critique of Konrad Lorenz's theory of instinctive behavior." Danny was already engaged in his ring dove studies and we used to meet once or twice a year, on one side of the Atlantic or the other, and would exchange data and ideas. We worked on related problems but were never in competition, and he became my closest friend. My relationship with him, and that with Jay Rosenblatt, which started a year or two later, not only provided me with marvelous opportunities for discussing data, but extended far outside the academic sides of our lives.

Another important influence in the fifties was Frank Beach. I first met him at a meeting at Harvard in the early fifties, organized, I believe, by Bill Verplanck. A few years later he invited me to a meeting designed to bring together ethologists and comparative psychologists held at the Behavioral Sciences Center in Palo Alto. It was a small group that included Lehrman, Rosenblatt, Harlow, Washburn, Baerends, Hebb, Vowles, van Iersel, Tinbergen, for a brief while Lashley, and others. We had several weeks in which to talk, with no publication requirement hanging over us, and Frank ran the meeting on the principle of "If you do not finish what you have to say today, there is always tomorrow," and discussion, anyway, spilled over to the darts board and Dinah's Shack. A few years later he organized a conference on "Sex and Behavior" in Berkeley that met for a second week after a year's interval. On the second occasion we were allowed to take our best pupil, and I took Pat Bateson—now a close colleague and friend—and I know the experience was important for both of us.

All the way through I was fortunate to be in on the early expansion of ethology. I learned much from the conferences I was able to attend, and made friends on both sides of the Atlantic. I was especially affected by an invitation to contribute a chapter on Ethology to Koch's *Psychology: A Study of a Science*—I suspect, another of Niko Tinbergen's rejects that I picked up. I was really much too unsophisticated to undertake such a task, but the terms of the request proved to be an important challenge. We were asked to write under such headings as "What are the independent variables in your system?"—suitable for the heyday of learning theories, but quite unsuitable for ethology. At that time I did not dare to reject the questions as unsuitable; the attempt to answer them made me really think about the nature of ethology, and to realize that its importance came not only from its status as a discipline in its own right, but from its contribution to other disciplines. A year or two later I started to write a book on animal behavior (initially in collaboration with Niko Tinbergen, but it later turned out that he had too much else to do). This became an attempt to link ethology with comparative psychology and physiology (1966). Subsequently I came to believe that some of ethology's most important achievements lay in its links with other disciplines. In collaboration with my wife, whose Ph.D. was in operant psychology, I became interested in the manner in which learning is guided by constraints or predispositions characteristic of the species or individual (1973). More recently I have emphasized the links between ethology and a wide range of biological and social sciences (1982).

The monkey colony at Madingley was financed, with the help of John Bowlby, by the Medical Research Council and the Mental Health Research Fund. Except for the actual bricklaying, a laboratory assistant and I built it ourselves. Digging the foundations, cementing, wiring, and making a road took the best part of a year. It was my good fortune that Thelma Rowell finished her Ph.D. just at the right time to help with the establishment of the animals, and that when she had to leave to go to Uganda her post could be filled by the late Yvette Spencer-Booth. Both of them were more perceptive about monkeys than I, and the work might have advanced faster if I had followed all their advice. Although monkeys were being widely used in psychological laboratory small-cage situations, there were few precedents for the work we wanted to do, and we had to learn how to keep monkeys, and how to record their behavior, the hard way.

In the early sixties, Louis Leakey was starting to sponsor research on the great apes. On the basis of my somewhat exiguous experience with captive rhesus, he asked me if I would supervise Jane Goodall's research on chimpanzees. This in time took me to the Gombe Stream, in Tanzania, and gave me the opportunity to supervise a number of students working in Africa, and later on Cayo Santiago, on monkeys, apes, and even elephants. Since the days when I was working on my doctorate, I had done no further fieldwork, and this opportunity to enjoy fieldwork at second hand has, I believe, helped me to become less narrow than I might otherwise have been. The students included another protegé of Louis Leakey's, Dian Fossey, now well-known for her studies of the mountain gorilla. Although I made only three visits to the camps of Jane and Dian, these provided me with breathtaking experiences—the orphaned chimpanzee Flint sitting by his mother's body; the incredible excitement of having a two-hundred-fifty-pound gorilla burst out of the foliage, lie down beside me, and play with my boot laces.

During the fifties and early sixties, Bill Thorpe carried all the high-level administration of Madingley, while I looked after its day-to-day affairs. I was thus practically never worried by policy decisions. In 1963 Bill proposed me for a Royal Society Research Professorship—a really marvelous appointment since it provided the freedom to do the research I wanted to do with practically no strings attached. When Bill retired, the Medical Research Council changed the grant support they had been giving the monkey work into a Unit, thereby giving it more continuity. This enabled me to go on working at Madingley without having to take over the administration of the laboratory. After a few years, Patrick Bateson succeeded to Bill Thorpe's job as director of the Madingley laboratory, now elevated from its original status as an "Ornithological Field Station" to a Sub-department of Animal Behaviour. This has been an excellent arrangement for me—Pat Bateson and I have worked very well together on administrative issues, and I have profited greatly from his developmental studies.

In the early seventies I gradually changed the focus of my work. There were a variety of reasons for this. I found it increasingly difficult to keep up with two literatures—one, relevant to the canary work, becoming increasingly physiological, and the other, relevant to the monkeys, psychological and psychiatric. The canary work had come to focus on factors influencing the behavioral effectiveness of steroid hormones; my colleague John Hutchison arrived at the same problem by a different route, and has subsequently pursued it with great success. Research on the determinants of the response of infant rhesus to a separation experience received a setback as a result of the devastatingly untimely death of Yvette Spencer-Booth, but Lynda McGinnis came to my aid, helping to run the colony and completing a series of experiments on the influence of the nature of the separation experience on the recovery of the infant.

One clear finding of that work was that the sequelae of separations varied with the nature of the mother-infant relationship beforehand. We were able to get correlations of 0.4, 0.5, even 0.6 between measures of the mother-infant relationship and the infant's distress after reunion—significant, but not so high as one would wish. I believed that we might get higher correlations if we could measure the mother-infant relationship better. Being a naive biologist, I supposed that guidance would readily be available in the literature of the human social sciences. I spent several years reading in social psychology, psychiatry, anthropology—and was frankly disappointed in what I found. There seemed to be a curious gap in the social sciences between the study of brief interactions and the study of groups— disappointingly little about the study of long-term dyadic relationships. As a biologist, I wondered whether one reason for this lack of a science of interpersonal relationships could be the lack of an adequate descriptive base, and I spent some time trying to see what form it would take (Hinde, 1979).

Around the same time, active research on interpersonal relationships increased at several other centers, and I found an atmosphere akin to that in ethology in the early fifties—an open field, no professional rivalry and a sense of common enterprise. I was surprised at the welcome an outsider received from such research workers as Harold Kelley, Anne Peplau, and George Levinger—they showed me a warmth that I felt was related to their intense interest in what they are doing.

I am now attempting to fulfill a long-standing ambition by concentrating my research on social behavior and relationships, and in particular those of three- to six-year-old children. I see this area as important not only for its practical implications—and after all, interpersonal relationships are the most important thing in the world for most of us—but for a special theoretical reason. The properties of every relationship depend on the personalities of the participants. At the same time, however, the behavior that an individual shows in the context of a particular relationship is influenced by his or her view of, and expectations for, that relationship. And in the longer term, the behavior he or she could show is profoundly influenced by relationships experienced in the past. The effects are of course most profound in the relationships of early childhood, but continue throughout life.

An understanding of relationships thus requires us to come to terms with a dialectic between personality and relationships. This view is of course intrinsic to much work on child development. It is also entirely in keeping with those of the symbolic interactionists, who see individuals as selecting from their repertoires of possible behaviors depending on whom they are with, and the interdependence and exchange theorists, who see each relationship as a product of its own past and the participants' expectations for its future. It is also entirely in keeping with current trends in personality theory. Earlier attempts to account for the consistencies we think we perceive in our own and others' behavior in terms of "traits" came under fire because of the limited cross-situational consistency of the traits. For a while, there was an excessive emphasis on the situational determinants of behavior. Now a more balanced view obtains, with the recognition that some people are consistent some of the time about some things, and the most important situational determinants of behavior of course involve interpersonal relationships.

Thus child development, the study of interpersonal relationships, and the study of personality have reached a common interest in the study of the dialectic between personality and relationships. This seems to me to offer enormous promise as an integrating glue that will help to bind together different subdisciplines within psychology, subdisciplines that at present are in danger of losing touch with each other.

I see no major discontinuity between this work and my work in animal behavior. My interest in interpersonal relationships stems directly from the finding that the consequences of separating a young monkey from its mother varied with the nature of the mother-infant relationship beforehand. Some of the dimensions I believe to be useful for describing relationships have come directly from animal work. My wife has used a rating technique derived in part from human personality inventories with some success for assessing the individual characteristics of rhesus monkeys, and her work influenced our current study of the relations between children's behavior and relationships at home and their behavior and relationships in preschool.

Nevertheless, although by origin an ethologist, I have some reservations about some human ethology. Too often it involves merely the application of techniques worked out with animals to our own species, where they can give only partial answers. Ethologists should not scorn the techniques already in use for the study of their own species on the grounds of "subjectivity" while in ignorance of the trouble taken to establish their reliability and validity. It is necessary to recognize that what an individual wants to do may be as important for some issues as what he actually does. Cultural norms, moral values, locus of control, and other issues important for understanding human behavior cannot be studied by purely ethological techniques. In our own work we find it profitable to use a variety of techniques—observation, interview, questionnaire, test—that partially validate and complement each other.

While I have given a logical account of my transition from animals to human subjects, I could also have written about it at other levels. Superficially, children of preschool age are delightful subjects to work with, much more fun than rhesus monkeys. At a deeper level, and probably closer to the truth, although I became an agnostic during the war, my evangelical upbringing left me with a need to apply my work to the human situation, and I find fulfillment in doing so. I miss sometimes the relatively neat experiments that were possible with canaries. I miss the illusion that one could sometimes have that one was getting closure on a problem, however trivial. But while intellectual satisfaction may be an adequate reward for research, in choosing between problems one criterion must surely be its relevance to human health and happiness.

This does not imply a belief that the study of animal behavior is becoming exhausted. Ethology has an excitement, interest, and value in its own right—an issue movingly discussed by Danny Lehrman. Ethology, and its subdiscipline of sociobiology, has much to contribute at an intellectual level to our understanding of our own nature. For example, functional considerations can provide a way to integrate otherwise diverse facts about human behavior—the nature of the constraints and predispositions that govern how we learn, or the nature of some sex differences in behavior. And, at a more practical level, ethological studies of animals still have much to contribute to our understanding of human behavior. This lies not so much in direct comparison with human behavior as with the elaboration of principles whose applicability to the human case can subsequently be assessed.

An obvious but neglected example is the "behavioral system." In the early days of ethology, Baerends was concerned with the relations between the different patterns of behavior in an organism's repertoire—with the structure of behavior. Unfortunately Tinbergen's hierarchical model of the organization of behavior became unpopular because of its association with energy models, but the concept has been kept alive through the painstaking work of Baerends and his colleagues. Bowlby's application of ethological concepts to the mother-child relationship clearly demonstrates the urgent need for further elaboration of this concept.

I have in fact continued to supervise the work of students working on the social behavior of monkeys (especially on Cayo Santiago) and other mammals. I believe that these studies are helping to formulate principles with rather wide applicability to the study of social behavior, and that the *relative* simplicity of nonhuman primates facilitates the distinctions between levels—behavior, interactions, relationship, social structure—and helps to specify the properties emergent at each level. I believe that an understanding of these distinctions, and of the relations between these concepts, are essential if social behavior is to be understood (Hinde, 1983).

Finally, while in recent years I have come to find more satisfaction in the study of humans than in that of animals, there is one problem that I have come to recognize as of overriding urgency. This is the absolute necessity to stop the fear-engendered power hunger of the United States and Soviet governments, their senseless stockpiling of nuclear weapons, and the irreversible contamination of the world consequent upon this and other follies. The governments of the United States, having led all the way in the arms race, must bear a large share of the blame. And the study of interpersonal relationships is not irrelevant to this issue. Just as the course of a relationship between two individuals is determined in part by the perceptions each has of the other's intentions, so is the arms race fueled by perceptions and misperceptions by each side of the other. Each new weapon, perhaps acquired with a genuinely defensive intent, is perceived by the other side as a further offensive threat, demanding countermeasures. That, however, is another story.

References

Hinde, R. A. 1952. The behaviour of the Great Tit *(Parus major)* and some other related species. *Behaviour* (Suppl. no. 2:1–201).
———. 1955–56. A comparative study of the courtship of certain finches. *Ibis* 97:706–45 and 98:1–23.

———. 1956a. The biological significance of the territories of birds. *Ibis* 98:340–69.

———. 1956b. Ethological models and the concept of "drive." *Brit. J. Philos. Sci.* 6:199–213.

———. 1959. Behavior and speciation in birds and lower vertebrates. *Biol. Rev.* 34:35–128.

———. 1960a. Energy models of motivation. *Symp. Soc. Exp. Biol.* 14:199–213.

———. 1960b. Factors governing the changes in strength of a partially inborn response, as shown by the mobbing behaviour of the chaffinch *(Fringilla coelebs)*. III. The interaction of short-term and long-term incremental and decremental effects. *Proc. Roy. Soc. B.* 153:398–420.

———. 1965. Interaction of internal and external factors in integration of canary reproduction. In *Sex and Behavior,* ed. F. Beach. New York: Wiley.

———. 1966–70. *Animal Behaviour: A Synthesis of Ethology and Comparative Psychology.* 1st and 2nd editions. New York: McGraw-Hill.

———. 1968. Dichotomies in the study of development. In *Genetic and Environmental Influences on Behaviour,* ed. R. E. Thoday and A. S. Parkes. Edinburgh: Oliver and Boyd.

———. 1970. Behavioral habituation. In *Short-term Changes in Neural Activity and Behavior,* ed. G. Horn and R. A. Hinde. Cambridge: Cambridge Univ. Press.

———. 1974. *Biological Bases of Human Social Behaviour.* New York: McGraw-Hill.

———. 1976. Interactions, relationships and social structure. *Man* 11:1–17.

———. 1977. Mother-infant separation and the nature of inter-individual relationships: experiments with rhesus monkeys. *Proc. Roy. Soc. B.* 196:29–50.

———. 1978. The influence of day length and male vocalizations on the estrogen-dependent behavior of female canaries and budgerigars, with discussion of data from other species. In *Advances in the Study of Behavior,* vol. 8, ed. J. S. Rosenblatt, R. A. Hinde, C. Beer, and M-C. Busnel, pp. 39–73. New York: Academic Press.

———. 1979. *Towards Understanding Relationships.* New York: Academic Press.

———. 1981. Animal signals: ethological and games-theory approaches are not incompatible. *Animal Behaviour* 29:535–42.

Hinde, R. A., ed. 1969. *Bird Vocalizations.* Cambridge: Cambridge Univ. Press.

———. 1972. *Non-Verbal Communication.* Cambridge: Cambridge Univ. Press.

———. 1982. *Ethology.* Glasgoe: Fontana Paperbacks.

———. 1983. *Primate Social Relationships: An Integrated Approach.* Blackwell: Oxford.

Hinde, R. A., and Bateson, P. P. G., eds. 1976. *Growing Points in Ethology.* Cambridge: Cambridge Univ. Press.

Hinde, R. A., and Davies, L. 1972. Removing infant rhesus from mother for 13 days compared with removing mother from infant. *J. Child Psychol. Psychiat.* 13:227–37.

Hinde, R. A., and Horn, G., eds. 1970. *Short-term Changes in Neural Activity and Behaviour.* Cambridge: Cambridge Univ. Press.

Hinde, R. A., and Roper, R. 1979. A teacher's questionnaire for individual differences in social behaviour. *J. Child Psychology and Psychiatry* 20:287–98.

Hinde, R. A., and Simpson, M. J. A. 1975. Qualities of mother-infant relationships in monkeys. In *Parent-Infant Interaction* ed. R. Porter and M. O'Connor (Ciba Foundation, 1974) Amsterdam: Associated Scientific Publishers.

Hinde, R. A., and Spencer-Booth, Y. 1971. Effects of brief separation from mother on rhesus monkeys. *Science* 173:111–18.

Hinde, R. A., and Stevenson-Hinde, J. 1976. Towards understanding relationships: dynamic stability. In *Growing Points in Ethology,* ed. P. P. G. Bateson and R. A. Hinde, pp. 451–79. Cambridge: Cambridge Univ. Press.

Hinde, R. A., and Stevenson-Hinde, J., eds. 1973. *Constraints on Learning.* London: Academic Press.

Hinde, R. A., Thorpe, W. H., and Vince, M. A. 1957. The following response of young coots and moorhens. *Behaviour* 9:214–42.

John A. King (b. 22 June 1921)

9

Those Critical Periods of Social Reinforcement

John A. King

Elkhorn, an extruded granite peak protruding from the ponderosa pines of the Black Hills, was the July afternoon destination of R. "Mike" Elliott, an older friend and adviser, and myself, as we scrambled over the boulders towards its base. We rested at intervals and talked of many things: birds, science, love, music, careers. On one shaded boulder, Mike drew an analogy from Darwin's descriptions on the origins of oceanic islands in order to contrast science with art.

Art is like a volcanic island that emerges from the seabed with the explosive force of creative genius. Science is like a coral atoll gradually rising above the sea's surface by the accretions of multitudinous, small organisms. To become an artist, an individual breaks through the surface by his own singular achievements. A scientist reaches sea level and above by the accumultive efforts of his fellows.

Although the analogy is not perfect, it was good advice to give to a youth torn between art and science. The analogy has also served to remind me of the sociality of science. Ideas, methods, goals, rewards, and frustrations derive from our epigenetic development, our culture, as well as from those scientists who have preceeded and are contemporary with us. A scientific autobiography could be fashioned from those social sources if they could be identified and pieced together with some accuracy.

I will attempt to trace the social sources of three areas of study that have given me satisfaction: (1) the descriptive study of prairie dog sociality that I did for my Ph.D. dissertation; (2) the experimental evidence I found for the interaction between genetics and early experience in deer mice and the concept that differential rates of development could be the mechanism for the interaction; (3) the recent study of deer mouse dispersal, which suggested to me that a search for mates was one reason for dispersal.

One of several social sources for these contributions is ontogeny, which must be omitted because early experiences are too complex and too remote to isolate and identify at this late age, except for one childhood realization: I like animals. Somewhere in the ontogeny of most people, there seems to be a switch that turns some, as children, onto animals and others away from animals. Rather than an on-off switch, the early experiences with animals resemble a throttle that is at rest in some children and pushed ever onward in others until their entire lives involve animals in some respect. My throttle is full speed ahead.

Like the epigenetic sources for these three contributions, the cultural sources are diffuse and indirect. Society rewards my scientific efforts by providing me a livelihood and a social environment in which science is accepted, rewarded, and encouraged. The most identifiable sources of social stimulation for my scientific efforts derive from other scientists, colleagues, professors, students, editors, and scientific authors. Thus, this autobiography will be largely a recognition of fellow scientists who have contributed ideas, criticisms, knowledge, skills, attitudes, stimulation, and, most of all, social reinforcement.

Living with Prairie Dogs

My Ph.D. dissertation on the social behavior of prairie dogs is probably my most often-cited study (King, 1955). The dissertation launched my professional career with vigor; it was internationally recognized; it rapidly established my reputation; it has been called a classic study; and it continues to bring me social reinforcement. In 1982 I was invited to present a thirty-year perspective of ground squirrel research, which my dissertation helped to initiate. It is probably the primary single reason for writing this autobiography. A doctoral dissertation is a significant document in establishing a research career. The confusion, anxiety, and procastination students often exhibit in choosing a dissertation topic is warranted by the perseverance of the dissertations' influence on the student's career.

My thesis was essentially that prairie dogs regulated their population density through their social organization and behavior. The thesis was supported by a description of their physical, biotic, and social environment. These large ground squirrels live in dense colonies on the open grasslands of the western prairie states. They interact with each other frequently. Members of small reproductive units, coteries, interact amiably with each other, but with hostility towards members of other coteries. Their alarm calls alert each other to predators and their burrows serve as communal refuges within a coterie. I postulated that they adaptively altered the vegetation by clipping it short, which permitted the visual detection of predators, by exposing the subsoils to the invasion of nutritious weed species, and by increasing the diversity of plant species.

The validity of these postulates has not been seriously challenged despite recent attention given to such adaptive strategies of their elaborate social behavior as, alarm calls, nepotism, and inbreeding avoidance. Even John Hoogland, who recently has made careful, long-term studies (1981) recognizes the complexities and perplexities of their social behavior. In a letter dated August 27, 1981, he says, "Whew, are prairie dogs ever complicated!"

The social stimuli leading to my investigation of the prairie dogs were diverse and intricate, like most other research undertakings. The most important single stimulus was Lee R. Dice, my major advisor. He was about sixty years old when I joined his group of ecologists and geneticists at the Laboratory of Vertebrate Biology. He had been curator of mammals in the Museum at the University of

Michigan, but for some unknown reason, he relinquished his curatorial responsibilities and established an experimental approach to evolution and systematics. In this age of the "New Systematics," so heralded by Julian Huxley, Dice was an active contributor with his studies on the genetics and ecology of deer mice *(Peromyscus)*. His days of collecting mammals from all over North America had ended and most of his time was spent writing in his office. His interest in the natural spacing of organisms took him occasionally into the field to record the distribution of various plant species. He also worked at intervals on electrical recording apparatus for mouse behavior. He kept regular hours and was always available for consultation, but he rarely invited it. In the late 1940s, Dice initiated a program of clinical consultations and research in human genetics with James Neel and Charles Cotterman. He also brought the anthropologist, James Spuhler, to examine the possibility of assortative mating in the local human population.

Dice, with his snow-white hair and carefully trimmed goatee, was a warm, soft-spoken, and reserved man. He was an originator and organizer with a subtle power that must have resided in his accomplishments. He was a political person without outward appearances of involvement, except in his ability to ally young, progressive scientists like Frances Evans, James Neel, and Clem Markert. At intervals, a deep-voiced visitor entered his office down the hall from mine. The visitor would often raise his voice and even pound the desk in a belligerent attempt to subdue Dice to his point of view. One never heard Dice during these encounters, but subsequent events revealed that the soft-spoken, white-haired gentleman had prevailed. The intensity of his feelings was obscured by his mild manners, was but vented in chronic stomach troubles that required a diet prepared carefully by his wife, Dora, of fruits and vegetables they raised in their garden on Day Street, Ann Arbor.

Dice taught a prosaic course on ecology that was the testing ground for his textbook *Community Ecology* (1952). He also conducted a seminar that was regularly attended by his graduate students. During one seminar, I reported on the mutation rate of a colonial species of primrose *(Oenthothera organesis)* (Lewis, 1948). The reason plants were discussed in a zoology seminar was their application to random genetic drift, which was then being championed by Sewell Wright. I had previously encountered Wright's ideas while aboard a Liberty ship on the way from Buenos Aires to Rotterdam. After my military service, I erased the memory of barracks' life by spending a winter alone in the Black Hills and later by a trip across Central and South America that ended in Buenos Aires. I returned to the United States by a circuitous route aboard a cargo ship. One book I purchased for the three-month voyage was *Evolution: A Modern Synthesis* by Julian Huxley, who embraced Sewell Wright's ideas of genetic drift with the same enthusiasm that W. D. Hamilton's ideas of inclusive fitness are accepted by current sociobiologists. If genetic drift occurs, most likely in relatively small breeding units isolated from each other except for rare dispersers, then a colonial rodent should be the best place to find evidence of it among mammals. Dice concurred that mammalian evidence was needed and he encouraged me to find a species.

If Dice was my academic father, William H. Burt, who bestowed many academic gifts upon me, was my academic godfather. When I was a freshman at

the University of Michigan, he offered me the opportunity to test his identification
key to the species of Michigan mammals at a museum desk. The following summer he arranged for me to accompany his recently hired assistant curator of
mammals, Emmet T. Hooper, on a collecting trip in northern Michigan. Again
after my discharge from the Army Air Force, he invited me to join a collecting
expedition with himself, Hooper, Norman Hartweg, Theodore Hubbell, and Hel-
mut Wagner to the vicinity of the recently erupted volcano Paricutin in Mexico.
Still many years later his skilled editorial eye provided the polish for the publication of the book I arranged on the *Biology of Peromyscus.* Burt's academic
munificence benefited everyone interested in mammals, from the wealthy stock-
broker William Harris to the naive undergraduate. Burt was about fifteen years
younger than Dice and working on his field guide when I asked for advice about a
colonial mammal to study for possible genetic drift. I see him now at his desk with
his reference cards at his finger tips smiling and saying, "Why not pitch a tent in
the middle of a prairie dog town and watch what goes? No one has ever really tried
living with a wild animal in their native habitat. That is the proper way to become
acquainted."

Dice accepted the prairie dog study with his customary subdued enthusiasm.
He suggested that we pursue this possibility at the next American Society of
Mammalogists meeting in Toronto. A former student, Victor Cahalane, who was
then chief biologist for the National Park Service would be attending and he would
know of a suitable prairie dog town. Calahane was responsive to the idea of the
study and recommend a hidden prairie dog colony in Shirttail Canyon, Wind Cave
National Park on the southern edge of the Black Hills. Perfect! I knew the Black
Hills from my hikes during the war years with my psychologist friend from Minnesota, Mike Elliot. Mike had advised me to spend a winter alone in the Black
Hills caring for a herd of Troy Parker's horses and trapping small mammals for the
university museum as a relief from army discipline. Now I would return to those
beautiful hills with a definite purpose.

In the summer of 1948, I lived among the prairie dogs as Burt had advised and
became enthralled by these delightful squirrels, whose social life was conspicuous. The small, semiisolated colonies that might have been suitable for studies of
genetic drift did not exist. Instead, the huge colony was organized into social
groups somewhat larger than a monogamous family. The genetics of this type of
breeding system might be interesting, but the social behavior was captivating.
When I returned to school and described the elaborate social dynamics of the
prairie dogs to Dice, he agreed that my future efforts should be devoted to behavior rather than genetics. Academic advisers are not always as flexible as Dice was,
but he quickly recognized the relationships between social organization and population genetics. The study of genetics could be approached from two directions:
how genes act upon the phenotype and how the phenotype, particularly behavior,
influence changes in gene frequencies.

Often the shape of one's science is molded by the family, particularly the
spouse. I can almost categorically state that the relationship with one's spouse has
a greater effect upon science than all the other social stimuli. This maritial influence is not unique to scientists, but probably affects any creative endeavor.

My second trip to study prairie dogs was a honeymoon. Joan McGinty and I were married before we departed for the Black Hills in a homemade house trailer. Joan was a freshman when we met on a bus returning to school from our homes in suburban Detroit. Her father was an endocrinologist with Parke-Davis, so she already knew about living with scientists. Besides, her keen, analytical mind intuitively made her a scientist, though her interests were humanistically oriented. She graduated with a bachelor of arts degree in sociology at the same time I received my Ph.D. in zoology, 1951. She later earned a master of social work degree and was a medical social worker at a Lansing hospital for eleven years. During our summers with the prairie dogs, we observed them, discussed them, raised some pets, and interpreted their behavior together. We have subsequently pursued our career activities independently, but we have always maintained a common interest in the little animals about us, including our two children.

With the postwar baby boom, society became concerned about the regulation of a geometrically expanding population. The wartime activities had reshuffled the population and initiated the shift from rural agriculture to urban industry and service. At the same time, research on rodent populations shifted from pest control to the quest for possible models for regulating human populations. The most undisguised promoter of rodent models for human populations was John B. Calhoun. He had joined David E. Davis and John T. Emlen in a study, done for public-health purposes, of the Baltimore rat populations.

It was found that the rats' social behavior severely restricted their movements. Calhoun further pursued the study of social behavior by observing enclosed populations of rats. He vividly described the emergence of "social pathology" as a result of the crowded conditions of increasing rat populations. The message was that either populations regulate themselves or social disorganization and behavioral dysfunction occur. Calhoun visited me at the prairie dog town and encouraged my social and population study of these free-ranging and readily observable rodents. The zeitgeist was population regulation through social mechanisms, while the genetic mechanisms were ignored.

Genetic mechanisms cannot be ignored for long in any biological system. They must be compatible with all other systems, including population regulation. The problem with my thesis was that there was a basic incompatibility between the ideas of cooperative self-regulation of the prairie dog population and of the individualistic struggle to be most fit by producing as many offspring as possible. This conceptual contradiction between altruism and selfishness was shared by many other biologists. It was not clearly exposed until V. C. Wynne-Edwards (1962) proposed group selection: with those populations regulating their numbers within the carrying capacity of the environment having a selective advantage over those populations that failed to regulate. Many of us had been thinking in terms of group selection without recognizing that it was antithetical to individual selection. Any individual that possessed the genetic capacity to relinquish its reproduction in favor of limiting population size would not leave descendants to maintain that genetic capacity. Conversely, any individual that cheated by reproducing beyond the curtailed reproduction of other individuals would have a good chance to increase the number of cheaters in the next generation. After W. D. Hamilton

(1964) provided the sociobiological concept of inclusive fitness, the conflict between individual and group selection was resolved. Selfish individuals may cooperate to the extent that they share more common genes than the population at
large. My thesis about cooperation among prairie dog societies died long before
my detailed description of their social behavior. Careful descriptions of natural
phenomena often endure beyond the fickleness of popular theory.

Nature-Nurture Intertwined

John Calhoun is a scientific visionary. After presenting a scheme for the North
American census of small mammals at the mammalogists meetings in Yellowstone
National Park, he stopped to see the Shirttail Canyon prairie dogs. He was returning to the Jackson Laboratory in Maine to continue his studies of mouse populations with J. P. Scott. He had joined the National Institutes of Health during the
bountiful years when his visions could become a reality. He had many novel ideas,
which explained how panic-stricken mice rush to fill a population vacuum created
by the removal of a natural resident population, or how subordinate species of
small mammals, became trappable only after the dominant species were removed.

I vividly recall his "Frontier of Science" paper presented at the Dallas AAAS
meeting in 1968 in which he envisioned a future species of human being. I had
become a concrete and down-to-earth person by the time he presented this fanci-
ful exposition. However, I concluded that science profits from these dreamers and
we are fortunate that society has the luxury of supporting individuals who are not
afraid of their imaginations. That hot summer afternoon when Jack Calhoun
visited me on the prairie, he suggested I apply for an NIH postdoctoral fellowship
to study at the Jackson Laboratory.

My introduction to the Jackson Laboratory in Bar Harbor, Maine, September
1951, was momentous. The staff behaviorists, J. P. Scott, John L. Fuller, and Emil
Fredericson, had organized a conference entitled "The Effects of Early Experience on Mental Health." Approximately twenty-five leading animal behaviorists
were invited to attend the four-day conference funded by NIH. For a newly
fledged Ph.D. to associate closely with Donald Hebb, Frank Beach, Ted Schneirla,
Karl Pribram, Leonard Carmichael, Richard Solomon, Calvin Hall, J. W. Whiting,
John Anderson, and David Levy was a wonderful and influential experience.
Since the Jackson Laboratory specializes in mammalian genetics, this conference
on early experience was directed toward the interaction between the genetic
predisposition of various genotypes and their vulnerability to modification by
developmental events. Epigenetics was resurrected, but with each participant
adhering to his own concept of the interplay between genetics and early experience. These concepts had been dealt with in my graduate studies, but I had never
wrestled with them as entities incarnated in the personalities of the famous men I
ate, drank, and talked with. My experience at the conference was stimulating and
humbling. I wanted to become involved in this intellectual resurgence of the
nature-nurture problem. Paul Scott and John Fuller warmly welcomed me.

The intellectual climate at Hamilton Station of the Jackson Laboratory during the 1950s was stimulating, exciting, and rewarding. Paul Scott and John Fuller were my seniors by about ten years and they complemented each other precisely in personality, in scientific skills, and in administrative ability. Paul was serious, contemplative, detached, and as concerned about human social interactions as he was about the social behavior of the dogs and mice he studied. One of his basic premises was that science was noncompetitive and that severe competion could disasterously impede the flow of scientific progress. Cooperation was the name of the game at Hamilton Station and Paul Scott was an altruist. It was a small, tight group that car-pooled the eight miles along the beautiful shores of Frenchman's Bay from town to Hamilton Station. We ate our lunches together, had weekly seminars, and planned the activities for the week each Monday morning. Scott usually conducted these meetings and I remember my frequent vexation following some meetings when problems remained unsolved and decisions were postponed. It took many years of maturing before I fully understood what Paul meant about the intrinsic value of communication.

In contrast to Paul's idealism and detached seriousness, John Fuller was realistic, involved, and playfully competitive. John and I often played chess and the accuracy of his scientific predictions was a game to him. He was critical of my ideas, helpfully steered me along a rational path, and taught me statistics. I probably talked more science with John than with any other person before or since our association. His office was across the hall and he was always available for consultation. Weekly we caught every dog in the kennel and inspected it carefully. It was always a pleasure to work, play, and talk with John.

Scott and Fuller (1965) were, each in his own way, attacking the problem of how experience interacted with the genetic differences among various strains of mice and breeds of dogs, but they were simultaneously influencing me. After completing my postdoctoral project on closed social groups in dogs, I studied the effect of early social isolation on the aggressive behavior of mice. Adult male mice readily fought when strangers met in a cage from which the partition that separated them had been removed. Their readiness to fight could be predicted from a knowledge of their genetic strain or from a knowledge of their previous social environment. The same early social environment did not similarly influence the aggression of two different strains of mice. It was possible to make a phenocopy of one genetic strain by environmentally manipulating the other strain. Instincts, innate behavior, genetics, or nature could be molded to some extent by early experiences.

The mechanisms of the genetic-environmental interaction during the ontogenetic development of behavior were just beginning to be explored by William Young, Frank Beach, Donald Hebb, Daniel Lehrman, and Jan Bruell. Ideas, theories, and recent data on the interaction were the trading commodity of the time just as foraging and reproductive strategies capture the attention of behaviorists now. My contribution to this flurry of intellectual speculation came from my experiments on the development of deer mice and from the ideas offered by C. H. Waddington's *The Strategy of the Gene* (1957).

I began the developmental studies of deer mice by trapping small mammals on Mount Desert Island. The local subspecies, *Peromyscus maniculatus abietorum* failed to reproduce despite my efforts to provide them the appropriate environment. So when William Prychodko, a former fellow graduate student, offered me stocks of *Peromyscus* he had rescued from Dice's disassembled colony, I took them into the confines of Hamilton Station of the Jackson Laboratory (such irreverence to strict quarantine could not be tolerated today).

The stocks of *Peromyscus maniculatus bairdi* and *P.m. gracilis* bred well and added a bit of wildness that was lacking in the inbred house mice. The two subspecies are interfertile, but distinct in morphology, development, and behavior. The species apparently had diverged around the Great Lakes basin during the Pleistocene. The northern branch adapted to the forests, while the southern deer mice were inhabitants of open prairies and beaches. Recently the two subspecies had migrated toward each other as the forests were opened for crops. Wherever the agriculturally created prairies join northern forests in Wisconsin, Michigan, and New York, both geographic races occur in their respective habitats. Now I had them together in my Maine laboratory.

The prairie race, *P.m. bairdi,* developed rapidly into an irritable, timid adult that cringed immobile one moment and burst into frenzied activity the next. The initial freezing probably enabled the deer mice to avoid detection by predators in the open fields. This behavior as well as the dull brown pelage, and short tail and ears contrasted sharply with the deliberate, cautious movements of its arboreal forest relative to the north, *P.m. gracilis.* This race lagged about four days behind *bairdi* in development, so at fourteen days of age young *bairdi* ran around like little miniature adults, while young *gracilis* still had their eyelids sealed shut and appeared infantile. The slow-developing *gracilis* became large, fawn-colored adults with "Mickey Mouse" ears and eyes, long tails, and a docile disposition. They clambered about my hand, up my arm, and around my neck as if I were a tree from its native habitat, while *bairdi* leaped senselessly from my hand held at any height. I measured the ontogenetic development of both subspecies in every conceivable respect and discarded over half of my attempts. It was a pleasure to work with these infant animals and I enjoyed handling them like I handled my infant children at home.

While studying the different rates of development in the deer mice, I could not avoid observing differences in my children. Kit at ten months, was walking, balancing wooden blocks stacked twelve high, and was swimming instinctively. But it took him three years before he was toilet trained and before he talked with facility. His young sister, Andrea, lacked the physical dexterity, but accompanied him through toilet training and talking despite a chronological gap of twenty-three months. Kit was active, almost hyperkinetic and Joan exhausted his energy by walking him to the shore each day, while pushing Andrea in a stroller. Kit could be held only when bounced or jostled before he scrambled restlessly out of your arms. Andrea cuddled cozily into your arms. Kit fought going to bed and awoke early, while Andrea enjoyed her crib.

The emotional and intellectual feedback from these human and rodent infants forced me to think about rates of development. I read Julian Huxley's *Problems of Relative Growth* (1932) and examined the figures of D'Arcy Thompson's *On Growth and Form* (1917): old books with perennially refreshing ideas. One mechanism for the genetic-environmental interaction could be allometric patterns of behavioral development that provided the rules for incorporating experiences into the adult character. This possibility needed further exploration, but information from another source compelled attention.

W. R. (Bob) Thompson had worked with Don Hebb at McGill and was establishing himself at Queens University through his genetic studies of behavior. Bob was a frequent, but temporary, resident at the Jackson Laboratory. I was attracted to his wit and jocularly exaggerated British mannerisms. We were about the same age and had similar interests, but our training and background differed. He was a psychologist with a cosmopolitan heritage and an Australian wife. I was a zoologist from the Midwest. On one of his visits, Bob teasingly mentioned the incredible findings of Weininger (1953), who found that rats increased growth and viability with handling. Louis Bernstein (1952) had independently found that handled infant rats learned more readily than nonhandled rats. He postulated that gentle fondling of the infant rats established a "relationship" with the human handler that enabled them to grow faster and gain more from their experiences. (Harry Harlow, who was pursuing similar ideas with rhesus monkeys, called the mouse and rat experimenters "rodent rubbers.") S. ("Gig") Levine and Vic Denenberg also experimented with infant mice during their summer visits to the Jackson Laboratory. In contrast to the social deprivation experiments I was doing, this group of contemporary investigators was overstimulating the mice and rats by handling and stroking them or merely by pulling the infants from their mothers as Levine and Denenberg did.

The prolific experiments of Levine and Denenberg revealed the relationship between infantile stimulation and developmental rates in rodents. I had found the closely related races of deer mice matured at different rates, which indicated a genetic contribution. The developing organism did not require specific genes for each behavioral attribute, nor did it need a critical experience. All that was required to form adult patterns of behavior was the proper meshing or timing of the exogenous stimulation with the inherent rate of development. If the appropriate rate of stimulation or reinforcement coincided with the related organic rates of development, the behavior could be canalized into adult patterns. A frequently employed behavior would be more likely to be reinforced by the environment, which would in turn increase the rate of the response constellation.

The idea was not new. Teachers of reading and foreign languages, for example, had recognized the value of providing essential language stimulation at the most suitable age. Waddington had conceptualized the process as an epigenetic landscape with genes and exogenous stimulation uniting to form physiological and behavioral patterns with the same startling results as the Cartesian coordinates that D'Arcy Thompson used to illustrate the effect of allometric growth upon adult morphology. My contribution was the exposure of mice with different genotypes to a variety of early experiences as a method of investigating the interaction. The next two decades of my life, roughly 1955 to 1975, were spent in attempting to exploit this preparation.

The first attempt was simply to combine the "rodent rubbing" early stimulation with the genetic differences of the two races of deer mice (King and Eleftheriou,

1959). The subjective element of human handling was replaced with a mechanical mouse handler, which gently compressed the infant mice between foam rubber pads several times each minute during the first weeks of life. Their adult tests for the later effect of this early "handling" were their performance in a Sidman avoidance conditioning task and their response to stress. The results were dramatic: the same experience enhanced the learning of one race and impaired learning in the other race. It was as if the early handling stimulation made one race more wild and intractable. (I recalled how my son squirmed out of my arms as an infant, while my daughter snuggled closer.) These results were rewarding. I will not soon forget the gentle approval from Gardner Lindzey, who was in some respects a protégé of my esteemed friend, R. M. Elliott (1952). Now I had to find the developmental processes of the mice that were so easily impressed by this experience. Here was the mammalian equivalent of avian imprinting that was concurrently and thoroughly being studied.

My experiments and speculations were stimulated by other social reinforcers. Jerry Hirsch invited me to a two-week-long conference at the Institute for Advanced Studies in the Behavioral Sciences; the paper I presented there was later published in *Behavior–Genetic Analysis* (King, 1967). I also contributed to *Early Experience and Behavior* (King, 1968), which was edited by G. Newton and S. Levine. About the same time, Frank Beach suggested that I apply for a National Science Foundation–National Research Council fellowship to study abroad. Both C. H. Waddington in Edinburgh and B. Rensch in Münster had influenced my thinking and I was eager to derive direct stimulation from them.

At the age of forty, nine years after receiving my Ph.D., scientific stimuli were converging at a rapid rate. I was excited. I was with exciting scientists. Funds were readily available from federal agencies. Behavior genetics and effects of early experience were "hot topics." The personal excitement I felt during the early 1960s was being felt by individuals in other scientific areas. It seemed as if science was about to replace religion in human belief systems. Each generation of scientists probably also feels that they are at the crest of intellectual achievement. Some individuals in each area ride to the crest of the scientific wave before it collapses. Some individuals leap from crest to crest, some rise again onto the next crest, and some reach shore in a trough. One advantage of aging is the opportunity long-lived individuals have to observe these changes, whether or not an active part is taken. In fact, witnessing historical events, personal or universal, becomes such a part of life that one hangs on tenuously, even when the events can be only dimly or distortedly seen.

A heady, initial excitement of even minor scientific progress rapidly dissipates into the pedantic, routine details of proliferating disciplines. By the time my theoretical chapters were published in the two books, the problem area I discussed had diverged into the brilliant studies on early learning of bird song (Marler and Mundinger, 1971) and the neural genetics of stridulation in crickets (Bentley and Hoy, 1972). Interest in the rate of early development was superseded by investigations on early endocrine alterations of sexual behavior development (Phoenix et al., 1959) and audition in prehatched ducklings (Gottlieb, 1971). Behavioral modification of the gene pool was eclipsed by sociobiology. My theoretical attempts went unrecognized and I abandoned future efforts in that direction, trusting in my empirical efforts to produce something more durable.

My experiments went in two directions: (1) genetically altering the rates of development by selectively breeding both races of deer mice, and (2) experimentally altering early schedules of reinforcement at appropriate times in development, which initially required detailed descriptions of behavioral development.

After much searching for an appropriate response to work with, I ended up with vision. The age of eyelid separation served as the index for developmental rates in the two subspecies of deer mice. Eye opening responded readily to selection, but was not an index of overall developmental rates since some mice opened their eyes as infants and others were active juveniles with their eyes sealed shut. This might have been an allometric advantage, if it meant that vision was also affected by selection. It was not. Infant mice with eyes wide open failed to exhibit an optokinetic response to revolving vertical stripes until they reached the age at which their controls normally opened their eyes (Vestal and King, 1968).

I also tried some light reinforcement experiments on mice reared in darkness and with different ages of eyelid separation (King, 1970, 1975). Mice reared in darkness were less reinforced by light than controls, but age of eye opening had no effect. Early exposure to light seemed to differentially influence later light preferences in various species and subspecies of deer mice, but any further pursuit of these results would only repeat the early "rodent rubbing" experiments. Furthermore, the experiments by Pettigrew and Freeman (1973) on the effect of early visual experience in kittens on electrophysiological recordings from the brain made my studies primitive by comparison.

The deer mouse selection lines gradually lost fertility from too much inbreeding, some expensive apparatus caused problems, and results from completed experiments were not very exciting. Other investigators were attacking similar problems with more elegance than I could provide. I was eager to get into the field again and used my 1970 sabbatical leave to examine nocturnal periodicity in desert deer mice in an attempt to relate periodicity to light reinforcement. The radioactive technique I used for tagging the mice failed to work on free-ranging mice, so I shifted my interest to visual orientation in nocturnal rodents. The orientation studies have been a series of on-off attempts over the last decade without adequate data to publish.

Thus, the sequence of my scientific efforts was as follows: the initial finding of a genetic-environmental interaction in deer mice, then the attempt to discover a specific response of the deer mice for the further analysis of the interaction. I tried the motor response of digging; then I went to vision in a nocturnal rodent, which was a mistake. Later I pursued the study of vision in light reinforcement, in periodicity, and most recently in orientation. I became overwhelmed by the technical problems, discouraged by variation in the results, and intimidated by the elegance of experiments in related areas.

When I was invited to write a review on the ecology of aggression for *Annual Review of Ecology and Systematics* (King, 1973), I eagerly welcomed the opportunity evaluate field studies again. My reaction to this review was that all the energy and manpower applied to most field studies failed to counterbalance the precision and control that can be achieved in laboratory experiments. The field studies that used aggression as an explanation for the regulation and distribution of rodent populations depended primarily on inference because aggression was rarely observed directly. Perhaps if I went into the field again, I could bring along my laboratory-learned respect for controls and replications.

The Ecology of Dispersal

Dispersal ranks with reproduction and mortality as one of the three fundamental properties of populations. Some individuals of all existing species disperse. The absence of dispersal leads to extinction. However, dispersal is a costly process that probably eliminates most dispersers. Still, the rewards are sufficiently great for those individuals that successfully establish themselves in a new locality. They can pass this attribute on to their descendants. Since some individuals do not disperse, this question arises: What causes some individuals to stay home and others to leave? The answer has been sought in classifying large samples of dispersing rodents according to age, sex, social status, or reproductive condition. The combined results of numerous studies led me to conclude that dispersers are pretty much a random sample of the entire population. Infants and their mothers are often absent from collections of dispersing rodents because they are restricted to the nest. The most likely answer to "Who disperses?" probably resides in an analysis of how individuals perceive their environment and how they make decisions based on those perceptions.

An inquiry into the perceptual and decision-making processes would be difficult enough without the additional impediment of the lack of a suitable way to measure dispersal. Laboratory attempts to study dispersal are not very convincing because dispersal is usually measured by spatial barriers involved in distance from the natal sites. Irv Savidge, a former graduate student, tried several barriers to dispersal and ended up with an electrified grid that was slightly noxious for the mice to cross (1974). Although his published research was largely ignored by field biologists, Irv's technique had merit.

If I was to study dispersal, a field study would be required to establish my credentials. My long-standing interest in the problem arose from my prairie dog research and from teaching the subject in my graduate classes. The study would also satisfy my desire to return to the field with greater regularity than my intermittent trips to the New Mexican desert to study orientation permitted. The dispersal study could be done in Michigan.

One thing twenty or so years in the laboratory taught me was the necessity of controls in behavioral studies. Control of all variables in the field is next to impossible, but room for improvement exists. Since a repeated observation on the dispersal of deer mice was that it changed seasonally, some control over seasonal variables in the field would be desireable. The resident population could be controlled by introducing a specified number of mice at a given age, sex ratio, and reproductive condition to an uninhabited area. Food, competition, and predation could be partially controlled by supplying food, and refuges from competitors and predators. We introduced eight mated pairs of deer mice with the female pregnant into an uninhabited area of about five acres. By allowing the population to grow only ten weeks in each season (spring, summer, fall), growth was limited to the one generation produced by the introduced mice.

These experimental procedures made it possible to control some of the seasonal changes that usually accompany dispersal: population density, reproduction, adult aggression, availability of food. One could argue effectively that dispersal was caused by any one or any combination of these correlated seasonal changes. Probably the most frequent explanation given was that during the breeding season many adult individuals aggressively defended their territories or nest sites. This reasoning associated the increasing population density with seasonal breeding activity and the repulsion of the new recruits from their natal homes and already occupied territories. Winter dispersal was caused by reduced food, which could also be a function of population density. The logic of these explanations is reasonable. The problem is that no one has observed young deer mice being driven from their native homes, so inferences were made from questionable laboratory observations on aggression.

The results of our study confirmed previous observations that most mice disappeared or dispersed during the spring and summer, in contrast to the fall when they remained in the population (King, 1984). Two hypotheses were eliminated by these results: that seasonal differences in dispersal were caused by adult breeding activity or by increasing population density. The adults continued to breed throughout all seasons and the highest population density occurred in the fall when there was reduced dispersal. Evidence for seasonal differences in competition, predation, or nutrition was lacking. The only significant change was the recruitment of sexually active juveniles to the population. Young mice born in the field reached sexual maturity in the spring and summer, but not in the fall. The seasonal differences in this one feature of sexual maturation suggested that the critical mechanism for dispersal was a search for mates. The search is apparently intensified by the recruitment of sexually maturing individuals and is little affected by breeding among the previously mated adults. This concept shifts the emphasis from aggression between generations of mice to a search for mates by most individuals, except those restricted to care for infants in the nest.

Other mechanisms for dispersal exist beyond the search for mates. I have stressed the search for mates when sexually maturing mice are being recruited in order to expose the myth of aggression. I refer to the "myth of aggression" because the scientific evidence for the role of aggression in dispersal is weak or lacking in most species of animals. Animals fight each other and the consequences are often dramatic, like wars among nations. Aggressive encounters draw the observers attention and their importance is easily exaggerated. Most social interactions among animals, like political relationships among nations, are resolved in the absence of overt aggression—or wars. Aggressive solutions to problems are so straightforward and definitive that we tend to prefer these solutions over the unsettling social ambiguities that are a more accurate representation of the real world.

Students and Other Support Systems

A popular observation is that too many Ph.D.'s are being awarded in some areas of biology for the number of positions available. This correct observation usually accompanies reproaches to the professor, who, it is assumed, aggrandizes himself with an overpopulation of graduate students (Janzen, 1982). Actually, this observation focuses on a fundamental educational problem. Is the Ph.D. exclusively a vocational education or does the training have values independent of stepping into a job? The same question can be asked of the baccalaureate degree, or of all education, if one chooses. Probably every educator anticipates that the training offered will aid the student in his/her pursuit of the life's activities, including jobs, recreation, and social responsibilities.

My answer is simple; the Ph.D. is not training for a specific job. It is training to investigate the unknown objectively. There are ample opportunities for objective investigations beyond the narrow confines of a specific career in science or academia. Most students recognize this as well, because they know the opportunities to pursue one's graduate training into a professional career are limited and becoming more so. It takes a special type of person to give up approximately four years of earnings to study on a graduate student's income without the promise of job commensurate with their education. When I meet such a person, I cannot deny him/her an opportunity to pursue graduate study. Anyone willing to sacrifice material gain for scholarship deserves moral support, encouragement, and a pleasantly exciting tenure in graduate school.

These "rare birds" who willingly trade earning power for the scholarly pursuit of their interests certainly deserve the freedom to pursue those interests without being funneled into my specialities. My students sacrificed precise instructions for the freedom to make mistakes. We often examined research problems together and ran the risk of arriving at the wrong solution. Most problems concerned something about the behavior of deer mice *(Peromyscus)* because these were the animals available both in the laboratory and nearby fields. Ed Price (1967) initially wanted to study Alaskan marmots, but settled on the problem of domestication as a result of breeding *Peromyscus* in the laboratory for many generations. Both Bedford Vestal (1973) and Kevin Murphy became involved with their work as research assistants. Bedford did a comparative study of visual acuity during ontogeny in two species of *Peromyscus* and Kevin pursued endocrine factors involved in dispersal in the field. Until Gene Brenowitz (1980) broke away from the study of *Peromyscus,* all students used *Peromyscus* as their research subjects. Gene studied mechanoreceptors in the paws of tree and ground squirrels. Then Suzette Davis Tardif (1982) went to Oak Ridge National Laboratory to investigate sexual maturation in a colony of marmosets they had there. Bruce Cushing is combining a study of weasels and deer mice by examining the effect of estrus in mice on prey selection by weasels.

Most students did their dissertation research in the laboratory, but my first, Dick Terman, (1961) studied social organization of deer mice in one of the first large mouse enclosures. Although Gale Haigh attempted to examine the effect of nest sites on population dynamics in white-footed mice, he later switched to study reproductive inhibition in *Peromyscus* (1983). Dick Tardif (Tardif & Gray, 1978) and Kevin Murphy (1982) combined both field and laboratory studies on dispersal in *Peromyscus*. Irv Savidge (1974) did his dispersal study exclusively in the laboratory.

Although I felt comfortable in advising students on the behavior of *Peromyscus*, many students went into areas beyond my competence. Jane Huff (1973) got involved with ultrasonic recording equipment and was helped by her husband in the recording of infant mouse vocalizations. Lincoln Gray (1979) probed deeply into psychophysical principles for his investigation of deer mouse foraging strategies. Bob Robbins (1978) delved far into the psychological literature on taste aversion learning, on which he did a comparative and developmental study. Lee Drickamer (1972) anticipated the current popularity of foraging behavior by examining olfactory conditioning on later food preferences and foraging behavior. Jim Hill (1977) convinced me that *Peromyscus* does not have territories like those of house mice or red squirrels, but other investigators cling to the old idea. Bill Barry (1976) investigated the neglected alternative to hibernation and migration during periods of food scarcity: food storage. Jim Joslin's (1977) studies on deer mouse orientation got me involved in that problem, which is more difficult to resolve than I initially thought. We learned a lot about *Peromyscus*, but many problems remain for another generation of students.

Outside of my family, until my retirement in 1984 students were the major social contributor to my university life. Usually we saw each other every day. We ate lunch together; we chatted with each other as went about our various activities; we shared information, news of events, and research progress in a casual way. Rarely did we have regular research meetings. Journal clubs and seminars were sporadic, depending upon the interests, personalities, and alliances among each contemporary group of students. Cohesion was dictated by the shared use of space, animals, and equipment.

I did not socialize much with students outside of the academic environment. Although I personally liked my students and enjoyed their company, our associations were primarily academic. They had top priority for my time at school, but not at home, where my family and privacy came first. My library of journals, books, and reprints were freely available to them. Their professional accomplishments provide one of the greatest rewards of my career. Those who have pursued nonacademic careers are also a continuing source of pleasure as they pursue their various goals. Their degrees should not straitjacket them for their entire lives, but should contribute to their performances in any occupation.

In addition to the daily social reinforcement of students, colleagues provide more enduring rewards and conflicts. Three animal behaviorists in the Department of Zoology have always been available for consultation, advice, assistance, and friendship: James C. Braddock, Martin Balaban, and Lynwood Clemens. Jim was responsible for bringing me to Michigan State as an ex officio member of Dick Terman's Ph.D. advisory committee and, in 1960, as a faculty member. He had offered a course in animal behavior for several years and found student enrollment increasing each year. Jim was a student of W. C. Allee at the University of Chicago and added a behavior course onto the already demanding lectures he gave in introductory zoology. I siphoned off some students into new courses, but Jim continued teaching the introductory behavior course. Jim had abandoned most personal research because of his heavy teaching load, but his graduate students cooperatively continued his work on the Siamese fighting fish. With his graduate students and close to a thousand undergraduates enrolled in his courses at any one time, he was always talking with someone in his office. Jim did a lot to promote the study of animal behavior and we worked harmoniously together as secretary and treasurer of the newly formed Animal Behavior Society, on summer institutes, and training programs.

Marty Balaban was another University of Chicago graduate, who received his degree with Austin Riesen in psychology. Marty was with Viktor Hamburger at Washington University when he joined us to take over my undergraduate teaching. A career development fellowship from the National Institute of Child Health and Human Development had released me from undergraduate teaching to pursue ontogenetic studies in several taxa of *Peromyscus*. Marty's interest in behavioral development could facilitate these efforts. Although Marty was about ten years younger, our interests, ideas, and approach to life meshed perfectly. We both hated conformity and hypocrisy, but, of course, had our share of both. Throughout most of our professional careers, we have grown together and influenced each other to a remarkable degree. Marty is more of a scientific philosopher than a practioner of science. He spends much time with his animals and in scholarly pursuits. However, rather than write about his subjects, he thinks about them, avidly searches the literature for novel approaches, and thus becomes so self-critical that he rarely exposes his ideas to public criticism. He believes scientific literature is already overburdened with old ideas and he refuses to contribute until he has something really valuable. Besides the common interests in our families, Marty and I have the same belief in the intrinsic merits of scholarship.

Lynwood G. Clemens joined Jim, Marty, and me a few years later in order to strengthen our NIH-funded training program in animal behavior. Lyn received his Ph.D. at Berkeley with Frank Beach and then had a postdoctoral fellowship with Roger Gorski at Los Angeles. Lyn added endocrinology, neural biochemistry, and a far more systematic approach to science than any of us. He rapidly built an active program with graduate students, postdoctoral fellows, and technicians who met weekly to discuss their current research activities and plans. Lyn is an organizer and team worker in contrast to our haphazard, individualistic groping. His youth, vigor, and directness contributed to the group and provided me with another and different type of friendship. After Jim retired, Marty, Lyn and I were united in the same building, where we see each other daily. We rarely talk science, but exchange personal notes and observations on the state of things locally, nationally, and internationally. Lyn is the pragmatic businessman. Marty is the philosophical scholar. I am the romantic aesthete. These ingredients make a delightful concoction to savor every day.

In addition to the support from students, colleagues, and family, the munificence of the government has contributed to my research. The federal government provided most support starting with a National Institute of Health postdoctoral fellowship, through research grants from the National Institute of Mental

Health, Child Health and Human Development, the Eye Institute, and a National Academy of Science-National Research Council fellowship. The government of Michigan has helped by providing my salary and by coming to the rescue in emergencies. The total dollar contribution is not large by comparison with many research budgets, but it is a huge personal debt that I owe society and it compels me to discuss how the debt has been repaid.

I would like to believe that I made some direct, practical contribution to mental health, child development, or vision, like some of my colleagues have when they have discovered ways to treat cancer or other illnesses. Other than a few citations in the clinical literature of some of my early studies, my contributions have been very indirect, like occasionally helping to advance some idea about animal behavior. The relevance of animal behavior to human welfare can be doubted because it is sometimes misused to establish codes of human conduct. However, this study, like most inquiries into the workings of nature, is a valuable and essential human attribute. Society wants, needs, and is willing to pay for a few of its members to explore the unknown without anticipation of personal gain. This human attribute is difficult for some politicians, bureaucrats, lawyers, college administrators, and even some scientists to understand. Society provides few refuges from the marketplace, where every person and every idea has a price. A few absolute values are needed and are often supplied by art, religion, science, or some personalized mixture of all three. This human demand for some priceless values in art, religion, or science is particularly strong in times of stress—financial, personal, political, military—when the inclination to abandon it is encouraged by practical and immediate needs. Still, society gains confidence and courage when it recognizes that a few individuals are pursuing truth without regard to personal hardship, ambition, aggrandizement, or financial gain. The price of these selfless attributes is submission to society's expedient needs.

References

Barry, W. J. 1976. Environmental effects on food hoarding in deer mice *(Peromyscus). J. Mammal.* 57:731–46.

Bentley, D. R., and R. R. Hoy. 1972. Genetic control of the neural generating cricket *(Teleogryllus gryllus)* song patterns. *Anim. Behav.* 20:478–92.

Bernstein, L. 1952. A note on Christie's "Experimental naivete and experiential naivete." *Psychol. Bull.* 49:38–40.

Brenowitz, G. L. 1980. Cutaneous mechanoreceptor distribution and its relationship to behavioral specializations in squirrels. *Brain, Behav. Evol.* 17:432–53.

Drickamer, L. C. 1972. Experience and selection behavior in the food habits of *Peromyscus:* use of olfaction. *Behaviour* 41:269–87.

Elliott, R. M. 1952. In *A History of Psychology in Autobiography,* vol. 4, ed. E. G. Boring, H. S. Langfeld, H. Werner, and R. M. Yerkes. Worcester: Clark Univ. Press.

Gottlieb, G. 1971. *Development of species identification in birds.* Chicago: Univ. of Chicago Press.

Gray, Lincoln. 1979. Feeding diversity in deermice. *J. Comp. Physiol. Psychol.* 93:1118–26.

Haigh, G. R. 1983. The effects of inbreeding and social relationships on the reproduction of *Peromyscus maniculatus bairdi. J. Mammal.* 64:48–54.

Hamilton, W. D. 1964. The genetical evolution of social behavior. I. and II. *J. Theoret. Biol.* 7:1–52.

Hill, J. L. 1977. Space utilization of *Peromyscus:* social and spatial factors. *Anim. Behav.* 25:373–89.

Hoogland, J. L. 1981. The evolution of coloniality in white-tailed and black-tailed prairie dogs (Sciuridae: *Cynomys leucurus* and *C. ludovicianus*). *Ecology* 62:252–72.

Huff, J. N. 1973. The "distress" cry of infant deer mice, *Peromyscus:* physical characteristics, specific differences, social function. Ph.D. diss., Mich. State Univ.

Huxley, J. S. 1932. *Problems of Relative Growth.* London: Methuen.

Janzen, D. H. 1982. Academic birth control: long overdue. *Bull. Ecol. Soc. Amer.* 63:5–6.

Joslin, J. K. 1977. Visual cues used in orientation by white-footed mice, *Peromyscus leucopus:* a laboratory study. Amer. Midl. Nat. 98:308–18.

King, J. A. 1955. Social behavior, social organization, and population dynamics in a black-tailed prairie dog town in the Black Hills of South Dakota. *Contrib. Lab. Vert. Biol.,* no. 67, Univ. of Mich.

———. 1967. Behavioral modification of the gene pool. In *Behavior-Genetic Analysis,* ed. J. Hirsch. New York: McGraw-Hill.

———. 1968. Species specificity and early experience. In *Early Experience and Behavior,* ed. G. Newton and S. Levine. Springfield: Thomas.

———. 1970. Light reinforcement in four taxa of deer mice (Peromyscus). *J. Comp. Physiol. Psychol.* 71:22–28.

———. 1973. The ecology of aggressive behavior. *Ann. Rev. Ecol. Syst.* 4:117–38.

———. 1975. Stimulus selection (light) in four subspecies of deer mice *(Peromyscus maniculatus). J. Interdiscipl. Cycle Res.* 6:267–78.

———. 1984. Seasonal dispersal in a semi-natural population of *Peromyscus maniculatus. Cam. J. Zool.* 61:2740–2750.

King, J. A., and Eleftheriou, B. E. 1959. Effects of early handling upon adult behavior in two subspecies of deer mice, *Peromyscus maniculatus. J. Comp. Physiol. Psychol.* 52:82–88.

Lewis, D. 1948. Structure of the incompatibility gene. I. Spontaneous mutation rate. *Heredity* 2:219–36.

Marler, P., and Mundinger, P. 1971. Vocal learning in birds. In *Ontogeny of Vertebrate Behavior,* ed. H. Moltz. New York: Academic Press.

Murphy, K. L. 1983. Dispersal in semi-natural populations of *Peromyscus maniculatus:* seasonal and hormonal relationships. Ph.D. dissertation, Mich. State Univ.

Pettigrew, J. D., and Freeman, R. D. 1973. Visual experience without lines: effect on developing cortical neurons. *Science* 182:599–600.

Phoenix, C. H., Goy, R. W., Gerall, A. A. and Young, W. C. 1959. Organizing action of prenatally administered testosterone propionate on the tissues mediating mating behavior in the female guinea pig. *Endocrinology* 65:369–82.

Price, E. 1967. The effect of reproductive performance on the domestication of the prairie deer mouse, *Peromyscus maniculatus bairdi. Evolution* 21:762–70.

Robbins, R. J. 1978. Poison-based taste aversion learning in deer mice *(Peromyscus maniculatus bairdi). J. Comp. Physiol. Psychol.* 92:642–50.

Savidge, I. R. 1974. Social factors in dispersal of deer mice *(Peromyscus maniculatus)* from their natal site. *Amer. Midl. Nat.* 91:395–405.

Scott, J. P., and Fuller, J. L. 1965. *Genetics and the Social Behavior of the Dog.* Chicago: Univ. of Chicago Press.

Tardif, S. D., 1983. Relationship between social interactions and sexual maturation in female *Saguinus oedipus oedipus. Folia Primatologica* 40:268–75.

Tardif, R. R., and Gray, L. 1978. Feeding diversity of resident and immigrant *Peromyscus leucopus. J. Mammal.* 59:559–62.

Terman, C. R. 1961. Some dynamics of spatial distribution within semi-natural populations of prairie deer mice. *Ecology* 42:288–302.

Thompson, D'Arcy W. 1917. *On Growth and Form.* London: Cambridge Univ. Press.

Vestal, B. M. 1973. Ontogeny of visual acuity in two species of deer mice *(Peromyscus). Anim. Behav.* 21:711–19.

Vestal, B. M., and King, J. A. 1968. Relationships of age of eye opening to first optokinetic response in deer mice *(Peromyscus). Develop. Psychobiol.* 1:30–34.

Waddington, C. H. 1957. *The Strategy of the Genes.* New York: Macmillan.

Weininger, O. 1953. Mortality of albino rats under stress as a function of early handling. *Canad. J. Psychol.* 7:111–14.

Wynne-Edwards, V. C. 1962. *Animal Dispersion in Relation to Social Behavior.* New York: Hafner.

Paul Leyhausen (b. 1916)

10

The Cat Who Walks By Himself

Paul Leyhausen

How It Started

I was a late child, born more than six years after the younger of my sisters. Thus I grew up being left very much to my own devices most of the time. But there were animals. My mother kept chickens, my father bred rabbits, and there was always a cat who, though left to come and go as she chose, was more of a family member than a mere pet. Later on, when a schoolboy, I adopted a stray dog and spent most of my vacations among the animals of a large farm.

Small wonder that from earliest times, whenever some inquisitive adult asked what my ambitions were, it was always something to do with animals: at five, I wanted to become a coachman—in the early twenties, most transport in my hometown of Bonn was still performed by horse-drawn vehicles. Between eight and twelve, the career of forester seemed immensely desirable; but as my sense of romance had led me to crave knowledge of and encounters with exotic animals, I also dreamed of traveling to the wilder corners of the globe and exasperated my geography teachers by knowing more about the plains, rivers, and mountains of East Africa than the immediate environs of Bonn. Finally, when entering university, I had sobered down somewhat and thought I might, with luck, become a zoo director.

My mother looked with indulgence on my infatuation with animals although, as well I knew, her secret wish was for me to become a priest. My father, though tolerant, looked on it with sorrow because, having had to fight economic vicissitudes and near-disaster himself all his life, he wanted me to acquire security by embarking on a civil service career. A prosperous uncle wanted me to become heir to his fortune and flourishing business. I could not find it in me to embrace any of these prospects.

On entering university, I found teachers who were sympathetic and helpful in every possible way. At the same time it was still possible—and in fact students were encouraged—to sample every topic the university offered, and so in addition to my chosen field of zoology I found myself attending classes in botany, geology and paleontology, ethnology, chemistry and physics, medical anatomy and physiology, neurology and psychiatry, and philosophy. After a few terms I concentrated on the first three of these, and eventually exchanged botany—a subject which wrongly and unfortunately held little interest for me at the time—for psychology.

Königsberg

After four terms, my zoology professor advised me, because of my proclaimed interest in animal psychology, to move to Königsberg University and become a student of Otto Koehler. I arrived there in January 1941. By an extraordinary stroke of luck, Konrad Lorenz had shortly before become director of the newly founded Institute of Comparative Psychology. Within a few days of my arrival in Königsberg I found myself firmly established in the new institute, which consisted of three small studies and a larger former seminary room in the process of being converted into an aquarium for the study of cichlid fish behavior. Arranging tanks, furnishing them, constructing filters and heating systems out of materials that were cheap and available in wartime, alongside Lorenz, whose only senior student I was, I had a free privatissimum course in emerging ethology practically all day long. It cannot be said, however, that at the beginning I was much taken by his ideas. My upbringing had been strictly Roman Catholic, and I was not ready to admit that similarities between human psychology and animal behavior were anything but superficial. Lorenz accused me of not believing one word he was saying, and he was quite right: I did not believe him. But in the end he convinced me.

As the only major surviving witness of that time in Königsberg, I think it befits me to say a few words about the often-alleged influence of National Socialist doctrine on German ethology. I also believe that I have every right to do so since my own record is untainted, through no merit of my own but thanks to the strict religious upbringing mentioned above: our national hero of the time was not Hitler or Göring but Clemens August Graf von Galen, Archbishop of Münster.

What made it appear, during the late thirties and early forties, that ethological findings and Nazi ideology might to some extent be convergent was the fact that as yet next to nothing was known about the intricacies of population genetics. Since the phenotypes of wild populations as a rule look far more uniform than do domesticated ones, most evolutionists concluded that the individuals of wild populations were homozygous to a very high degree. Lorenz's studies showed the disintegrating influence of domestication on complex behavior systems. Thus, apart from its undeniable uses to man, domestication was considered the destruction of the beautiful harmony of the animal with itself and its ecosystem, and in that same sense the so-called self-domestication of man must have evil consequences. The only apparent exception to this rule in domesticated animals were pure breeds, which in some instances at least seemed to regain harmony of even a higher order. Inevitably, the idea of eugenics held some appeal, and Lorenz expressed himself accordingly in several publications. Unfortunately, he used the term *racial hygiene (Rassenhygiene)* instead of *eugenics,* the latter term being banned since it had been coined by the Jesuit Father Muckermann. More accurate reading of these articles of Lorenz's than most of his critics felt necessary reveals, however, that Lorenz in no case extolled the virtues of the "Nordic race," or indeed of any other race, over another. His concern was not racist but humanistic. While fully aware that "self-domestication" was a prerequisite of cultural development and of all the more human and humane achievements, he also saw the dangers inherent in this process going too far, and he thought it his duty to warn against them.

It must, however, be borne in mind that the official Nazi attitude toward evolutionary theory and genetics was by no means clear-cut: Ernst Krieck, by the grace of Hitler professor of philosophy at the university of Heidelberg and, apart from Alfred Rosenberg, the foremost creator of Nazi ideology, called the theory of descent the "ape religion of the mentally deficient [*Affenreligion der geistig Minderbemittelten*]."

There really never was any question of bending ethology to suit Nazi ideology, but only a vague and—as things stood—ill-founded hope of applying the former to correct the latter.

In 1941 it was still possible to read all the more important ethological literature in a fairly short space of time. Also, Germany was cut off from free exchange of scientific publications with the rest of the world, particularly English-speaking. Thus the few that reached us via neutral countries, such as David Lack's "Behaviour of the Robin," had a tremendous impact. But on the whole, German ethologists had to work without such exchange of ideas and inevitably German ethology developed its own style or brand, which it has not quite lost to this day.

In Königsberg, investigations were concentrated on instinct movements (fixed action patterns), their differences in closely related species and the attempt to reconstruct their evolutionary history from the distribution of these differences within the given taxon, and on the release of activity through sensory mechanisms. The question of the heritability of behavior components, the "innate-acquired" controversy, and the search for methods to investigate the resulting problems were foremost in our discussions. I shall have more to say about this later.

It was the family of Cichlidae that Lorenz had chosen as the study object offering most favorable opportunities of approach to the problems mentioned. He himself has explained their relevant virtues many times. I need not recount them here.

One circumstance that fostered rapid contact and understanding between Lorenz and myself was that he had started out as an anatomist, while foremost among my zoological studies in Bonn was comparative vertebrate anatomy, to which I added experimental embryology in Königsberg. Hence, Lorenz's idea of applying the methods of comparative anatomy and embryology to behavior struck me as not only plausible but also promising. He never needed to explain to me his many "shorthand" formulations and terms which have so often been misconstrued by his later critics, many of whom to this day seem insufficiently conversant with what to Lorenz were such "matter-of-course" basics that it never occurred to him that his "working slang" might be ambiguous.

The Comparative Method and the Mammal

The blissful certainty that innate behavior components "behave like organs in the course of evolution" and the ensuing rather unsophisticated procedures were astonishingly successful in the beginning, and both the ducks and geese and the cichlid fishes proved to be extremely "cooperative": It was actually possible, with considerable accuracy, to reconstruct the probable evolution of complicated behavior patterns and to use them as taxonomic criteria with at least the same degree of reliability as anatomical traits (e.g., Lorenz, 1941). But alas, this happy state of innocence could not last forever. For, while it was not very difficult to find behavior elements characteristic of the species and fairly constant between individuals in insects, fish, reptiles, and birds, the far more variable behavior of higher mammals presented some problems. Indeed, many researchers firmly believed that all higher mammalian behavior was modified to a considerable degree during the life history of the individual. Because of this, it was generally agreed that the higher up the evolutionary ladder a mammal had climbed, the fewer instincts, innate behavior elements, fixed action patterns, and innate releasing mechanisms it could possess. Lorenz himself often quoted C. O. Whitman's statement that the breakup of instincts opened the gate "through which the great teacher Experience enters." If all this were true, then the use of the comparative method in the study of behavior would be limited to investigations on invertebrates and lower vertebrates. Any hope that ethological methods might be successfully applied to the study of human behavior would have been futile from the start. Precisely because of my initial doubts on the comparability of human and animal behavior, I was particularly interested in exploring such application, and I very early started in search of a mammalian taxon that would fulfill the following conditions:

1. The taxon should comprise a sufficient number of closely related species for comparison but not so many that it would appear hopeless to study them all, or at least most of them, in sufficient detail.
2. The species should display clear-cut differences in several areas of behavior.
3. The taxon should occupy an evolutionary level about midway between the "lowest" and the "highest" mammals.
4. Some at least of the behavior patterns should lend themselves easily to experimentation.

Small wonder that my lifelong association with cats and my romantic yearning for the exotic made me choose the cat family, and my youthful optimism allowed me to overlook the difficulty of meeting and studying its rarer members. So it took me a lifetime and still there are four of the forty species I have yet to meet: the Andean cat *(Oreailurus jacobita)*, the Borneo Red cat *(Pardofelis badia)*, the Gobi Grey cat *(Felis bieti)*, and the Kodkod *(Leopardus guigna)*.

In Königsberg, however, the Institute and its means did not allow for such an ambitious project. Apart from that, cichlids are indeed most fascinating objects for any student of behavior. They did much to sharpen my powers of observation and to impress me with the importance of environmental conditions that we now term *ecological.* The word *ecology* was not yet commonplace in those days, but I should like to stress here that from the beginning, ethology was *eco-ethology* by implication. Nothing else was meant when the early ethologists emphasized the necessity of first studying an animal in its "natural surroundings" before starting to experiment on it in the laboratory. The aquarium permits the recreation of a miniature ecosystem as close to the natural as possible.

Thus, I was not discontent with the cichlids, and if little ever came of my work with them it was because the war brutally interfered, transferring me to barracks near Berlin in December 1941, from there to North Africa, and then—in due course I might almost say—via the *Queen Elizabeth I* to a prison camp in Canada.

Yet my felid studies started in Königsberg. Through an extraordinary stroke of luck, the zoo of Königsberg possessed a female tigon, a hybrid offspring of a tiger and a lioness. A frequent visitor to the zoo, I soon noticed some peculiarities in this animal's behavior which distinguished it from both lions and tigers. I wrote a short note about it, which Otto Koehler tore to shreds, as was his way, while at the same time ordering me to find out whether there were similar hybrids in other German zoos. When I found out that there were two female ligers (offspring of a lion and a tigress) and a male tigon in Munich and another in Berlin, Koehler enabled me—no small feat in 1941 wartime Germany!— to travel to these and practically all other major German zoos to make a three-month study of the hybrids and of true lions, tigers, leopards, and jaguars for comparison. Most of the data and almost all the photographs were lost in Königsberg, where I left them when drafted into the army. But the remainder sufficed for my doctoral thesis, "Observations on lion-tiger hybrids, with some remarks on the systematics of the great cats" (1950). To my knowledge, this was the first investigation to establish that it was also possible with a group of highly evolved mammals to use behavioral traits along with anatomical characters to revise current systematics. My results were unorthodox in some respects, but were vindicated by later, more extensive research. (The tiger turned out to be least related to the lion, whose closest relative is the leopard.)

I may add that my interest in felid taxonomy and, more generally, mammalian taxonomy has never abated. All my comparative behavior studies on so many cat species were in part also governed by this interest, and I have, over the years, visited almost all the major museums in the world, measured and photographed felid skulls and studied the skins, and I wish here to thank the staffs of all of them for their unfailing hospitality. That same gratitude I also owe to the numerous zoological parks the world over where I was allowed to pursue my studies.

From the taxonomic point of view, one area of behavior proved particularly promising: vocalization. Being afflicted with a slight congenital hearing defect which grew worse with the years, I was never any good at identifying sounds and voices and miserably failed to identify bird songs on field excursions. Hence I also treated cat vocalization negligently until the availability of ever-improving tape recorders and microphones and the design of sonagraph machines made it possible to transform acoustical patterns into optical ones. From the mid-sixties, however, my wife and I began to assemble recordings of all available cat species, both in our own collection and in numerous zoos. We were joined later by a student, G. Peters, whose publications on the taxonomic use of felid vocalization have—I believe—already left their mark (1978, 1981). The vocal repertoire of mammals seems, unlike that of many song birds, to be completely innate. Only lately I began to suspect that house cats do indeed acquire personal dialects to some extent. But certainly all main species-specific cat vocalizations develop true to type whatever the circumstances under which a given individual is raised, even when it is congenitally deaf. There is still a wealth of material awaiting analysis. Hybrids have been particularly useful in this study, although there is the snag that there has never been a F_2 generation of any kind of felid hybrid. Females could be back-crossed in some instances, but so far the males were invariably infertile.

In May 1942 the army sent me to North Africa where I was captured one month later by the British and sent to Canada. After four years in Canada and eight months in Britain, I was repatriated in January 1947. With the aid of Otto Koehler, who had succeeded in fleeing from Königsberg and now occupied the chair of zoology at the University of Freiburg, I soon resumed my studies and acquired my doctorate just as Ludwig Erhard abolished ration cards and introduced the new deutsche mark. This left me, like some 95 percent of the Germans of Trizonesia (the combined American-British-French occupation zone, now constituting the Federal Republic of Germany), virtually penniless. After working as an unskilled laborer in several factories, I accepted an offer from the head of the psychology department of East Berlin University to set up an ethology unit there. I soon got into political difficulties, however, and had to leave suddenly after six months. Dr. Wolf, curator of mammals at the Museum Koenig in Bonn, then offered me a room at the Museum and a small hut with some adjacent cages in the Museum garden, and the Deutsche Forschungsgemeinschaft (something equivalent to the National Science Foundation in the U.S.) provided me with a small grant. It was here that my feline venture really began.

Early Problems in Behavioral Organization

The early years of Lorenzian ethology were almost exclusively taken up with establishing its basic concepts more firmly by accumulating factual evidence. We looked for more and more examples of instinct movements, taxes, and innate releasing mechanisms, trying to prove their "innateness." To some critics it seemed as if we disregarded, or at least vastly underestimated, the role of learning in behavior. However, the preoccupation with the innate arose out of the realization that research in this direction was sadly lagging behind and had been doing so for a long time, while research into the various kinds of learning had been flourishing and was still going strong. The aim was to draw even, not to relegate learning to insignificance. Hence, my first object in approaching the cat was also to find out whether there were behavioral components that would fit the definitions for the above-mentioned innate behavior categories. On the difficulties presented in this by both criteria and methods I shall have more to say later on.

But apart from establishing those innate elements, there was the question of how the elements organized themselves or were organized into complex, biologically efficient sequences or higher order patterns. Lorenz's first step in this direction was the concept of "intercalation" or "interlocking" of taxis and instinct movement *(Taxis-Instinkt-Verschränkung)*, as first exemplified in the classic study on the egg-rolling movement of the grey lag goose (jointly with N. Tinbergen, 1939). The orientation of the basically "aimless" instinct movement by means of one or several taxes was seen as the simplest form of striving toward an object or goal, that is, of appetitive behavior. Appetitive behavior, in those times referred to by Lorenz as the "non-analyzed rest" was made up of all forms of directed and undirected, learned and unlearned behavior leading up to the satisfactory release of the instinct movement which set it all in motion: the consummatory act, as it was called. This was all very vague and was more or less brushed aside in favor of following up ever more detailed analysis of components.

The first to tackle this all-important problem of higher unit organization more thoroughly was G. P. Baerends in his momentous study on the digger wasp, *Ammophila campestris* (1941). Under the term *hierarchy of moods,* he described how the readiness of the wasp to perform certain behaviors could be changed each morning by specific stimulus situations that would set the mood for a whole day. However, what the stimulus situation determined was not so much the level of readiness to perform individual instinct movements as their sequence and duration. Thus, although in each "mood" very much the same behavioral elements were combined, the difference in their arrangement led to different results: digging a new nest and providing it with a caterpillar on which to deposit an egg, or adding a single caterpillar to feed a young larva, or adding several caterpillars in the course of one day and closing the nest for good. Baerends's paper was the first to demonstrate clearly that, although it is of great importance to identify behavioral units at various levels of complexity, it is equally important to understand the ways in which they may be organized as to sequence, varying intensity, synergism, and antagonism. To this day there are few, if any, other papers which state the problems involved so clearly, analyze a given case so thoroughly and, on being reread today, open up so many still unsolved questions; in my opinion, it should be read and reread at least once a year by every ethologist interested in behavioral organization.

Only some ten years later (1950), N. Tinbergen published his model of the hierarchical organization of instincts. This differs from Baerends's model in requiring the performance of one specific activity before an activity next in sequence can be elicited. In this, it is closer to the old chain-reflex model. To some extent, the differences in the two models reflect only the differences in insect and vertebrate neural organization: while the "empty-handed" morning visit *(raupenloser Besuch)* to a particular nest decides the wasp's mood for a whole day, a vertebrate may alter its course of action at almost any turn. Yet both models have one essential feature in common: they are one-way systems proceeding from a more generalized level of mood or instinct to ever more specific levels, and the resulting hierarchy is thought to run parallel to the temporal order in which the consecutive activities manifest themselves.

Such was the authority of Niko Tinbergen at the time that his model was almost universally accepted without question. I myself tried hard to assemble the behavioral components of prey-catching, fighting, and mother-young behavior in the cat according to the model. In the first edition of my "Studies on Cat Behavior," I succeeded to some degree by presenting a highly idealized or typified account of these very complex activities (1956). Yet there were three areas where it was apparent that the model could hardly be stretched to cover the phenomena: in play; in the fact that some components of prey-catching could change their role under varying conditions of hunger and prey deprivation from consummatory to appetitive act and back again; and in the arbitrary use of components of threat and attack in social intercourse dependent on ranking and familiarity between any two animals. The significance of these incompatibilities with the Tinbergen hierarchy model had dawned upon me much earlier, but the evidence still seemed too incomplete to refute the model. I had realized that a hierarchy model following temporal sequence actually turned the original idea of Lorenz's upside down: according to

Lorenz it was the final or consummatory act that activated the guiding or steering
taxes and the sometimes long and involved chain of appetitive behaviors which
eventually led to, or produced a situation allowing, the discharge of the consummatory act. If this were true, then there had to be some kind of mechanism in the
nervous system which made it possible for a future behavior to influence, guide,
and organize the present, a mechanism capable of activating a variety of instincts
and learned behaviors in ordered and directed sequence. And behold, such a
mechanism was already well known, if ironically only by its sometimes bizarre
side effects.

At the 1952 Ethological Conference in Buldern, I presented a paper on the so-
called displacement activities (reprinted in English, 1973). The hypothesis then
favored was that an already activated instinct movement whose discharge was
impeded by either external obstacles or the simultaneous activation of another,
antagonistic instinct would "spark over" from its own "autochthonous" pathway
to an "allochthonous" pathway and thus drive another, unrelated movement
which was not inhibited by circumstances. Consequent on the hypothesis, it was
assumed that the discharge of the displacement movement would lower the energy
of the originally activated instinct(s). This, however, was not so in quite a number
of cases, where the activated instinct actually builds up while the displacement
activity is going on and often also increases in intensity. Moreover, a stimulus
often fails to elicit the adequate response to itself and a displacement movement
appears in its stead. The cats provided me with plenty of examples of both kinds.
My conclusion was that there was no 'sparking-over', but that the activated
instinct could disinhibit or directly stimulate the propensity component of another.

The disinhibition hypothesis later proposed by van Iersel and Bol (1958) tells
only half the story and cannot account for the cases in which the displacement
activity is increasing while the blocked activity which gave rise to it remains level
or even diminishes—when the displacement activity will often become "genuine"
(e.g., displacement feeding). In my interpretation that both disinhibition ("block
removing," so to speak) and facilitation (rise in excitation) were necessary to
account for all the observable phenomena, I was confirmed by von Holst's work
on relative coordination and the central nervous automatisms. He found that the
influence of one automatism over another can be inhibitory as well as facilitating,
depending on relative frequency and phase relationship. The influence affects
amplitude as well as frequency independently of each other: while the amplitude is
being increased, the frequency may be reduced, and vice versa. Thus the precise
ratio or relationship between the two (or more) automatisms at a given moment
determines the dominance of one automatism over the other(s) and in which way
that dominance manifests itself. My assumption was and still is that the mutual
relationship between instincts is determined by a neural/humoral set of mechanisms which work on the same principle as von Holst's central nervous automatisms.

Displacement activities are conspicuous for their "oddity"; they do not seem to
fit into the context in which they appear. However, what if the same kind of
mechanism would produce something that did not seem odd, which fitted
smoothly into the seemingly goal-oriented flow of actions and reactions? Clearly

what we then had would be "appetitive" behavior. My contention is that it is the same kind of mechanisms that normally initiate and maintain both appetitive behavior and, occasionally, something one might also call "freakish," which then, because it is so conspicuous and startlingly odd, often acquires a new function as a social signal.

I find it puzzling that so far no fellow ethologist has taken up this idea, nor has anyone felt the need to look for some other hypothesis on how a consummatory act not yet started is able to instigate, organize, maintain, and direct a chain of activities which sooner or later ends with that consummatory act. Conditioning, where it applies, is not the answer but only another way to formulate the same problem.

This way of reasoning necessitates a modification of the original Lorenzian definition of "instinct," which is of a narrowly defined motor pattern (the instinct movement or fixed action pattern) that creates its own "drive" energy. (To avoid confusion with older and infinitely varied drive concepts, I use the term *propensity*.) Lorenz claimed these to be inseparable, different aspects of the same device. If, however, the specific propensity which manifests itself directly in a specific motor activity has so many forms of manifesting itself indirectly, it must be essentially separate from the motor pattern. This is corroborated by the fact that cats who have never attacked and killed a prey animal, and who will not do so however adequate the hunger condition and stimulus situation, do so promptly and expertly when electrically stimulated in certain areas of the anterolateral hypothalamus. The motor sequence is latently present, fully programmed; but the propensity is lacking and is only momentarily replaced by the electrical stimulus. So far, no increase or continued repetition of the electrical stimulation has ever succeeded in setting the propensity to work of its own accord.

The Innate and the Gene

I think I should insert a few remarks here on the old nature-nurture controversy, the ethological concept of the innate and its relation to the heritability concept, and about the use and misuse of the concepts of homology, analogy, and convergence in ethology. Evolution produces a combination of genetic information and plasmon properties that, by interacting with the environment in which the combine finds itself, by accident or design produces functional structures which then perform those functions we call "life." The functions manifest themselves either permanently or at regular or irregular intervals. By and large, behavior belongs to the two latter categories. "Innate" is a characteristic that shows negligible variation despite considerable variability in some of the environmental factors with which the developing organism interacts. It is a (relatively) peristostable trait in the sense of classic Mendelian genetics and insofar "hereditary"; but this does not imply any assumptions about the underlying genes and the kind of phenogenetic developmental process which eventually produce it. Some critics of the concept of "innate" behavior have alleged that calling a behavior innate is tantamount to saying it has no developmental history. This was never implied, and is indeed nonsense. The instinct movements that in certain conditions produce "aggression" are certainly no more "in the genes" than the window of a house is in the architect's blueprint, and yet the blueprint determines what kind of window it is to be and where it is to be placed.

As the foregoing implies, each act, every smallest identifiable unit of manifest behavior has a threefold history: phylogenesis, ontogenesis, and actogenesis—the way in which the normally latent physiological mechanism produces observable behaviour. An "epigenetic" view (Kuo, 1967) is legitimate, but as little capable of making us understand the whole of behavior as either of the others when considered by itself.

Thus, when speaking of innate behavior, we are in reality referring to the phylogenetically and ontogenetically developed brain mechanism which at times produces that behavior. This mechanism, however, is never directly observable. The most refined methods of neurophysiology and biochemistry allow us at best glimpses of some part or aspect of it, but never the workings of the whole *at the same time.* This accounts for the difficulty in homologizing behavioral items: what we really want to homologize is the physiological mechanism responsible, which we can identify only by its manifestations. We can put an organ or a bone on a table, dissect it and compare it directly with others. The bane of the matter is that identical functions or results of functions are not proof that they are being produced in an identical way or by an identical mechanism. When a chimp, a rat, and a pigeon all learn the same type of maze, they produce very similar learning curves. But this does not mean that the learning mechanisms employed are similar or even identical. In fact, we know that they are not (Fischel, 1948; Bitterman, 1965). The performances of the chimp, the rat, and the pigeon are analogous, not homologous. Likewise, a morphological trait such as a jawbone or that elusive behavior-producing physiological mechanism that looks similar in related species may or may not be produced by identical genes and phenogenetic development. True, most of the time they are, but there are cases where the trait has been maintained while the genes were changed. Here it must be borne in mind that the homology concept was designed to deal with morphological characters and that its touchstone was not common genes but originally correlation with other characters and, later, common ancestry. When applying the concept to a wider range of phenomena, we must be careful to stay within the same level of system complexity. We may homologize genes or we may homologize traits, or developmental processes leading from the one to the other, but we must not confuse the levels.

When attempting to homologize behavioral items, again we must resort to the question of common ancestry in order to achieve more than speculative certainty that behavior similar in form—not function or result—is produced by homologous mechanisms. There is some justification for this assumption because the mechanism can be subject to natural selection only when and while it manifests itself. In those cases where natural selection favors certain functions or results, it may as easily be productive of analoga as of homologa. Form, however, the particular way a movement is performed, is strongly suggestive of corresponding properties of the underlying mechanism.

Very often one finds the homology issue confused with similarity. However, homologous structures and functional mechanisms are not necessarily similar. Where are the similarities between the gills of a shark and the parathyroid glands of a mammal? And the functional properties of a system can be changed out of all recognition by seemingly insignificant changes in the structure of the system—witness molecular structure. Hence we must be prepared to find homology in behaviors which are not at all similar in form. This has so far been shown only in some social releasers, but it may be so in many other cases where we may as yet—because of the dissimilarities of the behaviors in question—not even suspect the homology of the underlying mechanism.

I said that "innate behavior" means and always meant to me a relatively peristostable (sub)system. In this sense innate behavior components are "genetically determined"—but so, of course, are all peristolabile systems also; the range of modifiability as well as the kind of factors which may induce modification are preset genetically, and it must be borne in mind that modification by no means invariably implies adaptation. There are numerous modifications that are nonadaptive.

The peristostability-peristolability issue has unfortunately been most inadequately expressed and discussed in ethology (as in related disciplines) as either the "nature-nurture controversy" or the "learned-unlearned dichotomy." The nature-nurture discussion is rather nonsensical because nurture, where it occurs, is very natural indeed. It is logically inadmissible to start a dualistic conflict between two concepts which are not on the same level of complexity and one of which is contained within the other. Learned-unlearned is unfortunate because it considers only one particular set of mechanisms without identifying and investigating other ways of modifying behavior. In the heat of the argument, then, every observed modification was and often still is interpreted as "learned."

Methods and Ideas

In 1952 I joined the German Scientific Film Institute as head of the biology department. My work there widened my experience of techniques and problems, and I spent much time filming and studying vertebrate locomotion. My own cat research naturally had to be relegated to spare time. In 1958, at last, the Max Planck Society agreed that a small research unit should be built for me within the administrative framework of Lorenz's department in the Institute of Behavioral Physiology. For this I had to thank, besides Lorenz himself, the very active support of Erich von Holst and to some degree Alfred Kühn, then director of the Max Planck Institute of Biology in Tübingen, who was very interested in the progress of ethology. The unit was to be erected on the premises of a major zoological park to enable me also to study the larger cats and to command the services of an experienced zoo veterinarian for my own exotics. Eventually, we decided in favor of the zoo in Wuppertal.

While the unit was under construction, a grant from the American Air Force in

Europe enabled me to stay with Professor Walter Richard Hess in Zürich for two years to study the written and filmed records of his Nobel Prize-winning electro-stimulation experiments on cats.

In the course of thirty years these experiments had provided us with a better knowledge of the neural substrates of complex behavior in the domestic cat than we had of any other mammal. From the start this had been another factor influencing me to choose felids as my main study objects. The aim of my close scrutiny of Hess's material was to learn as much as possible about the method and its implications, but also to try and see whether the trained cat ethologist's eye might not be able to interpret some of the data in more detail. In a few minor instances it was. But on the whole the result was disappointing in this respect, partly because Hess had been a very shrewd observer and knew his cats extremely well, but also because he had had an exaggerated sense of economy and had always started his film camera too late and switched it off again too soon.

Hess guided and aided me in the most amiable way and I became convinced that from the ethological point of view, electrostimulation of the brain is the most valuable tool for investigating brain mechanisms of behavior. I hoped to start this line of research myself in the new Wuppertal unit, but soon found that I should never obtain the means and the personnel to do both comparative work on many cat species and electrostimulation experiments on a reasonable scale. After much heart searching, I decided in favor of comparative work, but never ceased to regret the necessity. My later friendship with John Flynn, however, provided some compensation when I was a guest in his laboratory.

My knowledge of the method and results of electrical brain stimulation provided me with a background against which to check my interpretation of how the observed behavior might be organized. The two behavior complexes I was particularly studying—prey-catching and agonistic behavior—were already well represented in Hess's material, although he himself failed to recognize the prey-catching elements. In later years, Flynn and several of his collaborators also made considerable contributions. Yet while electrical brain stimulation serves as an effective check on an all too free interpretation of normally observable behavior, the method itself is not without its limitations. Electrodes and the control of stimulus parameters may be infinitely more refined now than they were in Hess's day, but considering the structure and function of brain tissue this is still a brutally crude way of probing into it. The effects of brain stimulation may in part be due to the method rather than the brain's organization, and so they are in a number of cases. The observed behavior of the nonexperimental cat is the phenomenon to be explained. The experimentally induced artifact can often give us an idea of what the normal mechanism might be, but could also mislead us if accepted without discernment.

Following ethological principles, I began by observing my cats, identifying recurrent movements and their sequence, studying the complex situations which served to elicit, maintain, and end these activities, and eventually cautiously altering one by one the factors in the situation which could be manipulated. The other kind of experimental approach, still considered by some as the only truly scientific one—the laboratory experiment under strictly controlled conditions, with everything excluded or kept constant except one independent and one dependent variable—is of very restricted value when working with higher mammals. I was vaguely aware of this from the start and acted accordingly, but only in later years found out why it should be so.

The most obvious and distinct manifestation of what Lorenz called "action-specific energy" and I now call a "propensity" is the initiation and the maintenance of the specific activity by which it is identified and after which it is named. Another, less obvious function is to start and organize appetitive behavior when the immediate discharge of the specific activity meets with internal or external resistance or obstacles. A third function, according to Lorenz, is to reduce the selectivity of the perceptual apparatus employed in triggering any respective innate releasing mechanism. In translating Lorenz's original German term *angeborenes Schema* into innate releasing mechanism, the emphasis was shifted from the perceptual aspect to the mere triggering function. Since then, it seems to me, very few ethologists have given thought to the far-reaching significance of Lorenz's observation: The current level *(Aktualspiegel)* of a propensity determines specific perceptive processes not only quantitatively but also qualitatively. The rising of a current level does not just "lower the releasing threshold," it reduces the degree of similarity to the "adequate" sign stimuli which is required to release the specific activity; similarity of form, color, contrast, position and/or movement pattern (when a sign stimulus is a movement) cannot be assessed merely in quantities—"similarity" is essentially a qualitative category.

Too little notice has been taken of the fact that any change in innate releasing mechanism selectivity is only a special case of the more general influence the current state of the overall motivational balance of the organism exerts on the perceptual apparatus. Perception is not just recording objectively what the physical environment offers. It is an active performance of preset machinery that is scanning the environment for such items as are at the moment motivationally "required" or "expected." If or when the requirements are not met, then a variety of events are possible: (a) appetitive behavior is initiated; (b) substitute objects may be accepted and reacted to; (c) the scanning apparatus "runs wild," becomes disoriented and casts wildly around for what it "knows" ought to be there. The latter seems to me to be the case in many of the experiments under so-called strictly controlled conditions: a factor that is absent or being kept rigidly invariable is often not a controlled but a disturbing factor and enhances rather than diminishes the variance in the results of the experiment. Thus the "First Harvard Law of Animal Behavior," "Under the most strictly controlled laboratory conditions the animals behave as they damn please," originally meant as a joke, is the unavoidable outcome of a basically sound but in the case of mammals misplaced and generally overrated principle. The beautiful experiments of Gottschaldt and Young (1977a, b) have provided the final proof of my interpretation.

I have often been accused of setting too little store by quantitative methods and evidence. This is not correct. But first, one should not confuse "quantitative" with "statistical"; secondly, one absolutely must, before attempting quantification and most certainly before starting any kind of statistical treatment, exactly identify what there is to be quantified; and thirdly, it depends on the kind of questions one asks whether methods of quantification are of any great use. Examples:

1. After the last all-annihilating nuclear holocaust, the first scientists from outer space arrived on earth. In a heap of rubble one of them found a fairly intact four-stroke engine. By accident he set it in motion and saw that inside— one spark plug was missing—something went up and down, and that something on the outside had started revolving. So he took two measuring instruments, and since these were not very precise he found a ratio between ups-and-downs and revolutions of 0.9877. He concluded that *very probably* the ratio was 1:1. His colleague when told of this resolutely cut the crank case open, turned the machine *once,* saw piston, connecting rod, and crankshaft and said, "Indeed, the ratio *is* 1:1." This is quantitative but not statistics. It is the second type of information and verification I have always been after. This does not mean that I despise a statistical approach as long as I cannot do better. But when one is forced to have recourse to statistical treatment in biology, it is a sure sign that the knowledge of the functional principle underlying the phenomenon in question is insufficient. Quantity never explains a functional principle. In a few strokes I can sketch a diagram illustrating to the fullest the functional principle of the four-stroke engine. It is a matter of structure and design. Any quantitative factors necessary to understanding it are of a purely relative nature: the pressure in the combustion chamber must develop greater force than atmospheric pressure plus internal friction, and so on; exact ratios are not required.

2. Every chemist knows that he cannot start quantitative analysis of any matter before completing qualitative analysis. I submit that this also applies to any other analytical science. The pious hope that if only one amasses sufficient data stochastic treatment of a refined nature will sort them out later is quite unrealistic. The use of factor analysis to identify elements in human motivation failed because the "factors" obtained in this way had no functional reality in the subjects investigated. The high hopes initially raised by numerical taxonomy likewise ended in disappointment. There is no way around it: in ethology too, qualitative analysis must precede quantitative analysis. It is my contention that in most areas, the former has by no means reached a level which would make a quantitative approach more than a mere hazard. This is why I stayed with qualitative work.

3. I am not interested in "pushing behavior around" (Skinner's repeatedly used expression in a lecture I once attended). I want to understand how it comes about and how it works. Likewise, I have very little interest in predictability; as proof of any hypothetical assumption about the causes which lead to the predicted behavior, it is rather open to doubt (Leyhausen, 1965); where it means statistical predictability, I am always more interested in the statistical minority than any majority which fulfills the prediction. It is most often the minority that holds the key to the problem of interest to me: How does it work? In short, if I had been so lucky as to find the first archæopteryx, I should not have looked so eagerly for ninety-nine more of them (though it would, of course, be nice to have one hundred archæopteryxes) as for just one specimen

representing a slightly earlier or later stage in the evolution of birds, which I should want to reconstruct. For that very reason I also have some reservations as to the value of preclassifying behavioral units so that they may be recorded in such a way as to make the data fit for any stochastic treatment I might fancy. This always leads to omitting too much detail from the records, detail which could be essential for the kind of qualitative analysis I have in mind. It is the seemingly erratic detail in an otherwise unchanging sequence which must arouse attention instead of being dismissed, let alone overlooked.

All this is not, of course, to say that there are no questions requiring sufficient numerical data and statistical treatment for their answer, and I and my coworkers have employed the appropriate procedures in several cases (Verberne and Leyhausen, 1976, Schmidt, 1975, Peters, 1978, Leyhausen, in prep.). There are many other problems that I should have liked to handle in this way. But to obtain the necessary data on any complex behavior of a mammal such as the cat and to perform the necessary number of experiments in all the possible gradations and variations for the result to be conclusive requires personnel, facilities, and equipment on a scale I was never able to command. I shall give an example of this in the next section.

The Cat Grows Up

The only way to find out whether a behavior is innate is to test its modifiability. Peristostability can be understood as the expression of selective pressures directed toward ensuring that each individual of a species is equipped uniformly with certain traits. In other words, there must be a selective advantage in uniformity or "likeness" with respect to such traits. This means that peristostability must not be taken in any absolute sense but only relative to the environmental (including social) conditions which prevailed during the evolution of the species. Conditions that remained constant over geologically long periods of time offered selection no "leverage" to ensure continued trait stability in the event of their becoming unstable. Hence, when an organism is subjected in the laboratory to radical changes in such conditions, it cannot reasonably be expected that the trait under inspection will remain unaffected—although this turns out to be so in an astonishing number of cases. Peristostability can have been selected for only with respect to environmental factors which during the evolution of the species frequently or regularly fluctuated or were unevenly distributed over the range of a population or the ranges of the populations of a species. However, when a peristostable trait starts to yield under "unnatural" conditions, that is, conditions never encountered while the trait was evolving, such modification is normally nonadaptive and described as "freakish," a disturbance or malformation (most striking case of recent years: the thalidomide children).

The method thought to prove incontrovertibly whether a given behavior unit is peristostable is the deprivation experiment: A developing organism is deprived of those environmental factors or experiences that could possibly influence and modify the development of the characteristic under investigation. If the characteristic develops just the same and does not differ from "normal controls," peristostability with respect to the environmental conditions withheld is proven. A negative result, however, is inconclusive, as the deprivation may have caused more generalized damage to the developmental mechanisms. One point should, however, be made: If the result is positive in only one single case, this is sufficient evidence, as exemplified by the extraterrestrial scientist who cut the crank case open.

Deprivation experiments are notoriously difficult to make with mammals, and I know of not a single one that is free of all suspicion that the observed effects were more or less generalized and not at all specific to the behavior in question. I believe this to be particularly true of the once-famous deprivation experiments on rhesus monkeys conducted by the Harlows (1962a, b). But this need not be considered an impasse.

A behavior unit is suspect of being peristostable when it is performed almost identically by all observed members of a species. There are many such units in cats, and I proceeded to record the conditions under which these units develop in the young. Conditions that vary considerably from litter to litter or even from individual to individual obviously do not modify behavior developing to near-identical patterns. To make sure, however, I experimentally enhanced their normal variation in a number of cases beyond what is normally found. With respect to prey-catching, for instance, there is normally great variation in the time (age of young) at which the mother first introduces solid food, or live prey, in the kind and size of prey first offered to the young, in the way the mother handles the prey in front of the young, and many other factors. Factors which hardly vary in different mother cats and litters, nature has obligingly varied considerably in different cat species. Yet quite a number of behavioral units employed in prey-catching are virtually identical in most cat species. Even species and generic hybrids did not differ from the parent species in this respect, either in the finished product or in the developmental stages.

Because of all this I felt justified in claiming that these behavioral units are "innate," peristostable (Leyhausen, 1956), and still do.

This does not, of course, mean that the factors enumerated above and others as well are without influence: they do not alter the form, that is, the pattern of the movements in question, but they change the stage at which these first appear and the zest and persistence with which they may be performed. Under these circumstances I deemed it more profitable to describe single cases in great detail than to summarize all available cases. Again, a hundred archæopteryxes do not prove one iota more than one does.

My own investigation of the development of prey-catching in young cats was, of course, prejudiced by the previous work of Kuo (1930). From the ethological point of view, the obvious shortcoming of Kuo's procedure was his failure to observe first what happens when cats are allowed to rear a litter without human interference. I had done this from earliest youth dozens of times. It was inconceivable to me that anyone could expect to gain insight into the ontogenetic process of prey-catching development when rearing the kittens under such restricted and monotonous conditions that the introduction of almost any new object would of necessity elicit fear or even panic and thus overlay or distort the expression of any other motivation which might be aroused. I am not saying that the results of such experimental arrangements are worthless, but they can be assessed properly only when projected against the normal background. This, the first tenet ethology ever proclaimed, was scoffed at by experimental psychologists until in the end they found it to be a truism. To call something a truism is in some quarters a way of dismissing the truth as being irrelevant.

Apart from this, Kuo did not understand the significance of the archæopteryx simile. He concluded that young cats must learn to kill mice and rats, because out of twenty young "reared in isolation" (the rough approximate of a deprivation experiment), only nine became killers. He had not one but nine "archæopteryxes" but was blind to their message. If only one of the twenty had become a killer it would have been irrefutable proof that the influences withheld in the experiment could not be *necessary* to the development of killing behavior.

There is, however, a far more fundamental difficulty in any attempt to investigate the ontogeny of behavior. The following example may serve to elucidate its nature.

Baerends and Baerends (1971) were dissatisfied with my qualitative approach to behavioral ontogeny (which they thought to be merely speculative) and set about remedying this unhappy state of affairs by starting a series of experimental litters which they proposed to study quantitatively. However, although they present their results in a number of graphs and tables, they admit that their data do not suffice for subjecting to analytic stochastic treatment. So we are still left with qualitative interpretation.

But let us assume that statistical analysis had been possible. What insight into the formation of a behavior system would it have provided to learn that a certain item, say crouching before the jump, makes its first appearance between the ages of 2.5 and 6 weeks, with a mean on day 29 and a standard deviation of 4? Of course, we could then compute these values for other items and test whether they are correlated directly, inversely, or not at all. And what would we conclude?

Some would claim that such correlations, once established, would teach us something about the ontogenetic emergence of adult behavioral structure. Alas, I fear this to be a forlorn hope.

I have already explained why the temporal sequence of behavioral acts is no guide to their hierarchical organization. Likewise, the ontogenetic stage at which a certain behavior first manifests itself and the temporal order in which various behaviors first appear do not reliably reflect the course of behavioral ontogeny. The age-old belief of psychologists and biologists that this is so has been shown to be an illusion, an illusion which ethologists have shared. It should never have been allowed to arise because we knew that in many cases the motor pattern of a given behavior remains latent for long after it is fully functional. Lorenz has drawn attention to this in some of his earliest writings. It was proven in electrostimulation experiments. As overt behavior gives us only indirect access to the structures and mechanisms that produce it, and since *indirect* here also implies an incalculable time lag, overt manifestation does not mirror behavioral ontogeny. Many behavior patterns develop latently without needing any feedback from their overt activity, and so does much of their hierarchical organization—not, as we shall see, in the sense of a rigid order but as a more or less flexible and adaptable interdependence very similar, if not identical, to von Holst's "Relative Coordination." Thus only behavior manifestations during ontogeny which are dependent on external input and feedback of the kind described by von Holst and Mittelstaedt as the "reafference principle" (1950) reflect in part the ontogeny of the mechanisms of behavior themselves. *Maturation* is a word of the past. The phenomena for which it served as a label, however, still present us with methodological problems, some of which remain unsurmounted.

The Cat Walks—by Himself?

As I mentioned before, when writing my first account (1956) I attempted to interpret my data within the framework set by Baerends and notably Tinbergen. I knew I could do this only by ignoring certain incongruities (inconsistencies, discordancies), witness my 1952 paper. However, I was not sure then whether they were not of a superficial, accessorial rather than basic nature. The 1952 paper had not been well received by the ethological fraternity. Yet after completing the manuscript of "Studies on Cat Behavior" (1956), in 1954, I felt my doubts growing, and presented a paper "On the Relative Hierarchy of Moods in Mammals" to the 1955 Ethological Conference in Groningen. Again it fell flat. I felt deeply discouraged because I had thought that at least men like Niko Tinbergen and G. P. Baerends would see the possibilities of the new approach to the problem of hierarchical organization and the indubitable flexibility this organization exhibits in highly evolved vertebrates and particularly in mammals.

I took up other investigations and had nearly forgotten it all when almost ten years later I was faced with the need to write a paper to dedicate to Otto Koehler for his seventy-fifth birthday. I felt that none of the things I was working on at the time was really ripe for publication and in my despair I decided to write a purely descriptive paper in augmentation of the evidence and comparative material presented in the 1956 study, which had already been dedicated to Koehler. As record after record of species after species and individual after individual was scanned, the picture suddenly stood out in its entirety and in brilliant clarity. I could see the whole machinery working in every simultaneous and temporal detail. Every bit of a vast accumulated material with its often puzzling and sometimes seemingly contradictory evidence fell into place. Careful inspection of each piece of evidence and an exhaustive search for "misfits" followed. I did not find any and have not to this day. Thus the full theory of "Relative Hierarchy of Moods" was born.

It shows not only the way in which instinct movements change their role from consummatory to appetitive function and back and how they are integrated into various functional contexts, but also how they are joined up with acquired elements. The "learned" does not in any way modify or replace the "innate." The learned elements form a separately stored set of subprograms which may be "intercalated" with the innate ones under the organizing directive of the momentarily ruling appetence for the activity whose propensity is dominating the motivational balance. I cannot sum it up more concisely than I did in 1952:

The action-specific energies of an animal have a constant mutual relationship of inhibition and facilitation. Further, they can mutually raise or lower each other's internal threshold of elicitation. None of these mutual effects is determined for all time; they are all constantly changing, either gradually or abruptly, according to the "state of co-action" [von Holst, see Leyhausen, 1954a] of the system as a whole, at any given time. The influence exerted on a specific excitation and the associated threshold need not necessarily be oriented in the same direction.

It is apparent that my theory is based on Lorenz's old concept of action-specific energy. This concept has found less and less favor over the years, until some colleagues have mocked me as "the last believer in action-specific energy." There is, however, less of the believer in me than in most of those who so desperately try to explain away the very real phenomenon for which "action-specific energy" is just as good a descriptive term as perhaps half a dozen others. Animals do exhibit initiative. Certainly cats do, and in the absence of any relevant stimulus situation or its past traces in memory. For this there is irrefutable evidence, not only from direct and prolonged observation but also from physiological experiments (electrostimulation, differential effect of various tranquilizers on reactive and spontaneous fighting, etc.). Von Holst was the first to suggest that action-specific energy should be renamed "action-specific *substance*," and indeed there is a good deal of circumstantial evidence that the specific agent is humoral rather than purely neural. Recently it has been suggested that transmitter substances and their differential effects are what is responsible for the observable phenomena and not excitatory substances acting directly on or within the neurons' bodies. This could well be part of the story, but hardly all of it. Michael's findings (1960) point to a rather slow, cumulative process within the target area of sexual hormones for days after the hormone has left the brain tissue again.

"I am the cat who walks by himself," Rudyard Kipling lets his hero tell "first woman," and most of the time this is interpreted as referring to the supposed solitary, asocial nature of the cat. But the continuation, ". . . and all places are alike to me," reveals a second, less obvious meaning: places and situations are of no great significance, the cat is master of his own actions and intentions. He walks not by stimulus and response, he walks "by himself": he is an autonomous subject capable of initiative.

My position is, of course, open to four questions.

1. Is not the old "flush toilet" model Lorenz used to describe the functional properties of his action-specific energy + specific instinct movement system too simple altogether?
2. Why postulate a large number of specific propensities when a central general drive would serve the same purpose?
3. Why is it at all necessary to adopt an energy model of motivation?
4. What further approach would be available to test the validity of my theory?

1. The model appears simple, but its implications certainly are not. The trouble is that it has always been looked at singly. To my knowledge, nobody has ever attempted to trace all the possible interactions which might occur between only two systems working on the principle of the model, let alone three or more. Only recently (in press) I made a rather timid attempt to depict at least

partially the interactions of only two such systems. With three, the complexity of possible interrelationships becomes appalling; with a dozen they might defy any existing computer. My model of these interactions and phasic relationships is based on von Holst's findings on the functional properties of relative coordination between the central nervous automatisms of the spinal cord. Anyone wanting to understand what "Relative Hierarchy of Moods" is really about will have to read all von Holst's original work with painstaking attention to minute detail: unfortunately it is all in very difficult German (full bibliography in Lorenz & Leyhausen, 1973).

2. A "general drive" concept is incompatible with a number of facts. I have already mentioned that there are domestic cats who do not kill prey. If we were to assume a general drive, this failure could be due to two circumstances: either the motor pattern of killing would have to be absent or at least in some way disturbed or imperfect, or the channel into which the drive is normally released to activate the killing pattern might be blocked. The first hypothesis is wrong: when these cats are stimulated electrically in the right locus, they kill perfectly; the complete motor pattern is latently present. The second hypothesis is inadequate because both long-term observation and experiment show that the threshold and the "drive" vary independently of each other. In this, the internal and the sensory thresholds must not be confused. Sensory adaptation and sensitization are likewise independent of the internal releasing threshold. Lastly, if some "drive" were drained off or discharged in whatever channel happens to be opened up, the level of readiness to perform any other activity ought to be lowered quite unspecifically. In reality, the reverse happens more often than not.

3. Motivation sets behavior in motion. This is usually effected by the cooperation of a propensity (or a momentarily synergic cluster of propensities) urging from within and some motive (stimulus situation, past experience, etc.) from without (the influence of past experience being an outside influence when viewed from the instinct system). The ensuing behavior is then organized toward the consummatory act or situation. Neither the propensity's nor the motive's part can be accomplished without energy expenditure, however minute. The organization of behavior might perhaps (though I doubt it) be described or explained by using a nonenergy model; but organization by itself is dead. In short, he who wishes to do away with "energy" in motivational theory gainsays its essence. At best, he reduces the animal—and that includes humans—to the status of a somewhat advanced slot machine which is indeed unable to exhibit spontaneous initiative.

It has been suggested that we should be satisfied with a probability model of motivation. This would, of course, be some way of describing what is overtly observed, but it would also mean foregoing any aspiration to investigating the mechanics of motivational processes; it would not even allow the development of testable hypotheses about them. I have yet to meet a model of motivation that describes the observed facts of motivation more succinctly than the old Lorenz model. It has not yet outlived its heuristic usefulness by far, and as long as nothing manifestly better is offered, I shall continue to use it.

4. One way of testing my hierarchy-of-moods model is by direct multi-electrode brain stimulation. It has already been shown by von Holst and von St. Paul (1960) that certain assumptions of a hierarchical model can be tested and corroborated in such experiments. In the cat, the external situation as well as internal changes can alter the effect of an electrode from eliciting attack to eliciting defense, withdrawal, and even full flight. Now, it is largely the same motor elements which are employed in the course of attack as well as defense. The sequence in which they appear and their relative intensity determine whether the one or the other results—and this is relative hierarchical organization. A series of such experiments more directly aimed at the phenomena in question could provide some answers. A more elegant method of theoretically testing some of the implications of my model will be computer simulation. There are still some technical snags, but I am hopeful of presenting the result of such an attempt in the not too distant future.

The most violent objections to the spontaneity of propensities and the spontaneous arousal of appetences caused thereby have been encountered in the long-lasting discussions of Lorenz's theory of aggression. In my opinion, the critics are overlooking or neglecting four points that put a very different complexion on the whole question.

1. As Lorenz has pointed out from the very beginning, much of the specific energy of an instinct is expended in "small coin," abortive attempts and intention movements. A violent discharge in the absence of adequate stimulation is therefore rare and not the rule. "Dripping out" from the "reservoir" normally prevents too high a buildup of specific energy and keeps it at a more or less even level. This is particularly true of agonistic propensities, which easily find partial outlets in small display gestures.

2. Some propensities are subject to the effects of use and disuse. This is not introducing learning by the back door. The effect is "learned" as little as a muscle learns to grow stronger with use and weakens again with prolonged disuse. The killing bite of a cat is governed by such a trainable propensity: in ontogeny, it matures and reaches peak condition at about nine weeks. If not in use by then and not kept in training, it wanes away again. A three-month-old cat that has not yet killed is a hardened nonkiller. With some difficulty it can be turned into a killer later by a laborious training process (Leyhausen, 1965). An adult killer with no opportunity to meet live prey will show an increase in killing readiness for the first three months. This is demonstrated by a growing tendency to bite substitute objects of diminishing similarity to prey objects, eventually even "killing" lumps of clay. After three months, however, the readiness begins waning and has almost completely disappeared after a year. It can then be rekindled and slowly built up to its former level. It is possible to make this experiment several times during the lifetime of a cat, always with the same results. The fourth retraining in no way differs from the first. It is not a question of forgetting and remembering. No cat would forget something of such vital interest to it within a year—not in five years, as I have cause to know. It is strange that ethologists should so far have paid so little attention to this

phenomenon. This may be due to the prejudice that ascribes every modification of behavior to some kind of learning. There is good indication that some at least of the agonistic propensities of mammals are trainable in the sense described and that many of the results of experiments thought to prove that "aggression" is learned are really due to "trainability."

3. "Aggression" is not an instinct but a drive as newly defined by Lorenz (1963, 1978). It is a hierarchical system of agonistic instincts combined with appetitive behavior of varying complexity. This system is not permanent but arranges itself from time to time according to the principle of relative hierarchy of moods. Single agonistic instincts may show "vacuum activity"; "aggression" hardly ever.

4. In a propensity system governed by relative hierarchy of moods, propensity controls propensity. The hierarchy of the moment is not an undivided system. One emergent subsystem struggles with others and the one "winning" for the moment more often than not must compromise. Control of one subsystem by another is in part indeed a matter of learning. Besides the obvious eliciting elements of a situation, there are others latently present, the recognition and memorizing of which may be learned. In this way an antagonistic or even "multivectorial" control of one or more subsystems by one or more others can be conditioned. The mechanism thus established is what McDougall (1935) once sought to pinpoint by his concept of "sentiment." It is what I have called "stabilizing systems [*Verspannungssysteme*]."

These four points taken together ensure that aggression, and particularly violent aggression, is not inevitable, though at times it is certainly a tendency or even a temptation. And we do not always resist temptation—so there is, very occasionally, spontaneous aggression. War is an entirely different story again, especially modern war. A persistent war effort for years on end certainly cannot be motivated by a single drive, however complex. The war effort of a nation has many more and equally or even more powerful motivational sources besides and before any aggressive drive.

The Sociable Cat

Animals living in organized groups are usually thought to establish among themselves a rather rigid, durable, linear ranking order (dominance order, social hierarchy). For a long time I was unable to detect any clearly defined social order in my several groups of caged cats and contented myself with the assumption that cats, being solitary animals, were unable to form one. During that period, I was little interested in social organization and still very busy identifying and cataloging innate and acquired elements of social behavior, not their context.

However, while a prisoner of war in Canada (1942–46) I could not help casting an ethological eye over my fellow prisoners. This provided me with a fair notion of the effects of—among other things—crowding on social relations. Thus it was plain to me that some at least of the erratic social behavior of my caged cats was due to the effect of crowding. But only when we had assembled sufficient observations on the interactions of free-ranging cats were we able to understand what was actually happening in the cages.

The main clue had been provided by H. Hediger (1949) in a small paper on "Mammalian Territories and their Demarcation." Hediger pointed out that a mammal does not normally enter every square foot of its territory, but uses only certain pathways and places, *keeping to a more or less strict time schedule*. It was, of course, long known that in birds, territorial behavior underwent seasonal changes, but this is an all-or-nothing nature, the bird behaving territorially in one season and not at all in another. In mammals Hediger found a periodically changing use of certain parts of the territory, while territorial behavior as such remained. While there are certainly wide differences according to species, in felids, at least, the time of day (week, month) element in territory use is very pronounced. In addition, cats are never able to control all their territory simultaneously for the obvious reasons of terrain and time needed to travel from one part to another. This allows the partial overlap of neighboring territories.

Mammalian territories of this kind cannot be adequately described as a piece of land defended by the owner. A time element must be included in the definition. In practice, I found that the outcome of an encounter does not decide the issue once and for all, but only with respect to the place and the time of day (week or even month) where and when that encounter occurred. Thus territorial fighting will result not in fixed boundaries but in establishing right of way or use of locality respective to time. In other words, a ranking order between neighbors emerged which is dependent on locality and time—I called it "relative social hierarchy."

Returning to my caged cats with this in mind, I detected that their seemingly erratic behavior was a curious mixture of absolute and relative ranking. The more crowded the animals were, the more the balance shifted toward absolute ranking, with a consequent increase in mutual intolerance. It turned out that there are two social areas where absolute ranking is normal: within the litter and between males when they meet outside their territories. It is obvious that in a community of territorial neighbors, the proper balance between the two ranking principles and the predominance of the relative order prevents an overpoweringly dominant animal from becoming a tyrant: within its territory and at the proper time, each community member is superior—a kind of limited equal rights system. It enables the owners of adjacent territories to become truly a community. I call it a neighborhood system. So-called solitary animals may be aloof, but they develop their own kind of sociability and social organization.

On the basis of these findings and a great deal more detail than can be enlarged upon here, I described the social system of the free-ranging domestic cat. Its essentials can be found in the social systems of most other felids so far studied in the field. Yet I made one fundamental and inexcusable mistake: I took the form of the system as I found it in several urban and suburban house cat populations as universal.

It was long known that the social behavior and social structure of some mammals such as the wapiti and the hippopotamus vary according to different ecological conditions. I myself quoted these examples without quite realizing their full significance. In the lion, social structure ranges from presumably lifelong single-pair formation in desert habitats to the enormously complex system of the Serengeti population as described by G. Schaller (1972) and B. C. Bertram (1975), which is based on the interaction of female prides, male fighting "brotherhoods," and landless "nomads" (Leyhausen, 1979). All this should have warned me that in the domestic cat I had found only one of an unknown number of possible variations on the same theme. But since then several researchers have investigated the social organization of feral cats living under a variety of widely differing ecological conditions in Britain, Italy, Japan, Sweden, and Switzerland (Dards, 1978, in press; Izawa and Ono, 1982; Liberg, 1980, 1982; Natoli, 1983, 1984; Panaman, 1981; Schär, 1983). The patterns of social structure that emerge vary considerably, due largely to environmental factors, but genetic differences in individuals and populations also seem to be important (Schär).

In the meantime, I had continued my observations on the effects of crowding. But the really important contributions in this field were the experiments of J. B. Calhoun (1962a, b, 1963). Until he published his results, only one such effect had been discussed: It was said to increase aggressiveness. Calhoun, however, pointed out that the symptoms shown by overcrowded communities of rats and mice were not confined to increased aggression. Neglect of young, social indifference and apathy, neglect of self-maintenance (proper feeding and auto-grooming) were other such symptoms. They all occurred within the same overcrowded community, depending on the social status of the individual and on other factors, some of which one might almost label "cultural."

To cut a very long story short, I have now formed the hypothesis that the social system of a species is a range of potentialities that may strike a balance in a number of ways consequent on how environmental and social factors affect the various behaviors employed in social intercourse and in building a social organization. Any change in the factors may cause a shift in the balance. But systems often resist the pressures toward change; they exhibit a certain stability or inertia. However, when pressures grow to exceed the capacity of the system to strike a balance at all, serious disturbance or even total disruption will result. Overcrowding seems to be one such pressure with which the system of social behavior and structure seems unable to deal. Now, the limit of population density that determines "crowding" differs from species to species and it is to a certain degree dependent on various accessorial factors. Consequently the limit has a certain "band width," but it exists nevertheless. This limit determines what I have termed the *density tolerance* of a species. Because of the "band width," the density may exceed tolerance for some time, and latent damage to the social behavior and system of a population may reach considerable proportions before density itself is revealed as the cause of that damage.

Cats are Human and People Cattish?

Inferences from the behavior of other animals to human behavior are often questioned. Since we believe we are under stricter moral obligation to our own species than to others, we cannot apply to humans all the methods employed in research on animal behavior. A striking example is the deprivation experiment, ironically also known as the "Kaspar Hauser experiment." Thus we are deprived of one of our foremost tools when investigating instinct movements and innate releasing mechanisms in humans. However, the case is not entirely hopeless; the procedure is just more tedious and requires very great persistence. I have pointed out above that identical patterns in populations living under different ecological conditions can serve almost as well. This is one of the reasons why Eibl-Eibesfeldt is accumulating his cross-cultural documentation of human behavior.

It is often said that to homologize comparable behavior in higher mammals and man is merely speculative and that indeed there is no more than analogy. Those who argue thus—mostly sociologists and cultural anthropologists—succeed only because they always manage to have each case discussed in isolation from all others. If as many as possible were taken together, the ludicrousness of explaining them all by analogy would immediately be demonstrated. The sheer number of close conformities not only of elements but also of patterns excludes any notion of chance convergencies.

Another point concerns the basic functional principles on which the behavioral system is built and along which it works. Here we are on much safer ground than most ethologists seem to realize. Organs and functional systems are designed throughout the class of mammals on basically identical principles (cf. Eisenberg & Leyhausen, 1972). Whether we dissect a mouse or an elephant, a bat or a whale, a squirrel or a human, they all have a heart with two atria and two ventricles and completely divided pulmonary and systemic circulation. Of course, the apparent differences between all those hearts are enormous, but that in no way diminishes the fact that they are all, without even the remotest possibility of a doubt, hearts typical of a mammal. I think it incontestable that the motivational systems of mammals and particularly the propensities within those systems work on the same basic principles. However much the systems of various species may differ, the basic design of the system cannot but be mammalian—and that includes humans. I am reducing people to cat level as little as Lorenz is reducing them to goose level. I am not saying that if one electrically stimulated the antero-lateral hypothalamus of a man, that man would proceed to bite a mouse or maybe a pig or sheep in the neck and kill it. But if we are not to assume that the principle of Relative Hierarchy of Moods also applies to humans, we have no choice but to discard the whole notion of evolution along with it.

Likewise I would claim that the principle of density tolerance is as valid for man as for any other mammal. In fact, the principle extends over almost the whole animal kingdom; but I think that mammals have a basically identical pattern when faced with situations which tax their density tolerance beyond its limits. This pattern is recognizable in most present-day human societies and populations despite the enormous number of cultural and economic superstructures which greatly diversify and disguise the phenomenal symptoms. But the basic syndromes are there, exactly as they present themselves in Calhoun's mouse "universes." Please let this not be misunderstood: the symptoms of overtaxed density tolerance in a human population are human symptoms, as those in the mouse "universe" are murid symptoms—it is the pattern that is universally mammalian, not the individual symptom.

Sociobiology—Social Darwinism Resurrected?

Most cultural anthropologists would agree that all human behavior, and certainly that pertaining to crowded conditions, is of cultural origin and that the remedies for any resulting sociopathology must also be cultural. The rise of sociobiology has in part been a repercussion of this attitude. However much cultural traditions and mores may transform, cloak, or disguise behavior patterns, the patterns themselves are much older than any culture, most of them much older than man as defined in the genus *Homo*. Hence, if many of the phenomenal ailments of human societies seem to have cultural origins, the next question must, of course, be how it came about that the cultural conditions went so awry as to cause these ailments. This inevitably leads us back to the individuals who create and maintain culture, and to their behavior patterns. Something must have disturbed their system of social behavior, and it is this disturbed system which in turn creates unhealthy cultural developments. Derek Freeman, the Australian cultural anthropologist, once let himself be goaded into flatly stating: "Culture has no cause." How, then, can it become a cause? Robert Mayer might well start rotating in his grave!

It is fashionable today to regard the world of culture as a sphere by itself and to think in a dualism of the "merely biological" and the "cultural" as if they did indeed belong to separate worlds. Even biologists as great as Sir Julian Huxley and Theodosius Dobzhansky put forward the hypothesis that for all practical purposes in the human species "biological evolution" had ended and "cultural evolution" taken over. But in reality there is no such thing as "cultural evolution." For one thing, cultures do not "evolve": they have no young that are selected according to their relative success and then in turn reproduce. Cultures have no life but that of the individuals who build them, adopt them, maintain or change them, destroy them, and eventually forget them again—all in the course of relatively few generations. Cultures are developed by human individuals. Likewise it cannot be said that cultural factors have become the principal selecting agents. No single culture has ever existed relatively unchanged for a sufficient number of generations to exert any significant selective pressure in human evolution. Even after thousands of years of cultural separation, humans have remained just the one species *Homo sapiens*. The basic need of the species to develop and to "have" culture—any culture—is, of course, a different matter.

Thus *cultural evolution* is a misnomer and the term should be abolished. Developing a diversity of cultures is part of human biology, human behavior in particular. Any conceivable culture bears the marks of this. What sociobiology attempts to do is not revive social Darwinism, which anyway is not an invention of Darwin's or any biologist's but of some philosophers and sociologists of the day who did not understand what Darwin was about. Sociobiology attempts to put things in their proper historical perspective again. It really ought to please the cultural anthropologists had they not such a stunted notion of history. They are like a person for whom the history of the motor car begins with Carl Benz and who does not want to know about Mr. Otto, who invented the combustion engine, the prehistoric genius who invented the wheel, or anything in between. Likewise most of what makes humans human has a history much, much older than Adam. Sociobiology lays bare the roots of human social behavior, which are also the roots of all culture.

A very important difference between social Darwinism and sociobiology is, of course, that the latter stresses the survival value of altruistic behavior. This interest is genuine. When sociobiologists stress the fact that if viewed from gene level, altruism is "selfish," it must not be misconstrued to mean that they think altruistic behavior to be generally insincere and hypocritical; it is just a way of expressing that altruism must be of survival value to have evolved at all. However, in their endeavor to prove their point at the level of population genetics they fall into a trap.

They begin with the unanswerable premise that natural selection can work only on individuals. They proceed with a second, less unassailable premise that it is the differences between individuals that are inherited, not the characters which exhibit these differences. Here they confuse methodological requirement with the phenomenon to be investigated: It is the characters that are inherited, but the differences which can be measured. The measure is Janus-faced. If the different expression of a trait in any two individuals is due to genetic differences, measuring that difference does not tell us anything about heritability; if the difference is based on modification, heritability is the greater the smaller the mean of all the respective differences within the population. When a difference is genetically determined, they assume next, selection favors the more advantageous deviation and thus indirectly selects for the gene which determines it. Yet here is where the sociobiological argument makes a strange omission: In this way *any* gene is favored which brings about the favored trait difference, and often there is more than one that does so.

Fixated on the tenet that selection is individual selection, the sociobiologists renounce all possibility of selection for what benefits the species and not the individual. They cannot see how this could be. Here they are handicapped by their belief that selection works only through favoring advantageous variants which then, of course, would always result in enhancing a given difference. But since heritability is the stricter the smaller the differences within the population, we must assume that as often as not selection will also favor a reduction in differences, thus stabilizing a favorable trait within the population. In effect, the sum of selective processes reducing differences is what "benefits the species." It is very real and it is brought about entirely by individual selection.

If the model of the dynamics of population genetics put forward by the sociobiologists were the whole picture instead of just half of it, it would be unavoidable that each kin-selected clan would shoot off at a tangent from the rest of the species and there could never be anything like a species at all. I also fail to see how the "selfish gene" and "kin selection" hypotheses propose to account for the fact that despite all individual selection, allele distribution within a given population remains stable over long periods of time as long as the ecosystem remains stable. This is not to deny that at times conditions change in a manner which derails the homeostasis between stabilizing and innovating selective pressures in favor of the innovating ones. That is the way by which evolution proceeds. But it appears that these periods in the history of a species are short relative to the stable ones. Each burst of innovation is soon overtaken by the pressures that restabilize the system.

In short, sociobiology is basically sound, and it seems a pity that it should so unreservedly have adopted genetic models which, though interesting and stimulating, are rather too narrow in scope to give us a balanced picture of the various mechanisms of natural selection and the role played in them by the genes.

The Cheshire Cat

Looking at the world of cats for a lifetime, I awakened one day in 1969 to the realization that—apart from the ubiquity of the domestic cat—cats were a vanishing clan. But they do not vanish in a grin. Rather I should say it was a grim snarl. Not only were many of the larger cats traditionally regarded as bloodthirsty and dangerous brutes which it was meritorious to eradicate from the face of the earth by any, even the vilest means. Many of the smaller and quite harmless or maybe even useful species were also quietly disappearing because their habitats—the forests, savannahs, and even deserts and snow-covered mountains—were being invaded by more and more people who populated, cultivated, or otherwise exploited and thus destroyed them.

Consequently I joined the Species Survival Commission of the International Union for Conservation of Nature and Natural Resources and became chairman of a cat specialist group of the commission. In the years following, I was involved in many attempts at and projects concerning cat conservation, notably the Iriomote cat *(Prionailurus iriomotensis)* project on the Japanese island of Iriomote and Project Tiger in India. There were several aspects that made this work fascinating. The gratification of trying to preserve a large predator like the tiger lies in the fact that the animal is at the top of the food pyramid of its ecosystem. To preserve it means not just preserving a conspicuous animal that maybe takes the fancy of people who have no other cares. It means preserving an ecosystem in its entirety down to the smallest soil organism. So I looked around me, beyond the projects in hand and the sanctuary or national park, at the people and at what they were doing to themselves and to their own and their children's future. What I saw and experienced convinced me that nature conservation is by far the most useful and the most necessary kind of developmental aid one could conceive. Developing wildernesses as undisturbed future reserves, as regulators of climate, as the sole enduring guarantors of a country's water regime, as the most efficient and cheapest medium by far for recycling human-caused waste, as gene banks for future use, as places of recreation and research, will contribute more to a country's lasting stability, prosperity, and future happiness than any other kind of "development." Most developmental aid projects as practiced now will at best buy short-term relief, but then inexorably destroy first the receivers and then the suppliers.

However, I do not think that the greatest horror threatening man's future existence is the overexploitation and exhaustion of material resources. The continued and ever-increasing strain on what I called above "density tolerance" will cause far more formidable and—I fear—eventually irreversible damage of a psychosocial nature. Man is in great danger of becoming, by his sheer numbers and the intolerable levels of densities in many parts of the world, the only truly a-social mammal. Not to preserve resources, or rare plants and animals: no, to preserve his very own self man must curb his numbers. I suggest that four-fifths of the money now spent on so-called development be taken away from these projects and spent instead on promoting and implementing effective birth control programs, not only in the Third World but also in our own vastly overpopulated, industrialized countries. Overpopulation is not a problem of the developing countries alone, it is a world problem. Nothing is to be gained for future human happiness by further adding to our numbers; all may be gained by stopping the increase and even slowly but surely reducing numbers. It simply is no longer permissible to shy away from tackling this all-important problem and hide behind a screen of religious, economic, social, cultural, and political prejudices. If we do not tackle it soon and successfully, we may as well forget everything else.

There are many who provide sops to the apprehensive public. Man's cultural dimension, they assure us, will cope with our biological weakness; space and density will be sublimated from the merely physical onto a conceptual, spiritual level. But the fact that man has so far avoided total and final disaster is no guarantee that he will be able to do so again. Too often in the past he has managed to overcome his difficulties by a practice I call "ecological bill-jobbing" which is vulgarly described as robbing Peter to pay Paul. Our ecological debt has increased each time and apparently we are the generation who can find no more eco-credit. We are wasting the capital at an unprecedented pace. We devoutly hope to remedy the ailments due to past growth by further growth and do not wish to realize that in a finite world nothing can grow except at the expense of something else. So what are we to sacrifice next time, and the time after that, and the time after that. . . ? I am not pleading that all growth is bad. Only that for what we want to grow, we are sacrificing the wrong kind of things—those that are irreplaceable.

This is now leading us a long way beyond ethological science and its competence, or so it might appear. But it is our behavior that is at the root of the human problem. Not that the behavior as such is at fault in most cases. It is not so much what people do, but that there are too many doing it. To say that man must mend his ways is saying he should cease to be man. So it is our numbers and consequently our failure to redress the balance of birth and death that the advances in medicine and our increasing control of natural hazards have so profoundly disturbed.

Certainly much of what I conclude and recommend here is not entirely based on incontrovertible scientific proof. I am fully aware of this. But I am also aware that even a scientist is also a citizen of a human community and above this of spaceship Earth. Citizens nowadays expect, and rightfully so, that politicians and other decision makers base their decisions on the best available evidence. All too often, scientists when asked for information shrink from committing themselves and withdraw behind existing uncertainties and the need for further research. This is very respectable, but useless to the politician, who must decide without delay. As a citizen, the scientist in the case has a duty to set aside his scruples and give his résumé and advice on the basis of the best available evidence and in clear and decisive terms. I have done this and given my reasons in many publications in far more detail than is possible here—honi soit qui mal y pense. The quintessence:

ET CETERUM CENSEO POPULATIONEM HUMANAM ESSE DIMINUENDAM.

References

Baerends, G. P. 1941. Fortpflanzungsverhalten und Orientierung der Grabwespe Ammophila campestris JUR. *Tijdschrift voor Entomologie* 84:68–275.

Baerends-van Roon, J. M., and Baerends, G. P. 1979. *The morphogenesis of the behaviour of the domestic cat, with a special emphasis on the development of prey-catching.* Amsterdam, Oxford, New York: North-Holland Publishing Co.

Bertram, B. C. R. 1975. The social system of lions. *Sci. Am.* 232:54–65.

Bitterman, M. E. 1965. The evolution of intelligence. *Sci. Am.* 212:92–100.

Calhoun, J. B. 1962a. Population density and social pathology. *Sci. Am.* 206:139–48.

———. 1962b. A "behavioral sink." In *Roots of behavior,* ed. E. L. Bliss, pp. 295–315. New York: Hoeber Med. Div. of Harper & Brothers.

———. 1963. The social use of space. In *Physiological Mammalogy,* vol. I, ed. W. B. Mayer and R. G. van Gelder. New York: Academic Press.

Dards, J. L. 1978. Home ranges of feral cats in Portsmouth Dockyard. *Carnivore Genetics Newsl.* 3:242–55.

——— (in press): The behaviour of adult male cats in Portsmouth Dockyard.

Eisenberg, J. F., and Leyhausen, P. 1972. The phylogenesis of predatory behavior in mammals. *Z. Tierpsychol.* 30:59–93.

Fischel, W. 1948. *Die höheren Leistungen der Wirbeltiergehirne.* Leipzig: J. A. Barth.

Gottschaldt, K. M., and Young, D. W. 1977a. Properties of different functional types of neurones in the cat's rostral trigeminal nuclei responding to sinus hair stimulation. *J. Physiol.* 272:57–84.

———. 1977b. Quantitative aspects of responses in trigeminal relay neurones and interneurones following mechanical stimulation of sinus hairs and skin in the cat. *J. Physiol.* 272:85–103.

Harlow, H. F., and Harlow, M. K. 1962a. The effects of rearing conditions on behavior. *Bull. Menninger Clin.* 26:213–24.

———. 1962b. Social deprivation in monkeys. *Sci. Am.* 207:137–46.

Hediger, H. 1949. Säugetierterritorien und ihre Markierung. *Bijdragen tot de dierenkunde* 28:172–84.

Hinde, R. A. 1956. Ethological models and the concept of "drive." *Brit. J. Philos. Sci.* 6:321–31.

Holst, E.v., and Middelstaedt, H. 1950. Das Reafferenzprinzip (Wechselwirkungen zwischen Zentralnervensystem und Peripherie). *Die Naturwissenschaften* 37:464–76.

Holst, E.v., and St. Paul, U. v. 1960. Vom Wirkungsgefüge der Triebe. *Die Natur-wissenschaften* 47:409–22.

Iersel, J. J. A. van, and Bol, A. 1958. Preening of two tern species: A study on displacement. *Behaviour* 13:1–88.

Izawa, M., Doi, T., and Ono, Y. 1982. Grouping patterns of feral cats *(Felis catus)* living on a small island in Japan. *Jap. J. Ecol.* 32:373–82.

Kuo, Z. Y. 1930. The genesis of the cat's responses to the rat. *J. Comp. Psychol.* 11:1–35.

———. 1967. *The dynamics of behavior development: an epigenetic view.* New York: Random House.

Lack, D. 1939. The behaviour of the robin. *Proc. Zool. Soc. London A* 109:169–78.

Leyhausen, P. 1950. Beobachtungen an Löwen-Tiger-Bastarden, mit einigen Bemerkungen zur Systematik der Grosskatzen. *Z. Tierpsychol.* 7:46–83.

———. 1954a. Die Entdeckung der Relativen Koordination, ein Beitrag zur Annäherung von Physiologie und Psychologie. *Studium Generale* 7:45–60. (Engl. trans. in Lorenz & Leyhausen, 1973.)

———. 1954b. Vergleichendes über die Territorialität bei Tieren und den Raumanspruch des Menschen. *Homo* 5:116–24. (Engl. trans. in Lorenz & Leyhausen, 1973.)

———. 1956. Verhaltensstudien an Katzen. *Z. Tierpsychol.* (Suppl. 2): 1–120.

———. 1965. Über die Funktion der Relativen Stimmungshierarchie (Dargestellt am Beispiel der phylogenetischen und ontogenetischen Entwicklung des Beutefangs von Raubtieren). *Z. Tierpsychol.* 22:412–94. (Engl. trans. in Lorenz & Leyhausen, 1973.)

———. 1979. Cat Behavior. New York: Garland STPM Press.

———. In press. Antriebe, Motivation und Erleben. In *Vorträge der Klinik für Kinder- und Jugendpsychiatrie der Universität Essen,* ed. Ch. Eggers.

———. In prep. Revision of the family Felidae.

Liberg, O. 1980. Spacing patterns in a population of rural free roaming domestic cats. *Oikos* 35 (3): 336–49.

———. 1982. Home range and territoriality in free ranging house cats. Third Int. Theriol. Congr. Helsinki, 15–20. Aug. 1982.

Lorenz, K. Z. 1941. Vergleichende Bewegungsstudien an Anatinen. *J. Ornithol.* 89 (suppl. 3, Festschrift O. Heinroth): 194–293.

———. 1963. *Das sogenannte Böse. (Zur Naturgeschichte der Aggression).* Wien: Borotha-Schöler.

———. 1978. *Vergleichende Verhaltensforschung (Grundlagen der Ethologie).* Wien-New York: Springer Verlag.

Lorenz, K. Z., and Leyhausen, P. 1973. Motivation of Human and Animal Behavior (An Ethological View). New York: Van Nostrand Reinhold.

Lorenz, K. Z., and Tinbergen, N. 1939. Taxis und Instinkthandlung in der Eirollbewegung der Graugans. *Z. Tierpsychol.* 2: 1–29.

McDougall, W. 1935. The Energies of Men. 3rd ed. London: Methuen.

Michael, R. P. 1960. An investigation of the sensitivity of circumscribed neurological areas to hormonal stimulation by means of the application of oestrogens directly to the brain of the cat. *4th Intern. Neurochem. Symp, 1960*, pp. 465–80.

Natoli, E. 1983. Behavioural Evolution of the Feral Cat (*Felis catus* L.) as a Response to Urban Ecological Conditions. *Monitore zool. ital.* (N.S.) 17: 200–201.

————. 1984. Spacing pattern in a colony of urban stray cats (*Felis catus* L.) in the historic centre of Rome. Applied Animal Ethology (in prep.).

Panaman, R. 1981. Behaviour and Ecology of Free-ranging Female Farm Cats (*Felis catus* L.). Z. Tierpsychol. 56: 59–73.

Peters, G. 1978. Vergleichende Untersuchung zur Lautgebung einiger Feliden (Mammalia, Felidae). *Spixiana* (suppl. 1): 1–206.

————. 1981. Das Schnurren der Katzen (Felidae). *Säugetierkundl. Mitteilungen* 29: 30–37.

Schaller, G. 1972. The Serengeti Lion. Chicago: Univ. of Chicago Press.

Schär, R. 1983. Influence of Man on Life and Social Behaviour of Farm Cats. Poster, Int. Symp. on the Human-Pet Relationship, Vienna, 27–28. Oct. 1983.

Schmidt, W. 1975. *Qualitative und quantitative Untersuchungen am Verhalten von Haus- und Graugänsen.* Doctoral thesis, Dept. of Zoology, Univ. of Düsseldorf.

Tinbergen, N. 1950. The hierarchical organization of nervous mechanisms underlying instinctive behaviour. *Symp. Soc. Exp. Biol.* 4 *(Physiological mechanisms in animal behaviour):* 305–12.

Verberne, G., and Leyhausen, P. 1976. Marking behaviour of some Viverridae and Felidae: Time-interval analysis of the marking pattern. *Behaviour* 58: 192–253.

Konrad Lorenz (7 November 1903–27 February 1986), watching well-fed, large *Zanclus canescens,* a species difficult to keep in the middle of Europe

11

My Family and Other Animals

Konrad Z. Lorenz

When, as a scientist, one tries to get insight into the deeper roots of one's own life interest, it is quite instructive to delve into earliest childhood memories. As a very little boy, I loved owls and I was quite determined to *become* an owl. In this choice of profession I was swayed by the consideration that an owl was not put to bed as early as I was, but was allowed to roam freely throughout the night.

I learned to swim very early and when I realized that owls could not swim, they lost my esteem. My yearning for universality drove me to want to become an animal that could fly and swim and sit on trees. A photograph of a Hawaiian goose sitting on a branch induced me to choose a Sandwich goose as my life's ideal. I was not yet quite six years old when I was hit by the impact of Selma Lagerlöf's immortal book *The Journey of Little Nils Holgerson with the Wild Geese.* Consequently I wanted to become the sort of wild goose idealized by the Swedish poet.

Very slowly it dawned upon me that I could not *become* a goose, and from then on I desperately wanted at least to *have* one and when my mother obstructed this, because geese are too damaging in a garden, I settled for a duck. When a farmer in the neighborhood had a mother hen guiding a brood of day-old ducklings, I wheedled my mother into buying one for me. She did, in spite of the remonstrances of my father, who said that it was cruelty to animals to trust a six-year-old with a live duckling. He prognosticated its early demise. He was wrong. This duck lived to a ripe old age rarely attained by domestic ducks.

One interesting detail that, I confess, fills me with a certain pride, is that I chose the only wild-colored duckling in the brood. My future wife, who got another one for herself the next day, had to content herself with one that had a white wing. I don't know why we did not like white ducks, but admiring the wild and despising the domestic fowl seemed natural to me then, as it does now.

Our getting those two ducklings had, for its consequence, the discovery of imprinting. The process of imprinting is, in domestic ducks, not as clearly marked as in the wild mallard; in particular its time limits are not as strictly defined. Although they had already followed the mother hen, our two ducklings became tolerably well imprinted on both of us; mine, however, which I obtained nearly twenty-four hours earlier than my wife got hers, was clearly much more affectionate and much more closely attached to me than my wife's was to her—though she keeps denying this to this day. What we did not notice and what became apparent only many years later was the fact that I had, at that period, become imprinted on ducks. My undying love for ducks is a good illustration of the fundamental irreversability of the imprinting process. My wife is slightly older than I am: perhaps that is the reason she was not similarly affected.

Tribute is due, at this point of my life history, to Resi Führinger, who was my nanny. Possessed of the true peasant woman's deep wisdom, she taught me many things, among others how to keep a duckling well fed and, which is infinitely more important, well warmed. By helping those ducklings to survive, she probably exerted a more profound and lasting influence on my scientific career than any other of my most revered teachers.

After the acquisition of these ducklings, my future wife and I took to "pretending" again: we pretended to be mother ducks. We waded along the banks of suitable backwaters of the Danube, choosing pools with much insect life in them, watching gleefully how our ducklings enjoyed this natural food. They did, indeed, thrive most wonderfully and at least one reason for this was, besides Resi's teachings, that we had come to understand to perfection our duckling's utterances and expressive movements. We responded correctly when our charges gave the distress signal and we usually guessed correctly whether it was hunger, cold, or loneliness that made the bird cry out. We understood the "happy-feeding-note" and preferred puddles in which the birds uttered it, thus indicating that they had found palatable worms or Chironomid larvae in the ooze in which these prey remained invisible to us. When we heard their enchanting "going-to-sleep" trill, common to all Anatidae as well as to gallinaceous birds and they nuzzled us in the way they do when they want to be brooded, we instantly conformed and made a nice warm pocket in our sweaters for them to sleep in. We lived with them a complete duck's life and within a few months we were most thoroughly familiar with the whole repertoire of all the things a duck can do or say. We did not know that this was to be called an "ethogram" many years later. I remember in vivid detail everything we did with the ducklings, and what the ducklings did with us, in that happy summer of 1908. One evidently does not forget scientifically important memories, even if they are acquired seventy-five years ago.

The amount of important knowledge that two intelligent children can, without any special tuition, gain from two domestic ducklings bears witness to the fact that these birds are particularly instructive objects of study. Of course, my appetite for possessing—and knowing—more ducks was whetted. I wanted more ducks, my next aim being American wood ducks *(Aix sponsa)*, which A. E. Brehm's famous book said were easy to breed. My parents, to whom I owe an immense debt of gratitude, were singularly permissive toward my duck infatuation. The first pond was built, a second came along and finally a third. My passion for collecting was awakened, if, for the duration of my early childhood, unrequited. However, I could watch at the zoo those ducks that I could not own myself.

Even in that early childhood I lived a sort of double life, my interests being divided between my ducks and my aquaria. My first aquarium was brought into being by a female spotted salamander giving birth to forty-four larvae. Shortly prior to my going to elementary school, my parents brought back, from a Sunday walk in the Vienna woods, a spotted salamander. My father gave it to me with the injunction that we should liberate it again next Sunday at the place where it had been taken from. Reluctantly I agreed to this contract and was amply rewarded for my filial obedience. The salamander gave birth to forty-four larvae, all of which were swimming one fine morning in the small water container of the terrarium.

This event obviously, necessitated an aquarium and I had my first lesson about the metabolism of water-breathing animals. I wanted to keep the aquarium well filled up, but Resi Führinger convinced me that the larvae wanted very shallow water because soon after birth they would develop lungs, their external gills would be slowly reduced, and they would become dependent on air respiration. Owing to Resi's "green thumb" in keeping animals, we reared twelve of these forty-four larvae to metamorphosis.

A newly metamorphosed spotted salamander is among the most enchanting animal babies, being an exact miniature of the adult, showing the same bright black and yellow coloration and having very nearly identical body proportions. I remember as if it had been yesterday, how the young salamanders began to breathe air by moving the floor of their mouths, inhaling air, and pumping it into their lungs. At first these movements were irregular but soon they became as rhythmical as they are in most adult amphibia. I fully realized that the young salamanders will drown at that age if they are not able to crawl out of the water. None of mine did.

The salamanders turning into land animals were the cause of my building a beautiful large outdoor terrarium, which was occupied by these salamanders, and, additionally, by newts of different species and small frogs. I had become a collector and amateur of amphibia. I remember successes and failures, for example, the tragic failure in keeping small frogs alive through the winter.

When the last of my twelve salamanders had completed metamorphosis, the aquarium was vacant, so, of course, I began to keep freshwater fish, beginning with common native species. Even at that time I had a vague feeling that all spiny-rayed fish are much "nobler" than minnows. Later I specialized in North-American sunfish (Centrachidae) and later still on cichlids. Breeding cichlids and observing their parental behavior brought me some insight into the workings of instinctive activities. The most important discovery, however, was made by my friend Bernhard Hellmann. He was trying to breed the very aggressive South-American cichlid *Geophagus brasiliensis.* His male had killed the female and, after having been isolated for some time, incontinently proceeded to kill any new conspecific put into the tank, irrespective of sex. At this point Bernhard definitely had a stroke of genius. He bought a new female but, before putting it with the male, confronted the latter with a mirror and let the fish fight his own mirror image to complete exhaustion. After this, he put the female into the tank and the male immediately proceeded to court it. In other words, Bernhard Hellman, at age seventeen, must have had some inkling that what we now call "action specific energy," is "dammed up" during inactivity and is consumed by acting out a specific action pattern.

On finishing high school, I proceeded to study medicine at the Vienna University; it was my father's wish that I should do so. I really wanted to study zoology in general and paleontology in particular. I was obsessed by a burning interest in evolution. Again I very distinctly remember the awakening of this interest. We were sitting under the Caria tree in our garden having tea and a wasp alighted on a slice of bread and honey held in my father's hand. My mother wanted to chase it away but my father prevented her and explained that the wasp would never sting without being held fast. Then he went on to explain to me what an insect was and that it was so called because of the partitions of its body, head, thorax, and abdomen and he, furthermore, called my attention to the metameres of the wasp's abdomen sliding into each other like the parts of a telescope. Some days later I observed an earthworm creeping and saw all the metameres of the worm slide together and again away from each other as a certain part of the worm's body was contracted or extended. I was reminded of the body structure of the wasp and subsequently asked my father whether the earthworm was an insect. My belief in my father's omniscience suffered severely, because he obviously was out of his depth when I asked this question.

A few days later I discovered the picture of *Archaeopteryx* in a book by Wilhelm Bölsche called *Schöpfungstage (The Days of Creation)*. The text stated clearly that *Archaeopteryx,* with teeth in its jaws and the long tail with two feathers to each vertebra, represented a transition from reptile to bird. I immediately realized the explanatory value of evolution and at once generalized that, of course, annelid worms were the ancestors of insects!

Childhood memories are of the character of still pictures and not of movies. I know for sure where, on a walk in the woods near Altenberg, I explained my discovery of evolution to my father. Usually he was not too willing to listen to my prattle—my wife assures me that I was prattling rather excessively at that age. This time, however, my father listened very attentively and began to smile most benevolently. Then, quite suddenly, I realized that he *knew* all about what I was trying to tell him. I remember experiencing a deep resentment against my father for knowing something of such tremendous importance and not deeming it necessary to tell me all about it. I do not know when this conversation with my father took place though I know where it did. It must have been rather early in my life because I was not yet above the infantile play of pretending. In my enthusiasm about paleontology and that Bölsche book, my wife and I were inspired to play at pretending to be Iguanodons. I know for certain that we attached pieces of old garden hose to our backs for tails and walked around solemnly, keeping our hands in front of us and the thumbs extended upwards, as is shown in all pictures of *Iguanodon bernissartensis.* Infantile play persisted until very late in my life; nonetheless I don't think that I can have been much above six years of age when all this happened, because I distinctly remember having that Bölsche book read aloud to me by Resi.

My yearning to become a paleontologist was diverted by my study of medicine. One of the first lectures that I attended was that of the systematic anatomist, Professor Ferdinand Hochstetter. He was not only a brilliant comparative anatomist with an extensive knowledge of vertebrate zoology, but, what meant much more, his main interest lay in comparative embryology. He was a dedicated teacher of the comparative method and very intent on convincing the students of its value.

At that time I had reared not only ducks but a number of birds, fish, and amphibia and had studied their ontogenetic development of behavior, and I was also familiar with the behavior patterns of courtship, which, by converging evolution, are so strikingly similar in newts, fish, and birds. I would have had to be considerably more stupid than I actually was, not to realize at once that exactly the same methods of comparison could and should be applied to behavior. After all, Bernhard and I had already done this when we compared the artemia larvae's movement of its antennae with that of cladocera. Obviously, all I knew about waterfowl behavior should be evaluated in the same way.

This is actually the discovery on which all ethology is based. I certainly did not, at the time, realize the importance nor did I do so when, a few years later, I met the man who had long ago made the same discovery: Oskar Heinroth. Not only had he discovered that behavior patterns constituted just as reliable characteristics of species, genera, and higher taxonomic categories, but he had done so by studying the same group, Anatidae! Much later, after Heinroth and I had become close friends, we discovered the real pioneer in this field of comparative behavior study, Charles Otis Whitman who, ten years earlier, had fully recognized exactly the same facts—only with pigeons for a subject.

Neither Heinroth nor Whitman ever fully realized the far-reaching scientific consequences of their discovery. The fact that innate behavior patterns are clearly homologous in many species explodes certain doctrines to which quite a few psychologists adhere even now. Whitman and Heinroth remained unknown to psychologists; Whitman himself was never mentioned in American psychology. When I was studying in Vienna under Karl Bühler, who had many American psychologists for guests, I asked every one whether he or she knew of Charles Otis Whitman. None did. Not so many years ago I happened to meet his son, a very successful businessman. He had no idea of his father's importance. The only thing he knew about him was that he was "crazy about pigeons" and kept many aviaries full of these birds.

Let me put in, at this place, a word about the "amateur" or "dilettante." Amateur is derived from the Latin *amare*, "to love," dilettante from the Italian *dilettarsi*, "to delight in something." It is fashionable in science nowadays to experiment rather than to observe, to quantify rather than to describe. Yet descriptive science, based on plain, unbiased observation is the very fundament of human knowledge. As regards our knowledge of animal behavior, I contend that not even a person endowed with the almost superhuman patience of a yogi could look at animals long enough to perceive the laws underlying their behavior patterns. Only a person who looks with a gaze spellbound by that inexplicable pleasure we amateurs, we dilettanti enjoy, is in a position to discover that, for instance, the gruntwhistle is very much the same in many members of the genus *Anas,* but not in the Garganey-Shoveler group, or that the precopulation display is the same in swans and geese.

It was quite impossible that the homology of motor patterns should be discovered by anybody but "dilettanti." The same is true for the discovery of many other important laws of animal behavior. Karl von Frisch, one of the greatest biologists, made his most stirring discoveries on the honey bee, and he made them not in his famous laboratory but at his home in the old hamlet of Brunnwinkel on the Wolfgangsee, which has been in the possession of the Frisch family for many generations. I contend that this great experimenter would not have made his discovery if he had not, for many years, been taking a delight in bees.

We, the amateurs and dilettanti, can certainly claim a number of great scientists as members of our community and this is particularly true of those delighting in waterfowl. Not that an amateur need necessarily be a scientist; he is under no obligation to be one. However, if you are a true amateur, a true lover of some kind of fish, bird, or mammal, you cannot help becoming an expert. Again, the expert need not be a scientist, but the scientist is undoubtedly under an obligation to be an expert. If the expert so often tends to become a scientist this is, in some cases, brought about by the following sequence of events. The expert, who still does not consider himself a scientist, reads and hears what scientists of great renown have said and written about the subject with which he, the expert himself is thoroughly familiar. To his great surprise he finds that these great men have no idea what they are talking about and that they consequently talk nonsense.

This was my disillusioning experience when my teacher Karl Bühler made me read the works of the great vitalistic and purposivistic psychologists on the one hand, and of the great behaviorists on the other. None of them knew the phenomena that I was strenuously attempting to understand. One of the most encouraging facts in the study of animal behavior is that experts always agree on fundamentals. Whether I am talking to the nestor of waterfowl experts, Jean Delacour, or to one of the youngest, of which, luckily, there are several, there is never any discussion of those fundamentals that still are so controversial in psychology and even in ethology. All experts simply know what an innate behavior pattern is, often without ever having been told.

In 1922 Bernhard Hellman gave me as a birthday present Oskar Heinroth's classical work *Die Vögel Mitteleuropas,* which is neither more nor less than a book on the "comparative embryology of bird behavior." I realized that comparing the similarities and dissimilarities of living animals was a much better way to reconstruct their genealogy and to gain insight into the path evolution had taken, than the study of its fossil documents, which were too few and too far between. I suddenly realized that the sheer observation of what birds do, the occupation I had hitherto regarded as a mere hobby of an amateur, was indeed good legitimate science. The aims of my life as a scientist had become very clear.

During the last two years at high school—that is, from 1920 to 1922—Bernhard Hellman and I got intensely interested in an odd group of small crustacea, in phylopods. This interest stemmed from our collecting live food for our aquarium fish. An older friend, to whom I owe an undying gratitude, presented me with a small microscope and mere curiosity made me look at all the little creatures that I later fed to my fish. Confronted with the magnificent multitude of forms, I was caught by the devil of collecting. Maybe every zoologist in the making has to repeat the history of his science and pass through a phase of collecting.

The group that caught our fancy was the Cladocera, of which we collected preparations as well as photographs, which we made by taking the lens out of a very old large camera and mounting it directly over the microscope. We knew that

Cladocera belonged to the larger subclass of Phylopoda and we longed to study some representatives of this group.

I vaguely remember that in 1909 the water covering the fields during an inundation of the Danube had been teeming with free-swimming Crustacea much larger than Daphnia and we concluded retrospectively that these must have been Conchostraca and particularly Estheriides. The only Euphylopods that we could get hold of were *Artemia salina,* the eggs of which were obtainable in pet shops. When we observed the early nauplia stage of these creatures, and saw that they were scullying along by movements of their second antennae, we had the bright idea that Cladoscera were derived from Euphylopods by way of neoteny, which accounted for the reduction of their body segments.

We also formed the hyothesis that Cladoscera had developed di-phyletically, the bivalve forms from *Conchostraca,* while *Onychopoda* like *Bythotrephes* and *Polyphemus* had descended from forms similar to *Branchinecta* or *Branchipus.* It was not before 1937 that a severe and long persisting inundation of the Danube caused the meadows near Altenberg to be covered by water that evoked an eruption of thousands and thousands of Euphylopods of seven species, including the great *Triops cancriformis.* Bernhard Hellman had by then emigrated to Holland; this did not save him from being gassed by the Nazis during the war.

Even before I obtained my doctorate I became at first instructor, and later assistant in Hochstetter's department. Also I began to study zoology at the zoological institute of Professor Versloys and to participate in the psychological seminars of Professor Karl Bühler, who took a lively interest in my attempt to apply comparative methods to the study of behavior. It was Bühler who drew my attention to the fact that my findings contradicted, with equal violence, the opinions held by the vitalistic or "instinctivistic" school of purposive psychology on one hand, and those of the mechanistic or behavioristic school, represented by Watson and Yerkes, on the other hand.

Bühler made me discuss at his main seminar the most important books of the purposivistic school, W. McDougall's *An Outline of Psychology* and Edward Chase Tolman's *Purposive Behavior in Animals and Men,* and in a subsequent lecture, a book by Watson. By doing this, Bühler forced me to read these books thoroughly and in doing so I suffered a really shattering disillusion: none of these people really *knew* animals. None was familiar with them as Heinroth was or as even I was at the age of just over twenty years. I felt crushed by the amount of work that was still to be done and that obviously devolved on a new branch of science that, I felt, was more or less my own responsibility.

During the years from about the middle twenties to the middle thirties I seem to have done an amazing number of things at the same time. I again lived a double life, the less reputable side of which consisted in a rather violent passion for motorcycle riding. I rode a large, two-cylinder Brough Superior. Bernhard Hellmann and my wife each rode a 500 cc overhead-valve Triumph, and, together with our friend Willy Reif, we toured all over Europe during our summer vacations, visiting the Bretagne, Switzerland, and Italy. I was a pretty good rider and I accepted an invitation from the British Leyland Company to enter some road racing on a factory-owned machine. That was very good fun, but after I had one crash, fortunately, without injury to myself, my wife forbade further participation; she said that there were more rewarding forms of committing suicide.

Together with my friend Gustav Schmeidl, I bought a whaleboat, originally from the Austrian dreadnought *Viribus Unitis,* which had sunk and stayed on the bottom of the harbor of Pola since 1918. Into this well-preserved hull we built an ancient Mercedes motor of 13,000 cc's and, on this romantic vessel, successfully navigated the Danube. In order to do this I had to pass an examination that would permit me to pilot small steamers because a license to drive motorboats had not been provided for by the Austrian legislature. This was by no means the easiest exam I had to pass during my long life. On our farthest voyage, with my wife's brother and his wife for crew, we took our ship to Budapest and—which took far more time—back again.

All these not-so-commendable activities did not prevent me from being scientifically diligent. I published nine papers, one of which, "A Contribution to the Comparative Sociology of Colonial Nesting Birds," I read at the 8th International Ornithological Congress in Oxford. I kept birds in Altenberg, concentrating on social species, settling a colony of free-flying jackdaws in the roof of our house and a colony of night herons on the high old trees in our garden. At the same time, I did duty as assistant in the Anatomical Institute, I studied zoology and got my doctorate, I married, and I studied psychology at Karl Bühler's institute—at least enough to realize the tremendous importance that Bühler's theories on perception would have for a future epistemology.

I was deeply influenced by Egon Brunswick, at that time assistant to Bühler, who had just published his book, *Psychologie vom Gegenstand her,* which may be translated as *Psychology from the Point of View of the Perceived Objects.* What I learned from Bühler and Brunswick regarding the functions of perception was fundamental to the later development of my views on the epistemology of my late colleague Immanel Kant. (It was one of the virtually incredible acts of fate that wafted me, in the year 1940, to Königsberg to actually occupy the very chair of professor of philosophy held long ago by the greatest philosopher of all times. But I am getting ahead of my story.)

All my life I seem to have been persecuted by the goddess of good luck. I had had a sequence of good teachers beginning with Resi Führinger, proceeding to an excellent biology teacher, P. Philip Heberdey, a benedictine monk who taught us all about Darwin and natural selection, to Ferdinand Hochstetter, who taught me how to reconstruct the path phylogeny had taken by studying similarities and dissimilarities of living creatures, and to, last not least, Karl Bühler and Egon Brunswick, who kindled my interest in epistemology.

I decided that it was incumbent on me to read Immanuel Kant, which, in fact, is some undertaking indeed. With beginner's luck I struck on the *Prolegomena zur Kritik der reinen Vernuft* and had just finished reading it, when the goddess of fortune interfered. Two things happened in quick succession. I had applied for a lectureship in animal psychology at Vienna University and one professor raised objections on the ground that animals don't have a soul, another one raised difficulties about my becoming a lecturer at one faculty, while being assistant at another; psychology belonging to the philosophical and anatomy to the medical faculty. So I decided to leave my position at the anatomical institute, resigning myself to losing my salary and relying on that of my wife, who was at that time already responsible for a department in an obstetrical hospital in Vienna. I owe her great thanks for encouraging me to leave the anatomical institute, which was then under the directorship of another professor. People often asked her how she could stand the many birds and animals, including lemurs, capuchin monkeys, ravens, great crested cacatoos, to name only some of them, running and flying free in and around our house in Altenberg. Her habitual answer was that she did not mind because she hardly ever was at home, being on night duty in the hospital and earning the money necessary to feed the Altenberg menagery.

The climax to these fruitful and definitely formative years came when I gave a lecture on the concept of instinct at the Harnack House in Berlin and was invited to repeat it at a congress, entitled "Instinctuus," convoked by Professor Van der Claauw in Leyden. At both occasions I talked in detail about the fact, discovered before me by C. O. Whitman and O. Heinroth, that there are motor patterns, the similarities and differences of which, from species to species, from genus to genus, even from one larger taxonomic group to another, are retained with exactly the same constancy and/or variability as are morphological characteristics. In other words, these patterns of movement are just as reliably characteristic of a particular group as are the formation of teeth or feathers and such other proven distinguishing attributes used in comparative morphology. For this fact there can be no other explanation than that the similarities and dissimilarities of these coordinated motor patterns are to be traced back to a common origin in some ancestral form that also already possessed, as its very own, these same movements in a primeval form. In short, the concept of *homology* can be applied to them.

Neither Whitman nor Heinroth ever expressed any views concerning the physiological nature of the homologous motor patterns they had discovered. My own knowledge of the physiology of the central nervous system came from textbooks and lectures in which the Sherringtonian reflex theory ruled supreme and was regarded as the last word and the incontestable truth.

In this lecture I refuted the purposivistic opinion held by McDougall and E. C. Tolman that "instinct" as a supernatural factor directed animal behavior to its goal, as well as the behaviorists' doctrine that all animal behavior was formed by environment. I made it perfectly clear that any animal is perfectly capable of striving toward a purpose by goal-oriented and variable behavior, but that this purpose must not, as the purposive psychologists supposed, be equated with the achievement of the teleonomic function of the behavior pattern in question. The purpose toward which the animal, as a subject, is striving is neither more nor less than the run-through or discharge of that kind of innate behavior that Wallace Craig designated as "consummatory action" (1918) and that we now call "the drive-reducing consummatory act." Up to this point, what I said then is more or less what I believe today.

But what I had to say about the physiological nature of fixed action patterns was influenced by doctrinaire bias. Led by McDougall, the purposive psychologists had continued their battle against the reflex theory of the behaviorists and, quite rightly, had emphasized the *spontaneity* of animal behavior. "The healthy animal is up and doing," McDougall had written. I was already thoroughly familiar with the writings of Wallace Craig and, through my own research, I was well acquainted with the phenomena of appetitive behavior and of threshold-lowering for releasing stimuli—and I should have borne in mind a particular sentence of a letter Craig had sent shortly before, in which he had argued against the reflex concept: "It is obviously nonsense to speak of a re-action to a stimulus not yet received."

At that juncture mere common sense ought to have prompted me to put the following question: Innate motor patterns have apparently nothing to do with higher intellectual capacities, they are governed by central nervous processes that occur quite independently of external stimulation. Do we know of any other physiological processes that function in a similar way? The obvious answer would have been: Such motor patterns are very well known, for instance those of the vertebrate heart for which stimulus-producing organs are anatomically known and the physiology of which has been thoroughly studied.

I lacked the independence of mind and the self-assurance that would have been necessary to ask this question. My valid aversion toward the preternatural and inexplicable factors that the vitalists had summoned to interpret spontaneous behavior was so deep that I lapsed into the opposite error; I assumed that it would be a concession to the vitalistic purposive psychologists if I were to deviate from the conventional mechanistic concept of reflexes, and this concession I did not wish to make. During the course of that lecture I did cover completely, and with special emphasis, all those characteristics and capacities of fixed action patterns that could *not* be accounted for by the chain reaction theory. Yet, in my summary at the end, I still concluded that fixed action patterns depended on the linkage of unconditioned reflexes, even if the cited phenomena of appetitive behavior, threshold lowering, and vacuum activity would require a supplementary hypothesis for clarification.

Sitting next to my wife in the last row of the auditorium was a young man who followed the lecture intently and who, during the exposition on spontaneity, kept muttering "Menschenskind, that's right, that's right!" However, when I came to the concluding remarks described above, he covered his head and groaned "Idiot." This man was Erich von Holst. After the lecture we were introduced to one another in the Harnack House restaurant, and there it took him all of ten minutes to convince me forever that the reflex theory was indeed idiotic.

The moment one assumes that the processes of endogenous production and central nervous coordination of impulses, discovered by Erich von Holst, constitute the physiological basis of behavioral patterns and not some linkage of reflexes, all the phenomena that could not be fitted into the reflex theory, such as threshold lowering and vacuum activities, not only were easily explained, but became effects to be postulated on the basis of the new theory.

One important consequence of this new physiological theory of the fixed motor pattern was the necessity to analyze further that particular behavioral system that Heinroth and I had called the *arteigene Triebhandlung* (literally, "species-characteristic drive action") and that we had regarded as an elementary unit of behavior. Obviously, the mechanism that selectively responded to a certain stimulus situation must be physiologically different from the fixed motor pattern released. As long as the whole system was regarded as a chain of reflexes, there was no reason for conceptually separating, from the rest of the chain, the first link that set it going. But once one had recognized that the movement patterns resulted from impulses endogenously produced and centrally coordinated, and that as long as they were not needed, they had to be held in check by some superordinated factor, the physiological apparatus that triggered their release emerged as a mechanism *sui generis*. These mechanisms that responded selectively to stimuli and in a certain sense served as "filters" of afference, were clearly fundamentally different from those that produced impulses and from the central coordination that was independent of all afference.

This dismantling of the concept of the *arteigene Triebhandlung* into its component parts signified a substantial step in the development of ethology. The step was taken in Leyden at a congress called together by Professor Van der Claauw. During discussions that lasted through the nights, Niko Tinbergen and I conceived the concept of the *innate releasing mechanism* (IRM), although it is no longer possible to determine which one of us actually gave birth to it. Its further elaboration and refinement and the exploration of its physiological characteristics, especially its functional limitations, are all due to Niko Tinbergen's experiments.

The following summer, Niko Tinbergen, with his family, came to Altenberg to continue our discussions about instinctive motor patterns and innate releasing mechanisms and to cement what proved to be a lifelong friendship. At this period I was delivering a systematic lecture in Vienna twice weekly and Niko used to accompany me and to listen. When we were not discussing, we were digging a pond in the lowest parts of our garden in Altenberg—I was studying my first greylag geese at that time. One fine day, when the pond was progressing beautifully Tinbergen refused to come to my lecture; he preferred to complete a particularly intriguing part of the water conduit and he knew what I was talking about anyhow. Just that day a rather arrogant and slightly older colleague approached me and said in a somewhat supercilious manner, "I hear that Tinbergen is working at your station; what is he working on?" When I answered, quite truthfully, that, at the moment, Tinbergen was digging a pond, he thought I was pulling his leg and walked away offended.

Retrospectively, this summer with Niko Tinbergen was the most beautiful of my life. What we did scientifically had the character of play and, as Friedrich Schiller says, "Man is only then completely human when he is at play." Niko and I were the perfect team. I am, as described before, an amateur and prefer observing to experimenting. It comes hard to me to endanger the happy life of a brood of birds or fish in order to perform an experiment on them. Niko Tinbergen is the past master of the unobtrusive experiment, of asking a question of the organism without unnecessarily disturbing it. We published, in joint authorship, a paper that has become a classic.

I missed Niko Tinbergen most dreadfully when he left Altenberg in the autumn of 1937. Work went on, however, and the same situation, my wife earning the money and myself doing research, continued until after the beginning of the last war.

Then, in 1939, I received a call to the second chair of philosophy at the University of Königsberg. This surprising invitation was triggered by the fact that Erich von Holst was playing the viola da braccia in a chamber quartet in which Eduard Baumgarten played the first violin. Baumgarten was then professor of philosophy at Madison University and had just received a call to the first chair of philosophy at the University of Königsberg. Baumgarten, being a direct pupil of John Dewey and a dyed-in-the-wool pragmatist, was somewhat doubtful about settling in the cast shadow of Immanuel Kant. At one of the quartets, he casually asked von Holst, whether he perchance knew a psychologist with some biological, particularly evolutionary, background who might be interested in epistemological problems, particularly in the nature of what Immanuel Kant was calling the "a priori." Von Holst told him that by a rare coincidence he could furnish just such a rare bird, meaning myself. Von Holst and Baumgarten approached the zoologist Otto Koehler and the botanist Kurt Mothes, on whose authority the philosophical faculty of Königsberg invited me to the chair of psychology.

I don't think I am superstitious, but the incredible improbability of the coincidence of all these contriving factors impressed me as something like the "finger of God" and I accepted at once. As a professor of psychology, my biological background caused some controversy among my new colleagues on the several chairs of humanities. Interestingly enough, Professor Héraucourt, Anglicist and comparative linguist, at once claimed brotherhood of method and called attention to the fact that the methodology that I had learned from F. Hochstetter was identical with that used in linguistics in investigating the historical development of language. Other philosophers repudiated everything I said and did and asserted that by my occupying what had been administratively Immanuel Kant's chair, philosophy had not only gone to the dogs, but actually to the fish. This unkind cut referred to the many aquaria that I had installed at the rooms of my new department. I automatically became a member of the Kant Society, which convened every Monday evening, and I was seduced at an early date to air my rather immature views on epistemology.

In my opinion the recognition of new and important facts begins with a subconscious growth, not only in the mind of one single man but simultaneously in that of very many thinkers. Some ideas, already widely spread at a certain time in an unreflected subconscious manner, suddenly erupt in the form of a relevation in the mind of one who consequently considers himself and is considered by others as having wrought a great breakthrough. But, and this is the trouble, when an idea of this kind is mature enough to emerge, it is not only one man to whom it suddenly becomes clear, but very often to several at once, Wallace and Darwin being the classical example.

To any thinker thoroughly familiar with the facts of evolution, it is a matter of course that the organization which enables us to perceive the external world, the sensory organization as well as that of our central nervous system, is something real and has evolved in interaction with and in adaptation to the outer reality that surrounds us. I contend that a vast number of biologically informed scientists are holding this very same epistemological attitude even if they are completely uninterested in epistemology and never have given a thought to the problems of the Kantian *a priori*. Kant's question, how it is to be explained that our aprioristic categories of thought and forms of ideation fit adequately to the external world, receives an easy and even banal answer on an evolutionary basis.

For Immanuel Kant even these categories of thought and forms of visualization are *a priori* in the sense that they are there before any experience and must be there in order to make experience possible at all. According to Kant the *Ding-an-sich* is unknowable on principle. I never could quite understand how this can agree with his statement that our categories of thought and forms of ideation are "adequate" to make experience possible. I believe that my own understanding of this question was furthered by the paradigm of my old inadequate microscope, which adorned all objects seen through it with rainbow-colored edges. This made me realize that objectivity depends on a thorough knowledge of the apparatus through which we perceive the world. Otherwise it is not possible to avoid mistaking for a characteristic inherent to the object observed something that in fact results from the shortcomings of the instrument through which we perceive it.

I never believed that microscopic animals really had rainbow-colored edges but the great poet-philosopher Johann Wolfgang von Goethe committed just such an analogous error in regarding color qualities not as a product of our perceiving apparatus, but as physical properties of light itself. What I had learned from Egon Brunswick regarding the function of color constancy helped me considerably in understanding all this.

At an early age, thanks to the tuition of Karl Bühler, I had thoroughly understood that knowing an object results from an interaction between an organization within the observer and an object in the outer world, both of which are equally real. I sincerely believe that this fundamental truth is the solid base of all striving for objectivity in research. P. W. Bridgeman said as early as 1958, "The object of knowledge and the instrument of knowledge cannot legitimately be separated but must be taken together as a whole." Objectivity cannot be reached, as behaviorists think, by ignoring subjective experience, but by thoroughly studying what Bridgeman calls the instrument of knowledge, what Karl Popper called the perceiving apparatus, and what I have called the *Weltbildapparatur.*

Karl Popper writes with a remarkable casualness in *The Logic of Scientific Discovery,* "The thing-in-itself is unknowable; we can only know its appearances, which are to be understood (as pointed out by Kant) as resulting from the thing-in-itself and from our perceiving apparatus. Thus appearances result from a kind of interaction between the things-in-themselves and ourselves." Popper does not quite seem to realize that Kant himself, and some neo-Kantians even more so, would violently object to these statements. According to Kant's transcendental idealism, there is no correspondence between the *Ding-an-sich* and the way our a–prioristic forms of ideation make it appear in our experience. What we experience is, for Kant, in no way an *image* of reality, not even a crude or distorted image. He saw clearly that our forms of apprehension are determined by preexisting structures in ourselves and not by those of the object apprehended. What he obviously did not see is that these structures of our perceiving apparatus have something to do with reality. In paragraph 11 of the prolegomenon to the *Kritik der Reinen Vernunft,* Kant wrote:

If one were to entertain the slightest doubt that space and time did not relate to the Ding-an-sich, but merely to its relationship to sensuous reality, I cannot see how one can possibly affect to know *a priori* and in advance of any empirical knowledge of things, i.e., before they are set before us, how we shall have to visualize them as we do in the case of space and time.

Kant was obviously convinced that an answer to this question in terms of natural science was impossible in principle. He was right in contending that our forms of ideation and our categories of thought are not, as Hume and other empiricists believed, the product of individual experience. He found clear evidence that they are not individually acquired by learning.

A highly pertinent question is what Kant would have thought of the *a priori,* if he had been familiar with the facts of evolution. I assert that he would, without any objections, have taken the view held by what we call evolutionary epistemology. Karl Popper's statements quoted above put our views in a nutshell and Donald D. Campbell in his essay "Evolutionary Epistemology," has convincingly demonstrated why and how it is necessary for an understanding of our cognitive apparatus to know how it has phylogenetically evolved. It is an approach that has also received the express approval of no less a man than Max Planck, who wrote to me that it gave him "deep satisfaction that, starting from such different premises, I should have arrived at the same view on the relationship between the phenomenal world and the real world as he had done himself."

I regard evolutionary epistemology as superlatively important to our views of man and his relationship to the rest of creation. But I do not flatter myself to have made very important contributions. The time for the recognition of these epistemological revelations was simply ripe in our day and I do not doubt that besides Max Planck, Karl Popper, Donald Campbell, Rupert Riedl, and myself, many other thinkers have independently arrived at the same results. If I myself was the first to put it into words, it was because in my controversial position in Immanuel Kant's chair, I was exposed to a critique that simply forced me into counterattack.

As my knowledge of Kant's work was, at that time, infinitismal, this counterattack was an insecure undertaking. I relied, in regard to Kant's teaching, mainly on the utterances and particularly on the letters of my opponents. I owe great gratitude to the physiologist, H. H. Weber and to Annemarie Koehler, the first wife of my teacher and friend Otto Koehler, the zoologist. I often used the argument *hic dicat quispiam*—here the Kantian would say—and quoted literally what Weber or Frau Koehler had said originally. This is how my paper "Kant's Lehre vom Apriori im Lichte moderner Biologie" was written. This daring and even foolhardy attack on transcendental idealism was not published until I had paid off my indebtedness to my teacher Heinroth by publishing an elaborate paper "Comparative Studies on the Courtship Movements of Anatidae." Both papers were published after I had been recruited into the German army.

I was called up as a motorcycle dispatch rider in a motorized unit. My first days as a recruit, usually so disagreeable, were made easy by a piece of stunt riding. The sergeant in charge showed me a 600 cc Norton-licensed NSU with a huge sidecar, and asked me whether I was afraid of that machine. I said no, mounted it, and rode round the court on two wheels, balancing the sidecar very high in the air, which was difficult because it was on the outside of the circle. I let it come down with a thump in front of the sergeant and sat at attention. He at once deputized me as a motorcycle-riding instructor and left the locality. I was exempt from some very tiresome drill. Instructing, whatever it may be, is an amusing job.

It was not long, however, until it was discovered by the military authorities that I had been a professor of psychology. However, after a few weeks as a psychologist in Posen, which consisted mainly in administering routine psychological tests to aspiring officers, I lost my job because Göring abolished the institution of military psychology altogether. He had found out that according to previous psychological testing, Mölders, one of the greatest among fighter pilots, had been declared hopelessly unfit to ever become an airman.

During my occupation as a psychologist I had become acquainted with Dr. Herbert Weigel, who was in charge of the Department of Neurology and Psychiatry at the Reservelazarett 1 in Posen. He wanted me to come to his department, which, however, seemed impossible because Posen belonged to Wehrkreis 1 and I was recruited at the Wehrkreis of Königsberg, the number of which I have forgotten. Weigel applied for my transfer to the psychological department in Posen which, owing to Göring's intervention, did not exist any more. The higher levels responsible for the transfer were, as Weigel had forseen, quite unaware of this fact and I was duly transferred to Wehrkreis 1.

Immediately following my transfer, the medical section claimed me, as it was unlawful for a medical man to do duty of any other sort. I was included in Weigel's department, which again was one of my strokes of luck. Nothing is more humiliating than having to pretend to be expert on a subject of which one knows too little, and I definitely knew too little of medicine to fulfill the duties of a doctor. With the special subjects of neurology and psychology, on the other hand, I felt myself quite able to cope, as I knew enough of anatomy and of the physiology of the central nervous system, and also had some inklings of psychiatry acquired in mixed seminars held in Vienna by Professor Pötzl, the psychiatrist, and Professor Bühler, the psychologist.

Weigel was a good teacher and one of the few German psychiatrists who dared to confess that they took Freudian psychoanalysis seriously. During the two years of my activity in Posen, my chief occupation was the treatment of neuroses, mainly of hysteria and of compulsive neuroses. Much later this schooling proved important to me when I realized the degree to which neuroses have become epidemic and threaten humanity as a whole.

In 1944 I was sent to the front to a hospital in Witebsk, which was enclosed and beleaguered by the Russian army from the day on which I arrived. In the complete dissolution and disintegration in which the siege ended, I quite unwillingly found myself commanding a small group consisting almost exclusively of sergeants who had refused to panic. After waging two days of private war we were all taken prisoner. I was examined by a Russian major who spoke German tolerably well. When he found out that I was a university professor, he went into an elaborate propaganda speech, telling me that in Russia science was taken much too seriously to take university professors away from their posts and send them to the front. "But," he concluded, "*inter arma silent muse,* do you understand?" I did, and he seemed very impressed by it while I, on my side, received an entirely erroneous opinion of the erudition of the Russian army. In fact, I never again encountered a Russian, even among doctors, who knew as much Latin as that major.

My first appointment as a prisoner of war doctor was satisfying insofar as I really was able to save some lives. I was put in charge—of course under the supervision of Russian doctors, mainly women—of a neurological hospital in Chalturin full of more than six hundred patients, all of whom suffered from what by German army doctors was called *Feldpolyneuritis,* which is best translated by "front soldier's neuritis." It consists in an inflammation of the anterior roots, including the ganglia to which they belong, and is caused by the additive effect of cold, overexertion, and lack of vitamin C. Its symptoms are the disappearance of all tendon reflexes and, at later stages, in paralysis of striated musculature. At its worst the illness kills the patient by paralysis of the breathing musculature.

The Russians did not know this illness and the doctors in charge of the hospital all had thought that the patients were suffering from an epidemic diphtheria, which also produces areflexia. The treatment is extremely simple; it consists in keeping the patient quiet and warm and making him swallow great doses of ascorbic acid. The latter was available and so were some additional blankets and by this simple treatment I achieved what the Russians regarded as a miracle cure on all the patients except two, who, in the first days after my arrival died a horrible death by asphyxiation. An iron lung could have saved them, but none was available.

Another interesting story that happened in Chalturin is worth telling. The hospital often received patients emaciated and in the last stages of starvation. They always came from prisoner of war camps that were in an isolated position that made official supervision difficult. We received a young Austrian in this stage who, from pure weakness, had developed gangrene of the toes and the adjoining part of the right foot. I assisted my friend Hans Theiss, who was in charge of the surgical department, in amputating the forefoot in the Lisfrank's articulation. Unluckily, the gangrene crept upwards and a few days later I was asked to assist the Russian surgeon in amputating the leg at the knee joint. The Russian cut merrily into the bend of the knee and, against discipline, I could not help saying, "ostaroshno [look out] arteria poplitea." The Russian surgeon did not take offense but asked "shto takoi [What's that] Arteria poplitea?"—and cut it through in the next second. The artery squirted blood across the room, but only twice. At the next heartbeat the Russian had got the artery in his forceps and a second later had efficiently ligatured it. Obviously, manual dexterity can substitute for anatomical knowledge.

After this operation, the patient, who was a very neurotic personality, went on a hunger strike. The Russian prisoner of war hospitals took the most extreme pains to save their patient's lives. Russian nurses acted as donors for blood transfusions, and the man whose story I am telling, was offered the most incredible choice of food. However, no tidbit could tempt him and I was called in, as the case obviously came within the competence of a psychiatrist. Though the man came from Vienna as I did myself and thus spoke the same dialect, I found myself quite incapable of convincing him that the loss of his leg was only an advantage, as amputated prisoners regularly were sent home at once. He simply had had enough of it all and went into a hysterical state like a small child in a tantrum. He was in extreme danger and in desperation I resorted to the last means of treating hysterical reactions: I simulated extreme rage, roaring like a gorilla at the poor little remnant of a man, actually beating my breast and threatening to make mincemeat (in Austrian *Krenfleisch*) of him, if he did not eat this nice soup at once. Miraculously he did, and I felt tremendously relieved because I had had my doubts about the humaneness of my treatment.

It is characteristic that a person cured, or more accurately, "snapped out," of a hysterical reaction feels extremely grateful to the "snapper outer." From then on the patient would eat greedily, but only under one condition: I had to be present. The small starveling, who had looked about eighty years of age plumped out with amazing speed and turned out to be a quite good-looking young man in his early twenties. I had to look after my department consisting of over six hundred beds and I became tired of walking over, breakfast, lunch, and dinner, to the surgical department for his feeding. So at last I lost patience and told him that, unless he would eat his meals even in my absence, I would administer the beating I had threatened a few weeks ago. I concluded, "Do you think that I have the time to sit on your bed and feed you spoon by spoon?"

Weeks later I met the man again in Orithchi, where I was interned on a "vacation" after the hospital in Chalturin had been dissolved. He was due to be repatriated the next day and he offered with real heroism to take a message to my wife. He carried, in his cheek, the first message telling my wife that I was still alive—I had simply been reporting missing. The point of this story is the tale this man told to my wife. He said that he was lying in a corner uncared for, on the point of starvation and ready to die. Then he said that I had found him and had sat on his couch and saved him by feeding him spoon by spoon. When my wife told me the story later I did not, of course, remember at once who the man was until the story of my feeding him "spoon by spoon" rang a bell. The interesting thing is that the man knew, at heart, that I had saved his life but, for obvious psychological reasons, had displaced the disagreeable method by which I had done so.

Altogether I went through thirteen Russian prioner of war camps; the Russians shifted us about quite a lot. I was always working as a camp doctor. I strongly dislike the term *psychosomatic,* because there is hardly an illness that does not affect mind and body alike. Therefore, the keeping up of morale in a prisoner of war camp is at least as important as medical supervision and so I acted as a mixture of medical man, father confessor, and buffoon, the latter activity not being the least important.

I had some time on my hands, particularly when staying in smaller camps, so I started to write a book. I could get some ink but no paper. However, by bribing the camp's tailor with a few pieces of bread, I got him to iron out cement sacking, which I cut into suitable pieces as paper to write upon. It was quite a package when I had finished what ultimately became my book on evolutionary epistemology entitled *Behind the Mirror.*

Some of my prisoner friends thought this writing extremely dangerous and took a very glum view when, from the camp in Erewan, which sat at the foot of Mount Ararat, I, quite alone and convoyed only by one officer, was very suddenly transferred to Krasnogorsk in the precincts of Moscow. On the railway, my guard got a severe attack of malaria. When we arrived in the oil town of Baku he was quite unable to leave the railway car, so he gave me money and I had to walk quite illegitimately in a German uniform through the Russian town with nothing to prove that I was not an escapee. After having bought what I had been told by the convoy officer, I found the opportunity to wash myself at a fountain in the main square of Baku. While I did so, I suddenly saw approaching the intimidating figure of a one-legged Russian soldier with Mongolian features, waving in his hands a huge razor. I was reassured when he said in very bad and therefore for me very understandable Russian: "You have soap, I have razor, you give me some soap, I shall shave you." It is a very consoling memory to visualize a German soldier being shaved by a Russian invalid in the middle of a Russian city.

On my arrival in Krasnogorsk, my presence was made known to the *natschlanik lagera,* the commandant of the prisoner of war camp, by the spreading of the alarming news that one of the prisoners had gone crazy and went around catching flies and putting them into match boxes. The commander, being a highly intelligent man, at once realized that the professor had arrived. He had the amazing kindness to arrange that bird food be provided for me. In Armenia I had reared a starling that I could let fly free for most of the time, as it was securely imprinted on myself. Once it had flown away with a huge swarm of other starlings but had come back to me on my whistling when, later in the day, the swarm had happened to pass above the camp. This earned me the fame of being a magician. It was this fame that had preceded me to Krasnogorsk and the Russian lieutenant-colonel in charge of the camp had correctly associated it with the news of a madman catching flies.

In the camp of Kransnogorsk, I was asked to procure a typewritten copy of my manuscript and was promised that, after it had passed the official censor, I should be allowed to take a copy home with me. The date for the next repatriation transport for Austrians was approaching and my manuscript had not yet come back. The gloomy prognostications of my friends in Erewan threatened to come true.

One day before the transport was to start, I was suddenly summoned to the commander. This in itself was highly alarming for a prisoner of war in Russia but what followed was the most astonishing and, in fact, the most beautiful experience I had during the war.

When I entered the *natchalnik*'s office, he amazingly rose from his seat and bade me to sit down. Then, with a very serious face he said, "Professor, you are not a prisoner any more, nor am I your superior officer. Now, from man to man, I want to ask you a question: Can you give me your word of honor that your original manuscript which you have kept, contains nothing which is not identical with what you wrote in the copy submitted to censure?"

I did not understand in the least what he had in mind and I answered at once: "No, I have eliminated one chapter, enlarged another and generally improved on the style of the whole thing."

He laughed outright and said, "No, professor, you don't get my meaning. I am asking you whether your manuscript contains nothing besides your scientific work, any secret notes you took in some camp or other which are not contained in the manuscript submitted to censure."

It was then my turn to laugh and I answered that regarding this question I could indeed give him my word of honor [*chestnyi slowo*].

Thereupon he wrote a *propusk* that I was allowed to take with me, on the repatriation convoy starting the next day, "one manuscript, one bird cage and one wooden sculpture," the latter being a little wigeon duck that I had carved of hornbeam wood on my wife's birthday. Furthermore, he instructed the convoy officer that I was not to be searched, and that he should pass on this word to the convoy officer succeeding him and the latter should tell it to the next one, and so on. Of course, this exceeded by far the competence of a camp commander and he would have been in very serious trouble indeed if I should have been searched after all and should have been found in possession of any politically relevant notes. I do not think that I know of another example of one man utterly trusting another's word of honor. I am deeply moved whenever I tell the story and I find my eyes moist now, while writing it.

I came home in February 1949. My father had died that year at the age of ninety-one; the rest of my family was in good health. My wife had turned part of the tree nursery that she had inherited into a farm and herself from a gynecologist into a most successful farmer. So we had enough to eat but absolutely no money.

At the right time we received unexpected help from the English writer J. B. Priestley, who donated to the Austrian Academy of Sciences a considerable sum due to him for plays enacted in Austrian theaters, with instructions that it be used to support research in Altenberg. The first coworker to arrive at the new Station for Comparative Ethology under the Protectorate of the Austrian Academy of Science was Wolfgang Schleidt, now professor at the University of Maryland. Next followed Heinz Prechtl, now professor in Groningen, with his wife Ilse, a doctor of zoology in her own right, and, last, Irenäus Eibl-Eibesfeldt, who is now in charge of a department for Human Ethology of the Max Planck Society. The high quality of my early coworkers caused my wife to exclaim, many years later: "Funny, now all the boys are professors!" Of which I am admittedly proud.

The first non-Austrian ethologist who came to visit our station was William H. Thorpe of Cambridge, who had taken the immense trouble of getting a permit to visit us in the Russian-occupied zone. He remained a friend for life.

In 1950 I participated at a symposium of the Society for Experimental Biology in Cambridge and there I met Niko Tinbergen again. Though he had spent years in a German concentration camp and I even longer in a Soviet prisoner of war camp, we found that this had made no difference whatsoever, which Niko put in a nutshell by saying: "We have won."

We were living quite happily and very modestly at our station in Altenberg when the University of Graz proposed me unanimously to the chair of zoology as a successor to Karl von Frisch; Frisch was returning to his former chair in Munich. My friends in England, Thorpe and Tinbergen, had predicted that I would never get a chair in Austria, and indeed, I never did. The then minister of education refused to confirm my nomination, not because I had previously been at a Nazi university, but because, and this was explicitly stated, as a Darwinist and Evolutionist I was unwelcome.

I wrote this bad news to Niko Tinbergen and Bill Thorpe and they obviously worked a miracle because only a few weeks later I received a call for a lectureship at the University of Bristol. As an injunction it was added that I should work as an ethologist on the great collection of waterfowl at Slimbridge. This indicated that my friend Peter Scott also had a hand in that miracle.

I had already consented to accept the lectureship in Bristol when the Max Planck Society intervened. Erich von Holst had spoken to the president, Otto Hahn, and this great man acted at once. Unhesitatingly he exceeded his competence by asking me whether I would consent to stay in Altenberg, if the Max Planck Society were to pay me a salary of 1000 Austrian shillings monthly. Breaking my committment to Bristol University I at once accepted Otto Hahn's proposal.

However, these plans were soon superceded by other, more extensive ones. Late in 1950 the Max Planck Gesellschaft decided to install a station for ethological research at the castle of Buldern in Westphalia. In this location Baron Gisbert von Rhomberg had offered accomodation for scientists as well as some beautiful ponds for our waterfowl. The tremendous advantage of this offer was that I could give jobs to my coworkers from Altenberg; Eibl, Prechtl, and Schleidt became my assistants.

We worked in Buldern happily and satisfactorily; I myself was nominated professor at Münster University. My station was nominally under the jurisdiction of the Max-Planck-Institut für Meeresbiologie in Wilhelmshaven, at which Erich von Holst was working. In 1955 the Max Planck Society founded, for the two of us, the Max-Planck-Institut für Verhaltensphysiologie in Seewiesen. There followed a few—all too few—years of very fertile collaboration ended by Erich von Holst's tragically early demise. I myself continued working at the Max-Planck-Institut until my retirement in 1973.

During these years, ethology developed apace both in regard to results achieved and the number of research workers collaborating. A large store of data was laboriously assembled; many unique discoveries were made. If one chooses to criticize this period of felicitous research, it can be reproached for one-sidedness, even for a certain failure to think in terms of systems. This was inherent in an orientation that almost completely ignored *learning processes;* above all, the relationships and interrelationships that existed between the newly discovered inborn behavior mechanisms and the various forms of learning were barely touched. My modest contribution, which comprised a formulation of the "instinct-learning intercalation" concept, got no further than formulation; besides, the example on which the conceptualization—correct in itself—was based, was false.

In 1953 a critical study that had a behaviorist point of view but that did not come from a behaviorist appeared. In "A Critique of Konrad Lorenz's Theory of Instinctive Behavior," Daniel S. Lehrman dismissed, in principle, the existence of innate movement patterns and, in so doing, supported his argument substantially by using a thesis of D. O. Hebb, who had maintained that innate behavior is defined only through the exclusion of what is learned and, thus, as a concept was "nonvalid," that is, unusable. Drawing on the findings of Z. Y. Kuo (1932), Lehrman also asserted that one could never know whether or not particular behavior patterns had been learned within the egg or in utero. Kuo had already recommended abandoning the conceptual separation of the innate and the acquired. All behavior, in his opinion, consisted of reactions to stimuli and these reflected the interaction between an organism and its environment. The theory of a preextant relationship between the organism and the conditions of its environment is no less questionable, for Kuo, than the assumption of innate ideas.

My answer to Lehrman's critique was short and forceful but, at first, missed the most essential mark. The assertion that the innate in comparative studies of behavior is defined only through the exclusion of learning processes is entirely false: like morphological traits, innate behavior patterns are recognizable through the same systematic distribution of attributes; the concepts of innate and acquired are as well-defined as genotype and phenotype. The reply to the theory that the bird within the egg or the mammal embryo within the uterus could there have learned behavior patterns that then "fit" its intended environment was formulated by my wife with a single phrase: "Indoor ski course." I myself wrote at the time that Lehrman, in order to get around the concept of innate behavior patterns, was actually postulating the existence of an innate schoolmarm.

My formulation of the concept of the "innate schoolmarm" was clearly intended as a reductio ad absurdum. What neither I nor my critics saw was that in just this teaching mechanism the real problem was lurking. I took me nearly ten years to think through to where, actually, the error of the criticism and the counter-criticism was located. It was so very difficult to find because the error had been committed in exactly the same way by both the extreme behaviorists and by the older ethologists. It was, as a matter of fact, incorrect to formulate the concepts of the innate and the acquired as disjunctive opposites; however, the mutuality and intersection of their conceptual contents were not to be found, as the "instinct opponents" supposed, in everything apparently innate being, really, learned, but the very reverse. In fact, everything learned must have as its foundation a phylogenetically provided program if, as they actually are, appropriate species-preserving behavior patterns are to be produced.

Not only Oskar Heinroth and I, too, but other older ethologists as well, had never given much concentrated thought to these phenomena that we quite summarily identified as learned or as determined through insight and then simply shoved to the side. We regarded them—if one wishes to describe our research methods somewhat uncharitably—as the ragbag for everything that lay outside our analytical interests.

So it happened that neither one of the older ethologists nor one of the "instinct opponents" posed the pertinent question about how it was possible that, whenever the organism modified its behavior through learning processes, the *right* process was learned, in other words, an adaptive improvement of its behavioral mechanisms was achieved. This omission seemed particularly crass on the part of Z. Y. Kuo who had so expressly disassociated himself from every predetermined connection between organism and environment but who, at the same time, regarded it as axiomatic that all learning processes induced meaningful species-preserving modifications. As far as my knowledge goes, P. K. Anokhin was first among the theorists of learning to grasp the conditioned reflex as a *feedback circuit* in which it was not only the stimulus configuration arriving from the outside, but more especially the *return notification* reporting on the completion and the consequences of the conditioned behavior that provided an audit of its adaptiveness.

As in many other cases of erroneous reasoning, the behaviorists' absence of questions about the adaptive value of learned behavior may be traced to their emphatic antagonism to the school of purposive psychology. The latter's uninhibited commitment to behavior's extranatural purpose created in the behaviorists such antipathy to all concepts of purpose that, along with purposive teleology, they also resolutely refused to consider any species-preserving purposefulness, including teleonomy as defined by C. Pittendrigh (1958). This attitude, unfortunately, made them blind to all those things that could be understood only through a comprehension of evolutionary processes.

The "innate schoolmarm," which tells the organism whether its behavior is useful for or detrimental to species continuation, and, in the first instance reinforces and in the second extinguishes that behavior, must be located in a feedback apparatus that reports success or failure to the mechanisms of the first phases of antecedent behavior. This realization came to me only slowly and independently of P. K. Anokhin.

I published my theories on this subject in 1961 in my monograph "Phylogenetische Anpassung und adaptive Modification des Verhaltens," which I later extended and enlarged for a book in English, *Evolution and Modification of Behavior.* As I emphasized in that work, whenever a modification of an organ, or of a behavior pattern, proves to be adaptive to a particular environmental circumstance, this also proves incontrovertibly that *information about this circumstance* must have been "fed into" the organism. There are only two ways this can happen. The first is in the course of phylogenesis, through mutation and/or new combinations of genetic factors and through natural selection. The second is through individual acquisition of information by the organism in the course of its ontogeny.

Innate and *learned* are not each defined through an exclusion of the other but through the method of *entrance taken by the pertinent information* that is a prerequisite for every adaptive change.

The bipartition, the "dichotomy" of behavior into the innate and the learned is misleading in two ways, but not in the sense maintained in the behaviorist argument. Neither through observation nor through experimentation has it been found to be even in the least probable, still less a logical necessity, that every phylogenetically programmed behavior mechanism must be adaptively modifiable through learning. Quite the contrary, it is as much a fact of experience as it is logical to postulate that certain behavior elements, and exactly those that serve as the built-in "schoolmarm" and conduct the learning processes along the correct route, are *never* modifiable through learning.

But, on the other hand, every "learned behavior" does contain phylogenetically acquired information to the extent that the basis of the teaching function of every "schoolmarm" is a physiological apparatus that evolved under the pressure of selection. Whoever denies this must assume a prestabilized harmony between the environment and the organism to explain the fact that learning—apart from some instructive failures—always reinforces teleonomic behavior and extinguishes unsuitable behavior. Whoever makes himself blind to the facts of evolution arrives inevitably at this assumption of a prestabilized harmony, as have the cited behaviorists and that great vitalist, Jakob von Uexküll.

The search for the source of information that underlies both innate and acquired adaptation has, since those earlier years, yielded significant results. I will mention only the research done by Jürgen Nicolai with whydah birds (Viduinae) in which the information can be "coded" in such an intricate way: essential parts of the adult bird's song have been learned by monitoring the begging tones and other tonal expressions of whichever species of host bird by which the whydah happened to be hatched and reared.

Inquiry into the phylogenetic programming of the acquisition processes has proved to be important in many respects. Like imprinting, some acquisition processes are impressionable only during specific sensitive periods of ontogeny; a failure to perceive and meet their needs during those crucial periods in animals and humans can result in irremediable damage. Within cultural contexts the distinction between innate and the acquired is also significant. Man, too, and his behavior are not unlimitedly modifiable through learning and, thus, many inborn programs constitute human rights.

As early as 1916, Oskar Heinroth wrote in the conclusion of his classic paper on waterfowl:

> I have, in this paper, drawn attention to the behaviour used in social intercourse and this, especially in birds living in social communities, turns out to be quite amazingly similar to that of human beings, particularly in species in which the family—father, mother, and children—remain together living in a close union as long as, for instance, geese do. The taxon of Suropsidae has here evolved emotions, habits and motivations very similar to those which we are wont to regard, in ourselves, as morally commendable as well as controlled by reason. The study of the ethology of higher animals (still a regrettably neglected field) will force us more and more to acknowledge that our behaviour towards our families and towards strangers, in our courtship and the like, represents purely innate and much more primitive processes than we commonly tend to assume.

This early admonishment notwithstanding, ethology was curiously tardy in approaching humanity as a subject.

In the investigation of humans it is not easy to fulfill the primary task of ethology, which is the analytical distinction of fixed motor patterns. No less a man than Charles Darwin in his monograph "The Expressions of Emotion in Man and Animals" (1872), pointed out the homology of some human and animal motor patterns. The homology was convincing, but solid proof remained necessary.

Irenäus Eibl-Eibesfeldt was the first to afford this proof. He chose the same movements that Darwin had studied—those expressing emotions. For obvious reasons, the experiments involving social isolation that are generally used to prove a motor pattern to be independent of learning could not be used with humans, so Eibl fell back on the study of those unfortunates with whom an illness had already initiated this experiment in an equally cruel and effective manner: he studied children born deaf and blind. As he was able to demonstrate by means of film analyses, these children possessed a practically unchanged repertoire of facial expressions although, living in permanent and absolute darkness and silence, they had never seen or heard these expressed by any other human.

As a second route to approach, Eibl-Eibesfeldt used the cross-cultural method to study the expressions of emotions in humans. He observed and filmed representatives of as many cultures as he could, in standardized situations such as greeting or taking leave, quarreling, experiencing grief and enjoyment, courting, and so on. The essential patterns of expressing emotions proved to be identical in all the cultures he was able to study, even when the patterns were subjected to minute analysis by means of slow motion films. What varied was only the control exerted by tradition: this affected a purely quantitative differentiation of expression.

The most important result of Eibl-Eibesfeldt's extensive and patient research can be stated in a single sentence. The motor patterns shown undiminished by deaf-and-blind children are identical to those that, through cross-cultural investigation, have been shown to be inaccessible to cultural change. In view of these incontrovertible results, it is a true scientific scandal that many authors still maintain that all human expression is culturally determined.

A strong support for human ethology has come from the unexpected area of linguistic studies; Noam Chomsky and his school have demonstrated that the structure of logical thought—which is identical to that of syntactic language—is anchored in a genetic program. The child does not learn to talk; the child learns only the vocabulary of the particular language of the cultural tradition into which it happens to be born.

A surprising and important extension of ethological research was the application of the comparative method to the phenomena of human culture. In his 1970 book *Kultur und Verhaltensforschung,* Otto Koenig demonstrated that historically induced, traditional similarities on the one hand, and, on the other hand, resemblances caused by parallel adaptation—in other words, the reciprocal action between homology and analogy—are interacting in the development of human cultures in much the same manner as in the evolution of species. For an understanding of cultural history, the analysis of homology and analogy is obviously of the greatest importance.

Rather late in life this interest in human culture awakened my *medical* interest in my own species. New ideas arose from the association between the results of Eibl-Eibesfeldt and Otto Koenig and all that I had learned during my two years of activity as a psychiatrist. I do not think that without this schooling, as unwillingly as I had received it, I would have realized *how crazy* the collective behavior of humanity had become in our time. When, about twenty years ago, I heard a lecture by William Vogt, who was one of the first who had seen the immense dangers approaching, I was not impressed at all—as I must shamefacedly confess. It was Rachel Carson, whom I knew personally and respected highly as a marine biologist, who actually recruited me into the army of conservationists.

While still busily studying greylag geese at the Max-Planck-Institut für Verhaltensphysiologie in Seewiesen, I was writing the book *Civilized Man's Eight Deadly Sins,* which I dedicated to Eduard Baumgarten on the occasion about three years earlier, of his seventieth birthday. The book on greylags already exists in the form of an immense accumulation of data, but I am only just writing it.

Almost simultaneously with my retirement from the Max Planck Institut I received the Nobel Prize, sharing it with Karl von Frisch and Niko Tinbergen. Von Frisch ought to have got it much earlier, but the fact that Niko and I got it together is a matter of deep satisfaction. If ever two research workers depended on each other and helped each other, it is the two of us. I am a good observer, but a miserable experimentor and Niko Tinbergen is, as I have already said, the past master of putting very simple questions to nature, forcing her to give equally simple and unambiguous answers.

When I heard the news of the prize by telephone, my first thought was an objectionable one: That's one in the eye of behaviorism." The one excuse for this is that I was thinking not of myself but of the gain in respectability achieved by ethology as a science. My second thought, more commendable, was of my father, that is, what a pity it was that he did not live to hear this news. I can almost hear what he would have said: "It is incredible! That boy gets the Nobel Prize for fooling around with birds and fish." My father had been nominated for the Nobel Prize several times and had failed to get it by a very narrow margin. It is remarkable how strong the wish for a father's approval can still be in a man seventy years old.

An autobiography ought to end with the retirement of its author. I am afraid mine does not. In 1973 I felt that my work on the greylag goose, so far from being completed, left a surprising number of loose ends. The Max Planck Gesellschaft very generously agreed to finance further research, provided that I could find a place in Austria where I could continue it. By the mediation of my friend Otto Koenig, of the Austrian Academy of Sciences, of the Ministry of Science, and of the Duke of Cumberland it became possible to continue the work on greylag geese.

The Cumberland Foundation offered a gloriously suitable location in the Almtal in Oberösterreich and K. Hüthmayer, the head of the Foundation, worked miracles in the speedy construction of aviaries, ponds, and accomodation for scientists. In the year 1973 one hundred and forty-four greylag geese were moved to the Almtal. This uprooting and replanting of a colony of wild birds was a highly interesting experiment in itself. Among other interesting things, it turned out that the strongest factor keeping a free-flying wild goose from leaving the new location was its personal bonding to the human foster parent who had reared it, even if this had been two years ago.

The station in Grünau, first founded as a department of Otto Koenig's Institute for Ethology, has lately been turned into a separate institution by the Austrian Academy. The new Research Station for Ethology, Konrad Lorenz Institut of the Austrian Academy of Sciences, comprises at the moment research stations at three independent localities. The youngest one of these, the Biologische Arbeitsgemeinschaft Steiermark in Bruck/Mur in Styria, is occupied with research on the ethology and ecology of owls and raptors, with the aim of saving endangered species by breeding them in captivity.

The station in Grünau/Almtal continues its research on the ecology and sociology of greylag geese. The population of greylags in Grünau can claim to have been studied intensely and continuously for more than twenty years. There are, in the world, only two other populations of free-living undomesticated animals that can compete with this, the colony of *Macaca fuscata* in Japan, and the chimpanzees in the Gombe River Station studied by Jane Goodall and her coworkers.

The station in Altenberg consists mainly of an aquarium in which aggressive behavior of fish is being studied. Our most surprising result is that personal aquaintance constitutes the strongest factor inhibiting aggressivity. In *Zanclus canescens*, the formation of nonanonymous collective territory was proven and the same social structure was also found in a not too closely related species, *Zebrasoma veliferum*.

Concurrently a Konrad-Lorenz-Gesellschaft was founded, headed by my pupils Professor Antal Festetics and Professor I. Eibl-Eibesfeldt. These two men are a very comforting guarantee that my kind of work shall be continued even when I shall not be able to do so any more. So there is some hope that the research groups just mentioned will continue their work beyond the time at which I finally retire. Before I do so, I shall certainly have published the book at which I am presently at work, "Der Abbau des Menschentums, subtitled "und was man dagegen tun könnte [The Waning of Humaneness and what we should do about it]."

After having finished this book, which will be soon, I intend to "evaluate" the immense amount of data on the social life of greylag geese that we have collected through the years. I intensely hope that I shall be able to complete that book too. In case I should still be able to work after having done so, I intend to collect all I know about perchlike fish (Percomorpha) in a rather "pre-scientific" description, in which I plan to give a title culled from Heinroth's famous paper "Beiträge zur Biologie, insbesondere Psychologie und Ethologie der Anatiden," substituting fish for waterfowl.

References

Baerends, G. P. 1941. On the life-history of *Ammophila campestris* Jur. Nederl. Akademie van Wetenschappen, Proceedings 44: 1–8.

Bridgeman, P.W . 1958. Remarks on Niels Bohr's talk, *Daedalus* 87:85–93.

Brunswick, E. 1957. Scope and aspects of the cognitive problem. In *Contemporary Approaches to Cognition,* ed. J. S. Bruner et al. Cambridge: Harvard Univ. Press.

Craig, W. 1918. Appetites and aversions as constituents of instincts. *Biol. Bull. Woods Hole* 34:91–107.

Garcia, J. and Ervin, F. R. 1967. A neuropsychological approach to appropriateness of signals and specificity of reinforcers. *Proc. Intern. Neuropsychology Society Meeting.*

Hassenstein, B. 1965. *Biologische Kybernetic.* Heidelberg: Quelle & Meyer.

Heinroth, O. 1930. Über bestimmte Bewegungsweisen der Wirbeltiere. Sitzungsberichte. *Ges. naturforschende Freunde Berlin.*

Herrick, F. H. 1935. Instinct. *Western Res. University Bulletin* 22(6).

Hess, E. H. 1956. Space perception in the chick. *Sci. Amer.* 195(10): 71–80.

Holst, E. v. 1969. *Zur Verhaltensphysiologie bei Tieren und Menschen Gesammelt Abhandlungen, I and II.* Munchen: Piper.

Huxley, J. S. 1966. A discussion on the ritualization of behaviour in animals and man. *Philos. Trans. Royal Soc.* (London) 251B: 247–526.

Jennings, H. S. 1906. *The Behavior of the Lower Organisms.* New York: Columbia Univ. Press.

Kogon, Ch. 1941. Das Instinktive als philosophisches Problem (Kulturphilosophie, philosophiegeschichtliche u. erziehungswissenschaftl. Studien, Heft 16). Wurzburg: K. Triltsch.

Kummer, H. 1971. *Primate Societies, Group Techniques of Ecological Adaptation.* Chicago: Aldine.

Lehrman, D. S. 1953. A critique of Konrad Lorenz's theory of instinctive behavior. *Q. Rev. Biol.* 28: 337–63.

Lorenz, K. 1931. Beitrage zur Ethologie sozialer Corviden. *J. Ornithol.* 79: 511–19.

———. 1934. A contribution to the comparative sociology of colonial-nesting birds. *Proc. 8th Int. Ornithol. Congress,* pp. 207–18. London: Oxford Univ. Press.

———. 1937a. Uber den Begriff der Instinkthandlung. *Folia biotheoretia Serie B, 2, Instinctus,* pp. 17–50.

———. 1937b. The Companion in the Bird's World. *Auk* 54: 245–73.

———. 1939. Vergleichendes uber die Balz der Schwimmenten. *J. Ornithol.* 87: 172–74.

———. 1941. Kants Lehre vom Apriorischen im Lichte gegenwartiger Biologie. *Z. angew. Psychol. u. Charakterkunde* 59a: 1–81.

———. 1942. Induktive und teleologische Psychologie. *Naturwissenschaften* 30: 133–43.

———. 1950. The comparative method in studying innate behavior patterns. *Symposium of the Society for Experimental Biology 4, Animal Behaviour,* pp. 221–68. Cambridge: Cambridge Univ. Press.

———. 1951. The role of Gestalt perception in animal and human behaviour. In *Aspect of Form,* ed. C. C. Whyte, pp. 157–78. London: Bradford.

———. 1952. Die Entwicklung der vergleichenden Verhaltensforschung in den letzten 12 Jahren. *Zool. Anzeiger 1952 Suppl.,* pp. 36–58.

———. 1955. Morphology and behavior patterns in closely allied species. *Group Processes* (Transactions of the First Conference, Ithaca, N.Y., September 1954), ed. B. Schaffner. New York: Joshua Macy Foundation.

———. 1956. The objectivistic theory of instinct. In *L'Instinct dans le Compartement des Animaux et de l'Homme,* ed. P. P. Grosse, pp. 51–76. Paris: Masson et Cie.

———. 1959a. Methods and approach to the problems of behaviour. The Harvey Lectures, New York: Academic Press Inc. 60–103.

———. 1959b. Gestaltwahrnehmung als Quelle wissenschaftlicher Erkenntnis. Z. exper. angewandte Psychologie 6: 118–65.

———. 1959c. The role of aggression in group formation. 4. Conference on Group Processes. Princeton: Transactions of the Joshua Macy Jr. Foundation.

———. 1961. Phylogenetische Anpassung und adaptive Modifikation des Verhaltens. Z. Tierpsychol. 18: 139–87.

———. 1962. Kant's doctrine of the apriori in the light of contemporary biology. General Systems (New York) 7: 23–35.

———. 1963a. Haben Tiere ein subjektives Erleben? Munchen, Jahrbuch d. Techn Hochschule.

———. 1963b. A Scientist's Credo. In Counterpoint, Libidinal Object and Subject. New York: International Univ. Press.

———. 1964. Ritualized fighting. In The Natural History of Aggression, ed. J. D. Carthy and F. J. Ebling. London and New York: Academic Press.

———. 1965a. Uber die Entstehung von Mannigfaltigkeit. Naturwissenschaften 12: 319–29.

———. 1965b. Evolution and Modification of Behavior. Chicago: Univ. of Chicago Press.

———. 1966. On Aggression. New York: Harcourt, Brace, Jovanovich; London: Methuen & Co.

———. 1968. Die Entwicklung des Spiessens und Klemmens bei den drei Wurgerarten Lanius collurio, L. senator u. L. excubitor. J. Ornithol. 109: 137–56.

———. 1969. Innate bases of learning. In On the Biology of Learning, ed. K. H. Pribram. New York: Harcourt Brace & World.

———. 1970a. The enmity between generatiaons and its probable ethological causes. In The Place of Value in a World of Facts. Nobel Symposium 14 (Stockholm): 385–418.

———. 1970b. On killing members of one's own species. Bull. Atomic Scientists 26: 2–5.

———. 1971a. Der Sinn fur Harmonie. Kosmos 67: 187–91.

———. 1971b. Knowledge, beliefs and freedom. In Hierarchically Organized Systems and Theory and Practice, ed. P. A. Weiss, pp. 231–62. New York: Hafner.

———. 1972. The fashionable fallacy of dispensing with description. Naturwissenschaften 60: 1–9.

———. 1974. Analogy as a source of knowledge. Les Prix Nobel en 1973, pp. 185–95, The Nobel Foundation.

———. 1976. Die Vorstellung einer zweckgerichteten Weltordnung. Anz. phil. -hist. Klasse der Osterr. Akademie d. Wissenschft. 113, Jhg, 1976, So 2.

———. 1978. Behind the Mirror. New York: Harcourt, Brace, Jovanovich; London: Methuen & Co.

———. 1980. Die ethischen Auswirkungen des technomorphen Denkens. In Glaube und Wissen, ed. H. Huber and O. Schatz. Vienna: Herder & Co. Verlag.

———. 1981. The Foundations of Ethology. New York: Springer-Verlag.

Lorenz, K. and Kalas, S. 1980. *The Year of the Greylag Goose.* New York: Harcourt, Brace, Jovanovich; London: Methuen & Co.

Lorenz, K., and Rose, W. 1963. Die raumliche Orientierung von *Paramecium aurelia. Naturwissenschaften* 19:623–24.

Lorenz, K., and Tinbergen, N. 1938. Taxis und Instinkthandlung in der Eiroll-bewegung der Graugans, *Z. Tierpsychol.* 2:1–29.

Portielje, A. F. J. 1938. *Dieren zien en leeren kennen.* Amsterdam: Nederlandsche Keurboekerij.

Rasa, O. A. E. 1971. Appetence for aggression in juvenile damsel fish. *Z. Tierpsychol.* Beiheft 7.

Roeder, K. 1955. Spontaneous activity and behavior. *Sci. Monthly* 80:2362–70.

Aubrey Manning (b. 1930)

12

The Ontogeny of an Ethologist

Aubrey Manning

The opportunity to write a scientific autobiography was impossible to resist, although the closer the actual task of writing has approached the more obvious have become the pitfalls. It would be so easy to try to weave a pattern of coherent, sustained research that simply wasn't there, to be unconvincingly modest or unattractively self-congratulatory over favorite pieces of work. I will try to be honest without being solemn.

I do not regard myself as being in the front rank of ethological research workers although I think I have published one or two important papers. I believe also that I have assisted in the development of ethology in various ways and that this fact, as much as my research contribution, gained me election into this volume, which certainly does contain a high proportion of the major contributors to the first phase of ethology's history.

I came into biology along a route that is very well trodden in Britain—as a schoolboy I was a fanatical ornithologist. I was born in 1930 and brought up on the Surrey/Berkshire boundary, about twenty miles west of London—a region of brick houses, birch and pine heathlands with much rhododendron. During World War II it was artificially peaceful, for the cars and aeroplanes that now dominate this overdeveloped, overpopulated landscape were blissfully absent.

A school friend and I spent most of the leisure hours of daylight watching birds and my first article, on the breeding cycle of wood warblers *(Phylloscopus sibilatrix)*, was written with him. I have retained a love of birds, although it has become less professional over the years. It was always their behavior that captivated me and I read all the ornithological books I could get through the local library. I remember especially Lack's *Life of the Robin* and Kirkman's book on the black-headed gull *(Larus ridibundus), Bird Behaviour,* with its fascinating account of his early studies using model eggs to study their incubation responses.

I was fortunate in having an excellent biology teacher at school and from there I went up to University College, London in 1948 to read zoology. I had an inexhaustible desire to know more about animals and the UCL course exactly fitted my expectations. Its core was a two-year repeating cycle in which a year of vertebrates alternated with a year of invertebrates, so that in a three-year course one got a double dose of one part of the animal kingdom. I got two vertebrate years and have retained, even now, an affection for the acanthodian fishes, the apodan amphibia, the procyonid carnivores and other such vertebrate specialities. John Maynard Smith was an exact contemporary of mine at UCL and perhaps gives another view of the same course elsewhere in this volume.

Alongside our solid comparative anatomy, we had an idiosyncratic though interesting lecture course on comparative physiology from G. P. Wells, the son of H. G. It was possible to take a special option in physiology in one's final year, but I chose entomology. With hindsight this was probably an error for a behavior worker. I have never really acquired an easy handling of physiological concepts and in some part I trace this back to my solidly "whole-organism" training. Thus I cannot recall having had as an undergraduate any practical exercise involving live animals. But the entomology course was full of interest and I was all set for a career as an applied entomologist working in the tropical countries of the Commonwealth. However I got on well enough in my final year for my tutors to suggest that I should try to get a research grant to do postgraduate work. There was no doubt in my mind that I wished to study behavior and this feeling grew from my continuing fascination with birds. We had had no formal teaching in behavior at all, but someone had introduced us to the ideas of the new ethologists whose work was just beginning to make an impact. I can remember reading Lorenz and Tinbergen in the Society for Experimental Biology's 1950 Symposium volume during the summer of 1951, just before our final exams.

I obtained a research studentship from the Nature Conservancy and although, unimaginatively, I had wanted to stay on at UCL to do my postgraduate work, they wisely made the grant dependent on my being accepted at Oxford to work under Niko Tinbergen. I remember going to see him after I graduated and discussing projects. Although I had begun to think about applying the new ethological approach to insect behavior, I still found birds the more immediately attractive and I rather hoped that he would suggest I joined the group working on gulls. He, though, firmly identified me as an entomologist, the gull program was full at that moment, and he suggested a project on the foraging behavior of bees. I left more excited by the prospect of Oxford and research with the great man than by the project itself. I had the summer to spend reading up about bees—it took me very little time to become fascinated by them—and Tinbergen also gave me the manuscript of *The Study of Instinct,* due to appear in the autumn of that year, 1951.

Tinbergen himself is a contributor to this volume and it would be pretentious of me to try to interpret his research strategy, but it is valuable, I think, to set down how the group appeared to a young research student at this very active phase in the growth of ethology in Britain. Two other students arrived at the same time as I. Philip Guiton, who was to work on a motivational analysis of the male three-spined stickleback *(Gasterosteus aculeatus)* during the reproductive cycle, and Desmond Morris, who studied the ten-spined stickleback, *(Pygosteus pungitius).* We joined several others already in the group. Margaret Bastock, working on *Drosophila,* and two people working on black-headed gulls, Rita White and Martin Moynihan. During my time at Oxford others also arrived, among them Mike Cullen, Esther Sager (later Esther Cullen), Gilbert Manley, and David Blest.

Tinbergen was thus supervising seven to nine students at one time. I think most of us who have had graduate students would regard this as a daunting task. I can't believe it was an ideal situation for Tinbergen either, but he kept remarkably well in touch with all of us. We were quite a close social group and very conscious of the growth of a new approach to animal behavior. I remember Martin Moynihan called us "The Hard Core," and, over endless cups of coffee, we would discuss our results, very much within the framework of current ethological theory. We approached any new ideas and concepts from this conservative base but I think we were none the less constructive for this. During this time, three of the group wrote an important paper that certainly extended ethological thinking (Bastock, Moynihan, & Morris, 1953).

I think all of "The Hard Core" looking back would acknowledge that "Friday evenings" represented a key influence upon our early research careers. Each week we gathered after dinner in Niko and Lies Tinbergen's home to drink coffee and discuss ethology. I can see the room now, lined with books and huge photographs of Greenland scenery taken by Tinbergen on his visit there with Lies in the late 1930s; there was a decorated harpoon and other Eskimo artifacts. Much of the furniture was wooden, with rush seats, less uncomfortable than it looked. At these meetings students reported on their work and reviewed papers or ideas; over the years a huge diversity of topics was covered. Any visitors to Oxford—and they were frequent—were persuaded to talk; it was in this setting that I first met Danny Lehrman. Tinbergen would skillfully steer the discussion so that it was free, but not too discursive. He made sure that we had our say but kept to the point. Most importantly, he insisted upon understanding every step in an argument and never cared in the slightest if others appeared to have got there before him. Often, of course, they hadn't, but would have been prepared to let matters slide. We could rely on Tinbergen's persistent questioning to bring out the logic, or reveal that it was faulty.

Danny Lehrman's visit was a highly significant event for me. He came early in 1954 I think, not long after the publication of his famous "critique" in the *Quarterly Review of Biology*. We were all on the defensive. I have just mentioned that we—"The Hard Core," that is—tended to make our forays into speculation from what we saw as the secure home base of current ethological theory. Lehrman certainly appeared as something of a threat. I am speaking of the graduate student's response here—with hindsight I can now see that Tinbergen was way ahead of us. Of course he was critical of some of Lehrman's ideas and certainly ready to defend ethology, but he had more wisdom than we did and saw at once that ethology would benefit from having to take good criticism on board.

Lehrman's impact was even greater than we expected because he was so entirely different from the image we had formed of him. We were quite unprepared for his enthusiasm for animals in all their diversity and for his knowledge of natural history—he too was a fanatical birdwatcher. It rapidly became clear that he was in fact deeply sympathetic to the ethological approach and also hostile to much of white-rat experimental psychology. He wanted us to be more critical in our approach to behavioral development—that was really the crux of the matter—and who can dispute that he was absolutely correct.

Lehrman gave several talks that were characterized by the breadth of their coverage. He was familiar both with the ethological and the physiological literature—particularly as it related to hormones and reproduction. For me this diversity was most illuminating and I sensed all kinds of doors opening. Lehrman was the more effective because he did not assume that physiological explanations were any more interesting or important than behavioral or evolutionary ones—he interrelated all the levels.

Of course he also won us over by virtue of his personality. Danny was one of that rare type of human being that I would designate (following Evelyn Waugh in *Decline and Fall*) as a "Life Force." His warmth and kindness made all contacts with him a delight, while his enthusiasm and sense of humor were infectious. The story is that when Konrad Lorenz first met him, he exclaimed, "Ach, but you are a *large* man!" He too had probably expected a wizened laboratory psychologist, but Danny mostly weighed in at 250 pounds plus, I imagine.

I saw Lehrman regularly over the years, both here and in the States. While it is difficult to be specific about any effect he had on my research, I know that with Tinbergen, he was the most important influence in my career. I think his role in the development of modern ethology cannot be overestimated, and we still mourn his loss.

For me Tinbergen was an ideal supervisor; he didn't interfere much, but he was always available for discussion when you needed him. Probably the gull people got to know him best, because he was away camping with them in the field during March and April. This close contact was not without its trials, for Tinbergen was then a great believer in the virtues of the simple, not to say spartan life. He was at his most lively in the early morning, and some students found his high spirits in the predawn chill of a tent, before they set out for the first observations, rather hard to take!

The two main emphases of his work at this time were first, the evolution and motivational analysis of the displays used in communication by gulls and sticklebacks, and second, the investigation of the function of behavior in the sense of how it contributed to survival. Certainly these two lines overlapped extensively and both of them included issues concerning the stimuli to which the animals responded in natural situations. Tinbergen's own interests in behavior were very broad, however, and this breadth, coupled with the rich interactions within a large group of pretty diverse graduate students, made for a varied behavioral education. I have never since been attracted to the admittedly powerful research strategy that favors a concentrated team approach to one set of problems. I feel much happier with diversity around me, and the group at Edinburgh has evolved this way.

The problem Tinbergen set me concerned the stimuli from flowers to which foraging bees respond and the function of honey guides. Honey guides (nectar guides, Saftmale) are the patterns of lines or spots of contrasting color on the petals of flowers that mark out or appear to lead to the nectaries. It had generally been assumed that such patterns had evolved to guide pollinating insects. However, some fairly extensive recent work by a German botanist, Hans Kugler, had cast considerable doubt on their function. He claimed, for example, that bumblebees located the center of his model flowers just as easily when guiding patterns were absent.

Tinbergen found this hard to accept, taking the view that the patterns would not have evolved unless they increased the plant's survival. He wanted me to look again and use a more natural foraging situation than Kugler, who had his bees confined to cages. Tinbergen also suggested that I follow up some observations he had made with bees visiting hound's tongue *(Cynoglossum officinale)* in the sand dunes of the Netherlands, which suggested that they responded not only to the color and scent of the flowers, but to the general form of the plant itself. Both these problems would involve me in field experiments very much in the tradition and style of those that Tinbergen made famous with his experiments on the orientation of *Philanthus* to its nest and the pecking response of young gull chicks to their parent's bill.

Bumblebees are strictly seasonal animals in Britain, so after I arrived in Oxford to take up my research studentship, there was six months of waiting for my experimental material to emerge from winter hibernation. I spent the time reading the literature on the foraging behavior of bees. Most of it was in German, which I had laboriously and pretty ineffectively to learn from scratch. It was at this time that I first began to read Darwin seriously, particularly *The Effects of Self and Cross Fertilization* and *The Form of Flowers*. I was fascinated by his style and by the wealth of acute observation of insect behavior that was revealed. Darwin had already noted that bees may learn the general form of plants whose flowers they are visiting and his observations suggested ways I might proceed.

Tinbergen was away with the gulls as I began my first observations. I was fairly lost and had little real idea what to do first. This early bewilderment of mine has, I think, tempered my own responses with research students in their early stages of Ph.D. work. It was a formative period for me as I kept trying this and that, most of my efforts being concentrated on collecting bumblebee nests and trying to get bees to forage at small dishes of honey presented on artificial flower models.

Finally it began to work and I recorded their preferences for models of various shapes. The models were all quite large—5 cm across or more—and I noted that bees showed a strong tendency to react to the edges of the shapes, where there was a line of color contrast between the flower and its background. This suggested to me—and this was perhaps my first original idea in research—that the honey-guide patterns could operate by providing, in their turn, lines of color contrast at the center of the flower and hence attract the bee visitor away from the edge of the flower, to which it was responding initially.

This idea started me on a series of experiments, manipulating various patterns and shapes, comparing effects with flat and three-dimensional flower models, and I think I managed to demonstrate convincingly that my idea was correct (Manning, 1956a). At least it was correct for large flowers, (one can actually observe the "edge-effect" with bees visiting the huge flowers of magnolias or hollyhocks, for example);—whether they function in the same way on tiny flowers—some of which have very elaborate patterns—is hard to say. Of course the pollinators of such flowers are scaled down also and their responses may retain much in common with bees.

I have not kept up well with the literature on flower pollination, but I don't think much more attention has been given to honey guides since the neat work of Daumer (1958) revealed through quartz-lens photography how many more patterns were revealed when flowers could be observed by using ultraviolet as well as the wavelengths visible to us. Now that we know a great deal more about bee vision (e.g., Wehner, 1972), it would be possible to investigate the operation of "guiding patterns" in new ways.

The other part of my research with bees, as mentioned earlier, concerned their conditioning to the form of plants. It went far beyond this because I studied the behavior of bees foraging on the same group of plants over many days. Once marked and recognizable, each bee become an individual and it was fascinating to record their constancy day after day.

Bees learned the exact position of individual plants when foraging on *Cynoglossum* and visited them in a quite regular sequence. Occasionally a bee would leave the vicinity of familiar plants and make long sweeping flights over the ground nearby—flights that ceased when it encountered a new plant. On leaving newly discovered plants for the first time, bees sometimes made brief orientation flights and subsequently these plants became incorporated into their foraging area. If plants were removed, bees flew to their site and hovered there briefly before moving on.

It became clear that their foraging behavior involved far more than responding to stimuli from flowers or even learning to rely on the same area on each trip. I found that bees could learn a great deal more, but also revealed a subtlety of adjustment beyond this, for their responses varied according to the total stimulus situation provided by the plant they were currently visiting. Thus foraging on plants with large conspicuous flowers (the foxglove, *Digitalis*), they did not go through the elaborate orientation and learning that they did with *Cynoglossum;* they relied much more on direct visual stimulation from the flowers, (Manning, 1956b).

These observations represent for me, much more clearly now than they did at the time, a clear example of the way an animal's learning ability evolves to suit its life history. To use Seligman's (1972) useful terms, bumblebees come to forage with a number of "prepared" and "contra-prepared" responses. They learn no more than they have to—the association of certain stimui and reinforcement leads to learning only when this increases their efficiency in the location of food sources.

I have always hankered to get back to research on foraging, for it is a particularly agreeable field of study. It forces one to become something of a botanist as well as an ethologist and, since bees forage best in fine weather, one can excuse oneself for staying inside on cold, wet days. The delightful book by Heinrich (1979) describes a number of studies that have greatly extended my work on foraging, but I think there is still room for other ethological approaches. One of my most puzzling observations concerned the regularity with which male bumblebees, (especially *Bombus pratorum* males) would visit plants. They returned as regularly and appeared to feed as avidly as the workers, yet they have not been recorded as returning to the nest once they leave, and presumably have no more to do than feed themselves.

I have described my D. Phil. work because, although I have not worked in this field since, it sets the level of analysis at which I have always been happiest—the development and manifestation of behavior patterns within individual animals. This preference has shaped a good part of my research since.

The majority of my work has been in the field of behavior genetics and this also began at Oxford. During my second winter season, I analyzed data and wrote a report, but I also wanted to get involved with some laboratory-based research in the absence of the bees, and so I started collaboration with Margaret Bastock. She was studying the effects of the yellow mutant in *Drosophila melanogaster*. It had been known for some time that *yellow* males, although not disadvantaged in any obvious way, were less successful at inducing females to mate and this caused her to focus attention on the courtship behavior of both genotypes. Her study (Bastock, 1956) has become something of a classic and, although the answers she obtained have turned out to be incomplete, it was one of the first ethologically based studies in behavior genetics.

Tinbergen encouraged her in this at a time when behavior genetics as a field scarcely existed, but he saw it as a means of investigating the mechanisms of behavioral evolution. I began to think about the importance of relating what we could observe to be the behavioral effects of genetic changes to those we saw across the closely related species in the comparative studies that Tinbergen had running. *Drosophila,* with the large range of mutants available in *D. melanogaster* and a large number of closely related species and species groups with which to compare them, seemed ideal material.

However, Margaret Bastock convinced me that the first essential was to complete a detailed study of the courtship behavior of *D. melanogaster* and we worked on this. The Bible of *Drosophila* behavior at this time, and a work still to be consulted with profit, was Herman Spieth's (1952) monumental survey of some one hundred species, but, of necessity, he gave only the outline.

Margaret had developed a simple, but effective, means for recording courtship. Pairs of flies were watched in small cells, about 1.8 cm in diameter, covered by a glass microscope cover slip. They would be observed through a low-power binocular and she had evolved a system of time-sampling the behavior. A metronome kept the time, and the behavior of the flies, male or female, was called out onto a tape recorder every 1½ seconds—this interval was frequent enough to catch most of the changes in behavior and hence allowed a reasonable estimation of bout lengths for the different elements, while it was not too fast to handle.

We spent hour upon hour recording and transcribing the courtship of pairs of *D. melanogaster* and also pairs of its very similar sibling species, *D. simulans.* We produced a fairly detailed "ethogram" of their sexual behavior and attempted to account for the patterning of male behavior. This consists of several different elements that are to some extent superimposed upon one another, that is, simple orientation to the female, then orientation + wing vibration, then orientation + wing vibration + extension of the proboscis and "licking" of the female's genitalia, and so on. The elements have a preferred sequence (that given above), but this sequence can break off and elements may be repeated several times. No simple fixed sequence or reflex chain model can account for this pattern.

We developed a threshold model in which the second-to-second changes in the male's behavior depended on his level of sexual excitation. This level was determined by factors intrinsic to the male (his sex "drive" at the time) interacting with both positive and negative stimulatory factors from the female.

We described a number of graded "rejection" movements that a female made when she was unreceptive, and these we assumed to act negatively on the male's sexual excitation. The different elements of male sexual behavior, orientation, vibration, locking, and so on, have successively higher thresholds for their activation. Each is performed (and superimposed upon those already performed) as its threshold is reached. Fluctuations in sexual excitation are rapid and these account for the rapid changes from one pattern to another.

The paper that we published (Bastock & Manning, 1956) was certainly one of the first solid behavioral studies of *Drosophila* and I think it remains useful. Let me confess here that I have never opened the covers of the *Science Citation Index* but I would bet this is the most frequently cited contribution in which I have had a hand, largely because *D. melanogaster* has become such a popular animal for behavioral studies. To some extent we provided a kind of technical base for further work.

Our threshold model was attacked on the grounds that we had made invalid statistical deductions. Most probably this was so, but in fact we had other more telling evidence in its favor, which was there in the published data but which, for some unfathomable reason, we had not noticed at the time. In fact Andrew (1961) later drew attention to this in a paper in which he suggested a similar kind of threshold model to account for the varied alarm call sequences of blackbirds, *(Turdus merula)*. Tugendhat (1960) also used a similar model for the feeding responses of sticklebacks, *(Gasterosteus aculeatus)*.

It is interesting to remember that when we (or Spieth or Sturtevant before us) described the courtship behavior, we had no good evidence on what sensory modalities were involved. It seemed unlikely that wing vibration was a visual stimulus (*melanogaster's* mating speed is little affected in the absence of light) and sound or wafted scent seemed reasonable possibilities. The first good evidence for sound production came with Shorey's observations (1962); more recently my colleague Arthur Ewing (who was my first graduate student) has developed, with Henry Bennet-Clark, refined techniques for recording *Drosophila* courtship sounds. He has done some beautiful work surveying sound production through the genus, and gone on to study both its underlying mechanisms and their evolution (see Ewing, 1983).

After completing the *Drosophila* paper, I had a final season with bees and wrote up my D. Phil. thesis. In these days, one has become so accustomed to the four- or five-year Ph.D. that it is genuinely surprising for me to recall that mine was completed within three. I did have a fairly critical cutoff point to spur me on, because at the end of that third summer I had to go into the army for two years of National Service. Most people went in at eighteen when they left school, but I had had six years deferment for my undergraduate and postgraduate work. I did not relish the prospect. I was by now deeply attached to Oxford; I wanted to get a university job and continue with research and I regarded the two years ahead as a wasteful and unattractive interruption of my career.

Looking back on it now I wouldn't want to have missed the army. It must be remembered that this was peacetime and I had the pleasure of seeing a great deal of West German towns and countryside, for my regiment was part of the NATO forces there. I remember also a number of interesting people and perhaps, above all, I remember the exquisite agony of suppressed laughter. I seemed always to be in situations where the demands of the army as mediated through its rigid hierarchy and code of discipline expressed themselves in pure farce, but farce that could not be acknowledged by the least gesture or slip of the expression.

While away in Germany, I managed to keep fairly well in touch with Oxford, prepared my thesis for publication, and began to make inquiries about jobs. I drew a number of blanks and was particularly disappointed when I had no success with UCL but about six months before my service was due to end, I had a great stroke of luck.

This was through the good offices of J. B. S. Haldane, the professor of Biometry at University College, and his wife, Helen Spurway. They taught genetics to us at UCL and were very interested in ethology. I got to know them when I was a student and while I was at Oxford, we would meet on my visits back to UCL and talk ethology. Both were brilliant but in a manner so eccentric that a young and very conventional student like myself often had difficulty in coping. Nevertheless I was fascinated by their company and flattered by their interest in my work. They commented on early drafts of my bee work and helped me in numerous ways. Through all their eccentricities they were extremely kind and I remember them with great affection. They were fascinated by Tinbergen's approach and carried out some interesting experiments on the behavior of newts. Haldane also brought his mathematical analysis to bear on some of von Frisch's data on the bee dance.

Tinbergen found them quite a formidable couple and, I think, avoided meeting them when he could but Haldane sent manuscripts for Tinbergen to comment on. I happened to be with him when he received one such manuscript back. I should mention that Haldane had a most elegant prose style that combined simplicity and clarity. (His scientific essays are a delight still and his articles for the old *Daily Worker* in the thirties and forties remain some of the best popular science ever written). As Haldane flipped over the pages of his manuscript and took in Tinbergen's comments, I could see him becoming agitated. Finally he flung the papers down. "I will, of course, take note of his comments on the science. I will overlook the fact that he writes on the manuscript in ink . . ." (voice rising steeply) ' . . .but when he criticizes my English!" He positively shook with rage!

Haldane had visited Edinburgh and hearing that Michael Swann, the then Head of the Zoology department, was thinking of appointing an animal behavior person to the staff, he mentioned my name. I wrote, had an interview when next on leave from the army, and got the job. I was "demobilized" in September 1956 and went straight up to Edinburgh to take up the post of assistant lecturer. Twenty-eight years later, I am still there.

This might sound rather unenterprisingly static, but the fact is that both my wife and I like to establish roots in a place and Edinburgh has proved exceptionally good. Not only is it one of the most beautiful cities in Europe, it is surrounded by lovely and varied countryside in one of the last reasonably populated parts of Britain. Further, the university is excellent and formerly it grew at just the right pace to match my career—the jobs with suitable promotion fell vacant at just the right time (I reflect on this with wonder and humility when I consider the career prospects of a young person starting out in animal behavior research in 1984). Britain is such a small country that it is easily possible to keep in touch with colleagues anywhere else and I have had my fair share of travel abroad. I never had any real incentive to move.

I began work on *Drosophila* again, taking up a detailed comparative study of *melanogaster* and *simulans* (Manning, 1959). I was interested to discover how far the apparently sluggish nature of *simulans'* responses to a variety of stimuli could be related to the marked differences in the "tempo" of their courtship compared with that of *melanogaster*. Female *simulans* move much less than *melanogaster* when they are courted and as a consequence much more of the courtship is a static affair. The male vibrates rarely, but instead spends long periods facing the female, intermittently opening both his wings in a characteristic movement called "scissoring." A superficial description would tend to label vibration as the *melanogaster* wing display and scissoring as the *simulans* equivalent. However closer examination revealed that the differences between the species were reduced, if not eliminated, by altering the external conditions for the male. It proved to be very largely the level of the female's activity that determines the pattern of the male's courtship. Given a sluggish *simulans* female—and with patience one can usually get them to court the foreign species—the tempo of a *melanogaster* male slows down greatly and he shows occasional scissoring. Paired with an active *melanogaster* female, a male *simulans* becomes so highly aroused that his courtship becomes virtually indistinguishable in appearance from that of *melanogaster*. Indeed I often had to check my identification after a test to be sure I had the right males.

This means that during the divergence of these two species, the mechanisms controlling courtship in the male have not changed much. What has changed is the overall threshold of response to a wide variety of environmental stimuli as a result of which *simulans* is the far less active species. It is interesting to note that Spieth describes other sibling pairs in *Drosophila* that differ in a similar fashion. The lowered activity levels of *simulans* probably have ecological significance, for this species is known to withstand dessication at high temperatures better than *melanogaster*.

However, there have been some threshold changes within the courtship system too, for it appears that with equivalent levels of sexual stimulation, *melanogaster* produces more vibration and less scissoring. As we find so often from comparative studies, a group of species share a common repertoire of behavior patterns but perform them at different frequencies. These differences relate closely to the effects of small genetic changes revealed in studies of mutants, such as that of Margaret Bastock on *yellow*.

My interest in a more directly genetical approach led me to investigate the marked variation in mating speed of *Drosophila*. If one set up a hundred single-pair matings with flies as uniform in age and condition as possible, they nonetheless showed a remarkable range of mating speeds (the time from the introduction of the flies, until copulation began). Some mated in a few seconds, others were still unmated after many minutes, even though courtship was continuing for a good proportion of this time. I felt some of this variation must be genetic in origin and decided that artificial selection would be the easiest way to reveal the effects of genes operating on this trait.

I set up mass matings of fifty pairs of virgin flies in a brightly lit culture bottle from which copulating flies could be easily removed as they mated—drawn out into a "pooter." I could record the course of mating in the group and also select the fastest and slowest pairs to found lines selected for high and low mating speeds. In turn, fifty pairs from each of these lines were measured in the next generation, again selecting the fastest and slowest mating for breeding.

There is a particular fascination about selection experiments with a fast-breeding animal like *Drosophila*. The generations come round every two weeks, giving you rapid feedback. You are pressing enormously hard on the developing organisms, imposing huge selection differentials (20 percent in this case—it must be rare to reach 1 percent under natural conditions); how will it respond and—a behavioral question this—in what manner will such responses be mediated?

I got a rapid and substantial response to selection in both directions and soon had generated fast-mating lines in which 90 percent of the flies had mated before any flies of the slow lines had begun. I began some quantitative genetic analysis of these results, recognizing, of course, that all the concepts of quantitative genetics relate to measurements made upon individuals. Mating speed depends on the interactions between two individuals and my estimates of the heritability of the trait (about 30 percent) are of doubtful significance. However it was possible to demonstrate unequivocally that natural populations of *D. melanogaster* carry many genes affecting mating speed.

As often with selection experiments on behavior, the correlated responses were of particular interest because they reveal more of the way by which selection has operated to change the behavior and match the selection criteria. In brief, I found the most important behavioral change was to activity and general responsiveness to disturbance. The slow-mating flies were hyperactive and took ages to settle down when introduced into the mating chamber. Consequently it was several minutes before any sexual behavior could begin. The fast maters were much more sluggish, but courtship began within seconds. There were also changes to the detailed pattern of courtship of males—fast males showed more vibration and licking—and to the receptivity of females.

I think this (Manning, 1961) was an interesting and useful experiment. I was fortunate in having one of the world's great genetics laboratories on my doorstep and I was most grateful to Douglas Falconer and Forbes Robertson for their help and advice. I took the trouble to select replicate lines in both directions and to measure unselected controls each generation. The importance of controls is revealed by the remarkable overall fluctuations in mating speed: in some generations everybody mated much faster or slower although keeping their relative positions. The fact that controls did the same shows that such fluctuations must be of environmental origin and although they are commonplace in *Drosophila* work, we still have no idea as to their exact source.

The changes in the courtship and general activity thresholds went in opposite directions within the selected lines, but in controls both thresholds are low. It seems clear they are independent of each other and this is one example of the way that studying the effects of genetic changes helps to reveal something of behavioral organization. The distortions of the normal behavioral associations reveal an independence that might otherwise be unsuspected—this aspect of behavior genetics has been referred to as "parsing the phenotype."

At about this time, I spent a sabbatical year in the U.S.A. in Vince Dethier's lab at the University of Pennsylvania. I greatly admired his beautiful experimental approach to the feeding behavior of flies—work that was on the borderline between behavior and sensory physiology. He put me on to a nice problem concerning food deprivation and activity in *Phormia* in which I made no progress at all— largely because for some reason, now unfathomable, I stuck myself with a quite unnecessarily complex and precise method for measuring activity. I could never get it to work reliably.

However I learned a tremendous amount from Dethier and the rest of his group. There was a weekly "Feeding Seminar" and here the biologists met with physiological psychologists like Al Epstein, Phil Teitelbaum, and Elliot Stellar—it was an excellent and lively group that argued about concepts of hunger and the control of food intake in flies, rats, monkeys, and humans.

I spent a good deal of time in the States thinking about the questions I mentioned above concerning the effects of genetic changes on the thresholds of behavioral responses and the link between these effects and the observed changes that are the result of the microevolution within a species group. I completed a review of my ideas, which in a modified version (Manning, 1967) appeared in the symposium volume *Behavior-Genetic Analysis* edited by Jerry Hirsch.

At the end of my year in Philadelphia, Margaret (Bastock) and I (we had married in 1959 and she started on her book on courtship [1967] while in the States) drove across to California to take part in the first of the two meetings in Stanford on which this volume was based. This session and the one that followed it in 1962 provided an excellent opportunity to review the whole growing field of behavior genetics. There were powerful injections of psychology from people like Howard Hunt and Gardner Lindzey and Bob Thompson, with equally strong genetical inputs from Crad Roberts and Ernst Caspari. Altogether it was a splendid group in which to discuss the links between genes and behavior at many levels, with little of the polarization between wholistic and reductionistic approaches that are now so evident. I think the book that resulted is still one of the best multi-author surveys.

Back in Edinburgh, I began a fairly fruitless excursion into the physiology and behavior of cockroaches, although I hasten to add that a Ph.D. student and later colleague, Leonie Ewing, picked up one problem concerning their social organization and made really important contributions. I continued work on *Drosophila* too and completed a number of other selection experiments. These refined some of the earlier results and enabled me to focus upon particular effects of genetic changes.

Of course in selection experiments one is always dealing with the effects of many genes and thus the establishment of causal links is very difficult; nevertheless it is possible to establish some important facts on the general nature of gene action. In one case I obtained a line of *D. simulans* selected for slow mating that was different from all others I had worked with because the effects were entirely confined to the females. Males from this line had normal mating speeds with other genotypes of females and, conversely, changing their male partners had no effects on the receptivity of slow-line females. I was able to show that the normal onset of receptivity with maturation did not occur in these females and could go one stage further in identifying the nature of this genetically controlled developmental effect (Manning, 1968).

In an earlier study (Manning, 1967), I had discovered that on eclosion, females were quite unreceptive to males, although they were certainly attractive to them and elicited good courtship. About thirty-six to forty-two hours of age (in flies rendered arythmic by rearing for generations in constant light), a rather rapid change took place (I called it "switch-on") and they began to accept males after normal periods of courtship. On the second day from eclosion all females were receptive and, if unmated, remained so for several days until the effects of old age began to creep in after a week or so.

Switch-on coincided with a time of very rapid growth of the female's ovaries. Females hatch with very few mature eggs but feed avidly and their ovaries enlarge dramatically. It was known that the incorporation of protein into the ovaries was controlled by juvenile hormone secreted from the reactivated corpora allata. Insects show an admirable parsimony in some features of their endocrine system. Juvenile hormone is secreted actively around the time of each larval molt and prevents metamorphosis from taking place prematurely. Naturally its secretion must be suppressed for the final molt, so the corpora allata are inactive during pupation, metamorphosis and at eclosion. However insects have also pressed juvenile hormone into service as a kind of gonadotrophin that acts in young adults, so following eclosion the corpora allata have to be reactivated.

There was a possibility that this hormone might have behavioral effects too— this had already been demonstrated in the grasshopper *Gomphocerrippus* by Loher (1962). I could show that it was certainly not ovarial growth per se that changed behavior in *Drosophila* (one could imagine possible feedback from the distention of the abdomen with growth, for example) as starved flies whose ovaries remain tiny nevertheless show behavioral switch-on.

At this time it was almost impossible to get good hormone preparations from insects, so I tried dissecting out the corpora allata of mature females and injecting them whole into the pupae of other females just prior to eclosion. Such females showed precocious receptivity and this result was a further clear demonstration of a hormone affecting sexual behavior. It may also work in males, but if so it must work more rapidly than in females because males will show adequate courtship behavior at about twelve hours from eclosion in *D. melanogaster.*

Returning to the unreceptive line of *D. simulans,* I was able to take the analysis of gene action one stage further. The females showed no behavioral switch-on (it was a regrettable necessity, that to keep the line going at all I had to rely on effectiveness of males in forcing their attentions on the unwilling females), but since their ovaries grew normally, this strongly suggested that it was not hormone deficiency that eliminated their behavioral response. I proved this by showing that corpora allata taken from the unreceptive females would induce precocious receptivity when injected into the female pupae of normal genotype, and that no amount of corpora allata from normal females would bring about normal receptivity in the unreceptive-line flies. The only conclusion is that somewhere down the line some target organ(s) upon which corpus allatum hormone operates must be unresponsive and it seems likely that this will involve neural mechanisms in the female's brain.

I thought it was interesting to be able to suggest genes affecting the threshold of neural target organs in *Drosophila,* since analagous effects had been demonstrated in mammals such as guinea pigs and mice. The next stage of analysis for the *Drosophila* would be to investigate brain neuroendocrinology and neurochemistry. Certainly this would be worth doing, but it didn't interest me to go down that path, for my real interests have always remained at the gross behavioral level and there are problems enough there.

Jerry Hirsch spent six months in my lab at this time and we carried out a fairly detailed genetic analysis of the slow-mating *D. simulans* line—it was only a single line, for its replicate selected in exactly the same way showed virtually no response. At first sight it appeared that a single major recessive gene was responsible for the effect on female switch-on—we got ratios of receptive and unreceptive females that were close to simple Mendelian ones. However, as Sewall Wright had demonstrated fifty years previously, it is possible to get clear segregation of this type from a polygenic system operating about a threshold. With a genotype up to the threshold value, development proceeds along one pathway; beyond this value the alternative pathway is activated. Repeated backcrossing revealed that this was the situation for the control of receptivity in the *D. simulans* populations we had been using (Manning & Hirsch, 1971). Artificial selection must have rendered the slow line homozygous for a number of recessive genes that affect the response to hormone.

While pursuing this work on *D. simulans,* I had been writing an extended review on *Drosophila* and the evolution of behavior, in which I tried to draw together the genetic and comparative studies in this genus. I felt that the subtle qualitative differences between the so-called fixed action patterns of display postures and movements that give them such a species-typical form could also be readily accounted for by gene-controlled changes in threshold within the motor system that must be responsible for their production. In considering such mechanisms I was greatly influenced by my old college friend from UCL days, Graham Hoyle, who by this time was already established in Eugene, Oregon. I think his early contributions to the field now called neuroethology were of the very greatest importance and I benefited very greatly from discussions with him.

My review (Manning, 1965), published in a series called *Viewpoints in Biology* that rapidly became defunct was, I think, a useful survey. I was tremendously pleased when Danny Lehrman wrote and asked me for several copies, as, he told me, he made it compulsory reading for all his graduate students.

Since being appointed at Edinburgh I had been responsible for developing the animal behavior teaching. This gradually increased in scope and by the mid-sixties, we had a pretty broadly based course, taking an ethological viewpoint certainly but trying to link with psychology through discussions of learning and the physiology of motivation on the one hand, and to evolutionary biology and genetics on the other.

I had had some ideas about using my lecture course as the basis for a book but always put it off, until one day a publisher must have caught me during a sensitive period. I had always pictured scientific authors rather like first-time novelists, desperately hawking round their precious manuscripts and sinking into despair as they covered their walls with rejection slips. Nothing could be further from the truth; any scientific writer with a half-respectable manuscript will find himself avidly being courted in a seller's market. Edward Arnold skillfully signed me up, and my *Introduction to Animal Behavior* first appeared in 1967.

There was then no introductory text strong on ethology and the book obviously filled a gap. It was very successful and represents one way in which I feel I have been useful. Nothing has given me more pleasure than to hear from teachers that their students have liked the book. It has gone into three editions (genuine new editions, which involved new material and extensive rewriting) and been translated into eight languages. I was especially pleased that a cheap, subsidized edition is produced for use in developing countries. It is with wonder and mortification that I recall that the first edition contained no proper account of social organization. My priorities were straightened and my education completed when John Deag, fresh from field studies on the Barbary Macaque *(Macaca sylvanus)* joined Arthur Ewing and I in Edinburgh.

The book appeared just before I set off to go to the 1967 International Ethological Conference in Stockholm and it was following that meeting that I was first elected to the International Ethological Committee. These biennial conferences have been a powerful influence in ethology and form a great tradition that has strong devotees and and not a few strong critics. They have their origin in the meeting that Lorenz called in Buldern in 1951, which almost everyone in the infant field of ethology attended. The next conference was in Oxford in 1953 while I was a graduate student, and about eighty attended as I recall. Tinbergen and his students ran the whole thing and it had all the advantages of a small meeting that could be run on totally informal lines.

During the early years of these conferences, the tradition was for a single session only, so that nobody need miss anything. The program was pretty loosely organized—I don't remember much fuss over time limits—and two-way translation, German/English and sometimes French/English, was provided by Lorenz, Tinbergen, and Gerard Baerends. People would stop every five or ten minutes to have a section translated. It was a mammoth task for the Grand Old Men (although I hasted to add they weren't *that* old at the time) particularly as they were effectively trapped into having to attend the presentation of every paper. If Lorenz or Tinbergen were missing, the young research workers reading their papers would be deeply hurt.

Meetings went on for eight or nine working days and often until ten at night. I can recall reading a paper at the Freiburg meeting in 1957 quite late one evening with Lorenz and Tinbergen absolutely grey with fatigue but still heroically there in the front row. A merciful tradition calls for a day off at some point during such meetings. These were originally called "migraine days," for Tinbergen, who is a migraine sufferer, had—understandably—to retire into a silent, darkened room halfway through the Oxford meeting.

The Oxford conference closed somewhat unconventionally. I cannot remember exactly how the idea for a spoof lecture and demonstration evolved but certainly Tinbergen and Desmond Morris were the ringleaders. Morris, David Blest, and myself constructed a ludicrous contraption of flasks, tubes, balloons, levers, chains, and so on, so linked as to represent a psycho-hydraulic machine (psychohydraulics was the irreverent name given to Lorenz's model of motivation). This machine moved colored liquids through itself, generated smoke, and spurted a great deal of water, while Morris and I milked it for every possible sexual and scatological belly laugh (Morris gives a good account of its operation in his autobiographical book *Animal Days*).

More memorable, though, was Tinbergen's introduction to our display that evening. The whole affair had been kept secret and the machine, draped in cloth, stood in the lecture hall. Tinbergen began—as previously announced—by summarizing the work of the conference. He spoke perfectly seriously and said some important things. Then he began to digress to discuss areas of ethology, not well covered at the meeting, modeling being one of them. He maintained a perfect deadpan style and to us, waiting in the wings, as it were, it was fascinating to see how preposterous he could become before anybody felt it was safe to laugh. Social psychologists know all about such constraints on human behavior at public meetings of this sort. I have often promised myself I'll try it out when the opportunity arises.

After joining the International Ethological Committee in 1967, I was asked to organize the 1971 Conference in Edinburgh and subsequently became secretary-general of the IEC. Gerard Baerends, my predecessor, had managed to establish a system that operated flexibly, with just the right blend of autocracy and democracy. I tried to maintain and develop this, as the influence of the ethological approach extended and ethologists from more countries became involved. The term *secretary-general* has a strong UN flavor about it and I must say the job did often have a diplomatic as well as a scientific side. Nationality and personality had to determine the nature and tone of much of the correspondence during my eight-year stint.

The IEC came to have the characteristics both of a senate, with each ethological nation represented, and of a congress, with some recognition of the size of the ethological community in each country. For years, the IEC, and the great majority of the conference members, strongly adhered to the idea of "closed" conferences. This rule kept numbers down to about four hundred and fifty, but at the price of considerable resentment from those excluded. The secretary-general tried to keep a reasonable "quota" of invitations for each country to match its ethological pressure and we succeeded in keeping a balance between young bloods and established workers. The system held up well until the 1980s, but in the end a majority felt that the dangers of losing touch with important developments on the margins of core ethology outweighed the problems presented by large numbers, and future ethology conferences will be open. From about fifty at the Buldern meeting in 1951 we shall have evolved to twelve hundred or more in the 1980s. Lorenz and Tinbergen may reflect on the infectiousness of their approach and the offspring it has produced!

During the late sixties and early seventies, I spent an increasing proportion of my spare time with what is popularly called "the environmental movement." This commitment grew very directly out of my delight in the natural world of plants and animals. In common with many other biologists, I have long felt that the world is going to hell in a handcart and this feeling was quite unrelated to the most immediate threat that faces us now—the nuclear holocaust. Indeed during the decade from 1966 to 1976, the nuclear threat seemed to recede a little, while the erosion of natural resources, the obvious increase in pollution and environmental degredation of all types began, at last, to attract attention.

I had a somewhat lower threshold for such features than most and from about 1966, I think, I began to see the growth of human populations as the driving force behind all our problems of getting into balance with the planet that supports us. The Conservation Society in Britain has a slogan, "Whatever your cause, it's a lost cause unless we limit population." Nobody pretends that population control on its own will restore a sustainable society, but without it the task is absolutely impossible. This conclusion is as true for the rich North as for the poor South, indeed much of the North in general and Britain in particular are already grossly overpopulated.

For a time such issues were "popular" in the sense that every organization, dining club, youth group, etc. wanted someone to come and talk. As I stomped about Scotland and a good bit in England and Wales too, I sometimes saw myself as the poor man's Paul Ehrlich. Certainly I identified myself with a substantial majority of his views. I feel that he and Garret Hardin will prove to have been important figures in forming late-twentieth-century attitudes.

The "environmentalist's" view is now recognized, even if it is always identified as antidevelopment. The tragedy is, as Robert Allen points out in *World Conservation Strategy,* that scarcely ever have conservation and economic development been planned together—they are erroneously seen as being always in conflict.

I still remain an active, and I hoped informed, ecofreak, and try to convince administrators at all levels that our obsession with economic growth of the conventional type only delays our real, sustainable recovery, our adaptation to a postindustrial society of a type that can survive. Perhaps the present dreadful employment situation for biologists may force some of them into business, administration, or even politics. If so we can only gain. The huge majority of administrators and politicians across the world are totally ignorant of the basic principles by which the earth supports life. They do not even begin to acknowledge the real problems.

There are so many obvious, intense, and immediate human problems in the world that people with these obsessions can be accused of inhumanity in our concern for other species and for curbing human reproduction. I deny this charge, for I regard the growth of our numbers as the greatest long-term threat to human dignity and freedom. In any case we need *some* people to speak up for the planet and its other inhabitants. I think of Robert Burns who could address a small rodent thus:

> I'm truly sorry man's dominion
> Has broken Nature's social union,
> And justifies that ill opinion
> Which maks thee startle
> At me, thy poor earth-born companion,
> An' fellow-mortal!

It was during this period of intense environmental activity that my own micro-environment was changing too. Margaret and I had lived for fifteen years in a top flat in Edinburgh's late Georgian New Town. It was a splendid place in most ways—five minutes' walk from Princes Street—and one of the most heartening things about Edinburgh is that so many people live close to its center. However we hankered after some land to work with.

In 1972, together with four other families, we bought a share in about thirty acres fifteen miles to the southeast of Edinburgh. This was pasture and neglected woodland with a solid old coach house and some stables and other earlier buildings that had been farm workers cottages. We restored these, making three houses, and two new houses were built close by. We shared the costs of services, and so on.

We moved in in 1974 and ten years later we are all still friends. We share land and some machinery but we have our separate homes into which we can retreat. Our house was made from the oldest building that remained. It is the base of a small fortalice or tower house, dating from—probably—the fifteenth or sixteenth century. The walls at the ground-floor level are six feet thick and there is a delightful spiral stair to the upper floor—all of stone.

We have found it marvelously refreshing to live surrounded by woodland, yet so close to Edinburgh. The group of us is trying to restore the woodland by planting a diversity of tree species and thus create a richer habitat for wildlife. When research or administration are going badly, I reflect that probably by far the most useful thing I have done in my life is planting some thousand trees—that at least is important and lasting! I have a simple four word formula for the easier salvation of our planet, and our species with it: more trees, fewer people.

My research interests continued to center upon the ways in which the effects of genetic changes within a population or genetic differences between population could throw light upon behavioral organization itself. Just before the Edinburgh conference, Tom McGill from Williams College, who had worked extensively on the genetics of mouse sexual behavior, spent a sabbatical year with me. We began some collaborative work on mice, in particular the relation between testosterone and sexual behavior in three inbred lines, C57 Black 6, DBA/2, and Balb/c.

McGill had previously been skeptical of the common assumption that the sexual behavior of male rodents was completely hormone dependent. He suggested that this assumption was based upon evidence from highly inbred mice and was not an adequate comparison with carnivores or primates, where hormone dependence—

or in fact greater hormone independence—was estimated from heterozygous out-bred stocks. Certain hybrid mice had also shown greater hormone independence.

We investigated this in more detail by studying the retention of sexual behavior in castrated males of the three strains mentioned above and all their F_1 hybrids. All the males were allowed to mate and had equal experience of sexual behavior prior to castration at around ten weeks of age.

The tests revealed some striking differences. Two of the inbred lines, DBA and Balb, behaved according to the old assumption. The sexual behavior of most males disappeared totally and almost immediately following castration. The third line, C57B1.6, showed a less-rapid decline, but it was still complete within a week or two. Hybrids between DBA or C57 and Balb showed very little heterosis for this character but those between C57 and DBA were remarkable animals. These BDF_1 showed scarcely a falter in their sexual motivation following castration. Months or even a year later, a good proportion of them were ejaculating regularly. Old age often claimed them before their sexual behavior declined (one memorable ancient male, well over two years old, and of incredibly decrepit appearance collapsed and died in mid-copulation!). McGill and I tried to analyze in more detail the nature of BDF_1's hormone independence. By this time I had been joined by Mike Thompson from the University of Georgia, who brought both surgical and behavioral skills to help us. We got some way (e.g., McGill & Manning, 1976; Manning & Thompson, 1976); in particular we found that the hormone-independence only related to a mature mouse whose sexual behavior mechanisms had developed under the influence of androgens both perinatally and at early puberty. Males castrated just before puberty did not develop sexual behavior unless treated with androgens. However they proved to be amazingly sensitive to androgens administered in adulthood and began to show sexual responses after only two or three injections of $25\mu g$ of testosterone propionate—a dose far lower than that required by other strains. Interestingly enough, in an unpublished student project, Janson-Smith showed that female BDF_1 were similarly responsive to estrogen. Minute doses of 250 ng estradiol benzoate were enough to restore reasonable levels of lordosis in ovariectomized animals.

The editor will rightly not approve of using this volume to present hitherto unpublished results, but let me just finally add here that we found that the androgen-independence of BDF_1 males was at least partly specific to sexual behavior. Their aggressive behavior declined quite rapidly following castration, which suggests—not surprisingly—that sexual and aggressive systems are affected differently by the combination of genes that come together in BDF_1.

Such a result also aligns with the previous conclusions for rodents and other vertebrates (also for my slow-mating *Drosophila* line described above), that genes more readily change behavior via changes to the target organs upon which hormones act than by affecting hormone levels themselves. Of course there is genetic variation in endocrine function, but we know little of its significance for behavior, it any.

Jennifer Batty did her doctoral work on questions concerning hormone levels and sexual behavior in our different mouse strains (Batty, 1978a, b). She found, in fact, a negative relationship between circulating levels of testosterone and levels of sexual behavior, but the high behavior lines proved to be very sensitive to androgens. There was also an association between the intensity of sexual responses and the effects that such behavior had upon subsequent endocrine secretion. We don't know how plasma testosterone levels relate to the uptake of hormone by the hypothalamus, but Batty's work, and that of others, strongly suggests that in male mice, hormone levels in all strains are well above that needed for expression of sexual behavior. With testosterone at permissive levels we can observe genetic variation affecting sexual behavior via other systems.

I remain attracted to the mouse for behavior-genetic studies, although it is not often easy to come up with a situation that enables one to probe beyond the simple association of genetic and behavioral differences. Perhaps one of the more promising possibilities is that now being explored by a graduate student, Bob Goupillot, who is following up an earlier study (Stewart, Manning, & Batty, 1980) on the effects of the Y chromosome from different strains upon the aggressive behavior of males carrying them. Alastair Stewart has worked extensively on the role of the Y chromosome and set up a breeding program to eliminate the confounding of maternal effects with Y chromosome type that can arise with simple hybridization between two strains. The effects on aggression are quantitative and probably not large under most circumstances—certainly genes carried on the autosomes have greater effects. However, the simple inheritance of the Y, which acts as a single genetic unit, will, we hope, give us a good mammalian "preparation" in which to study gene/environment interaction in the development of a male's aggressive behavior.

I consider that the action of genes during behavioral development is one of the most challenging problems facing ethologists at present. It is made the more so by recent sociobiological theorizing concerning the evolution of behavior. The earlier dichotomizing of genetic and environmental determinants of behavior has clearly proved unsatisfactory. Unfortunately there has been a tendency to substitute a wishy-washy kind of interactionism in which genes and environment are seen as being in complete interplay with everything affecting everything else. Development cannot be like this. It must have a *structure* in which there is a sequence of events (gene/environment interactions if you like) that run through the course toward adulthood. The overall program in many organisms must be under genetic control but in the absence of an adequate environment, distortions or malfunctioning will occur.

We have a number of clear examples of the effects of environmental determinants that the developing animal can utilize only at particular stages (sensitive periods)—with some bird songs, for example, or the bias of orientation to the visual units of the mammalian striate cortex. Deprived of the normal environmental inputs, the young animal may latch on to other normally ineffective inputs and move on to the next stage of development with an altered pattern. Alternatively it may exhibit a capacity to "catch up" and adjust back to the normal pattern at a later stage. The nature of its responses to changed environments may yield clues as to the normal mechanisms of development.

Bateson (1976a,b) has provided a stimulating discussion of behavioral development along these lines and I particularly like the way in which he utilizes and extends Waddington's old model of the "epigenetic landscape" and the arguments put forward in Waddington's old model of the "epigenetic landscape" and the arguments put forward in Waddington's book *The Strategy of the Genes*. It is certain that the genetic strategy unfolds through the successive activation of different subsets of genes. If we could identify subsets involved in behavioral development, this would not only be of intrinsic interest, but would also provide clues to the way behavioral potential is organized in genetic terms.

It is a very long shot but one that is worth pursuing, and there are techniques that would allow us to test the "genetic architecture" at different stages of development, provided we can first isolate two strains or populations that differ in the trait to be examined. A change would indicate that different genes are involved. There have been some attempts in this direction already that look at the genetic background for different stages of learning a task—we can regard this as one form of behavioral development. One of my long-term aims in research is to pursue just this point in collaboration with Patrick Bateson who in his studies of sexual imprinting in Japanese quail may have suitable material.

I began to speculate about this type of approach to development in the course of writing a couple of reviews (1975, 1976) that had some overlap. Both tried to review the achievements of behavioral genetics and to place this field in relation to the study of behavior generally. The first of these reviews was for a volume of essays on function and evolution that Gerard Baerends, Colin Beer, and I edited as a *Festschrift* for Niko Tinbergen's seventieth birthday. Editing this was a labor of love and, as it happens, the essays were written just before the upsurge of interest in sociobiology and I think they are none the worse for this. The volume includes a diverse collection of overviews by Tinbergen's students and close associates dealing with an approach to the survival value of behavior and its evolution that represents one of Tinbergen's great contributions to our field.

My second review focused less upon evolution and more on possible uses of genetics to investigate behavioral organization—such as in the suggestions for developmental studies just considered. I presented this as a paper for the twenty-fifth birthday celebrations of the Madingley subdepartment of Animal Behaviour at Cambridge. This was a delightful occasion, based at King's College, with plenty of good company, food and drink. The success of Madingley has consistently been one of the high points of European ethology. Bill Thorpe, Robert Hinde, Patrick Bateson, and many others have secured its reputation and influence through their consistently innovative work.

I had an opportunity to try out my ideas before the Madingley meeting for I was delighted to be invited to give the 1976 keynote address to the summer meeting of the Animal Behavior Society, being held that year at Wilmington, North Carolina. I was fascinated by the local environment—the furthest southeast I had ever been in the States. I remember the excitement of watching boat-tailed grackles and fishing black skimmers from the balcony of my hotel.

The ABS meeting was huge—about nine hundred I think—and thus about twice the size of an ethology conference. The circumstances of my address were scarcely conventional. It was scheduled to take place in a large hall on the campus of the university following the conference dinner. I suppose about half the delegates came to the dinner, which took place in a barnlike seafood restaurant about ten miles outside Wilmington. The food was good, when one eventually got it, and there came the snag. For some reason there was no set dinner as is usual on such occasions, we all ordered separately and the staff were totally unable to cope with serving 400 people. It took absolutely ages before food came and meanwhile there was nothing to do but drink. From somewhere came an inexhaustible supply of canned beer. American beer is weak stuff, but consumed in quantity it makes it mark and as tottering pyramids of empty cans piled up on the tables, conversation rose through the various levels of volume and fluency toward downright hilarity. I became dimly aware that the time at which my keynote address should begin approached rapidly, then was passed, then was passed by a considerable amount.

Coaches eventually took us back to the campus and a large and very good-humored crowd piled into the hall to find that a large number of other delegates were already seated and had been seated for the previous one and one-half hours. Jerry Hirsch introduced me and there before me ranged my divided audience; about half had (as the Scots would say) "a drop taken" and were probably best prepared for a cabaret turn, the other half were stone cold sober and furious at the preposterous delay.

It was difficult to know the best tone in which to frame my opening remarks. I hoped that my own alcohol intake had not been too incapacitating and plunged in. I remember beginning by opening yet another can of beer to toast the audience and then telling a story about circus life in ancient Rome under the emperor Caligula. But I did go on to review behavior genetics with reasonable sobriety and the audience were uniformly courteous despite their diverse preparation for my lecture.

To conclude by coming up to date, I don't feel my enthusiasm for animal behavior has waned in the slightest over the years. I have less time than I would wish to "work at the bench" and one has constantly to take care to keep *something* going because it is so easy just to get out of the habit of doing research. I do keep pretty good contact with the lively group around in Edinburgh. I am helped in this by the fact that Margaret* has, over the years, turned herself into a full-time research psychologist. Beginning with her acute observations of our own two children, she moved on to study the aggressive behavior of nursery school children, and she has held a number of research grants. Her work seems to me to be an important blend of the ethological and psychological approaches to human behavior and she has developed an excellent framework for understanding young childrens' behavior in social situations. She has developed good collaboration with psychologists and psychiatrists in Edinburgh.

There is a weekly seminar on behavior—ranging in subject matter from ecology to psychiatry—and we often muster twenty to twenty-five people from diverse departments. I try to recapture some of the feeling of those Friday evenings thirty years ago. I think of Tinbergen and ask elementary questions when I don't understand.

Apart from the mouse Y-chromosome project mentioned above, I am also in-

---

*deceased 1982

volved (with Candace Lawrence and John Deag) in a study of maternal behavior and the development of cats. Then there is a rather delectable *Drosophila* project which I want to get down to, concerning the persistence of sexual arousal after males have lost contact with the females they've been courting. A cynic has suggested that each research worker does only one experiment, just repeating it with minor variations throughout his or her career. As I collect virgin flies and sort them into vials, I know what he meant.

I shall not really be content until I've made some attempt to get to grips with the problem of genes and behavioral development. I am convinced some of us will need to try this for behavior itself and not all rush into a reductionist approach. We know behavior evolves at its own level and that the course of development also evolves to match the life requirements not just of the adult but of the young animal. Accordingly we need to know how the genotype is organized to enable the expression of large-scale units to evolve without necessarily affecting every aspect of behavior. There is no evading recognition of how daunting this task will be, but it will not be solved by turning behavior genetics into a branch of developmental neurobiology.

Even though I have been trying for years to discover how genes affect the expression of behavior, I nevertheless think it is extremely important to recognize their limitations—particularly in ourselves. So commonly, genetic factors are equated with inevitability of expression even by some who should know better. As I get older, I am coming to the reluctant conclusion that the nature-nurture controversy will be with us for some time yet. The idea of fixed genetic determination has a consistent appeal to a strong fatalistic view of life that many cultures support. If it does nothing else, it seems to provide reassuringly simple answers.

The only way past this mental block is to concentrate on development and try to discover some of its basic grammar. One of the referees for my most recent experimental paper (a study on mouse sexual development with Bruno D'Udine in Rome) wrote, "I often wish that research in behavioral genetics produced conclusions of greater significance." Oh how I agree! However intractable the problem and the material, we must try harder. Perhaps, also, I might offer a conventional but not completely meaningless defense—we hope we are now beginning to ask the right questions.

References

Andrew, R. J. 1961. The motivational organisation controlling the mobbing calls of the blackbird *(Turdus merula)* IV. A general discussion of the calls of the blackbird and certain other passerines. *Behaviour* 18:161–76.

Bastock, M. 1956. A gene mutation which changes a behavior pattern. *Evolution* 10:421–39.

Bastock, M., and Manning, A. 1955. The courtship of *Drosophila melanogaster. Behaviour* 8:85–111.

Bastock, M., Morris, D. J., and Moynihan, M. 1953. Some comments on conflict and thwarting in animals. *Behaviour* 6:66–84.

Bateson, P. P. G. 1976a. Specificity and the origins of behavior. *Adv. Study Behav.* 6:1–20.

———. 1976b. Rules and reciprocity in behavioural development. In *Growing Points in Ethology,* ed. P. P. G. Bateson and R. A. Hinde, pp. 401–21. New York: Cambridge University Press.

Batty, J. 1978a. Plasma levels of testosterone and male sexual behaviour in strains of the house mouse *(Mus musculus). Anim. Behav.* 26:339–48.

———. 1978b. Acute changes in plasma testosterone and their relation to measures of sexual behaviour in the male house mouse *(Mus musculus). Anim. Behav.* 26:349–57.

Daumer, K. 1958. Blumenfarben, wie sie die Bienen sehen. *Z. vergl. Physiol.* 41:49–110.

Ewing, A. W. 1983. Functional aspects of *Drosophila* courtship. *Biol. Rev.,* 58:275–292.

Heinrich, B. 1979. *Bumblebee Economics.* Cambridge, Mass.: Harvard University Press.

Hirsch, J. 1967. *Behavior-Genetic Analysis.* New York: McGraw Hill.

Loher, W. 1962. Die Kontrolle des Weibchengesanges von *Gomphocerus rufus* L. (Acridiinae) durch die Corpora allata. *Naturwissenschaften* 17:406.

Manning, A. 1956a. The effect of honey-guides. *Behaviour* 9:114–39.

———. 1956b. Some aspects of the foraging behaviour of bumble-bees. *Behaviour* 9:164–201.

———. 1959. The sexual behaviour of two sibling *Drosophila* species. *Behaviour* 15:123–45.

———. 1961. The effects of artificial selection for mating speed in *Drosophila melanogaster. Anim. Behav.* 9:82–92.

———. 1965. *Drosophila* and the evolution of behaviour. *Viewpoints in Biology* 4:125–69.

———. 1967a. The control of sexual receptivity in female *Drosophila. Anim. Behav.* 15:239–50.

———. 1967b. *An Introduction to Animal Behaviour.* New York: Addison Wesley.

———. 1967c. Genes and the evolution of insect behaviour. In *Behaviour-Genetic Analysis,* ed. J. Hirsch, pp. 44–60. New York: McGraw-Hill.

———. 1968. The effects of artificial selection for slow mating in *Drosophila simulans.* I. The behavioural changes. *Anim. Behav.* 16:108–13.

———. 1975. Behaviour genetics and the study of behavioural evolution In *Function and Evolution in Behaviour, Essays in Honour of Professor Niko Tinbergen, F.R.S.,* ed. G. Baerends, C. Beer, and A. Manning, pp. 71–91. Oxford: Clarendon Press.

———. 1976. The place of genetics in the study of behaviour. In *Growing Points in Ethology,* ed. P. P. G. Bateson and R. A. Hinde, pp. 327–43. New York: Cambridge University Press.

Manning, A., and Hirsch, J. 1971. The effects of artificial selection for mating speed in *Drosophila simulans.* 2, Genetic analysis of the slow mating line. *Anim. Behav.* 19:448–53.

Manning, A., and Thompson, M. L. 1976. Postcastration retention of sexual behaviour in the male BDF$_1$ mouse: the role of experience. *Anim. Behav.* 24:523–33.

McGill, T. E., and Manning, A. 1976. Genotype and retention of the ejaculatory reflex in castrated male mice. *Anim. Behav.* 24:507–18.

Seligman, M. E. P. 1970. On the generality of the laws of learning. *Psychological Rev.* 77:406–18.

Shorey, H. H. 1962. Nature of the sound produced by *Drosophila melanogaster* during courtship. *Science* 137:677–78.

Spieth, H. T. 1952. Mating behavior within the genus *Drosophila* (Diptera). *Bull. Am. Mus. Nat. Hist.* 99 (7):401–74.

Stewart, A. D., Manning A., and Batty, J. 1980. Effects of Y-chromosome variants on the male behaviour of the mouse *Mus musculus. Genetical Research, Cambridge* 35:261–68.

Tugendhat, B. 1960. The normal feeding behavior of the three-spined stickleback (*Gasterosteus aculeatus* L.) *Behaviour* 15:284–318.

Wehner, R., ed. 1972. *Information Processing in the Visual Systems of Arthropods.* Berlin: Springer-Verlag.

Peter Marler (b. 24 February 1928). Ingbert Grüttner photograph

13

Hark Ye to the Birds: Autobiographical Marginalia

Peter Marler

If as a child I had believed in reincarnation, I would undoubtedly have chosen to be reborn as a bird with the next turn of the wheel. I have been fascinated by birds as long as I can remember. As an eight-year-old, I was already an avid bird-watcher, but the first explicit announcement of ornithology as an avocation seems to have been made at age eleven. I was invited to vacation with a fond uncle and aunt on a farm in Somerset, in England's West Country, as a reward for passing the elitist rite of passage from elementary school to grammar school, known as the 11 + exam. The year was 1939. Everyone was preoccupied with the threat of war, which was to cast a shadow over my youth, much as the First World War had done over my father's.

I was still in Somerset when war was declared, and we began excavating an abandoned root cellar as an air-raid shelter. Back home in Buckinghamshire, my parents were already digging, in our suburban garden, a pit to be covered with sod, after lining it with the sheets of corrugated steel that the government provided for the purpose. As an industrial town, Slough was a potential bombing target, and by the time of the Battle of Britain, it was bristling with antiaircraft guns. We three children spent many nights underground with our parents listening to the muffled drone of German planes and the booming guns. We didn't actually see much damage, although Slough was bombed several times. Once, at the height of the bombing, my father took me up to London. We were awed by the wasteland around St. Paul's Cathedral, but the tragedy did not really sink in until we saw a man searching for his belongings in still-smoldering rubble that had been an apartment building the day before.

Despite frequent air-raid warnings the entire family still assembled most weekends for the Sunday excursion, embarking on foot from the town suburbs into the Buckinghamshire countryside. I was always intensely preoccupied on these trips, collecting flowers to key out at school, adding new birds to my list, and occasionally supplementing my growing collection of bird eggs. Often we walked out to Stoke Poges, where everyone would sit in the churchyard to read the elegy inscribed in the wall of Thomas Gray's tomb. I was usually more interested in a pair of bullfinches that lived in the yew trees nearby. I owe more of my love of natural history than I ever appreciated at the time to these nature walks, which provided my father with a much-needed change of pace after laboring all week in a factory as a toolmaker.

He enjoyed his work, however, and was always good with his hands. A shed in the garden was fixed up as a workshop with a jeweler's lathe and all kinds of beautifully kept hand tools. At first I was only allowed to work there under supervision, but later he gave me free access and even allowed the invasion of a collection of noxious and in some cases corrosive chemicals that I began to experiment with as a teenager.

In grammar school I enjoyed all subjects, though there was never any doubt that I was to be a scientist, but I could not imagine how even the most benign society would employ me as an ornithologist. Instead it seemed more practical to take up chemistry, a decision that created new problems for the family. Reluctant aunts were coaxed into subscribing for homemade cosmetics, tactfully accepting with equanimity, and in some cases actually using, hand lotions that were more depilatory than emollient. Some relatives were so selfless as to order a regular supply, thus unwittingly underwriting my less well publicized diversions into pyrotechnics. The air-raid shelter was ideal for detonating homemade fireworks. My parents' patience and understanding must have been limitless, for I cannot recall once having been reprimanded by them for explosive and inflammatory experimentation, although neighbors were less inhibited.

I matriculated at the Slough Grammar School at the age of fifteen, and most of my friends left school to earn a living. I stayed on for what was called the Higher School Certificate, specializing in science, but I gradually realized that my mathematical aptitudes were not up to the abstractions of physical chemistry. Instead I came under the influence of an outstanding biology teacher, and with his collusion, I decided that a botany degree was probably my most appropriate preparation for employment. Again ornithology was relegated to the back seat, but it persisted. My interest in egg collecting waned, and a few experiments with a friend's airgun convinced me that I was more devoted to birds alive than dead. I began to keep birds at home, and to breed them, first domesticated species, and later wild birds taken as nestlings and reared by hand—a pursuit I find fascinating and instructive to this day.

With a modest scholarship from the Buckinghamshire County Council of £100, income from sundry odd jobs, and unswerving parental support, I registered as a day student in botany at University College London, with minors in zoology and chemistry. As a student at the end of the war I just missed the draft, though I was a member of the Slough branch of the Air Training Corps, the aeronautical equivalent of R.O.T.C. I even put in a few apprehensive sittings at the controls of a sedate "tiger moth" biplane, a matter of some pride until my younger brother was called up and became a fighter pilot at Biggin Hill. After my early morning newspaper round, I sprinted each day for the 8:12 train from Slough to Paddington, to attend lectures by such luminaries as J. B. S. Haldane; D. M. S. Watson, the paleontologist who was chairman of Zoology; G. P. Wells, the expert on *Arenicola* behavior; J. Z. Young in anatomy; and above all, W. H. Pearsall, my botany professor. My first publications, all concerned with plant distribution and ecology, derived from expeditions as a botany undergraduate.

During a summer studying the ecology of mosses near Cape Wrath in Scotland, I came to know and treasure the friendship of James Fisher. He was on a trout-fishing vacation in Sutherland at the time with Stephen Potter, who proved to be as witty in the flesh as you might have expected from his books on the arts of one-upmanship. I had already met James Fisher as founder and editor of the New Naturalist series, setting new standards for natural history writing in Britain. He arranged an interview for me with Nicolaus Pevsner who was producing a series of King Penguin booklets on natural history, delightfully illustrated with water-colors. I had painted a collection of seashore plants on the Norfolk coast, but they did not make the grade.

The botany department was situated in an annex above the famous Slade School of Art, which no doubt explains my ventures into painting and drawing. Although I was never very good at either, both gave me great pleasure and relaxation. Again my parent's patience was monumental, as I brought home not only oil painting paraphernalia but also the more sinister impedimentia for egg tempera, a technique that makes use of egg yolks as a pigment vehicle. Some years later when Desmond Morris introduced me to his unique brand of surrealist painting, he enthused over a new textural quality created directly on the canvas, using egg tempera as a culture medium for mold. I was already well acquainted with it.

As a day student at University College, much of my life was still centered at home. I joined forces with friends to establish the Slough Natural History Society, transformed and enlarged a couple of years later into the Middle Thames Natural History Society. Weekends became a ferment of feverish activity, surveying migrant birds at local reservoirs and sewage farms, organizing fungus forays in Windsor Great Park and botanical excursions to the Chilterns. Britain is remarkable for the energy and erudition of its amateur naturalists, and the Thames Valley had more than its share. One of our founder members, A. J. Balfour, was a distinguished botanist who took me to meetings of the Royal Horticultural Society at Kew Gardens. Guy Mountfort was another early member, a superb field ornithologist, subsequently to coauthor the standard *Field Guide to the Birds of Britain and Europe*. Derek Goodwin, now an authority on the biology and behavior of crows and their relatives, was a founder member.

In a more professional setting, some meetings were held at the Institute for Animal Populations, based in Slough, with as one of its staff the Australian-born insect ecologist, M. Solomon. Another member who became a close friend was Kenneth Alsop, an ex-fighter pilot who had lost a leg in the war. He was a disciple of the eccentric British animal novelist, Henry Williamson, known to a small devout readership for such books as *Tarka the Otter*. Alsop wrote several books in this vein while earning a living as a star reporter for a London newspaper. I still receive the Society's annual report, with its litany of concerns for local conservation.

Society excursions enriched my experience as a naturalist, but nothing could match the inspiration of a field trip with W. H. Pearsall. He had an almost mystical appreciation of the relationships between rocks and soil, plants and animals, and the debt of the present to the past. I used to return from courses in plant ecology at Blakeney Point, on the coast of Norfolk, with its bird colonies and salt marshes, deeply envious of those who were lucky enough to devote their lives to the study and nurture of plants and animals in wild places. Other heroes from this period were Frank Fraser Darling and Robert Lockley, appealing as much for their philosophy of living as for their scientific reputations. This emotive feeling for natural history was eventually balanced by intellectual preoccupations that, in retrospect, were surprisingly slow to emerge.

Another formative experience in my years as a botanist was participation in David Lack's Christmas bird conferences at Oxford. These remarkable gatherings organized more or less single-handedly by David Lack at the Edward Gray Institute for Ornithology at Oxford, brought together thirty or forty students, mostly undergraduates, to present papers and engage in discussions that, despite our garrulous inexperience, he managed to imbue with high scientific standards. My first serious bird research, on song variation and niche expansion in the Azores, was inspired at the first of these conferences I attended, and presented at the second. It eventually saw the light of day as my first publication on bird behavior, coauthored with fellow U.C. botanist Derek Boatman.

At Oxford I savored the delights of a lecture by Niko Tinbergen, recently arrived from Holland to take up his position as reader in animal behavior at Oxford University. Another fellow student was Mike Cullen, erstwhile Tinbergen student and colleague, now professor of zoology at Monash University in Queensland, Australia.

On graduating with my botany degree I would probably have joined a forest survey team in East Africa if Pearsall had not offered me a graduate studentship. I enjoyed working as a teaching assistant, especially on field trips. Discipline was lax enough to permit a good deal of bird-watching on the side, especially since one of the zoology undergraduates, Aubrey Manning, was as eager for the diversions as I was. My research assignment was to work in the Lake District on the ecology of plant succession in Esthwaite Water. Pearsall was on the advisory board of an organization called the Fresh Water Biological Association, housed in a counterfeit castle on the shore of Lake Windermere. A group of brilliant researchers worked on problems in limnology and fresh-water ecology. I spent long hours drilling mud cores in the postglacial deposits of the Esthwaite Water bogs and subjecting them to chemical analysis. Alas, I was less inspired by the chemistry of mud than by the remarkable diversity I found in the song dialects of the chaffinch in the surrounding valleys. At this stage, however, my chaffinch studies were still aesthetic rather than scientific, more on a par with my excursions into the Vedas and The Cloud of Unknowing. I was enthusing over George Fox's journals at the time, and it was moving to visit the Quaker Meeting House in Ulverston, where Fox had often preached.

The fact that my research enthusiasms were divided caused a good deal of frustration on everyone's part. I suspect that there were sighs of relief all around when I finished my London Ph.D. in short order and took a job with the newly formed Nature Conservancy, surveying possible conservation sites in Scotland. Although my twenty-first birthday had come and gone, I was still in some sense drifting. This is not quite the appropriate word because I was always deeply involved in something, but it changed from month to month. I enthusiastically scoured Scotland for the dwindling examples of primal plant communities, though I have no idea how many of the recommendations made by my fellow botanical colleague, Heather Salzen and I were ever acted upon. I began research on peat bogs and their progressive decay, convinced that it resulted from the changes in chemistry and structure caused by the five-yearly burning of the heather, by which farmers maintain good forage for the sheep. My colleague Sidney Holt did his best to instruct me in the finer points of statistical theory, leavened with Marxist interpretations of history and aesthetics. Painting and life classes at the Edinburgh Art School were evening occupations.

Despite this rich diet of distractions, I became increasingly preoccupied with bird behavior, and my growing conviction that birdsongs must be learned. I pressured the long-suffering directors of the Nature Conservancy to let me launch a research program on the developmental basis of chaffinch song.

I got my first intoxicating taste of ethology from Margaret Morse Nice's 1943 monograph on the behavior of the song sparrow. The many insights into bird behavior were interleaved with her interpretations of ethological theory, derived from Konrad Lorenz's 1955 "Kumpan" papers, which she had translated and republished in *The Auk* in 1937. Equally momentous for me was the publication by the Society of Experimental Biology of the proceedings of a conference held at Cambridge in July 1949 on "Physiological Mechanisms in Animal Behavior." Only a few of its authors were known to me at the time, though I came to revere many of them either by proxy or in the flesh.

Backed by the warm and charismatic director of the Nature Conservancy in Scotland, John Berry, a fountainhead of ethological knowledge and hilarious anecdotes about his beloved ducks and geese, my proposal for birdsong research found a sympathetic ear in one of the organizers of the Cambridge conference, William Homan Thorpe. By a heaven-sent coincidence, he was in the process of establishing the Ornithological Field Station at the University of Cambridge for research on bird behavior, focusing especially on birdsong and its development. In a display of the ineffable flexibility and understanding that often lurks below the forbidding exterior of the British Civil Service, the Nature Conservancy arranged for me to transfer to the University of Cambridge in 1951, complete with a research fellowship. I owe this good fortune in large part to the eminent amateur ornithologist and conservationist E. M. Nicholson, who had pioneered studies of birdsong in collaboration with the BBC and Ludwig Koch, the founder of its by now vast collection of natural sounds.

Thorpe welcomed me warmly in his laboratory, equipped with the first Kay Electric Company sound spectrograph available in Britain for the analysis of animal sounds. Bill had watched one in operation in the National Physical Laboratory in Teddington and immediately saw its revolutionary implications for the study of animal communication. It was developed by Bell Laboratory scientists in the United States as a tool for speech analysis, especially to help deaf children to learn to speak. It turned out to be of limited value for this purpose, but as a method of visualizing the structure of complex sounds, easier to read than oscillograms, it was an ideal tool for the new science of bioacoustics.

Although basically simple in conception, the sound spectrograph was temperamental to operate, and obtaining reproducible results was an art. Once we had figured out how to use it, there seemed no limit to its revelations, especially since Bill Thorpe had free access to the rapidly growing birdsong library at the BBC. We also had high ambitions for our own recordings, but in those days equipment problems were serious. To begin with, Thorpe used to record songs by laboriously cutting original phonograph discs. Tape recorders, especially portable ones, were hard to come by and less than reliable. One of the more basic early recorders, the hand-cranked Magnemite, had a large cast-iron flywheel on the recording deck to stabilize tape speeds; this had to be unscrewed before you could change tapes. Recordings from those days are sadly deficient by modern standards, and I still get twinges of embarrassment listening to them.

As a new student in Cambridge, I was invited by Thorpe to be a member of Jesus College, where he was a senior tutor. In my first encounter with Oxbridge chauvinism, I found my London Ph.D. unacknowledged. It also seemed inauspicious to launch my new career as a zoologist on the strength of botany degrees, so I decided to reregister for a Cambridge Ph.D., this time in animal behavior.

My true initiation into ethology fell largely into the hands of Robert Hinde, recently appointed to direct Thorpe's newly founded Ornithological Field Station in the village of Madingley, outside Cambridge. As a graduate student with David Lack at Oxford, Hinde had begun research on the behavior of the great tit. But Lack's primary focus was in ecology, and when a year or so after he began, biology at Oxford was enriched by Tinbergen's arrival, Robert Hinde struck an immediate chord with him and transferred. In Cambridge I had the double benefit of Hinde the disciple, and Hinde as critic and innovator. Every new idea, and each word I wrote was subject to the closest analytical scrutiny. He was generous in sharing his own ideas, and several of the dominant themes in the field study of the behavior of chaffinches in Madingley Wood that became my zoology Ph.D. thesis are direct reflections of his current enthusiasms, especially ethological drive theory and the insights it provides into the structure and evolution of display behavior.

My conversion was completed by the Second International Ethological Congress held in Oxford in 1952. It was intellectually inspiring and, as with all good conferences, it served to cement what were to become lifelong friendships with students of Otto Kohler, Gustav Kramer, and Konrad Lorenz in Germany, of Gerard Baerends in Holland, with Eric Fabricius in Sweden, and Holger Poulsen in Denmark, and especially the coterie of Tinbergen students who ran the conference. As I recall, the New World was hardly represented, apart from a few early converts such as Eckhard Hess, who was already taking up the theme of imprinting in his research. Nowadays Americans make up a majority of the attendance at ethological conferences. Present enthusiasm runs so high that many are now turned away, for which I, serving as the secretary-general, tend to be held personally responsible. Hopefully our efforts to throw the conference open to all comers will be successful, replacing the old system of national quotas that caused so much bad feeling.

In 1952 there were no such problems. The high point of the Oxford meeting was the entertainment on the last evening by Desmond Morris, Aubrey Manning, and

David Blest. As nostalgically described by Desmond in *Animal Days,* the entire front desk in the main lecture hall was covered with a Rube Goldberg apparatus illustrating Lorenz's "water closet" principle of action-specific potential. At the climactic moment of the "Ubersprungbewegung," bulging water-filled balloons, assaulted by sperm launched on wires from the back of the lecture hall, exploded and drenched the front rows of the wildly applauding audience.

Back in Cambridge, I now had a complete picture of the vocal repertoire of the chaffinch, based on field studies in Madingley Wood. I began to question Lorenz's assertion that animal signals are arbitrary in structure. It seemed inevitable that the evolution of calls would be influenced by the transmission properties of the environments in which they are used. I also became convinced that problems of sound localization held the key to understanding the peculiar acoustic structure of some animal calls.

In the breeding season, male chaffinches have a thin, high-pitched whistle that they use in extreme danger, especially when a hawk is nearby. I found the "seet" call to be strangely ventriloquial, so that the calling male was hard to track down. Moreover, the process of scanning to and fro to home in on the immobile, cryptically crouched vocalizer often revealed, not a chaffinch, but a great tit. Not only do these two species have virtually identical hawk alarm calls, but I found that they also engaged in mutual exchanges, along with several other species living in Madingley Wood. Unlike many vocal signals, with purely private functions to serve, these alarm calls also provided a public interspecific service, hence their lack of species-specificity. Several songbirds appear to have sacrificed specific distinctiveness in the interest of the mutual communication of extreme danger. The varying selection pressures for divergence and convergence in the course of vocal signal evolution were the topic of my first paper in the journal *Behaviour* in 1957, at that time the major outlet for ethological publications.

To explore the implications of sound localization for the evolution of vocal signals, I needed to review the acoustic and psychophysical principles involved. John Pringle in the zoology department was eager to discuss his new findings on the evolution of sound signals in cicadas, and the peculiar operation of the abdominal tymbals by which sound is created. Armed with a primer from him on the physics of sound, I sought out Donald Broadbent, working on human sound localization at the Medical Research Council Unit on Applied Psychology. He took time to explain and demonstrate the intricacies of human sound localization.

Suddenly it was obvious that chaffinches and great tits had hit on a way to design a ventriloquial alarm call, minimizing cues for sound localization by using a narrow bandwidth and a gradual onset and termination. By contrast, many other birdcalls were adapted to maximize localizability, with a wide bandwidth and abundant sharp discontinuities, rich in transients. Reasoning that birds probably localize sounds by the same principles as mammals do, I published a paper in *Nature* in 1955 pointing out that certain aspects of the physical structure of many animal sound signals could be understood in terms of selection either to maximize or to minimize the cues for localization, the emphasis depending on the caller's vulnerability to sound-oriented predators, and the balance of benefits and disadvantages of easy localization for the signaler and its communicants.

More than twenty years elapsed before proof was forthcoming that my hypothesis was valid. By the late seventies it had become clear that in bird hearing there is a greater emphasis on pressure-gradient reception than in mammals, implying rather different procedures for sound localization than with true pressure reception, which is the traditional, somewhat idealized view of human hearing. For a time my theorizing was in doubt, until Charles Brown of the University of Missouri demonstrated that the "seet" alarm call is in fact hard for hawks to locate, and that adaptations for facilitating or hindering sound localization by pressure-gradient reception are not so different after all.

Mark Konishi had made a similar point a few years earlier, also showing that the high pitch of the call does not discourage localization by predators, as I had thought, but rather hinders detection by the predator, due to rapid attenuation with distance. When I first presented the alarm call story at a seminar in Cambridge, Laurence Picken, equally at home in biology and oriental music, recalled an old Chinese expression for "a floating note" with a structure that perfectly matched the chaffinch hawk alarm call. I was especially pleased when the dean of German scientific ornithology, Erwin von Stresemann translated my 1955 paper and published it verbatim in the *Journal für Ornithologie*.

In the mid-fifties, J. B. S. Haldane became concerned with problems in animal behavior, and I met him for the first time since my anonymous presence in his genetics lectures for undergraduates at University College. He liked the alarm call story, and urged me to get it into print, after incorporating it in his Christmas lectures on animal behavior at London's Royal Institution. We came to know him and his wife, Helen Spurway, and to admire especially their capacity for vintage cider, at The Mill in Cambridge, and for beer at various London pubs. Spurway is well equipped vocally, and became even more loquacious as evenings wore on. We had to leave one pub hastily, in the face of verbal abuse, when she threatened to drown out the resonant voice of Winston Churchill coming through the radio on the bar. Much of the time with them I was content just to sit and listen to Haldane's displays of erudition. It was a sad day for British science when he and Spurway emigrated to India, to the accompaniment of scurrilous verses in *Punch* celebrating their departure.

The possibilities for winter field work in Cambridgeshire were limited, so I turned to the extensive aviary facilities at Madingley. Studies of aggression in winter flocks demonstrated that the critical distance at which one bird must approach another before aggression is triggered can be precisely measured. It was also satisfying to discover that the red breast of a male chaffinch is a classic example of a Lorenz/Tinbergen releaser, modulating the distance at which another bird would no longer be tolerated. Males always dominated females in the winter hierarchy, and I discovered that faking male coloration on a female's breast feathers with red ink guaranteed her high rank when placed in a group with other females. The annual program we all participated in for rearing chaffinches by hand for the song development studies also permitted me to explore the developmental basis of female responsiveness to the male red breast, which proved to be innate.

My motivation to explore the basis of animal aggression stemmed in part from the little book, *Frustration and Aggression,* that one of my future colleagues at the

Rockefeller University, Neal Miller, coauthored with several others in 1939. Impressed though I was by their attempt to create a comprehensive explanation for the basis of human aggressive behavior, I was sure that the stimulus triggering of aggression had been neglected in their account, at least as far as animals are concerned. The American ethologist Wallace Craig was right to stress the stimulus-dependence of some activities, notably escape behavior, and their apparent lack of an endogenously motivated appetitive phase except under special conditions. I felt the same to be true of avian aggression, and I assembled some supporting evidence. I thus found myself diametrically opposed to the view of Konrad Lorenz, so eloquently expressed some ten years later in his book *On Aggression,* that there is an endogenous appetite for fighting. I am still inclined to favor Wallace Craig's position.

Paradoxically, my ambitions to work on song learning were still somewhat frustrated. Although I was involved in many of the experiments on song development, Thorpe made it clear that vocal learning was his domain. When I came to submit my 1958 paper on "The voice of the chaffinch and its function as a language," one of the first to catalogue and describe the structure of the entire vocal repertoire of an animal, there was some question as to whether sound spectrograms of song should be illustrated at all. The problem was solved diplomatically by keeping the figure small. I felt no inhibitions about working on calls, but Bill, as the boss, exercised territorial rights when it came to song.

In 1955 I joined the competition for a much-coveted Research Fellowship at Jesus College, and to my delight I was successful. One member of the review board was the physicist Dennis Wilkinson, who had an interest in animal behavior, and collaborated with G. V. T. Matthews on a device attached to a homing pigeon that would tell you how much of the time taken returning home was actually spent in flight. He questioned me closely about my aggression studies in the interview, clearly interested, and that probably tipped the balance in my favor. As the junior fellow of the college, my main duty was to serve the port after high table, doing my best to avert the disapproval of the assembled fellows by serving them all equally from the bottles provided without leaving any dregs.

Thorpe's rooms in Jesus College were the rendevous for an exciting seminar that he organized with the chairman of the psychology department, neurolinguist Oliver Zangwill. The small group encompassed a wide range of interests, including Horace Barlow and Richard Gregory on visual perception and physiology, Larry Weiskrantz, David Vowles, and Zangwill on neuropsychology, Donald Broadbent on learning and perception, and Hinde and Thorpe on ethology. Other participants between 1953 and 1958, when the meetings terminated, included Richard Andrew, David Blest, John Crook, Peter Klopfer, Geoffrey Matthews, Hugh Rowell, Thelma Rowell, Bill van der Kloot, Margaret Vince, and Martin Wells. A book called *Current Problems in Animal Behavior,* published in 1961, summarized many of our discussions, which I found stimulating and provocative. In later years I had several requests to reprint my chapter "The Filtering of External Stimuli during Instinctive Behavior," but the editors refused permission, wanting to retain the integrity of the original book.

A small fraternity at Jesus College kept a program going in the arts, and we had a couple of shows of paintings, including one at Heffer's Gallery. I never sold anything, no doubt because it was obvious to everyone except me that my apocalyptic creations were pale shadows of the Blake paintings from which they were all too obviously derived.

In 1954 I married Judith Golda Gallen, an alumnus of the Slough High School for Girls, strategically placed to be as far from the Grammar School for Boys as it could be while still inside the town boundaries. In those days, women were virtually nonpersons in Cambridge college life, and we took the opportunity to travel whenever we could. Jurgen Nicolai, one of the great aviculturalists, was studying cardueline finches, and we visited him and his wife for several weeks at the old castle in Buldern in Westphalia, where Konrad Lorenz's institute was housed before being moved to more luxurious quarters at Seewiesen in Bavaria.

Lorenz was a cordial and generous host, but he was distracted at the time by his conviction that the baron had shot one of his dogs. Since relationships with the landlord were strained, castle upkeep was minimal. A rat plague had occurred a week or two before our arrival, and had been combated by poisoning; most of the rats had succumbed deep within the castle walls, and their fate was only partially obscured by an abundance of potted hyacinths in full bloom.

We assembled regularly for afternoon tea, and had endless discussions about ethology with Irenäus Eibl-Eibesfeldt, deeply immersed in his already classic studies of mammalian behavior, Helga Fischer, Lorenz's long-term colleague in the research on goose behavior, and Beatrice Oehlert, student of cichlid behavior, who subsequently married Lorenz's son Michael.

To gain access to species that were critical for our planned survey of the ethology of cardueline finches, Nicolai and I dreamed of an expedition to the Himalayas, to be launched with a Land Rover we planned to drive out from Britain, with our wives, by way of the Middle East and Afghanistan. We had already applied to the Royal Society for funds when I got wind of a job possibility at the University of California in Berkeley and the trip was never made. The fact that we even contemplated such a journey is a reminder of how much the world has changed since.

The Cambridge zoology department was a hotbed of creative research during this period. Victor Wigglesworth's empire was on the top floor, and as an erstwhile entomologist, Thorpe lingered there as an avian intruder. Downstairs in Carl Pantin's laboratory, Adrian Horridge was in active pursuit of issues in invertebrate neurophysiology. Hans Lissman's revolutionary investigations of electric fish and their abilities to locate objects by the electric sense were in full spate, and the basement aquarium was home for several huge *Gymnotus* he had brought back from West Africa. We became close friends with the keeper of the aquarium, Ken Klose and his wife Rita, and often drove out on the motorcycle to their cottage in Haslingfield, nominally for German lessons, but more to listen to his stories about the Spanish Civil War and many other adventures besides. Meanwhile I began to plan my first course in animal behavior.

I approached the prospect of teaching with some trepidation. My only serious prior experience with lecturing in zoology came from evening classes in Cambridge for adults. The prospect of confronting large undergraduate classes was daunting. I began a feverish revision of all that I had learned and forgotten, and much that I was still innocent of, in basic vertebrate zoology, my other major teaching assignment, along with animal behavior, in Berkeley.

I gave a lot of thought to how I would structure my animal behavior course, and on another trip to Germany I discussed the alternatives with several people. One was Bernhard Hassenstein, then an assistant under Franz Möhres at Tübingen, who was in favor of a physiological approach to behavior, using a homeostatic model of interactions between the behavior organism and the environment.

Otto Koehler, with a professorship in Freiburg that Hassenstein was eventually to occupy, entertained us in the garden of their home with his beautiful young wife Amélie, daughter of Erwin von Stresemann. He was more inclined, as I recall, to favor a psychological approach, in line with his own research on the cognitive and communicative capacities of animals. We discussed the alternatives over tea, at which my wife and I were regaled for the first time with *waldhonig,* the strong-flavored honey gathered by bees from the honeydew of aphids, living in pine trees in the Black Forest. I always pick up a jar when I visit Germany, even though we now have beehives of our own.

Graduate student Franz Sauer's Freiburg laboratory was an education in itself. His interest in song development in warblers had already drawn us together. Observing their restless activity at night, he was convinced of their ability to orient by star patterns in the night sky. His findings were radical enough to cause some consternation, especially since they contradicted the prevailing views of the reigning monarch of bird orientation studies, Gustav Kramer, and Sauer's career in Germany was hampered as a result. He later joined us in California to work with Bill Hamilton on night orientation, using the planetarium of the California Academy of Sciences in San Francisco. Eventually he was successful in competing for a professorship in Bonn. We were grieved to hear of his unexpected death in 1980.

The process of appointment to the Berkeley position was somewhat unorthodox. It had been offered first to David Vowles, but his wife did not want to leave Britain. I was next on the list, but because either time or funds were short, the zoology department decided to hire me sight unseen. Gordon Orians, who was a graduate student at the time in Frank Pitelka's Berkeley laboratory, on leave for a year at Oxford, stood in as a job interviewer, and we got on famously. My three letters of recommendation were from Haldane, Lorenz, and Thorpe. Chairman at the time was Richard Eakin, the embryologist of pineal fame, also a superb teacher. He was the first to receive the outstanding teacher award when it was instated some years later on the Berkeley campus as an early sign of the coming student revolution.

Eakin began the process of acculturation to Californian informality by insisting that our correspondence should be on first name terms from the outset. It took several letters and some anxious consultation before we succumbed. Cambridge traditions were quite otherwise, at least in relationships with senior faculty. Bill Thorpe and I had known each other for four years before he invited me to use his first name, when he came personally to tell me about my Jesus fellowship.

We traveled on a freighter from Glasgow to San Francisco by way of the

Panama Canal and Los Angeles. Our baggage included a large cage containing a dozen jackdaws we had reared by hand a year before. The cage was too large to go inside our taxi across London, so we had to strap it on the roof. As the taxi began to move, the breeze brought the jackdaws into full preflying display, that included a vocal chorus, which they kept up continuously from Liverpool Street to Paddington, audible a block away.

Once on the boat, the crew encouraged us to exercise the birds, which was fine as long as we released them one at a time. They became increasingly adventurous as the days went by, and on more than one occasion in mid-Atlantic the captain had to slow the boat down so that a desperately flapping jackdaw, struggling against a head wind, could catch up with us.

Judith and I both took to life on board, delighting in the freedom to relax and read entire books from cover to cover at one sitting. I worked on lectures, and had a dozen or so prepared before we made a landfall in Long Beach, California. We had our first unexpected experiences of Mexican hot peppers and California smog on the same day.

Californians were more impatient with decorum than Cantabrigians, as was appropriate in a state with such a mobile population, few of whom were natives. People in the San Francisco Bay area were relaxed, kind, and generous and we soon came to love Berkeley as a home, and California for its extraordinary natural beauty and grandeur, all within a day's drive. The zoology department was overcrowded in the monolithic Life Science Building, an erstwhile W.P.A. project, and it was not until the third year that I graduated to my own permanent quarters. Despite the separation in other departments of the botanists, biochemists, and molecular biologists, the zoology department was large, well balanced, and barriers to communication were minimal. The major strengths of the department were in cellular and organismal biology. One of several eminences grises was the human geneticist Curt Stern who immediately made my wife and I welcome, and made a special point of introducing us when distinguished visitors such as Dobzhansky or Goldschmidt were around. Max Alfert kept an interest in the arts alive, and we attended life classes for several years, as long as there was time.

From the moment we arrived, the personal hospitality of the faculty was unbounded. The incoming chairman Morgan Harris and his wife Marge met us at the boat and gave us free run of their house for several weeks. The only charge was that we take care of the garden, an assignment that escalated into a running battle between us and the local gophers about who had primary access to the rose bushes.

Such colleagues as the Balamuths, the Berns, the Hands, and the Mazias vied with one another in the splendor and bonhomie of their dinner parties. The Smiths always asked us to join their New England-style family gathering on Christmas Day, to feast on Todd's provender, and to savor Ralph's latest advice on about how best to bivouac if you happen to get caught on a mountain top in a blizzard, skiing in the middle of winter.

When the Berkeley fog was too much for us, there was always an open door at the Pearson's out in Walnut Creek, and we often gathered there with the Quays to talk about Payney's latest paradox, such as why mice have whiskers if they survive better without them—presumably because they stay in their burrows until their whiskers regrow?

Martha and Ted Bullock, on sabbatical in Berkeley to put finishing touches to the monumental "Structure and Function in the Nervous Systems of Invertebrates" with Adrian Horridge, engineered our most explicit trans-Atlantic initiation with a round of triple-decker banana splits on Fisherman's Wharf. With all rites properly accomplished, Marge Harris and Anita Pearson accompanied us to the Oakland courthouse some years later to vouch for our moral fiber and good standing, when we decided to become U.S. citizens.

The major focus of interest in ethology, however, was in the Museum of Vertebrate Zoology, closely integrated with the zoology department, under the broad-minded directorship of Alden Miller. Frank Pitelka, curator of birds, had an outstanding program in ecology and population biology in full spate, and many M.V.Z. students like Jerram Brown, Bill Hamilton, Jack Kaufmann, Gordon Orians, Dick Root, Bill Thompson, Ed Willis, and Larry Wolf were well informed about ethology, and eager to learn more. Several served as teaching assistants in my laboratory course in animal behavior, often teaching me as much as I taught them. I always tried to attend their weekly luncheon, to keep up with the results of the latest field projects, and to find out what Seth Benson, Ned Johnson, Starker Leopold, and Bill Stebbins were up to.

We enjoyed many pleasant weekends with John and Betty Davis at the Hastings Reservation in the Carmel Valley, run as a field station by M.V.Z. Keith and Martha Dixon were also there on sabbatical in the first year. Bill Lidicker was working on a catalogue of mammals of Aguascalientes and Judith and I spent a memorable summer with him and his wife Naomi camping in Mexico. We set small-mammal traps every night, and while Bill was skinning his specimens in the morning, I recorded birdsongs, searching for potential subjects for research on song development. After experimenting with house finches and juncos, the final choice was the white-crowned sparrow, common on the Berkeley campus and, most importantly, displaying an elaborate system of song dialects throughout California. With a Berkeley graduate student, Miwako Tamura, as a research assistant, I spent several years working out the sensitive period for song learning and the structure of the innate song, also demonstrating their ability to reject sounds of other species in the song learning process.

Although my own research and teaching were firmly planted in the zoological tradition, I valued the presence of the psychology department in the Life Science Building. A couple of years later it moved into new quarters in Tolman Hall, decimating the library in the process. Mark Rosenzweig had soundproof chambers in the basement for work on psychoacoustics that I inherited when psychology moved out. Gerald McClearn, the behavioral geneticist, was another close friend, and we taught courses together in comparative psychology.

Once Frank Beach arrived from Yale, I found myself teaching animal behavior to increasing numbers of psychology graduate students, including Norm Adler, Burney LeBoeuf, Lyn Clemens, Don Dewsbury, Ben Sachs, Donald Sade, Del Thiessen, and Nick Thompson to name only a few. More often than not, half of the students were nonzoologists, and even included occasonal geneticists and entomologists, such as Maurice Tauber, who still maintains interests in insect behavior as chairman at Cornell. Hybrid vigor was rampant both in my animal behavior course, and in the weekly seminar that was held in our house in the Berkeley hills, overlooking Tilden Park.

In addition to psychologists there was a regular contingent of anthropologists, urged by Sherwood Washburn to extend their education into the biological domain. Kroeber was still active, and much interested in evidence of protoculture in animals. He invited me to lunch to tell him of the latest developments in song learning. The three departments, led by a committee of Beach, Washburn, and myself persuaded the University of California to give us some precious land up in Strawberry Canyon for construction of an animal behavior station with NSF funding. One of the first to work there, on skunk behavior, was a postdoctoral fellow from Australia, John Nelson, who fascinated the entire zoology department with his films of cunnilingus in fruit bats.

My own group of graduate students grew into an independent organism with its own way of life. John Eisenberg filled the greenhouses in the Life Science Building courtyard with an endless procession of small mammals, some local, the others exotic, housed in glass-sided burrows modeled on those I had seen Eibl using in Seewiesen. John's recent book *The Mammalian Radiations* records the fruits of a twenty-year program launched in those early days. Edith Neal worked on the behavior of cichlid fishes, Julia Dewey on tarantulas, and Keith Nelson on the time structure of the peculiar courtship antics of glandulocaudine fishes. His house was full of eccentric pets, including a parrot with an expressive vocabulary, and a coati-mundi that terrorized zoology department picnics. Keith was a convert from architecture, erudite in literature and the arts. After the evening seminar in animal behavior, held at our home, Eisenberg, Nelson, and I often talked on into the small hours, much to the impatience of my wife, who had to cope with the children in the morning.

Keith was cofounder of a student drinking club called the Peter Artides Society, named after the Dutch ichthyologist who met his insobrietous demise one night after misplacing the bridge across a canal in Leiden. When the Ethological Congress was held there in 1964, a Berkeley delegation paid homage at the commemoration plaque on the wall marking the place. Cofounder of the club was Berkeley graduate student and icthyologist Robert Behnke, who brewed most of the beer himself and stored it under the floorboards of his house. As the party proceeded, he would raise one board after another until the youthfulness of the most recent brew brought proceedings to a natural close. I acquired the habit of brewing from him, and I have made my own ever since.

There was a ferment of work on birdsong. Father James Mulligan, a Jesuit who introduced the Artides Society to some of the more robust altar wines produced by the Order, worked out the basis of song development in the song sparrow. George Hersh brought his skills as an acoustician to bear on the functioning of the avian syrinx. By persuading white-crowned sparrows to sing normally in helium air he demonstrated for the first time that they make sounds according to different principles than the human vocal tract, as Crawford Greenewalt showed by other means in his fascinating book *The Physiology of Birdsong.*

In 1958 a student from Japan, Masukazu Konishi, joined the group. He immediately impressed everyone with his good humor and brilliance. His demonstrations of the impact of deafening on vocal development in songbirds provided an elegant complement to my own ontogenetic studies. His standing with Berkeley faculty was high by the time he left for postdoctoral work in Germany. His preliminary examination committee, consisting of Frank Beach, Ledyard Stebbins, Sherwood Washburn, Ralph Smith, and myself, predicted that he would go far. He was the first of my students to face the physiological implications of ethology squarely, and he set out to acquire the necessary skills for what was to become a new beachhead in neuroethology, now under full sail in his laboratory at Caltech.

Encouragement to develop the interrelationship between behavior and physiology derived not only from me, but also from a new member of the Berkeley faculty, Donald Wilson, whose demonstrations of the capacity of the central nervous system of the locust to generate flight rhythms endogenously revolutionized thinking about the origins of patterned motor activity. We joined forces to coordinate teaching in neuroethology and collaborated on a training grant. Our students exchanged ideas and facilities freely. Ingrid Waldron discovered *Drosophila* songs in my laboratory, before transferring to Wilson's group to study insect flight.

After we left Berkeley, Wilson moved to Stanford for another burst of creativity before a tragic drowning accident in the Snake River in Idaho ended his meteoric career. He was a radical, deeply involved in the student unrest in the mid-sixties. I saw relatively little of that era because of a sabbatical in Africa in 1964–65. I came back to find one of my students, Mildred Eley, who had been working on the behavioral significance of the pecten in the eye of birds, now in court and threatened with jail. Her studies were never completed, and her name next appeared as treasurer on the masthead of *Ramparts,* a radical political magazine. Other promising scientific careers changed course at that time. The operant psychologist Marilyn Milligan was working with me then as a postdoctoral fellow, trying to gain operant control over the responsiveness of birds to song. She and her husband, a biochemist, were so affected by the repressive actions against Berkeley students that they joined the counterculture, and never returned to science.

My own fragmentary recollections from that era are mixed. The Aquarian age had its attractions, especially for anyone with bohemian inclinations, but I found much of the political activity frightening, if not Hitlerian in some of the tactics invoked to get everyone involved. The town of Berkeley never recovered from the excesses of that period.

Over the years I had supplemented the Thorpe and Tinbergen texts with lists of supplementary reading so voluminous that the reserve shelves in the library threatened to overflow. It seemed high time for another textbook, and I set to work on *Mechanisms of Animal Behavior,* asking Bill Hamilton to help me with the chapters on orientation. The book was written from the viewpoint of an organismal biologist, and could not comfortably accommodate much in the way of psychological thinking. Its dominant theme was the interplay of endogenous and exogenous influences in the control of behavior. The preoccupation with proximate factors was paramount, and the disciplines with which we sought to link ethology in this book were sensory physiology, endocrinology, and neurobiology, rather than population and evolutionary biology which, as we indicated in the preface, called for a very different treatment.

A couple of years later, in 1969, Howard Bern invited me to speak in the President's Symposium at the annual meeting of the American Society of Zoologists on the theme of "The Interface between Population and Organismal Biology." I argued for renewed attention to the relationship between behavior and population genetics, singling out W. D. Hamilton's two 1964 papers on the genetical evolution of social behavior, among others, as "containing the seeds of a revolution in our thinking about social organization in animals and its evolution." That was in 1969. Within five years the sociobiological revolution was upon us, and the theorizing of Alexander, Brown, Maynard-Smith, Trivers, and Wilson made it possible for the first time treat the ultimate factors bearing on the evolution of animal behavior in an organized and coherent fashion.

Primatology tends to be the province of physical anthropologists and psychologists and, except for a few courageous mavericks such as John Emlen, most zoologists have been reluctant to sponsor research on primate behavior, let alone do it themselves. I was still living in the era of C. Ray Carpenter and Solly Zuckerman when modern primatology arrived on the Berkeley scene, in the person of Sherwood Washburn. An intellectual innovator from the outset, Washburn, in the face of departmental opposition, forcefully introduced animal behavior to the curriculum of his graduate students. I found myself closely involved as a teacher, a participant in preliminary examinations, and as a thesis examiner of such students as Richard Lee, Jane Lancaster, Donald Sade, and Suzanne Chevalier-Skolnikoff, and a graduate student from Chicago, Irven DeVore, who became a close friend and colleague. As always with such interdisciplinary encounters, I learned an enormous amount. One of the rewards of participating in preliminary examinations was the chance to see Sherry Washburn take an apparently anonymous bone fragment, that had left a hapless student totally nonplussed, and explain how muscles were inserted on it, gradually bringing to life its role in a posture of a pattern of locomotion.

In 1962, in collaboration with David Hamburg in the Department of Psychiatry at Stanford, Washburn organized a major conference on primate behavior at the Institute for Advanced Study in the Behavioral Sciences in Palo Alto. It gave rise to a book, *Primate Behavior: Field Studies of Monkeys and Apes,* published in 1965, that effectively launched a new science of behavioral primatology, shifting emphasis in primate studies from the laboratory to the field, and stressing the importance of ecology in understanding the behavioral subtleties of monkeys and apes. In planning the conference, the primary focus was on new information on behavior in the field, almost virgin territory prior to 1960, apart from the monumental contributions of C. R. Carpenter. Partway through the conference, the need emerged for more synthetic treatments, pulling the field data together and interpreting them in general terms. Bill Mason, a close friend thereafter, was invited to review social development against the background of a generation of laboratory studies, especially those of Harry Harlow and his students. Jarvis

Bastian and I were also asked, at rather short notice to consider problems of communication; Jarvis in relation to human language and me in relation to other animals. I managed to squeeze several visits to Palo Alto into a crowded schedule.

The preparation of my chapter on "Communication in Monkeys and Apes" prefaced a new phase in my ethological career. The deluge of new facts was an education in itself. I became more conscious than ever before of the logical and practical problems of defining signal categories, especially when there is continuous intergradation of signal features, as sometimes occurs in higher primates. I was forcefully struck by the implications for the evolution of signal structure of using auditory and visual displays in close concert, generating a high degree of redundancy. This occurs commonly in communication within large and highly organized primate social groupings, a far cry from the more dispersed social organization of the songbirds with which I was familiar. At that time we were virtually unaware of such avian social complexities as group territories, helpers at the nest, and the other forms of social cooperation, topics for an entire plenary session at the 18th International Ethological Congress in 1983 in Brisbane, Australia.

With a sabbatical coming up a year later, I succumed to the temptation to take a year off from bird studies for a trip to Africa to look at the vocal behavior of forest monkeys, about which virtually nothing was then known. I wanted to see whether the relationships I had described in birds in 1957 between call function and rates of evolutionary divergence could be generalized to primates. As it turned out, the predictions were nicely confirmed. When I compared the vocal repertoires of two related and cohabiting East African monkeys, the blue and the red-tailed monkeys, the alarm calls were less divergent than those concerned with such intraspecific functions as inter-troop spacing and rallying of the social group prior to movement.

Aided by a Guggenheim Fellowship, Judith and I and our son and eldest daughter, Chris and Cathy, aged five and three, embarked for Africa (Marianne was in embryo before we returned). Through the courtesy of Professor David Wasawo, Zoology chairman at Makerere University in Kampala, Uganda, an appointment was arranged, with a few teaching duties. Thelma Rowell, on the staff there, was immersed in studies of baboon behavior under both field and captive conditions. She provided us with much-needed primatological advice and hospitality. Her husband, Hugh, another ethological alumnus from Cambridge, now diverted from displacement activities into neuroethology, was trying to determine the direction in which messages were passing through the nerves that control locust locomotion. We had many memorable trips together, some exploiting his extensive knowledge of African music. Like all of the staff there, the Rowells were deeply committed to education in Uganda, although both eventually ended up on the faculty at Berkeley.

In the selection of a study site, another Makerere staff member, Dennis Owen, was immensely useful. I needed a forest habitat with several monkey species living in sympatry, in reasonable abundance. Washburn suggested the Kibale Forest in Western Uganda, in retrospect perhaps a better site, but logistics favored the Budongo Forest in north central Uganda, a bit closer to Kampala.

Commuting betweeen Kampala and Budongo, I managed to get a lot of research done, although there was less field time than I had hoped. Despite all the help we got beforehand from experts like Stuart Altmann and Irv DeVore, the rigors of expeditionary research were a new experience, and a lot of time was wasted getting projects started. We were also distracted by events in the outside world.

Berkeley was in a turmoil, and letters from home gave us an inkling of the pressures students were under. We were further unsettled by two tentative job offers, different enough from Berkeley to be intriguing. The New York Zoological Society, under its visionary and mercurial president, Fairfield Osborne, decided to establish a research institute for work on the ethology of animals in the collections. Bill Conway, the erudite and efficient director of the Bronx Zoo, offered full access to the zoo facilities, to be complemented by field research possibilities on the scale of those we were pursuing in Africa.

Then in midyear, Ernst Mayr wrote to ask if I would be interested in being considered for an Agassiz Professorship in the Museum of Comparative Zoology at Harvard. On the way back from Uganda we passed through Cambridge. It was not the first time we had overlooked the special significance of the Fourth of July but, inopportune though our timing was, Ernst and his wife entertained us warmly, at their country place. Much at Harvard was attractive and challenging. The prospect of a long period of new construction for the Biological Laboratories was somewhat daunting, but more than outweighed by the fabulous opportunities, not only within biology, but also for interaction with psychology. George Miller, Chairman of the Department of Social Relations, was enthusiastic about inter-departmental possibilities, and so was Irv DeVore, who already had an active group in primate behavior. The seeds of the coming sociobiological revolution were germinating in Ed Wilson's mind and, much though we loved Berkeley, we felt ready for a change.

Several new developments followed in short order. Unbeknownst to us, Don Griffin, chairman of Biology at Harvard, was advising the NYZS on the new Institute for Research in Animal Behavior (IRAB) and we now learned that he might be tempted to join himself. For him, as for us, the lack of an academic affiliation was a serious deterrent, however. Don solved this problem by involving the Rockefeller University. Both he and I were now possible candidates for professorships at the Rockefeller, with joint appointments at the NYZS, and research facilities in both places. Such a conjunction of attractions was irresistible, and the magnetism of Detlev Bronk's personality when I met him with Don Griffin on a hastily arranged rendezvous in Maine on his yacht, the *King Haakon,* finally tipped the balance. By the end of October, 1965, three months after returning from Africa, we had decided to move to New York. A meeting with Chancellor Heyns failed to dissuade us, though talk of a new Berkeley Institute of Neurobiology and Behavior gave cause for thought. So far as I know, nothing came of it.

As director of the new Institute for Research and Animal Behavior, Don Griffin was generous to a degree, both with funds and facilities, in all the administrative help he provided, and above all with his scientific stimulation and advice that, although never forced on anyone, was always wise and creative. His radar tracking studies of night-migrating birds, and work on the echolocation of bats flourished, and engaged the first in a line of R.U. graduate students, Jack Bradbury, who analyzed the physical basis of a bat's ability to discriminate between three-dimensional shapes on the basis of echolocation. His firm grounding in the physical sciences and mathematics was valuable to me on more than one occasion, especially while preparing my 1969 paper on the physical basis of tonal quality in birdsong. Later, as a postdoctoral fellow with me and, after a stint at Cornell, when he returned as a faculty member, Jack launched his comparative studies of bat social organization, providing the basis of modern thinking about the evolution of polygyny in mammals. Kathy Ralls, who had worked with me in Berkeley on the role of Jacobson's organ in pheromone perception in mice continued work on mammalian chemical communication at IRAB, making use of the thriving colony of mouse deer in the Zoo.

In those early halcyon days at IRAB, Don and I each had funds for two junior faculty members. He invited Richard Penny, an expert on penguin behavior and navigation, and Roger Payne, who was a priceless source of advice and enthusiasm. Roger was studying prey catching in owls, but not long after, his life took a sea change when he discovered the song of the humpback whale, and he has devoted himself to the behavior and conservation of these wonderful creatures ever since.

I chose Fernando Nottebohm and Tom Struhsaker to join me as assistant professors. Both had been students in Berkeley. Tom was originally a student of the late wildlife biologist and conservationist Starker Leopold, in the Museum of Vertebrate Zoology at Berkeley, but Starker was diffident about supporting the plan for thesis research on the behavior and ecology of vervet monkeys in Africa. Tom appealed to my newly found enthusiasm for work on primate communication, and I took him on. Maneuvers for funds from a variety of sources eventually yielded enough for a plane ticket, and after a couple of false starts, he began what was to prove a historic field study of free-ranging monkeys at the Masai-Amboseli Reserve in Kenya.

The site was well chosen, both for the animals, and the presence of Stuart Altmann, from the University of Chicago, working on the behavior and ecology of the olive baboons of Amboseli. Struhsaker benefited from the rigor that Stuart and his brilliant wife, Jeanne, brought to field studies, to say nothing of the value of friendly colleagues in combating the ennui and isolation that bedevil field researchers in remote areas.

Tom made classic contributions to the understanding of primate territoriality and the use of visual signals in communication. Above all, he prepared the first complete catalogue of the vocal repertoire of a monkey. His work set new standards for the quantification of field research data and for reporting what was actually seen, rather than an interpretation of it, as was all too common among ethologists and anthropologists alike.

Having accepted my invitation to join the Rockefeller, Tom immediately launched into new projects on West African monkeys. Only later did he return to East Africa to establish a long-term study site in the Kibale Forest, to which I had made a fleeting visit in my sabbatical year, to make the first recordings of the fascinating graded vocal system of the red colobus, later the subject of a monographic study by Tom. In Uganda he pioneered studies of the behavioral ecology of an entire primate community managing to survive even the deprivations of the Amin era, to become a world authority on the behavioral biology and ecology of forest primates. Throughout this period he has kept alive a rich vein of research of primate behavior in my laboratory, running field classes, advising graduate students, and helping to keep us up-to-date with the rapid development of primate studies throughout the world. You only have to read the 1965 Primate Behavior book to see how much improvement there has been in the rigor and quantification of primate field studies, partly attributable to students of mine like Cheney, Green, Seyfarth, and Waser in whose training Tom played a role.

I offered the second junior faculty position to Fernando Nottebohm. After graduating with the Departmental Citation in zoology at Berkeley in 1963, Fernando approached me about graduate work, in a discussion conducted over a bottle of Argentinian wine at the animal behavior station. Konishi's work interested him, and he was especially intrigued by the contrast between the effects of early and late deafening on white-crowned sparrow song. The former resulted in a highly degraded song, whereas postponement of deafening until song had matured had almost no effect. Had proprioceptive control taken over, or an endogenous motor tape? Under Konishi's tutelage, Fernando quickly gained the necessary surgical skills, for which both had a natural talent.

The first subjects were ring doves, but auditory feedback proved unimportant in their vocal development. Anticipating my absence in Africa on sabbatical, Fernando applied for a fellowship to work with Bill Thorpe for a year in Cambridge. He chose as subjects my old favorite, the chaffinch, which he deafened at various stages of song development, showing that they can retain much of the song structure already achieved once a certain point is reached. In 1966 we moved East, and Fernando and his wife Marta joined us soon afterwards. After a spell of field research in Trinidad, working on vocalizations of the Amazon parrot, he resumed his old interest in proprioception and its role in vocal control. As the fates would have it, that question remains unanswered because of diversion by a strange discovery.

Proprioceptive afferents from the syrinx pass along the hypoglossal nerve, and the first step was to see what effect cutting this nerve had on mature song. Again chaffinches and white-crowned sparrows were the subjects. Study of effects of bilateral section was impossible because of interference with respiration, but he noticed that even with one side cut, there was some disruption of song pattern, though the effects were variable. It gradually became clear that much of the variation occurs because the effects of left hypoglossectomy are more drastic than those on the right. The discovery that song production is lateralized convinced him that it was time to focus an intense research effort on the neural basis of singing behavior.

Some five years earlier I had imported an inbred strain of German Wasserschlager canaries, and we already had a good picture of the basis for song development. As a species easily bred in the laboratory they were ideal subjects for research on the neuroethology of birdsong. Song control proved to be lateralized in the canary, as in the chaffinch and the white-crown and once more the left side was dominant.

Analogies with human vocal control, always lurking in the back of our minds, now forced themselves upon us. Would the lateralization extend up into the left hemisphere? Before this could be answered, the basic connectivity of brain circuits for song control had to be worked out, an exercise accomplished with the invaluable aid of Rockefeller colleagues. With graduate student Arthur Arnold, Fernando discovered the remarkable sexual dimorphism of brain centers controlling song. The lateralization of control did indeed extend centrally.

The canary was a lucky choice because, unlike many birds, males change their songs from year to year. A striking seasonal pattern of growth and shrinkage of dendrites in song control areas was found, apparently correlating with the change of repertoire, preparing the ground for an attack on the most burning question of the day in neurobiology, the neural basis of learning.

Another research colleague in the laboratories of the Institute for Research in Animal Behavior at the Bronx Zoo, was Paul Mundinger, a postdoctoral fellow, who had found at Cornell that in addition to song, calls are learned in some birds, especially cardueline finches. During the breeding season, each pair of goldfinches has its own distinctive flight call pattern. The sharing apparently serves to maintain the bond between them. Together we studied song learning in the red-winged blackbird. Unlike the white-crowned sparrow, redwings were as ready to accept the song of orioles as their own, when given a choice over a loudspeaker. This is perhaps a case where visual reinforcement plays a role, in harmony with the rather dramatic plumage differences in male icterids. For several seasons we adorned tame males with artificial, velvet, red, or black epaulettes and exposed young to them during song training, but to no avail. They continued to learn unselectively, and I reluctantly gave up on this problem. Why redwings accept oriole songs for learning in the laboratory but not in the field remains a mystery.

Long-term field projects were difficult for me to conduct with the Bronx and Manhattan as a base. Shorter projects were more feasible, and an exciting summer was spent in 1967 recording the vocalizations of Jane Goodall's study population of chimpanzees at the Gombe Stream in Tanzania. Had I found dialects or other signs of vocal learning in the chimpanzee, I probably would have changed the main course of my research and concentrated on laboratory experimentation with primates, but we found none. I remain skeptical about whether any nonhuman primate displays learned local dialects in its vocal behavior, although Steven Green found some indications that I might be wrong.

For his Rockefeller thesis, Steve analyzed the vocalizations of Japanese monkeys, collaborating on a project that John Emlen had launched in Japan with a student at the University of Wisconsin, Gordon Stephenson. The problem of the graded vocal signals of macaques and their meaning lay untouched until Steve grappled with it. He showed that the variations in vocal morphology are by no means random, but constitute a highly ordered system, with different variants produced in distinct circumstances. Evidently the effective repertoire size was much larger than anyone had supposed. Steve stayed on as an assistant professor and worked in India on the seriously endangered lion-tailed macaque, becoming more and more committed to the cause of primate conservation in the process. Meanwhile, the Japanese monkey story unfolded further.

At a meeting of the advisory board of the Primate Research Center at Duke University in 1974 I met William Stebbins, an expert on psychoacoustical research and director of the Primate Laboratory of the Kresge Hearing Research Institute at the University of Michigan. He complained about the sterility of tones and clicks as the traditional stimuli in psychometric studies. I tried to convince him that new revelations would ensue if, as in studies of speech perception, natural vocal stimuli could be substituted. Before the site visit was over, we had developed plans for a collaborative project.

With Steve Green's connivance, another postdoctoral fellow, Steve Zoloth, trained by Norm Adler at the University of Pennsylvania, edited out a large sample of field recordings of one part of the vocal repertoire of the Japanese macaque, the "coo" system, where subtle acoustic variation in the position of a frequency inflection appeared to have meaning to the monkeys. Stebbins assembled an impressive team of young collaborators led by Michael Beecher who showed that Japanese monkeys do in fact process their vocal signals in a different way than other species when confronted with exactly the same sounds. The study was the first demonstration of perceptual specializations for the processing of conspecific signals using animal subjects. As a new development it was surprisingly slow in coming, considering how logically it follows from ethological innate release mechanisms.

This study also yielded the first clear evidence of cerebral lateralization of perceptual processing of vocal signals in a monkey, directly paralleling the right-ear dominance that we display in the perception of speech. One of the Michigan students, Mike Petersen, had the inspiration to present test stimuli to the Japanese monkeys monaurally, changing randomly from ear to ear. This made it possible to review the distribution of errors in the judgments of call type made by the right and left ears. The monkeys displayed a strong right-ear advantage when required by the experimentor to classify calls by a feature that we thought was "linguistically" relevant. The advantage disappeared when the task was based on a "linguistically irrelevant" cue. Other monkey species, for whom these calls were meaningless, showed no such effect. The parallels with human specializations for the processing of speech sounds are another reminder that the roots of language reach further back into our primate ancestry than has been supposed.

Along with the difficulties of conducting long-term birdsong studies at IRAB, there were also growing pains from the increase in staff and budgets, as I soon discovered when I took over from Don Griffin as director in 1969. The Rockefeller offered to take over the Institute entirely, and after an extended search, we found a new home ninety miles north of New York City in Dutchess County. With the aid of a generous grant from the Cary Trust, a station was established on several hundred acres of land completely under our control. Roger Payne and Tom Struhsaker stayed behind with the NYZS to form the nucleus of a new Animal Research and Conservation Center, and were joined shortly thereafter by George Schaller; all of them kept up the ties with the Rockefeller.

The rest of us began the laborious process of moving house and building new laboratories, with the final move in spring 1972. Meanwhile several more students graduated and took teaching positions at other universities. Alan Lill worked on manakin leks in Trinidad as a postdoctoral fellow, defining for the first time the female role in the mating system more fully than ever before; later he left us to join the faculty of Monash University. After a year at the Harvard Medical School, Haven Wiley also changed course to work on the evolution of polygyny in birds. At that time, there was not even an adequate description of how matings occur in a polygynous bird. After a careful search the sage grouse was chosen, and two years of field work in Wyoming served to define the mating system and show that it is indeed highly assortative, with a tiny proportion of males doing most of the mating in any one year. Wiley also set new standards in defining the nature of territorial aggression.

At the University of North Carolina, where Wiley now teaches animal behavior, a talented undergraduate, Meg McVey was fired by his enthusiasm to pursue the evolution of mating systems further. I had a growing sense of involvement in an extended family when she came to the Rockefeller, for a brilliant thesis on the mating system of dragonflies. Wiley and others showed that lifetime reproductive prospects must be assessed, rather than just those of one season, if any sense is to be made of the selection pressures on animal societies that favor monogamy, polygamy, or promiscuity. The short reproductive life of dragonflies made them ideal for further pursuit of this problem.

Carl Hopkins switched from biophysics with Keffer Hartline to work on social communication in electric fish. I had been witness years earlier in Cambridge, to Hans Lissmann's demonstrations of the role of the electric sense in object location, but its communicative role was still obscure. Too little was known of the natural history of electric fish to pursue the problem effectively in the laboratory. It was not even known whether the signals of male and female were different. Hopkins made himself an expert on their taxonomy, and established a field laboratory in Guyana where they breed, at the height of rainy season. The staunch aid of his wife, Kathy, who had previously guided the running of my laboratory with a firm hand, helped overcome logistic problems that would have daunted weaker souls, to say nothing of leishmaniasis and other rainy season hazards. Carl found that the electric sense serves as a highly sophisticated communicative system and, with the connivance and good counsel of Ted Bullock and Walter Heiligenberg, electric fish have become an ideal neuroethological preparation, demonstrating many sensory specializations for the processing of species-specific communicative signals.

Other primatological postdoctoral colleagues from that period were Carolyn Ristau, and Elizabeth Missakian who did excellent work on the supposed incest taboo in the rhesus monkeys of Cayo Santiago. She showed that the supposedly forbidden mother-son matings do in fact occur, though so secretively that they are easily overlooked. Soon after, she became so dedicated to support of the drug rehabilitation programs of Synanon that she gave up her academic career to join them full time.

Carolyn studied effects, or the lack thereof, of early deafening on vocal development in squirrel monkeys, in a project in which she was unfortunately scooped by Detlev Ploog and his group, who also demonstrated normal vocal development after deafening. Steve Gartlan also worked with us for a time, collaborating with Tom Struhsaker on difficult field studies of behavior of drills in West Africa.

Tom Struhsaker was now more preoccupied with social organization and feeding behavior than with vocal communication, but in his early vervet studies, he had described a system of alarm calls, apparently specific to different types of predators. He convinced me that if one wanted to grapple with the difficult problem of animal semantics this would be an ideal place to begin. Don Griffin was already extolling in Rockefeller seminars his convictions, formulated more fully in his 1976 book *On Animal Awareness,* that such issues as internal representation and mental processing in animals were ripe for reexamination.

In 1975 I was approached by Dorothy Cheney and Robert Seyfarth, a husband and wife couple who had just finished Ph.D.'s with Robert Hinde in Cambridge. I invited them to join me to see whether, taking a leaf from Peter Waser's book, vervet monkeys would respond to playback of their alarm calls under field conditions. First as postdoctoral fellows and subsequently as members of the faculty, they threw themselves into the project with typical energy and imagination. The experiments were successful beyond our expectations, demonstrating that the monkeys respond differently to eagle, leopard, and snake alarm calls, apparently irrespective of the context, in the absence of any predator. Moreover the subtleties of the responses, such as looking into the sky for the eagle that the tape recorder told them must be there, and into the long grass for the nonexistent snake, convinced us that Don Griffin had been right to chide us about neglecting notions of mental imagery in animals.

By this time Griffin's campaign for cognitive ethology was well under way, and we found ourselves as prime illustrations of the insights that flow from admitting that, with appropriate caution, mental processes of animals are indeed amenable to study. The result has been a new approach to communication in higher animals, in which central representations of experience become a key component.

Having established the semanticity of alarm calls, we looked at calls carrying social information. Picking up another theme from my 1965 review, Sarah and Harold Gouzoules returned to the study of rhesus monkey vocalizations, twenty years after Robert Hinde and Thelma Rowell had characterized them as an extreme case of signal intergradation. At least with scream calls, this turned out not to be the case. At least five discrete calls were found, all used in the process of recruitment of allies during agonistic encounters. Again the evidence from playback experiments points to a representational function. The calls appear to label different classes of opponent as defined by dominance rank and matrilineal relatedness, helping adults to decide at a distance whether or not to intervene in the quarrels their offspring become involved in.

The result was yet another exception to the old view of animal sounds as signs of emotion and nothing more. Nor can one fall back on purely reflexive processing, such as might have been inferred from early ethological conceptions of stereotyped and unchanging releasers and innate release mechanisms. Instead, we find something closer to processes of linguistic naming than had previously been suspected. In the fullness of time, we may discover that this is true of birds as well as monkeys.

From the early days at IRAB, canaries were a favorite subject because they are so easily bred, and they continue to be prime targets in Fernando Nottebohm's work on brain mechanisms and song learning. Canaries learn well both from a live tutor and a loudspeaker. Carrying the analysis of auditory feedback a step further, we found that they will breed even when exposed to white noise so loud that it masks their hearing of their own voice. Application of this technique threw new light on the question of innate auditory templates that Mark Konishi and I had invoked on the basis of the almost total degradation of songs of some species when deafened early in life. Mark had found some exceptions to this rule, however, and Fernando's chaffinch studies suggested that variation in the amount of predeafening experience of auditory feedback from vocal production might be crucial. However Mary Sue Waser and I found that canaries reared in white noise and immediately deafened still developed a normal phrase structure in their songs, paralleling a more recent finding with song and swamp sparrows as subjects. It begins to look as though auditory templates interact in vocal development with motor programs that define the gross lineaments of the vocal behavior endogenously. This might well be true of speech as well as of birdsong.

Ethological fieldwork is perhaps more demanding of time than any other type of research. Even at the Rockefeller, where nonresearch burdens are light, it was hard to keep up, while still maintaining laboratory programs in both Manhattan and the Bronx. With the move to Millbrook, however, I found myself closer to ideal circumstances than ever before since leaving Cambridge. With field sites and laboratories side by side, it was possible to work outdoors for several hours and still put in a full day on the multitude of chores that occupy a senior scientist's time.

Millbrook was soon a ferment of activity, with Don Griffin and Ron Larkin tracking night-migrating birds by radar, Joe Torre-Bueno flying birds in a wind tunnel, and Jim Gould working toward his brilliant resolution of the honey bee waggle dance controversy. Directorship of the new Field Research Center for Ecology and Ethology took up a good deal of time, but I have been blessed throughout my career with a series of outstanding and devoted research assistants, especially Susan Peters, to whom I owe much of the progress in understanding the song learning process over the past eight years.

The plan after the move to Millbrook was to identify a pair of species breeding in the area, preferably within earshot of one another, to become the targets of a comparative program of research on song ontogeny. To complement field studies, both species would be reared in the laboratory under identical, acoustically controlled conditions, with species differences providing a window on genetic contributions to song development. Study of their abilities as fledglings to reject songs of the other as models for the learning process would tell us whether something like "innate release mechanisms" are involved and if so, how they interact with experience in the learning process. At first, field and chipping sparrows looked promising but, as on other occasions, my own students and colleagues showed me otherwise.

Years earlier in Berkeley, Jim Mulligan had shown that the song sparrow can come closer to normal song when reared in isolation than most birds, although there are still abnormalities, as Don Kroodsma went on to demonstrate. Even more to the point was the fact that, reared in species isolation from the egg, innate song sparrow songs were very different from those of the closely related swamp sparrow. It became clear from Kroodsma's intensive fieldwork on variation in swamp sparrow song that, with the song sparrow, this was the ideal species pair for my purpose.

Don continued on his own creative trajectory as a faculty member, laying the groundwork for modern interpretations of the significance of song repertoires, and revolutionizing our thinking about the lability of sensitive periods for avian vocal learning. Blessed with a green ethological thumb for rearing difficult species in the laboratory, he had ventured into study of the structure, development, and functional significance of some of the most complex birdsongs known to us, especially those of warblers and wrens, the latter following up earlier studies made by Jerry Verner while he was a postdoctoral fellow with me in Berkeley.

Meanwhile, I played it safe with sparrows, which are easy to raise, at least when you are as expert as my wife Judith, who has managed our hand-rearing program at home for years. The sparrows have taught us many new facts about song development, not the least remarkable being the great overproduction by the swamp sparrow of song material in subsong and plastic song, that is later winnowed down to size as mature song crystallizes, by a process reminiscent of the emergence of speech from babbling in children. The choice of songs for learning did indeed prove to be innately guided. The sparrows provide a satisfying illustration of the interplay between nature and nurture, which Konrad Lorenz first demonstrated in imprinting, that probably pervades much more than we think of the learning process as it takes place in nature, in species-typical environments. The analogies between song learning and speech development become increasingly compelling, as my psychological colleagues often point out, though there are regrettably fewer of these now than when I first joined the Rockefeller.

In 1965, Detlev Bronk appointed psychologist-sensory physiologist Carl Pfaffmann as vice-president for academic affairs, to advise in adding a roster of behavioral scientists to the faculty. I could not believe my good fortune when we were joined by such luminous figures as William Estes and Neal Miller, each with a cadre of outstanding colleagues. The arrival of George Miller from Harvard was a special blessing for me, opening my eyes to a new universe of research findings on the nature of language, and the development of speech behavior in infancy. Common ground between biology and psychology was further extended by seminars on the biology of speech, which we organized together, that introduced me to such authorities on speech behavior as Alvin Liberman at the University of Connecticut. Contacts with Al have grown closer over the years.

With retirements and departures, the Rockefeller faculty has shifted its center of gravity in a neurobiological direction that is stimulating in a different way and compatible with the new directions in the birdsong work, but I sometimes miss the contributions of experimental psychologists. On occasion I have the impression that the research literature on birdsong gets a closer reading from them than from zoologists, where the fashion has shifted toward population and evolutionary biology, rather than to the analysis of behavioral development at the organismal level. On the other hand, birdsong studies have a role to play there, as I found when Myron Baker joined my group.

The functional significance of the dialects in white-crowned sparrow song was unclear. I had speculated about the possible impact on the genetic constitution of local populations, but tests were not forthcoming. During vacations in his native Argentina, Nottebohm made the relevant discovery that in the closely related chingolo, the size of dialect areas varied in direct relation to the homogeneity of the habitat. Out on the open pampas, dialect areas were huge, whereas on mountain slopes, where life zones change rapidly, they were small, illustrating something like Julian Huxley's "step clines," with song dialects limiting gene flow from one to another. To tackle this problem directly Myron (Mike) Baker joined us, first as a postdoctoral fellow and then as a faculty member.

Applying electrophoretic techniques, Mike obtained unequivocal evidence that song dialect boundaries hinder gene flow between populations sufficiently to be compatible with Nottebohm's original hypothesis. Donning his other hat as a population geneticist, Mike related the ethology of song variation and learning to population structure, obtaining the first reliable estimates of deme size in a free-ranging songbird. It is something of a paradox that while sociobiological theorizing is making such rapid advances in understanding the genetic basis for social evolution, few vertebrate ethologists have the skill and inclination to gather the relevant genetic information directly, as Baker is doing.

Bill Searcy maintained a similar tradition as a postdoc and faculty member in behavioral ecology, dividing his time between the mating system of red-winged blackbirds, which he studied in a productive collaboration with postdoctoral fellow Ken Yasukawa, and a study of mate selection in our sparrows, illuminated by an imaginative series of playback studies to males and to estradiol-treated females, using both natural and synthetic songs. Before moving on to join the faculty at the University of Pittsburgh, he was able to complete a satisfying synthesis of many of the ideas and experiments that first germinated while he was a graduate student with Gordon Orians at the University of Washington. His field studies were complemented in the laboratory by Bob Dooling, with whom Margaret Searcy worked as a talented laboratory assistant for several years.

Joining us from the Central Institute for the Deaf in St. Louis, first as a postdoc and then as a faculty member, Bob Dooling brought the special insights of the psychoacoustician to bear on the perception of complex sounds by birds, especially songs. Following up on the efforts of Jeff Baylis, a brilliant student of George Barlow who ushered us into the age of the computer, Bob initiated us into the mysteries of synthesizing birdsong, a valuable tool in establishing which elements are crucial in eliciting responses. He has an extraordinary knack for getting wild birds to submit happily to a variety of audiometric conditioning procedures with the result that he has learned more than anyone about hearing in birds and the special predispositions they bring to bear on the perception of song. He now has his own laboratory for avian psychoacoustic studies at the University of Maryland.

Each generation of students brings a new perspective on what animal signals mean, how they develop, their effects on population structure, and how the brain sustains these marvelously intricate patterns of behavior that are among the most complex known to us from the animal kingdom. For me the wheel has taken another turn recently, when in 1981 Fernando Nottebohm took over the directorship of the Rockefeller Field Station in Millbrook and its attendant burden of chores and headaches. My gratitude to him is unbounded as I anticipate the freedom to participate more fully in what modern ethology has to offer, including the exciting new developments in etho-endocrinology that John Wingfield has recently introduced to my laboratory.

There are already indications that neuroethology is destined to play a major role in new developments in the neurosciences. I am equally convinced that ethology has unique contributions to make to population biology as well. I know that Ed Wilson will forgive me for accusing him of poetic license in predicting cannibalism of ethology by population biology on the one hand, and neurobiology on the other. While I believe that experimentation is destined to supercede observation as the method of choice in ethology and behavioral ecology, I am also convinced that future investigators ignore at their peril the ethological incantation that, when in doubt, take another look at the behavior, preferably in its natural environment.

Ethology has that quality of steering research in profitable directions, as a direct consequence of sensitivity to the biological context in which a given behavior has evolved. This is the credo I strive to observe in my own work, moving to and fro between the inspiration of field study and observation, and the rigor of testing by experiment, either in the laboratory or, if we have the courage, even in the field. The jury is still out on whether the approach is as scientifically fruitful as I believe it to be, but I can vouch for the fact that it is a lot of fun!

References

Gould, J. L., and Marler, P. 1984. Ethology and the natural history of learning. In *The Biology of Learning,* ed. P. Marler and H. S. Terrace pp. 47–74. Berlin, Heidelberg, New York: Dahlem Konferenzen: Springer-Verlag.

Gouzoules, S., Gouzoules, H., and Marler, P. 1984. Rhesus monkey *(Macaca mulatta)* screams: Representational signalling in the recruitment of agonistic aid. *Anim. Behav.* 32:182–193.

Green, S., and Marler, P. 1979. The analysis of animal communication. In *Social Behavior and Communication, Handbook of Behavioral Neurobiology, vol. 3,* ed. P. Marler and J. Vandenbergh, pp. 73–158. New York: Plenum Press.

Marler, P. 1952. Variation in the song of the Chaffinch, *Fringilla coelebs. Ibis* 94:458–472.

———. 1955. Characteristics of some animal calls. *Nature* 176:6–8.

———. 1956a. The voice of the chaffinch and its function as a language. *Ibis* 98:231–61.

———. 1956b. Behavior of the chaffinch. *Behaviour Suppl.* 6:1–186.

———. 1957. Specific distinctiveness in the communication signals of birds. *Behaviour* 11:13–39.

———. 1959. Developments in the study of animal communication. In *Darwin's*

Biological Work, ed. P. R. Bell, pp. 150–206. Cambridge: Cambridge Univ. Press.

———. 1960. Bird songs and mate selection. In *Animal Sounds and Communication,* ed. W. N. Tavolga, pp. 348–67. A. I. B. S. Proceedings.

———. 1961a. The filtering of external stimuli in instinctive behavior. In *Current Problems in Animal Behaviour,* ed. W. H. Thorpe and O. L. Zangwill, pp. 150–66. Cambridge: Cambridge Univ. Press.

———. 1961b. The logical analysis of animal communication. *J. Theoret. Biol.* 1:295–317.

———. 1964. Inheritance and learning in the development of animal vocalizations. In *Acoustic Behavior of Animals,* ed. M. C. Busnel, pp. 228–43. Amsterdam: Elsevier.

———. 1965. Communication in monkeys and apes. In *Monkeys and Apes: Field Studies of Ecology and Behavior,* ed. I. DeVore, pp. 544–84. New York: Holt, Rinehart & Winston.

———. 1968. Aggregation and dispersal: two functions in primate communication. In *Primates: Studies in Adaptation and Variability,* ed. P. Jay, pp. 420–38. New York: Holt, Rinehart & Winston.

———. 1969a. Tonal quality of bird sounds. In *Bird Vocalizations,* ed. R. A. Hinde, pp. 5–18. Cambridge: Cambridge Univ. Press.

———. 1969b. Of foxes and hedgehogs: The interface between organismal and population biology—II. *Amer. Zool.* 9:261–67.

———. 1970a. Birdsong and speech development: Could there be parallels? *Amer. Sci.* 58:669–73.

———. 1970b. A comparative approach to vocal learning: Song development in white-crowned sparrows. *J. Comp. Physiol. Psychol.* 71 (2, monog.):1–25.

———. 1973. A comparison of vocalizations of red-tailed monkeys and blue monkeys, *Cercopithecus ascanius* and *C. mitis,*in Uganda. *Z. Tierpsychol.* 33:223–47.

———. 1976a. On animal aggression: The roles of strangeness and familiarity. *Amer. Psychol.* 31:239–49.

———. 1976b. Sensory templates in species-specific behavior. In *Simpler Networks and Behavior,* ed. J. Fentress, pp. 314–29. Sunderland, Mass.: Sinauer Assoc.

———. 1976c. Social organization, communication and graded signals: The chimpanzee and the gorilla. In *Growing Points in Ethology,* P. P. G. Bateson and R. A. Hinde, pp. 239–280. Cambridge: Cambridge Univ. Press.

———. 1977a. Development and learning of recognition systems. In *Recognition of Complex Acoustic Signals,* ed. T. H. Bullock, pp. 77–96. Berlin: Dahlem Konferenzen.

———. 1977b. Primate vocalization: Affective or symbolic? In *Progress in Ape Research,* ed. G. Bourne, pp. 85–96. New York: Academic Press.

———. 1983. Some ethological implications for neuroethology: The ontogeny of birdsong. In *Advances in Vertebrate Neuroethology,* ed. J.-P. Ewert, R. R. Capranica and D. J. Ingle, pp. 21–57. London, New York: Plenum Press.

———. 1984a. Animal communication: Affect or cognition? In *Approaches to*

Emotion, ed. K. R. Scherer and P. Ekman, pp. 345–365. Hillsdale, N.J.: Lawrence Erlbaum Assoc.

———. 1984b. Song learning: Innate species differences in the learning process. In *The Biology of Learning,* ed. P. Marler and H. S. Terrace, pp. 289–309. Berlin, Heidelberg, New York: Dahlem Konferenzen: Springer-Verlag.

Marler, P., & Boatman, D. J. 1951. Observations on the birds of Pico, Azores. *Ibis* 93:90–99.

———. 1952. An analysis of the vegetation of the northern slopes of Pico—the Azores. *J. Ecology* 40:143–155.

Marler, P., Dooling, R., and Zoloth, S. 1980. Comparative perspectives on ethology and perceptual development. In *The Comparative Method in Psychology: Ethological, Developmental and Cross-cultural Viewpoints,* ed. M. Bornstein, pp. 189–230. Hillsdale, N.J.: Erlbaum.

Marler, P. & Hamilton, W. J. III, eds. 1966. *Mechanisms of Animal Behavior.* New York: Wiley & Sons.

Marler, P., and Hobbet, L. 1975. Individuality in a long-distance vocalization of wild chimpanzees. *Z. Tierpsychol.* 38:97–109.

Marler, P., Konishi, M., Lutjen, A., and Waser, M. S. 1973. Effects of continuous noise on avian hearing and vocal development. *Proc. Nat. Acad. Sci.* 70:1393–96.

Marler, P., and Mundinger, P. 1971. Vocal learning in birds. In *Ontogeny of Vertebrate Behavior,* ed. H. Moltz, pp. 389–450. New York: Academic Press.

———. 1975. Vocalizations, social organization and breeding biology of the twite, *Acanthus flavirostris. Ibis* 117:1–17.

Marler, P., Mundinger, P., Waser, M. S. and Lutjen, A. 1972. Effects of acoustical stimulation and deprivation on song development in red-winged blackbirds *(Agelaius phoeniceus). Anim. Behav.* 20:586–606.

Marler, P., and Peters, S. 1977. Selective vocal learning in a sparrow. *Science* 198:519–21.

———. 1980. Birdsong and speech: Evidence for special processing. In *Perspectives on the Study of Speech,* ed. P. Eimas and J. Miller, pp. 75–110. Hillsdale, N.J.: Erlbaum.

———. 1981. Sparrows learn adult song and more from memory. *Science* 213:780–82.

———. 1982. Long-term storage of learned birdsongs prior to production. *Anim. Behav.* 30:479–82.

———. 1982. Developmental overproduction and selective attrition: New processes in the epigenesis of birdsong. *Developmental Psychobiology* 15:369–78.

Marler, P., and Pickert R. 1984. Species-universal microstructure in the learned song of the swamp sparrow *(Melospiza georgiana). Anim. Behav.* 32:673–689.

Marler, P., and Sherman, V. 1983. Song structure without auditory feedback: Emendations of the auditory template hypothesis. *J. Neurosci.* 3:517–531.

Marler, P. and Tamura, M. 1962. Song "dialects" in three populations of white-crowned sparrows. *Condor* 64:368–77.

———. 1964. Culturally transmitted patterns of vocal behavior in sparrows. *Science* 146:1483–86.

Marler, P. and Tenaza, R. 1977. Signaling behavior of apes, with special reference to vocalization. In *How Animals Communicate,* ed. T. Sebeok, pp. 965–1003. Bloomington, Ind.: Indiana Univ. Press.

Marler, P. and Vandenbergh, J., eds. 1979. *Social Behavior and Communication, Handbook of Behavioral Neurobiology, vol. 3.* New York: Plenum Press.

Marler, P., and Waser, M. S. 1977. The role of auditory feedback in canary song development. *J. Comp. Physiol. Psychol.* 91(1):8–16.

Searcy, W. A., and Marler, P. 1981. A test for responsiveness to song structure and programming in female sparrows. *Science* 213:926–28.

Searcy, W. A., Marler, P., and Peters, S. 1981. Species song discrimination in adult female song and swamp sparrows. *Anim. Behav.* 29:997–1003.

Seyfarth, R. M., Cheney, D. L., and Marler, P. 1980. Vervet monkey alarm calls: Semantic communication in a free-ranging primate. *Anim. Behav.* 28:1070–94.

Waser, M. S., and Marler, P. 1977. Song learning in canaries. *J. Comp. Physiol. Psychol.* 91(1):1–7.

Zoloth, S. R., Petersen, J. R., Beecher, M. D., Green, S., Marler, P., Moody, D. B., and Stebbins, W. 1979. Species-specific perceptual processing of vocal sounds by monkeys. *Science* 204:870–73.

John Maynard Smith (b. 6 January 1920). Photograph from the School of Biology, University of Sussex

14

In Haldane's Footsteps

John Maynard Smith

I was born in London in 1920, and lived there until I was eight years old. In 1928 my father, a surgeon, died, and my mother, my sister, and I moved to the country. My father had little influence on me, other than a genetic one, since I had seen him perhaps once a week. Also at the age of eight I was sent to a single-sex boarding school, first a "preparatory" school, and later a "public" school, Eton. Apart from being extremely unpleasant, neither of these schools taught me any science, let alone any biology. Eton did, however, have compensations. I was taught mathematics extremely well, usually in a class of fewer than ten boys, and largely through the medium of a weekly problem paper, on which I would spend up to ten hours, partly out of a spirit of competition with two of my contemporaries as to who could solve most problems.

The other compensations at Eton were the existence of an admirable school library, and adequate spare time, so that I was able to teach myself quite a lot of science. I remember reading Jeans, Eddington, Einstein and Infeld, Julian Huxley, Darwin's *Origin,* and the *Science of Life* by Wells, Huxley and Wells, as well as Marx and Haldane, of whom more later. Consequently, my knowledge of science at the age of eighteen, and, indeed, until I was almost thirty, consisted in an excellent training in solving problems—including problems in "applied mathematics," which in those days meant Newtonian mechanics—and a fair understanding of some of the more fundamental theoretical ideas in physics and biology. It included no practical or experimental work, and no chemistry or geology.

In the school holidays we lived in the country, Berkshire in winter and Exmoor in spring and summer. My mother's family were well-to-do Scots who had moved south. We owned horses, and hunted with enthusiasm. On Exmoor, red deer are still hunted with hounds, and the first hymn in church on the first Sunday of the stag-hunting season always was, and for all I know still is, "As pants the hart for cooling streams, when heated in the chase. . . ." However, I had acquired a passionate interest in animals before we left London in 1928. I can remember regular visits, on my insistence, to the Zoo and the Natural History Museum, and hours spent with picture books of animals. Later, I preferred reading stories about animals to stories about people; a major thrill when I first visited North America in 1960 was to see the animals that decorate the margins of Ernest Thompson Seton's stories. When we did move to the country, I spent hours watching birds, rather ineffectively because at first I had neither binoculars nor a field guide, and keeping pond animals.

I am unable to account for this absorption in animals, which has remained with me. It was not derived from any obvious adult influence; my family rode horses, but were not naturalists. It predated our move to the country. It also predated my entry to boarding school. I was lonely at school up to the age of about sixteen, and I think that children who are lonely, for whatever reason, are often attracted to animals and to natural history. Thus, although my interest was probably strengthened by loneliness and by life on Exmoor, its origin lies elsewhere.

Competence at mathematical modeling, and love of natural history, are two of the roots of my interest in evolutionary biology. The third is a curiosity about the nature and origin of things. I have thought about philosophical issues for as long as I can remember. I was also much influenced by science fiction, particularly Wells and Stapledon. I cannot remember when I first came across the idea of evolution, but I do remember that it fairly soon became clear to me that evolution was incompatible with the rather simplistic version of Christianity in which I was raised. I abandoned Christianity after some struggle, but ultimately with great relief. At the same time—approximately, at age sixteen—I ceased to be merely miserable at Eton and became actively hostile to it. I noticed that there was one person who attracted the particular hatred of several of my teachers. This was J. B. S. Haldane, himself an Old Etonian, who had betrayed his class and religion, and who lost no opportunity of attacking everything Eton stood for. Thinking that anyone they hated that much could not be all bad, I sought out his books in the school library; it is to Eton's credit that the books were there.

I found Haldane's mind immediately sympathetic. Today I find it hard to distinguish resemblances between us that arose by independent convergence, and those that arose because I have copied him. His method of thinking was an extraordinary combination of the abstract and the particular. Faced with any problem—the action of enzymes, evolution, protection against air raids—his habit was to make a simple mathematical model. Having reached conclusions from such a model, he would seek concrete and familiar illustrations of those conclusions. Haldane's ideas were clear ideas; he never left you wallowing in a sense of misty profundity.

On leaving Eton at the age of eighteen, I went to Cambridge to read engineering. This may seem an odd choice. My family expected me to enter my grandfather's stockbroking firm, but I knew that was impossible. All I had read of physics was the work of Newton, Bohr, and Einstein, and I knew I could never do anything like that. I was innocently unaware that one could earn one's living in biology—after all, Darwin didn't. So engineering seemed a possible outlet for mathematical ability.

Academically, I got less out of Cambridge than I had out of Eton—there, at least, I learned some mathematics. This time, however, the fault was partly mine and partly Hitler's. It was hard, in 1938, to take either academic work, or one's own future, seriously. I joined the Communist party in 1939, and most of my time and energy at Cambridge were spent in political activity, as was most of my spare time after leaving the university, until about 1946. This is not a political autobiography, so I will say no more than that this seemed to me the right thing to do at the time. It did, however, have two consequences. One is that between the ages of eighteen and twenty-six, when most scientists are identifying, if not solving, the problems on which they will subsequently work, I was otherwise engaged. The second is that I acquired a familiarity with Marxist philosophy that is uncommon among Western scientists.

After 1946, I was politically inactive, although I did not finally leave the Communist party until the Russian invasion of Hungary in 1956. During that decade, I became increasingly disenchanted with the Communist party. One reason for this was the behavior of the Soviet government during the Lysenko affair. This forced me to think harder about the Weismann-Lamarck issue than most biologists bother to do. It also forced me to think about the relationship between philosophical views in general—and the Marxist one in particular—and the practice of science.

On leaving Cambridge, I worked in an aircraft design office, first at Armstrong Whitworth's and then at Miles Aircraft. On the whole, I enjoyed this and was good at it. It proved further training in problem solving. I also acquired the ability, rare among biologists, to perform massive numerical calculations, with no aid other than a slide rule, and without making mistakes; a mistake could mean that someone got killed. This skill has been rendered largely obsolete by digital computers; however, I found that the ability to compute translated rather quickly into an ability to program a computer.

By 1947, I had decided that I did not want to spend the rest of my life designing airplanes, mainly, I think, because my poor eyesight means that I cannot hope to fly airplanes, and hence I do not much like them. By that time, I knew that I would like to get into academic science. Theoretical physics still seemed too difficult, and experimental physics was ruled out by my inability to use tools or operate machines. I therefore applied to University College London to read an undergraduate degree in zoology. I am not certain that it was really necessary to start again at the beginning, but in retrospect I am glad that I did. I chose UCL because I knew Haldane was a professor there. Whether I would have had the courage to do so if I had known how explosive a character he was, I don't know, but again I do not regret the decision.

At UCL I met ethological ideas for the first time, but again—a recurring theme—not as part of the formal curriculum. However, at that time the ideas of Lorenz and Tinbergen were in the air. Two of my fellow undergraduates were David Blest and Aubrey Manning, and they were quicker to appreciate these ideas than I was. At first, I was unsympathetic. A standard text, and an excellent one at that time, was Frankel and Gunn's *Animal Orientation*. Its mechanistic approach appealed to my engineering background. I was slow to accept the obvious fact that there must be structures in an animal's head capable of producing a complex response to a simple stimulus.

My major interest, then and always, was in evolution. The course provided two relevant bodies of material. The first was the course in vertebrate zoology, for which D. M. S. Watson gave the lectures. I have mixed feelings about this. Watson's lectures were a joy to listen to, and we did learn a lot about vertebrates. But the intellectual core of the course was the search for homologies, and I increasingly came to feel that this search is sterile. No doubt I was influenced by an intense dislike for the smell of formalin, and my inability to dissect a vertebrate without puncturing the vena cava. But I had more rational grounds for thinking that arguments about whether X is homologous to Y are often unsettlable, and would shed little light on the mechanism of evolution even if they were settled. I was, and am, more interested in knowing how X works, and why it evolved. I am therefore somewhat puzzled that Lorenz should regard it as a triumph of ethology that it resembles comparative anatomy in recognizing homologies between patterns of behavior.

The second input relevant to evolution was the teaching of Haldane, and his wife Helen Spurway, on genetics, ecology, and systematics. Most of what I have published since is based on that teaching; imitation is the sincerest form of flattery.

When I graduated, in 1951, I stayed at UCL to work in Haldane's laboratory (not that we ever allowed him into the lab if we could help it). A year later—and I feel ashamed when I think of the struggle people have today to get a teaching post—I was offered a lecturership in the zoology department, a post that enabled me to go on working with Haldane. By that time, Watson had retired as professor of zoology, and had been replaced by Peter Medawar, who, after Haldane, has been the major influence on the way I see science. He is a militant Popperian, and has a knack for thinking up experiments that actually do corroborate or falsify theories. This is not an ability I share, but it is one I admire and envy.

Almost all my research has been directed toward understanding the mechanism of evolution; I have studied behavior because it raises special problems for an evolutionist. After some initial work on the effects of inbreeding and outcrossing, the major part of my time has been spent on three problems: the evolution and physiology of aging, the evolution of sex and of breeding systems, and the evolution of behavior. In each of these three cases, I was first attracted to the topic because it was difficult to find an explanation that made sense in terms of natural selection. Senescence is puzzling because an animal would be fitter if it did not deteriorate with age. The evolution of sex faces the major problem of the twofold cost of meiosis, and a number of subsidiary problems. The problems raised by behavior are discussed in more detail below, but essentially they concern "altruistic" and "conventional" actions that, at first sight, seem not to contribute to the success of the individual.

From 1951 to around 1965, most of my time was spent on genetic and physiological experiments with *Drosophila subobscura*. Increasingly, since 1965, I have switched to mathematical and theoretical work. There are a number of reasons for this change. When starting in research, I had several theoretical papers (on animal locomotion) rejected by journals, and I found this discouraging. I was aware that Haldane could solve theoretical problems in a fraction of the time I could, and since he was just across the corridor it seemed easier to leave such things to him; after he departed for India, I found myself spending more time on theoretical work. In 1965, I left London to become the first "dean" (departmental chairman) of biological sciences at the new University of Sussex, and this involved a lot of administration, which does not mix well with the daily routine of *Drosophila* work. Perhaps most important, I began to acquire the confidence that I could contribute to evolutionary theory.

My first behavioral work came about by accident. I was trying to find out why most of the eggs produced by inbred lines of *D. subobscura* fail to hatch. One possible reason was that the female had not mated. The easiest way to find out was to watch. This led to a study of sexual selection in *subobscura*, suggesting that the side-to-side dance of the females enables them to distinguish between males. These experiments illustrate nicely the advantages and disadvantages of working without training or supervision. The drawbacks are obvious; had I had some training in the techniques and outlook of ethology, my descriptions of the courtship displays would have been incomparably better. But I might not have discovered anything. At that time, it was traditional to explain differences in behavior in terms of "motivation." I might, therefore, have concluded that inbred males do not mate because they lack motivation. As it was, I concluded that they did not mate because they could not keep up with the female, no matter how hard they tried. I don't think I fully proved my point. However, I did try to make the distinction between an animal not doing something because it couldn't, and because it didn't want to. This distinction, although obvious, was not often made by ethologists, partly, I think, because of the absence of a nonsubjective way of stating it. It becomes crucial if one attempts to analyze contest behavior.

In 1963, two events took place that were to be decisive in turning the attention of ethologists to functional explanations—that is, to discovering the selective processes responsible for the evolution of behavior. These were the publication of Hamilton's idea of inclusive fitness (described in more detail in his 1964 papers), and Wynne-Edwards's book *Animal Dispersion*. I encountered Hamilton's ideas when I refereed his 1964 paper for the *Journal of Theoretical Biology*. The central concept was not wholly new to me, because Haldane had published a crude version of the idea in 1955, and I had heard him discuss it. Haldane, however, did not work out the mathematical details. In particular, I don't think he asked what would happen if the gene for "altruism" was not rare. More important, he did not see the idea as the basis of a program for analyzing the social behavior of animals in the light of the genetic relationships between members of social groups.

I did not at the time follow the mathematical details of Hamilton's paper. What persuaded me that he was on to something was the wide range of applications he was able to suggest, and in particular his discussion of the relevance of coefficients of relationship in haplo-diploids.

At the time, however, I was more interested in Wynne-Edwards than in Hamilton. To understand this, one must know that the central problem of ecology in 1960 seemed to many of us to be the regulation of population numbers. Wynne-Edwards was proposing an explanation in terms of group selection. I had learned the weakness of such explanations from Haldane when an undergraduate. His method of teaching this was to write the words *Pangloss' Theorem* in the margin of an essay in which I had made, unconsciously, a group-selectionist assumption. The meaning of this comment was not at once obvious, but once understood it was never forgotten. Wynne-Edwards's thesis, however, was widely accepted, and there seemed to be a need to contest his views. This seemed particularly necessary because group-selectionist views were widespread, although less clearly stated, in other important fields. Darlington's *Evolution of Genetic Systems* had given a group-selectionist tone to most discussions of this topic, and similar thinking was present in ethology, particularly in Lorenz. The relevance of Hamilton's work in this context was that he had proposed a plausible alternative explanation for some traits that had previously been explained by group selection. It seemed important therefore, that in criticizing group selection I make it clear that Hamilton's mechanism was exempt from the criticism. It is for this reason that I coined the term *kin selection*, and that I have been critical since of attempts to argue that the two processes are not different.

My main research interests during the sixties, however, were in the physiology of aging and in the genetics of patterns, and, toward the end of the period, in various theoretical issues in population genetics, in particular the significance of sex in evolution. I was aware that it was unreasonable to criticize Wynne-Edwards for explaining population regulation by group selection if one assumed, as most evolutionary biologists did at that time, that sexual processes had evolved by that means. I have continued to work on this problem, which I regard as still partly unsolved.

In 1970 I spent three months in the University of Chicago, with the committee on mathematical biology. While there, I learned some game theory, with a view to applying it to animal behavior. The stimulus to do this came from George Price, who, a year or two earlier, had submitted a long paper to *Nature* arguing that the usual "good of the species" explanation of conventional fighting behavior in animals must be wrong, and suggesting that the reason animals did not always escalate their fighting during contests was that if they did, their opponent might retaliate. This paper was sent to me to referee. I reported that the paper contained an interesting idea. Unfortunately the manuscript was far too long for *Nature*, so I suggested that Price be encouraged to write a shorter version for *Nature*, and/or submit the original manuscript to some other journal.

While in Chicago, I developed the idea of an evolutionarily stable strategy, or ESS, and applied it to Price's problem by developing the Hawk-Dove-Bully-Retaliator game. I also invented the "war of attrition" game, and found its ESS. I should add that Hamilton, in his 1967 paper on "Extraordinary Sex Ratios," had used the concept of an "unbeatable strategy," which is formally similar to an ESS. I knew Hamilton's paper, but was not consciously influenced by his ideas, although, oddly enough, I was aware in 1970 of the similarity between what I was doing and the method MacArthur had used in 1965 to find the stable sex ratio.

When I had written up this work, I tried to find Price's paper so that I could quote it, but found that it had not been published. It turned out that Price was living in London and working on theoretical biology, initially without financial support, although later he did get an SRC grant. After some discussion, we agreed to publish a joint paper, which finally appeared in 1973. It is ironic that, although I think that Price's idea about retaliation being responsible for conventional behavior is often correct, and although I also think that evolutionary game theory is the clearest way of analyzing such questions, we did not do a very good job of applying game theory to retaliation in our 1973 paper. Indeed, Gale and Eaves later pointed out that we had missed an alternative ESS of the game we published.

Since 1970 I have spent a lot of time trying to clarify the concepts of evolutionary game theory, and to apply it to specific problems. I soon became aware of Parker's work on dung flies, and on animal contests in general. In his Ph.D work he had interpreted the stay times of male dung flies at cowpats in terms that closely resemble my war of attrition. In 1974, we both published papers on contest behavior in the *Journal of Theoretical Biology*. Parker's paper introduced a sharp distinction between motivation and "Resource Holding Power," or RHP. This was parallel to the distinction I had tried to make, between "won't" and "can't," when working on sexual selection in *Drosophila*.

In 1967 Parker and I published a joint paper on asymmetric contests. The origin of the idea of an asymmetric contest is curious. In the war of attrition, each contestant can, in effect, choose the cost he is prepared to pay to gain a given reward. It turns out that, at an ESS, the cost chosen is exponentially distributed. More relevant, the average gain to an individual, per contest, is zero; the costs incurred exactly cancel out the gains. Thinking about this, it occurred to me that two human beings playing this game would surely agree to toss a coin to decide who should get the reward. Then each could expect to gain in a series of contests. This seems paradoxical; the introduction of a randomizing device has improved the prospects of both contestants.

At first I concluded it was just bad luck for animals that they cannot toss coins. However, all that a coin toss does is to introduce an asymmetry into a previously symmetrical game. I was therefore led to analyze animal contests in which there were such asymmetries, of size, ownership, and so on. It is then easy to see that, at an ESS, asymmetries will be used to settle contests conventionally, even if they are, like coins, uncorrelated with either payoffs or RHP. At first, I thought that this conclusion was of logical and mathematical interest only; increasingly, I have come to think that almost all contests between just two individuals should be treated as asymmetric ones.

During the past ten years, I have been interested in extending the range of application of the ESS concept. One major field of application is in sexual allocation theory—that is, sex ratios; resource allocation in hermaphrodites. This field has been mainly developed by Charnov, but he and I did independently obtain the same basic prediction, published in 1976, about resource allocation in hermaphrodites. Another field of application, still largely unexplored, is in life history theory.

When I first started thinking about evolutionary games, I did not clearly distinguish between contests involving just two individuals (e.g., the Hawk-Dove game), and games in which an individual is playing against the rest of the population, or some subset of it. An example of "playing the field" in this way is the sex ratio game in a random mating population; that is, Fisher's sex ratio problem. Once this distinction is clearly understood, and it is appreciated that the concept of an ESS can easily be extended to games against the field, it should prove much easier to apply the method.

One other extension of evolutionary game theory has given me a good deal of pleasure, although I have not been an active contributor. Several groups of mathematicians have seen that the game theory method can readily be replaced by a set of differential equations describing how the frequencies of different strategies change in time. The equations obtained are identical to those used by Eigen and Schuster to describe the evolution of replicating molecules during the origin of life. The identity is not surprising, because the equations merely describe the evolution of populations of entities with heredity but no sex. I suspect that this approach will be more important for chemical evolution than for the evolution of behavior. My pleasure comes from seeing the same mathematical structures emerging from such diverse problems as the evolution of behavior and the origin of life.

It will be apparent that I have not seen myself as an ethologist, but as a student of evolution. If I have contributed to ethology, it has been largely by accident. My central concern has been to think more rigorously about the ways in which natural selection molds phenotypes.

References

Maynard Smith, J. 1958. *The Theory of Evolution.* London: Penguin Books.

———. 1978. *The Evolution of Sex.* Cambridge: Cambridge Univ. Press.

———. 1982. Evolution and the Theory of Games. Cambridge: Cambridge Univ. Press.

Curt P. Richter (20 February 1894–21 December 1988). Photograph by Fabian Bachrach

15

It's a Long Long Way to Tipperary, the Land of my Genes

Curt P. Richter

Introduction

For this occasion I have had for the first time to review my scientific career with relation to the course of the rest of my life. I realize now how one event completely and permanently changed my life pattern, changed it from a long period of inactivity to a career of intense scientific activity, particularly in the field of "animal behavior."

In my efforts to explain this great change, I have tried many different theories and have finally decided that ethology may give the best explanation. On the following pages, I will try to describe (with apologies to ethologists) how my life and career were determined by what I call "the release of my gene."

Life in Denver from Birth 1894 to Departure for Dresden, Germany in 1912

My parents came from a small village in the mountains in the province of Saxony in Germany. Both my father's and my mother's families had lived there for many generations.

My father spent two years or more in the German army. After that he took a degree in a Saxony engineering school. The only pictures I have of him in his youth in Saxony show him in various military uniforms, so I take it that he enjoyed the military life. My earliest recollections of him are of his putting me through exercises to produce a very straight posture. Later, in grade school, teachers often used me, to my embarrassment, to illustrate good straight posture. My mother and her sisters had the usual rigid German Hausfrau training. Saxony was very densely settled and life was not easy. Many enterprising young men and women left to come to America. My father arrived in Denver in 1890. He sent for my mother in 1892. I was an only child.

Denver was then just starting its great growth, and my father started a firm for the construction of iron and steel parts of buildings. It apparently did fairly well from the start.

My father was a definite disciplinarian, very strict and demanding. I can still

357

remember his administering severe whippings with a leather strap on several occasions.

His mind ran strictly to physical and mechanical phenomena. I do not remember his ever showing any interest at all in biological phenomena.

Father's Death

Near the end of my eighth year, my father died after a hunting accident. This left my mother and me literally to shift for ourselves. We got some help from my father's lodge (Woodmen of the World) but otherwise there were no helpful friends. My father and mother had not lived in Denver long enough to have made more than superficial aquaintances.

Fortunately the firm had done well enough during the first nine to ten years of its existence so that it did not collapse after my father's death. Though she had no experience, either technical or financial, my mother stepped in, and managed somehow to get good men to help run the firm. She took over the job of secretary and treasurer. However, for the next five years we literally lived from hand to mouth.

My mother had to spend almost the entire day at the factory and I had to take care of myself as well as I could. For a number of years, I prepared my breakfast and lunch and occasionally dinners, but my mother took care of my clothes and always managed to make certain that I was simply but well dressed.

As an only child, I had for many years to spend much of my time alone. Somehow I managed to take care of all the necessary chores and duties.

"Play" Period—Until after My Father's Death

Before I was three years old, I used my time in the factory to learn about tools and machines. Before long I was capable of handling hammers, pliers, sheers, and every kind of tool that was in use. It gave me a great deal of satisfaction to heat a one-half-inch soft iron rod in a forge, first to a red-hot glowing condition and then to beat it into different shapes by pounding it on an anvil with a heavy hammer. I became familiar with the operation of most of the machines in the factory. The workmen were very friendly with me and showed me how to do things whenever I put questions to them. At home I spent a lot of time working with different tools, or with small steam engines or electric motors that my father had given me. I spent a lot of time with a magnet, which somehow fascinated me, and I tested its strength in picking up screws and nails and all types of pieces of iron.

I was stimulated at one point to run an experiment on the magnet—actually the first experiment in my life. I tried to find out whether I could increase the strength of a magnet by adding a nail every day to the armature. I added a nail to the Bull Durham tobacco bag that hung from the armature to determine whether in this way, after a long period, the magnet would become stronger than it had been before. Unfortunately our moving to another house after several months interrupted this experiment and so I never managed to get an answer.

I should mention also some constructive activities I carried out at the age of five on the tin bathtub I had used as a baby. My mother had discarded this tub and left it in the backyard. I decided to rebuild it to make a blast furnace out of it. I turned the tub upside down and cut a hole at the large end to make a door and a hole on the top to hold a small length of ordinary stovepipe. When I filled this little tub with paper and wood, I managed to start a roaring fire which delighted me and made me feel that I had some big furnace going. After a week or more, mother finally made me get rid of it because of fire hazard.

I spent a lot of time working on locks and clocks—taking them apart and putting them back together. I also spent a lot of time taking my bicycle apart and putting the wheels and gears back in place. I managed somehow to keep the bicycle in good running order after all of these dismantlings.

I also spent a lot of time taking care of my pocket knives. In those days all the boys had to have at least one or two pocket knives and they had to be kept well sharpened and cleaned for making whistles from twigs and for carving our initials on school desks and on trees.

After my father's death the "play" period lay dormant until after 1919—when my gene was finally released. This early non-verbal "play" period during which I scarcely learned to read had definitely become an important part of my life.

Public Grade School

As I look back now on that time after my father's death, I feel certain that the most important thing that happened to me was that a redheaded boy in my class at school asked me to spend a Saturday playing at his house. This started a friendship that lasted until his death about twenty years ago. The boy was M. E. Traylor and his name later became widely known nationally in financial circles as that of the president of the Massachusetts Investment Trust in Boston, one of the highest-paid men in the country, and also the father of eight children. Ed's father ran a grocery store, his mother a boardinghouse—that for years was my second home. We were very good athletes and very competitive. We both developed special skills in playing marbles and tops for "keeps," so that between us, we practically cleaned out all the boys in our school, and managed to build up large collections of tops, agates, and marbles, which we kept in his cellar.

In the lawn cutting of spring or summer, he and I again joined in partnership and did very well by ourselves—with me pushing the lawn mower, and him trimming the edges of the lawns with clippers.

I mention our various early partnerships here, because later in life when Ed had well advanced his career as a broker and investment expert, he frequently begged me to go into partnership with him. For reasons that will become clear later on, I held out for some unknown goal that would not become known to me until after the release of my gene.

In our last year in grade school, the class elected him president and me vice-president. We both were having fun.

For a number of years after my father's death, I worked on a farm for at least one or two months each summer. I look back on this farm experience with special satisfaction. For one thing, it taught me "patience." When weeding long rows (in some instances a mile or more) of corn on hot days, I had to stick at it, row after row, for hours. This patience has often stood me in good stead in later life in my science career. Life in these dry farms was not easy so there was almost no time for relaxation. However, collecting the cattle on horseback from among the sand dunes on the "range" in the late afternoons gave me great pleasure.

Coyote hunting constituted a rare but exciting pleasure. Once we picked up the trail of a coyote, our dogs, two hounds and a bull terrier, would put on a great display of teamwork. The hounds going at great speed would come up on either side of the coyote. One hound would pull the coyote down with a firm bite on one side of the throat and the other hound would sink its teeth in the throat from the other side. The hounds would hold their grips on the throat until the slow waddling bulldog would catch up to dispatch the coyote with one crushing bite. Since learning more about coyotes and learning to like and respect them, I no longer like to think of these hunting expeditions.

I did not do well during my eight years at the Whittier Grade School. Each year I just barely passed to the next grade. I can remember at least three occasions when the teacher passed out the sealed certificates and, in my anxiety about passing, I took mine to the alley in back of the school before mustering up enough courage to open it. Suffice it to say that at no time during the eight years in grade school was the existence of biology ever hinted at. Nevertheless, I have the happiest memories of the teachers at the Whittier School; they were mainly maiden ladies, who inspired us to do our best. In return we treated them with great respect.

Public High School

Ed Traylor and I now went on to high school, which meant about a two-mile walk each day going and coming from school. Ed had to spend more time in his father's store, so we did not see as much of each other as before.

The high school offered all kinds of sports that were much more definitely organized than in grade school. We had to "go out" for baseball, football, track, or some other sport. Ed and I both "went out" for cross-country running. Ed did very well and won many of the long races. I had to stop because of a heart condition and I finally ended by spending most of my spare time in basketball and track.

I do not have any clear recollection of what my schoolwork was like at that time—all I know is that I did not do very well, but I managed to stay in school and advance each year to the next higher level.

Social life connected with fraternities played an important part of particularly the last three years of high school. Members of the same fraternity tended to gather into cliques and fraternity meetings and initiations were important to us. Initiations often involved unmerciful beatings with shillelaghs on the hamstring muscles of plebes.

Now as a part of more formal high school athletics, I spent much time on practicing the high and broad jumps and shot puts in my own backyard, and I kept fairly good records of my performances. I also installed a horizontal bar in my backyard. The entire setup, which I designed and built myself, gave me a great deal of pleasure for a number of years.

One of the members of my fraternity, Henry M. Winans, became a close and lifelong friend who had an important influence on my life. He went to Stanford University and later studied medicine at the Johns Hopkins University, where he graduated at the top of his class. He was the last man at the Hopkins Medical School to have to give up his internship because of marriage. In our junior and senior years in high school, we spent many Sunday mornings after church in his father's library discussing all sorts of topics. In contrast to me he had read widely in English and American literature, but he was not in any sense a bookworm. We kept in close touch with each other long after we both left Denver, right up to his death. In the election for the presidency of our class of 400, held at the beginning of the senior year, Henry made the nominating speech that won me the election. This time Ed Traylor was elected vice-president.

My election as president of the class of 400 students and my captaincy of a championship basketball team kept me very active throughout my senior year. In addition, I organized all kinds of class parties and dances.

My senior class work brought me into close contact with Dr. W. Smiley, a Harvard man, principal of the East Denver High School, affectionately known to many generations of East Denver graduates simply as Bill Smiley. We formed a close relationship that undoubtedly had much to do with my going to Harvard several years later, after my return from three years in Germany. During these three years, we kept up a regular correspondence. Bill finally became superintendent of education for all the schools of the city of Denver. We met again later as fellow members of the Cactus Club there.

Almost immediately after graduation, I set out to carry out my father's wish, first made when I was four years old, that I study engineering in Germany. When I look back it becomes clear to me that my father's plans for my future, particularly his ideal of my becoming his partner and going to Germany to study engineering, had the effect of a posthypnotic suggestion. I never questioned the plan, even though after my father's death I had progressively lost contact with his firm and with engineering in general. I had developed quite different interests. Anyway, shortly after graduation from high school I set off for Germany. I had not even inquired whether I could get good, or better, engineering training right here in America, for instance at M.I.T.

After graduating I was able for the first time to get a better perspective of my successes—my term as class president, my involvement in all kinds of student activities—and of myself as a person with a promising career ahead of me as a student of Colorado University. I suddenly made an important decision for my future life—I would never go in for that kind of political activity again. It was all fine when I did it and I was successful, but it was not anything that I would ever want to do again. Throughout the rest of my life I have held strictly to the decision. By then I had already clearly begun to know what things I did not want to do in life, but had no idea at all of what I really wanted to do.

Again I must emphasize that during high school, I had no exposure to biology in any form. It was not only that I did not get exposure to biology; somehow I had developed a really poor opinion of biology. I can well remember one day in my senior year when in talking about one of my classmates, I finished him off by remarking, "Oh, he takes biology."

Dresden, Germany 1912–15

My father had not specified where in Germany he wanted me to study engineering, so I decided to go first to Dresden, where several of my closest relatives lived. That decision turned out to be very lucky for me, since at that time Dresden was one of the most beautiful and interesting cities of Europe and also the site of one of the best German technical schools (Technische Hochschulen) on a par with our M.I.T.

I never will forget my first view looking down the main street, Prager Strasse, on emerging from the train depot in the late morning in June. As far as I could see flower boxes lined both sides of all four floors of the buildings, a really beautiful sight.

I submitted my Denver High School credentials at once to the "Hochschule" secretary and received acceptance on the condition that I get intensive tutoring in mathematics.

The "Hochschule" did not open until October, so I used the summer months to explore the art museums and other collections and the countryside around Dresden; also I spent much time in churches, getting used to spoken German; and I spent much time in reading German and in visiting relatives, particularly my grandmother, who still lived in the village where my father and mother had been born. It was the site of an old castle, Augustusberg, which I and my mother had visited when I was six years old. The village, perched up on a hill near the castle, could be reached either by a cable car or by walking.

I was still in fine shape from my Denver athletics and deeply tanned from the boat trip so one day I decided to walk up. When I got near the top, I saw a group of boys of about my age who were putting the shot on a flat space of ground. Since putting the shot was one of my favorite track events I stopped to watch their performance. I saw right away that they were novices and not doing very well. After quite a while, when the shot happened to roll near my feet I picked it up and walked over to the starting block and put the shot myself. Without much effort I put it about twice as far as the best put by any of the boys. After a few minutes of total silence every last boy quietly disappeared leaving the shot and me there alone. Then I went on with my walk up the hill. The next day my aunt learned from the local gossip in the bakery shop that a *Käse-Gesicht Japaner* ("cheese-faced japanese) had suddenly appeared in the village and she quickly surmized from the description that it was me. My deep tan, very healthy appearance, and the long shot put apparently had a "man from Mars" effect. Some of the boys probably were relatives of mine, since my family had lived in the village so many generations.

When the fall term opened, I found that I was the only American in the school, but there were many Norwegians, Danes, Swedes, Finns, Swiss, and Austrians, and a few Bulgarians and Turks. The school did not have any facilities for bringing students together for social contacts, so for a few months life was a bit grim.

However, it turned out that students in Germany had very special privileges to attend all governmental functions, and performances of the Royal Opera, and Royal Theatre. During the season, the Royal Opera presented a different opera every night and the Royal Theater a different play. Thus, during the year we could have heard almost all of the well-known operas or seen plays by all the world's playwrights. At the last moment on any afternoon the school porter could get us choice seats in the balcony just above the Royal Box for the equivalent of a quarter. It was truly a wonderful privilege, especially for me, who had never heard an opera or seem more than a few plays acted at local theaters in Denver.

Life brightened up for me after a few months when one of the students addressed me in English and said that his name was William Dunkel and that he was an American of sorts since his mother had been born in New York. It turned out that he himself had been born in Buenos Aires, had studied in Lausanne, in Switzerland, and Paris. He spoke five languages, was widely read in the world literature, and liked sports. We got along very well and later had joint lodgings. He was studying architecture at the Hochschule; later in Zurich he became one of the leading architects of modern housing in Europe.

In the beginning of 1913 we were both elected to the Akademishcher Sport Verein, an organization of students set up in opposition to the German Fencing Corps. The membership consisted of 30 percent Germans, 70 percent Scandinavians, Swede, Danes, Finns, Norwegians, and a few Austrians, and Swiss. We had our frequent special evenings sampling new wines, and so on. We also had our own hockey field, twelve tennis courts, and a field for track events. We all got to know each other very well, so the club in itself offered an education.

The track season later in the spring and early summer gave me a great opportunity to shine as an American. I was still in good shape after leaving Denver, even though I had had no opportunity for exercise. In the club track meet that had ten events, I won eight firsts—running, jumping, shot put, and others. I starred especially in one event—throwing a hocket ball. As an old baseball player, I threw the ball about twice as far as any other member. Later in the first track meet ever held between the Dresden Hochschule and Leipzig University, I again won almost all events, for which the minister of education of Saxony gave me an award of a silver lion desk set piece. Now that the East Germans have become such great world athletes, I would not be able, so to speak, to get to first base in any event. In 1912 when I left Denver, I had never even seen a ski, but on my arrival in Dresden I found that practically everybody skied. On moonlit nights several of us took an evening train from Dresden to the skiing areas and skied the entire night, then catching an early morning train back to Dresden in time to make a 7:00 A.M. class.

Through my interest in tennis I met Dr. Herman Rau, a lawyer and a member of the Akademishcher Sport Verein. We played tennis fairly often and took long skiing trips to the Giant Mountains in Silesia. Bill Dunkel often joined us in these longer skiing trips. Rau later moved to Berlin where he became a member of the

Rot-Weiss Tennis Club and Manager of the German Davis Cup team. I visited him in Berlin many times between the wars. He has since died. Dunkel is still alive but is no longer active.

Dresden offered so much for students in the way of entertainment, especially in the pre-Lenten season, called "Fashing," when the entire city practically gave up to a period of having a good time for four of five days.

Later on, in 1913, a truly memorable event occurred in Dresden on the occasion of the Royal Maneuvers at the Hellerau Infantry Training field near Dresden. This brought the crowned heads of all the countries in Europe to Dresden, coming and going from Hellerau through the narrow Prager Strasse, some on horseback, some in beautiful equipages, along with the brilliantly dressed cavalry of Germany, the Kaiser, the Crown Prince, the King of Saxony, King of Bulgaria, the Czar of Russia, and many others. It was breathtaking and gave us all the feeling that the world would go on this way forever. Little did we realize that the "Guns of August" were even then gradually being brought into position all over Europe.

Many other amusing and interesting things happened during my stay in Dresden, but I shall mention only two of them. First I shall describe the rituals and formalities associated with the so-called preliminary examination "Vorpruefung," given at the Hochschule at the end of one and one-half years. For each subject, students were examined in groups of four by the head professor. The students had to appear in full dress—white tie, top hat and all—beginning in most instances at 7:00 A.M. The professor also had to appear in full dress. A very formal atmospere persisted, but during the hour, the professors usually asked very gentle questions. When I appeared on Prager Strasse in full dress and top hat and white tie at 7:00 A.M., people did not know whether I was coming or going. Anyway, I managed to pass the exams.

The other event I can never forget. Early in July 1914, I went to a tailor to have my first fitting for an English homespun suit. I had several fittings before starting off on a surveying trip in a district located about thirty miles from Dresden, and the tailor promised me to have the suit ready on my return from the trip. Residents in the surrounding area put up the forty students for private lodgings and an inn provided meals. We spent the evenings pleasantly with beer drinking and a little singing. Usually we all met for breakfast before starting our surveying practice. But on one morning when I went down for breakfast I could not find a soul. I thought at first that I had overslept, but it was still early. Finally, I found an old waiter who told me that during the night Germany had declared war and that the students, all officers in the German army, had left at once to report to their units. All means of transportation had also disappeared so I had somehow to find my way back to Dresden, where the war fever had already reached a high pitch. An endless line of soldiers marching and singing passed through the Prager Strasse on the way to the rail station. On my walk through the streets I could see that all "English spoken" signs on windows had already been removed. And later, newspapers carried accounts of searches for spies and incidents in which spy suspects had been pummeled to death with canes.

The war between Germany and England somehow did not for the moment upset me too much. My main concern was to make my way to my tailor and my new homespun suit. I found my suit was packed and ready for me to take along.

In my room I tried it on and was much pleased with what I saw. I decided then to take my suit for a walk along the Prager Strasse and further on to the River Elbe. By then all of the marching and singing soldiers had departed.

On the way back from the Elbe, I suddenly became aware of half-audible whispering and talking in back of me and looking back through the reflection in shop windows I could see that a crowd of people was following along about thirty yards behind me.

When I got about halfway on one side of the Altmarkt—an open market area about six blocks in each direction—four burly Germans approached me with raised canes and started shouting *spione, spione,* "spy," "spy," and with that the entire Altmarkt, now almost all filled with people, took up the cry *spione.* These shouting people completely surrounded me and I expected any moment to have them start pummeling me with their canes. I kept my cool and kept saying *Tch bin ein Amerikaner.* Fortunately a man in the crowd recognized me and said that he knew my relatives. Someone sent for the police, and after ten tense minutes they marched me off between two rows of cops to the police station, where the officer released me.

I guess it was a narrow escape. I am certain that if one man had taken a swing at me with his cane they would all have followed and made me into mincemeat. However, my homespun suit survived many years and gave me much pleasure.

In the fall, only a few foreigners were left for classes. I had to make up my mind soon about staying to see whether the war would come to an end quickly or to go home. I finally decided to return home. But, most importantly, I decided to give up my engineering training. With the posthypnotic suggestions still urging me to go on, I had difficulty in breaking with my father's wish, but I had long since lost all interest. What else could I do?

On the basis of a short, fifteen minute sketch that Dr. Smiley had given the senior class at East Denver on Harvard and what it meant to him, I made up my mind to go home to Harvard and try to find some other profession. I had little experience in any other work; my only skills were in athletics, and I had no interest in business.

Since it took some time to work out all the arrangements for my return passage, I did not actually start home until late April and did not arrive in Cambridge until the middle of May. Toward the end of my stay in Dresden, after I had left engineering, I attempted to sum up what I learned from my three years experience at the Hochschule. On the surface I certainly did not seem to have learned very much that could help me in finding a new career. However, looking back on my work at the Hochschule, I have come to realize how much I learned that has helped me in the running of my Laboratory of Psychobiological Research. Certainly, the experience gained from making the many charts and graphs used in "Mechanics" helped me in many ways in working up results in my biological studies.

On the other hand, in the laboratory, in a course in physics, we were not allowed to set up the experiments for ourselves. They were all set up by the "Dieners" before we arrived. All we had to do was to make the readings and add up the results. When I tried to set up the experiments on my own, they would have no part of it. So I really did not enjoy the course. In those days the Hochschule did not have any courses in biology of any kind. I am sure that now they have a number of courses, they may even have a biology department.

Boston and Cambridge

When my boat from Rotterdam landed in New York, I caught the first available train for Boston, where I arrived on a beautiful sunny day in May.

It gave me a wonderful feeling to get home to America after the last few months in wartime Germany. I decided just to wander around Boston before taking the subway to Cambridge. Luckily, almost at once I found a game of baseball in session on an empty lot in the Boston Commons between fifteen boys, twelve to fifteen-years-old. They had picked sides and the game was in full swing. With my love of baseball, this could not have been a greater treat for me. I stood there for almost an hour. What interested me in particular was the way the boys settled what seemed at first to be knockdown arguments about some of the plays— whether a boy was "out" or "safe" at first, for instance. Hard words flew back and forth but invariably the argument was settled and the game went on. In my three years abroad I had seen that most Europeans had not learned to settle their differences in our way—to go on with the game.

When I presented my Dresden credentials to the secretary at Harvard, I did not get a very friendly reception—for even at that early date, sentiment in Cambridge had definitely turned away from Germany. I had hoped to get credit for the three years spent at the Dresden University, but I got credit for only two years. This meant that I would have to take an extra year to graduate. After completing arrangements for my admission to the college I went home to Denver to visit my mother and friends.

My friend Ed Traylor had risen to the top of a successful brokerage firm and was obviously well on my way to his important positions in New York and Boston. On learning that I had given up all ideas of joining my father's old iron-steel firm in Denver, he begged me repeatedly to join him in his brokerage firm. Although we still were close friends, I refused, but not on the basis of having any other career in view.

My other friend from East Denver High School, Henry M. Winans, had returned to Denver for the summer holiday. He was completing his work at Stanford in preparation for studying medicine at the Johns Hopkins Medical School and had his whole career well mapped out. He talked with great enthusiasm about the Johns Hopkins University and its high ideals of research and scholarship, and I think that in a very subtle way he helped me later on to decide to go to the Johns Hopkins University myself.

Our old firm, in which my mother still had financial interest, did not in any way attract me. I did not have the least desire to become a part of the firm.

I returned to Cambridge from Denver with even firmer conviction about several occupations that I did not want to follow as a career. Clearly I had to make an entirely new search for a career.

What made it so difficult to find some new interest is that from the ages of six to twenty-one, the period of my commitment to engineering, I had not considered any variety of possibilities, the way most boys do. I had absolutely no ideas about where to start. I bought a Harvard catalog of courses and read it straight through.

International Diplomacy

As incredible as it may seem now I settled first on a course in international diplomacy given in that summer school by Professor George Graften Wilson, an ex-ambassador. It seemed to me that the great mix-up between nations that I had witnessed at close hand in Europe needed men whose mission it would be to try to bring the various nations to overcome conflict and somehow to learn to settle arguments after the fashion of the boys playing baseball in the Boston Commons. I started the course with great expectations but soon became completely disillusioned when Professor Wilson spent most of the time quoting poems from Kipling, in particular the one on "Boots, Boots, they go marching on." After a couple of weeks I gave up to concentrate on a course in French that I had also signed up for. That settled "International Diplomacy" for me.

Economics

Then when college opened in the fall, I decided to try a course in economics. Many of the students that I had met during the summer recommended an introductory course, Economics EcA, given by Professor Frank Taussig, one of our leading economists. I found however, that in that particular year Professor Taussig would not give the course; I went back to my Harvard catalog to look for a course in economics that might take the place of the introductory course and decided on a graduate course given by Professor Taussig on "John Stuart Mill, David Ricardo, and Adam Smith." Only about fifteen students signed up for the course, most of whom, I found out later, had taken the introductory course. They all seemed like very bright young men. I saw very soon that I would have difficulty in keeping up with them. Also, I found it very difficult to follow the reasoning of economics. So after about six weeks in the course, one morning I found a card from Professor Taussig in my mail box: "Mr. Richter, I do not think that economics is your forte. Please drop the course at once." That cut off one more career for me. Looking back on that situation I now realize that Professor Taussing did me a great favor, for I am sure that the struggle with economics would have completely floored me in the end.

The sequel to this experience is that many years later, Professor Taussing's daughter Helen (of Blue Baby fame) and I became good friends and laughed over what her father had done to me. In my short stay in the course I was befriended by one of the brilliant students, Jacob Viner, who turned out later to be one of the leading economists in the country and a successor to Stephen Leacock, in Toronto. In 1951 when I was spending a year at the Institute of Advanced Study in Princeton, I found that Professor Viner was head of the economics department at Princeton. I got in touch with him and we renewed our old friendship. By that time I was happily settled in an entirely different career.

European History

After my debacle with economics, I decided, possibly on the basis of my having lived in Germany for three years and having seen in 1914 most all of the crowned heads of Europe at one time in the narrow street of Dresden, that I might know enough history of Europe to be of help to someone. The young instructor in the course thought next to nothing about my knowledge of European history and would not even give me a passing mark.

A Course on the Genetics of Eye Color in the Fly Drosophila

Although I had never taken a course in biology in Denver or Dresden and actually disparaged biology in general, I decided to take a preliminary course in genetics of the fly *Drosophila*. Professor William E. Castle, a leader in genetics, gave the course, part of which consisted in observations on the effects on eye color produced by breeding flies with different eye colors.

Somehow the lack of all biological knowledge made it impossible for me, no matter how hard I tried to become interested in breeding experiments with these flies. When after a weekend spent in New York City I returned to find all my flies had died, I knew that I was in trouble. Professor Castle asked me to drop the course. Many years later—in 1954—at a symposium on twenty-five years of progress in genetics and cancer held in Bar Harbor, Dr. Castle and I were members of the panel of six speakers. I reported results of some of my studies on the effects of domestication and selection on the behavior of the Norway rat. Dr. Castle spoke well of my paper and I never reminded him of our earlier meeting many years before. Since then of course I have become very interested in genetics.

About this time I joined the R.O.T.C. I was promptly made a first lieutenant, possibly because of the straight posture my father had insisted on in my early years.

Lectures on the Philosophy of Nature

I did not select these lectures, given by Professor E. B. Holt, on my own. I had not even heard of Holt until some students told me that he gave very interesting and exciting lectures. I sat in on a few lectures, then signed up for the course. I knew at once that what he was talking about interested me. A recent biography of Walter Lippmann states that Lippmann found Holt one of two of the most stimulating lecturers in the college. In looking back, I know that these lectures did several important things for me. In the first place they introduced me to Freud and psychoanalysis. In quick order I read many of the writings of Freud, Ferenczi, Jones, Brill, Adler, and others. These men seemed to open up an entirely new world for me. At the same time, I read some of Havelock Ellis's books. Holt's lecture also mentioned conditioned reflexes, which I had not heard of. The lectures stimulated so much interest that the students, sometimes twenty or more, stayed for a long time after the end of the lecture arguing over his ideas.

Holt's lectures also stimulated me to sign up for a course in psychopathology given by Professor Southard, then director of Psychopathic Hospital at Harvard. He took us on several trips to visit patients in state mental hospitals. These trips greatly extended my knowledge of what can happen to man. Southard took little stock in psychoanalysis and spent most of his lecture hours trying to demolish Freud.

Most important for me is that Holt's lectures stimulated me to sign up for a short experimental course on insect behavior, given by Professor Robert Yerkes. This course, though short, made me feel that at last I had found something that really interested me. Further, Professor Yerkes thought that I had done quite well and gave me an A in the course, the only A that I ever managed to get during my two years at Harvard, or for that matter, elsewhere.

Yerkes recommended that I read a book, *Animal Behavior* by John B. Watson, professor of psychology at the Johns Hopkins University. After having read only snatches here and there, I became convinced that I should try to go to work with him, if not right then, then at a later date. I must point out here that at that time, and for a long time later, I had no idea about what Watson meant by "behaviorism". For me "behaviorism" simply meant "behavioural."

My important problem at this point concerned my graduation from college. At the end of my first year and a half in college, my record did not look very encouraging, so the authorities put me on probation. At the end of the second year, the year for my graduation, possibly as a result of my A from Professor Yerkes and my active part in R.O.T.C. work, the authorities let me graduate.

On the whole, I had a very interesting but not very happy time in Cambridge. I think I was in a mild depression much of the time. Coming in as an unclassified student, I did not have any classmates, I did not have any friends from Denver, and my association with Germany did not help the situation.

However, on several occasions I had the good fortune to be invited to have lunch with L. J. Henderson, professor of biochemistry and W. M. Wheeler, professor of biology and a great authority on ants. These conversations had a great and lasting effect on me, though I never had courses with either of these professors.

In addition, during my two years at Cambridge I had managed to keep up with my sports by leaving an hour free from classes in the late mornings to row on the Charles River in a single shell. This managed to keep me in fairly good shape.

I had also, during my stay at Cambridge, become an active reader of *The Nation, The New Republic* and *The Masses,* as well as the books of H. G. Wells and George Bernard Shaw.

Little did I suspect that in another ten years, "biology" would take over my life practically to the exclusion of everything else.

Service in Army, 1917–18

After my graduation from Harvard, I went to the Second Officers Training

Camp at Plattsburg and received my commission about three months later. I
served six to eight months in a heavy artillery batallion on Fort Andrews, an island
in Boston Harbor, during a cold winter when the harbor remained frozen for many
months.

Nothing much of interest happened except that my appendix ruptured during a
stay in Boston. A severe septicemia kept me in bed with drains in a small hospital
in Cambridge for eight weeks, but I finally made a full recovery.

In the spring, a visit from Professor L. J. Henderson of Harvard and his wife
constituted the high point of interest during my stay on the island. He brought me
a number of books.

From the island the army transferred me to Boston where for several months I
was in charge every other day of the military police of Boston. This meant that on
alternate days I had to inspect, at irregular intervals, all of the sentinels on duty
throughout the main part of the city.

Periodically, the men on the island were transferred to Camp Eustis, Virginia, to
await shipment to Europe. It was there that the end of the war found me.

Johns Hopkins, 1919

After I was mustered out of the service, I began to think about my next venture—the move to Baltimore to work on my doctoral thesis with Professor
John B. Watson. I realized that I knew very little either about Professor Watson or
about The Johns Hopkins University. Professor Yerkes at Harvard had urged me
to go to work with Watson and my high school friend in Denver, Henry N. Winans,
had spoken frequently about the great research interest at The Johns Hopkins
University.

I did not know that Hopkins was a very young university, started in 1878; that
its first president, Daniel Coit Gilman had inspired many very capable young men
to join the academic and medical parts of the university and had inspired those
men with high ideas of research and, particularly, of the freedom of research.
Gilman's inaugural address had made a great impression on these men who
formed the first members of the staff.

I found Dr. Watson's laboratory on the third floor of the Phipps Psychiatric
Clinic, one of the main buildings of The Johns Hopkins Hospital. Little did I think
that morning, while walking up the three flights of stairs to Dr. Watson's laboratory, that I would still be walking up those same three flights of stairs every
morning sixty-three years later and that from then on the Johns Hopkins Hospital
and Medical School would become my permanent base. Watson's laboratory
looked very new and unused, and I later learned that this was because he had
moved over to the hospital from the academic part of the Hopkins at Homewood
only a short time before and had not yet had a chance to use it.

After exchanging a few pleasantries, he assigned me to one of the rooms in his
laboratory and told me where I could find lodging and meals with medical students.

Then came what was to be the great turning point in my life and the start of my scientific career. Watson said, "I want you know that I am only interested in getting a good piece of research. You do not have to take any courses or attend any lectures. You are strictly on your own." From that moment on I knew that I had made the right choice in coming to The Hopkins, and I know now that Watson's remark was pure vintage Gilman.

This clearly was the great event of my life. Suddenly for the first time I knew what I wanted to do and was set to do something about it. What a wonderful feeling! I had real freedom of research. I could now go directly to what I wanted to do without getting entangled in all the academic red tape. How to explain this great change in me? I have thought of various possible explanations and have concluded that Watson's "freedom of research" announcement had released my innate gene and that this had resulted in an overwhelming change in my whole being—freeing my initiative and self-confidence and restoring or renewing my early enthusiasm.

My freed gene could also have brought out specific characteristics that I had inherited from my father, that is, an interest in experimenting, using his hands and head, and readiness to deal with new adventures as he had to do when he left his native bureaucratic Saxony to settle in the open spaces of free America.

This burst of energy got me started very quickly, setting up experiments of various kinds that resulted, over the next sixty years or more, in the publication of more than 375 papers, a series of lectures, and a book. This stood in sharp contrast to the long years in Denver, Dresden, and Cambridge in which I had not even written a line.

Watson's part in this remarkable transformation of my life needs some comment. He released my gene for freedom of research. He did this as a member of President Gilman's young staff, clearly not as a "behaviorist" who did not believe in heredity.

After our first meeting, Watson took me to another part of the Phipps Clinic to introduce me to Professor Adolf Meyer, the director of the Phipps Psychiatric Clinic and one of the top psychiatrists in this country. Dr. Meyer had invited Dr. Watson to shift his laboratory from Homewood to the Phipps Clinic. After Watson left us, Dr. Meyer and I talked for some time about our common interest in the study of behavior and the opportunities offered by the Phipps Clinic for Behavior Studies. I quickly realized that he and Watson had much the same ideas as Gilman did about giving full freedom to research workers. He offered to let me attend any of his lectures that might interest me or attend his ward rounds on patients. This was the start of a wonderful friendship that lasted through ups and downs until his retirement in the forties. Watson left the university about a year and a half after I got there. Not long after I got my degree, Dr. Meyer put me in charge of Watson's lab and until his retirement he supported all of my many projects without question.

Fortunately, during our first conversation, Watson had not asked me what I planned to work on for my thesis, for at that moment I did not have the least idea. In fact, I had come without even having given it any thought at all, for I had not expected to be put on my own at such very short notice, if at all.

After a week or more of browsing around the hospital and medical school, I found a cage with twelve rats in my room. I was not sure whether Watson had sent them to give me some ideas about what I could do for a thesis, or simply to give me some company while I was making up my mind.

Watson did not know that I had not even handled a white rat, in fact, that I had not even seen a white rat close-up. I had never worked where rats were kept. None of my friends in Denver had ever had white rats as pets and none of my courses at the Hoschschule in Dresden or at Harvard College had dealt with white rats.

Now, after a sixty-year period, during which I handled many thousands of rats and operated on several thousands, I find it difficult to recall the fear and anxiety that I experienced in trying to get myself to handle the rats that Watson had sent.

However, from a cursory observation of the rats, I decided that I might work on their spontaneous activity—what makes them active. This spontaneous activity definitely fascinated me and appeared to be one possible problem to start with.

Then I considered a second possibility—the grasp reflex. Watson was making observations almost daily on the grasp reflex of newborn babies. It fascinated me to see a baby hanging by one hand, entirely unsupported, from a pencil or a rod. I could think of many different studies that I might make with this phenomenon, but this was Watson's research, and so it was out of the question for me, at least for the time being. Later on, after Watson's departure, I published a paper on the "Natural History of the Grasp Reflex from the Cradle to the Grave," and about twenty additional papers.

Only a few days after my arrival at the Hopkins, Professor Meyer had told me that some years before he had purchased one of the most recent models of the Hindle String Galvonometer for his assistant Dr. Stanley Cobb, who was going to carry on studies of muscular tension, but that Cobb had left to accept a professorship at Harvard before he could do anything with it. Dr. Meyer wanted to know whether I could use the instrument. Not knowing a thing about galvonometers, but definitely interested, I told him that I would like to try.

The galvanometer turned out to be a very impressive looking instrument about twelve feet long, with an enormous magnet, a lantern and camera, and all kinds of other equipment. Dr. Meyer had it sent to my room and I started to work with it almost at once. Before long I was well acquainted with all its parts, particularly with the almost imperceptible fine "string" that had to be suspended between the walls of the magnet. It became clear to me that it had great possibilities for what I would like to do sometime, but not just then. This ruled out the use of this instrument for an immediate research project, but later on I used it for well over one hundred papers on the functioning of the sympathetic nervous system, in normals and various types of psychiatric patients as well as on many monkeys and cats.

Thus I was left with Watson's rats—and the problem of the nature and origin of spontaneous activity.

Well, I did finally get myself to handle rats. I put them in separate cages. During this getting acquainted period, I became even more impressed with their spontaneous activity, jumping around and climbing the walls of the cages. Before long it occurred to me that it might be a good idea to find out whether this spontaneous activity is constant or irregular; and what causes it. I decided to construct a simple cage for obtaining continuous records on even the slightest movements of the rats. Here my "play experience" came in handy.

Stationary Cage—Tambours and Smoked Drums

With my proposed experiment in mind, I collected a few pieces of equipment that I found in Watson's laboratory and then built some cages out of wire cloth and sheet aluminum purchased at a shop in town. In the laboratory I found rubber tubing and glazed kymograph paper for smoking and shellacking. I learned how to smoke glazed paper, and how to get a record on the kymograph. The cage rested on tambours covered with rubber dam and connected by rubber tubing to a small recording tambour. Even the slightest movement of the rat in the cage resulted in a mark on the smoked paper. A chronometer registered time on the chart.

Each morning before putting a rat in the recording cage, I gave it a dish of bread and milk to satisfy its hunger. In the recording cage it had access only to a makeshift inverted water bottle attached to the cage. Then I recorded its activity for the next ten to twelve hours.

In less than four weeks after my conversation with Watson, this experiment demonstrated that spontaneous activity of the starved rat occurred in regularly recurring periods of activity and total inactivity at intervals of about ninety minutes. After each totally inactive period, the rat became active again, at first in very small amounts, then in a great burst that lasted until the rat suddenly became totally inactive again. The finding of the periodicity—the existence of these recurring periods of activity and inactivity gave me great pleasure and started a lifetime interest in periodic phenomena.

In the following experiment, I decided to determine the relation of feeding times to these spontaneous activity periods. I designed and constructed a different type of cage made up of two independent parts: one for recording only gross bodily spontaneous activity, the other containing a food cup for registering feeding periods. The two cages rested on separate sets of tambours which recorded separately on a smoked drum. This setup made it possible to record activity separately in the main cage, and in the food cup cage.

In this experiment I could not use bread and milk which rats could carry back into the main cage, and so I replaced that food with a finely ground mixture prepared according to a formula provided by Professor E. V. McCollum, (then one of our leading nutritionists). I used this diet for all of our experiments not only on rats, but, on hamsters, gerbils, chipmunks, desert rats, and ground squirrels, from then right up to the present time. I must mention here in passing that the contact made at this early stage with Professor E. V. McCollum blossomed into a friendship that lasted throughout his life. During his active career, he read and criticized almost all of my papers dealing with nutrition. In our later years both of us "Westerners" became interested in archeology and spent many happy hours together discussing various problems in that field. I inherited his collection of books on archeology.

In the double-cage setup just described, an ordinary tin cup contained the McCollum diet. Even the slightest touch of the powdered food by the rat registered on the smoked drum, which was, of course, completely separate from the drum measuring activity in the main cage.

I found that with food available all the time, intervals between active periods became longer, averaging 3.6 hours. Just as in the previous experiments, active periods were separated by sharply defined inactive intervals. Activity in the living part of the cage started very slowly, then increased quite rapidly until it reached a high level when suddenly the rat shifted to the food compartment and began eating. It ate for twenty to twenty-five minutes, then returned to the living compartment where it soon became totally inactive again. Clearly before the rat entered the feeding cage, it became progressively more active, but it never stopped to sample food in the food cup until the time of the single entrance to the area of the food cup.

In the early twenties, after I had demonstrated the periodic nature of the spontaneous activity of the rat, I did a lot of exploring with different ideas about behavior of the rat and also, trying out different kinds of cages for carrying on experiments. The most interesting experiment I did at that time demonstrated that the rat eats for calories. I did that in this way; I kept rats on 8 or 16 percent solution of alcohol as the only source of their water intake. I found to my great interest that the rats reduced their food intake in each instance in direct proportion to the calories received from their alcohol. This meant of course that with all concentrations of alcohol the total caloric intake remained the same. Later I was able to confirm this observation in many other experiments. This indicated to me that the rat might regulate intake of individual components of its diet.

This observation encouraged me then to go on to do more experiments on rats in a more systematic way. I thought at first that I might make some observations on the Norway rat in the wild state. I gave up this idea very quickly after considering that wild rats spend most of their life in burrows where they cannot be seen and where no measurements or studies can be made on their behavior.

I decided then to carry out experiments in the laboratory where the animals could be followed throughout the entire day and night under controlled conditions. I decided to build a small cage in the laboratory in which I could record all of the important parts of the rat's behavior and in which the rat could live through a normal lifespan and remain healthy and active, but would still be constantly accessible for inspection and handling, for weighing, and so on.

I built various types of cages in which each one of the rat's activities could be measured and recorded. Such a cage would have to keep a record of the daily food intake, water intake, and running activity in the revolving drum under conditions in which the rat could be observed throughout the twenty-four hours.

I decided to incorporate everything that I had learned during my previous three years and to build a large battery of such cages that would be kept in a fairly small space. The important part was, of course, to know what materials to use for the cages to make them firm and easy to handle and easy to keep clean.

I finally settled on a model that has required almost no change throughout the following sixty years. After about thirty years, I did add a few new features on the cages but their essential parts have remained the same.

It is important here to give a full description of the cages, since they were used exclusively in all of our experiments. We decided to build 80 cages to start with, 5 stands, each with 16 cages. The stands would be all metal and easily moved around in an ordinary sized room. Each cage consisted of two parts separated by a strong metal partition. The partition had a 3-inch opening that connected the two parts. The living compartment contained a food cup and a graduated inverted water bottle; on the other side of the partition, a revolving drum was built with the best type of bicycle hub and an axle that extended through the central partition over the living compartment and connected there with a cyclometer that recorded revolutions of the drum in either direction.

An eccentric hub and microswitch on the distant end of the axle of the drum made it possible to register single revolutions of the drum in either direction. The individual revolutions then registered on an Esterline-Angus recorder in an adjoining room. Wiring of the water bottles made it possible to register every lick that the animal made in drinking. These drinking times were recorded on the Esterline-Angus recorder. A simple device on the food cup registered eating times. Every pawful of food that the rat removed from the food cup made a mark on the recorder.

Daily records of the total number of revolutions on the drum in the cage made is possible to follow the presence of the four- to five-day cycles.

Each rat had a chart—on coordinated paper, with 10 spaces to the inch—showing daily total revolutions of the drum, intake of food in grams, intake of water in milliliters. Each rat's record showed twenty-four-hour photographic Esterline-Angus records of running times, eating times, and drinking times.

I kept the stands of rats in a large room under standard controlled conditions. The lights were turned on at 6:00 A.M. and off at 6:00 P.M. The records of water intake, food intake, and running activity were taken daily from 8:00 to 9:A.M.. Vaginal smears were taken daily on each female. All of the rats were weighed every ten days.

The Esterline-Angus records were serviced daily; the Esterline-Angus sheets were cut into small strips, one for each rat, and pasted on large ruled charts weekly. At the end of the week, the large charts were photographed and reduced to a small size for filing with the rest of the rat's records.

All activity, food, and water intake records were plotted on Friday. This meant that records of all types were ready for viewing Saturday morning.

Each rat had its own number and corresponding ear clips.

After completion of experiments, charts for each rat were indexed, filed, and registered in our library of experiments. This made it possible to retrieve charts for each rat used in special experiments at any time.

The records for each rat gave a full account of the animal's total behavior in the cages, throughout its entire stay in the lab.

A variety of other types of equipment served as accessories to the 80 main activity cages. They served to subject the rats, temporarily removed from the 80 cages, to various short forms of controlled stress.

1. Swimming tanks in which rats could be forced to swim in water of controlled temperatures for fixed periods of time.
2. Fighting chambers in which rats (mainly wild rats) were stimulated by electrical shocks to fight one another for fixed periods.
3. The same cages used with only one rat stimulated by an electrical current.
4. Equipment for administering electroshock to the brain with the same instrument used on human patients.
5. Large closed boxes for observing pairs of rats removed from the stands for mating trials.

Some ethologists apparently have a prejudice against the use of cages for the study of animal behavior, largely, undoubtedly, because of the use of cages for conditioning experiments. Cages described here have served repeatedly for the study of innate behavior of rats, particularly on problems concerned with the twenty-four-hour clock. The use of our cages makes it possible to put very definite questions to the rats, and to get definite answers. It is important, of course, to put questions in a form that the rats can give a definite answer to and that questions must involve survival of the gene.

Summary of Observations on My Gene

1. Watson's freedom of research instructions released my gene to start me on an active career in the study of animal behavior and on clearly related biological areas.
2. For long years I knew very definitely what I did not want to do in life; but I had no idea what I actually wanted to do. For a long time my gene must have somehow withheld the secret.
3. I never knew until the gene was released, what an overwhelmingly great effect it had in starting me to work.
4. Without release of my gene I might have had to go on throughout the rest of my life without ever developing any kind of special behavior.
5. During the first twenty-five years of my life, gene activity did not show itself except possibly in the "play" or "learning" period during my first nine years of life.
6. My first acceptance of an "animal behavior" career did not depend in any way on "behavioristic mechanisms."
7. My gene might have manifested itself earlier in my life if my father had not died during my eighth year, leaving me to fall in with conventional school, social activities, and sports.
8. Existence of my gene probably was disclosed by presence of the "play."
9. Release of the gene had overwhelmingly emotional effects on my whole life. The effects probably resembled in some ways those produced by starting life in a new country.

10. President Gilman, by his strong belief in freedom of research, was undoubtedly responsibe for release of my gene.
11. My gene must have had much in common with that of my father, who sought freedom from his native bureaucratic Saxony.
12. Early manifestations of effects of my gene: the decision on graduating from high school (a) to give up all political activities; (b) not to accept very attractive financial offers of partnership in his brokerage firm by my friend Ed Traylor.
13. There is probably real irony in the fact that Professor Watson, the great advocate of behaviorism and the negator of all hereditary beliefs, should be the one who released my gene, by entirely hereditary mechanisms.
14. How can we account for my inability before release of the gene, to tell myself or any one else what it was that I wanted to do in life. Does the gene have some inner control over the inner workings of the mind?
15. Could a release of Darwin's gene, after his struggle with medicine and religion, account for the sudden appearance of his wide range of mental activities after joining the *Beagle?*

Research

My doctoral thesis, completed in 1922, was my first publication in animal behavior and had the title, "Behavioristic Study of the Spontaneous Activity of the Rat." From the beginning of my studies on animal behavior in Watson's lab, my interest focused entirely on what animals do on their own, that is, their innate behavior, not on what they can be taught to do. *Behavioristic* was thus most definitely a misnomer (I am happy not to have had to make this correction during Watson's lifetime, for I held him in great affection and esteem). On the following pages, I will describe (in general terms) some of my more important research, and provide references for those who wish to read further.

Norway Rat's Qualification for Behavior Studies

At this point I should like to call attention to the advantages that the rat offers for behavior studies as well as for scientific studies in general.
1. Its dietary needs are very nearly the same as man's. This explains how observations made on it can readily be extrapolated to man.
2. It is a very stable and reliable animal.
3. It reproduces readily.
4. It is just the right size for all kinds of anatomical and physiological studies.
5. Its short lifespan of about three years makes it possible to study factors involved in growth, development, and aging.
6. It is a very clean animal. Given the opportunity, it will keep itself perfectly clean.

7. It is well equipped with tools for manipulation of it's environment: with very sharp and strong teeth, very dexterous paws and feet, and strong hind legs.
8. It is well equipped with sensory organs: for tasting, smelling, seeing, hearing, and tactile orienting (whiskers and tail).
9. It is available in large numbers, in both the domesticated and wild forms.

Studies on Cycles Undertaken on the 80 All-New Metal Activity Cages

After receipt of these new cages in 1925, I started to work almost at once on the study of cycles that appear normally in rats and other animals. Next, I became interested in bringing out cycles by various forms of experimental interference, for instance, by inducing changes in the endocrine glands or creating brain lesions. We have continued these studies almost uninterruptedly right up to the present time and have used not only rats, but hamsters, desert rats, chipmunks, ground squirrels, gerbils, and some other small animals.

Of interest here is that these animals thrive on the regular McCollum diet that we used for rats. Furthermore, our food cups prevented them from spilling or scattering food. This made it possible to get very accurate food-intake records. This was important to me because I believe that studies of any kind should start, first of all, with good records of food intake and water intake.

The studies constituted the start of a lifetime interest in cyclic periodic phenomena: (1) of cycles present in normal rats and rats with various types of experimental lesions; (2) of cycles present in other small animals after various forms of experimental interference; (3) of cycles present in normal and abnormal patients in mental hospitals. I also undertook a series of studies on the twenty-four-hour clock of rats and other small animals including squirrel monkeys after treatment with heavy water, which has a remarkable effect on the twenty-four-hour clock.

The following bibliography lists some of my studies on cycles or timing devices.

Richter, C. P. 1922. A behavioristic study of the activity of the rat. *Comp. Psychol. Mongr.* 1:1–55.

———. 1927. Animal behavior and internal drives. *Quart. Rev. Biol.* 2:307–43.

———. 1930. Biological approach to manic depressive insanity. *Proc. Assoc. Res. Nerv. & Ment. Dis.* 11.

———. 1938. Two day cycles of alternating good and bad behavior in psychotic patients. *Archives of Neurology and Psychiatry* 39:587–98.

———. 1953. Behavior cycles in man and animals. *National Academy of Sciences Meeting*, 27–29 April, 1968, Washington, D. C.

———. 1955. Behavior and metabolic cycles in animals and man. Proc 45th Ann. Mtg. of the Amer. Psychopathal. Assoc. In *Experimental Psychopathology,* ed. P. H. Hock and J. Lubin. New York: Grune & Stratton.

———. 1955. Experimental production of cycles in behavior and physiology in animals. *Acta Medi. Scandina.* (supplement 307, 35). International Society of the Study of Biological Rhythms, September 1953, Basel Switzerland.

———. 1957. Hormones and rhythms in man and animals. *Recent Prog. in Hormone Research* 13:105–59.

———. 1958. Diurnal cycles in man and animals. *Science* 128:1147–48.

———. 1958. Neurological basis of responses to stress. A Ciba Foundation Symposium of the "Neurological Basis of Behavior." In *Commemoration of Sir Charles Sherrington,* ed. G. E. W. Wolstenholme and Celilia M. O'Connor, pp. 204–17. London: J. A. Churchill, Ltd.

———. 1960. Biological clocks in medicine and psychiatry: Shock-pause hypothesis. *Proc. Nat. Acad. Sci.* 46:1506–29.

———. 1964. Biological clocks and the endocrine glands. *Proceedings of the Second International Congress of Endocrinology,* 16–21 August, 1964, London.

———. 1967. Sleep and activity: Their relation the the 24-hour clock. In *Sleep and Altered States of Consciousness.* Assoc. for Research in Nervous and Mental Disorders, vol. 45. Baltimore, Md.: Williams & Wilkens.

———. 1968. Clock mechanisms esotropia in children. Alternate day squint. *Johns Hopkins Medical Journal* 122:218–23.

———. 1968. Inborn nature of the rat's 24-hour clock. *Proc. Nat. Acad. Sci.* 61:1153–54.

———. 1968. Inherent twenty-four hour and lunar clocks of a primate—the squirrel monkey. *Comm. Behav. Biol.* (part A) 1(5):305–32.

———. 1971. Inborn nature of the rat's 24-hour clock. *J. Comp. Physiol. Psychol.* 75:1–4.

———. 1977. Discovery of fire by man—its effects on his 24-hour clock and intellectual and cultural evolution. *Johns Hopkins Med. J.* 141:47–61.

———. 1977. Heavy water as a tool for study of the forces that control length of period of the 24-hour clock of the hamster. *Proc. Nat. Acad. Sci.* 74:1295–99.

———. 1978. "Dark-active" rat transformed into "light active" rat by destruction of the 24-hour clock and synchronizers. *Proc. Nat. Acad. Sci. USA,* 75:6276–80.

———. 1978. Evidence for existence of yearly clock in a surgically and self-blinded chipmunks. *Proc. Nat. Acad. Sci. USA* 75:3517–21.

———. 1979. *Biological Clocks in Medicine and Psychiatry.* Springfield, Ill.: Charles C. Thomas.

———. 1980. Growth hormone 3.6-h pulsatile secretion and feeding times have similar periods in rats. *Am. J. Physiol.* 239 (Endoc. Metab. 2): E1–E2.

Richter, C. P., and Rice, K.K. 1956. Experimental production in rats of abnormal cycles in behavior and metabolism. *J. Nerv. Ment. Dis.* 124:393–95.

Richter, C. P., Honeyman, W., and Hunter, H. 1940. Behavior and mood cycles apparently related to parathyroid deficiency. *J. Neurol. Psychiat.* 3:19–25.

Richter, C. P., Jones, G. S., and Biswanter, L. T. 1959. Periodic phenomena and the thyroid. *Arch. Neurol. Psychiat.* 81:233–55.

Neuro-Endocrine Study of Levels of Spontaneous Activity

We were able to produce not only cycles but levels of hypo- or hyperactivity in rats and other animals by many different forms of interference with the endocrine glands by lesions made in different parts of the brain. For these experiments we used not only rats, but monkeys and cats.

Along with these studies on the animals in the new activity cages, I continued to collect from the literature case histories of all patients, medical or psychiatric, who showed regular cycles or changes over long periods, or great changes in activity levels. This collection now numbers well over 1800 cases.

The following list contains a few of our publications on production of energy levels of activity by various forms of interference.

Langworthy, O. R., and Richter, C. P. 1939. Increased spontaneous activity produced by frontal lobe lesions in cats. *Am. J. Physiol.* 126:158–61.

Richter, C. P. 1933. The role played by the thyroid gland in the production of gross body activity. *Endocrinol.* 17:73–87.

———. 1935. The effect of early gonadectomy on the gross body lesions of rats. Second Internat. *Neurol. Congress.*

Richter, C. P., and Eckert, J. F. 1936. Behavior changes produced in the rat by hypophysectomy. *Proc. Assoc. Res. Nerv. Ment. Dis.*, 17:561–71.

Richter, C. P., and Hawkes, C. D. 1939. Increased spontaneous activity and food intake produced in rats by removal of the frontal pole of the brain. *J. Neurol. Psychiat.* 2:231–42.

Richter, C. P., and Hines, M. 1937. Increased general activity produced by prefrontal striatal lesions in monkeys. *Trans. Am. Neurol. Assoc.*, 63rd Meeting, pp. 107–9.

———. 1938. Increased activity produced in monkeys by brain lesions. *Brain*, 61:1–16.

Richter, C. P., and Wang, G. H. 1926. New apparatus for measuring the spontaneous motility of animals. *J. Lab. Clin. Med.*, 12:289–92.

Surgical Operations

Soon after release of my gene and after I had gotten started on the study of the part played by neuroendocrine factors in the control of spontaneous activity, I began to learn various surgical techniques, especially those concerned with the removal of the endocrine glands.

I began with removal of the ovaries and testes, the adrenals, the parathyroids, and ended with the pituitary. Then I started on removal of the different parts of the brain and nervous system. Clearly, I had some native surgical talent for operating on these small animals, for I did very good operations almost from the very first time. Experience from my early "play" period may have helped. Over the years I worked out more than forty operations, chiefly on the rat. The artist, Mr. P. J. Malone, who made many beautiful drawings of the operations, and I hope to publish a volume of these drawings before long. Later I did many operations on monkeys, hamsters, chipmunks, and other small animals.

Thus, surgical operations undertaken as part of the study of animal behavior constituted one of my lifelong projects. We paid close attention to postoperative care. I do not remember that any of the thousands of animals operated on at any time showed or developed an infection after an operation.

Behavioral Regulation of a Constant Internal Environment by Maintenance of a Constant Caloric Intake after Forced Alcohol Ingestion—Rats Eat for Calories

Richter, C. P. 1926. A study of the effect of modern doses of alcohol on the growth and behavior of the rat. *J. Exper. Zool.* 44:397–418.

———. 1941. Alcohol as a food. *Quart. J. Studies on Alcohol.* 1:650–62.

Experiments on Homeostasis

When I first heard that adrenalectomized rats lose salt and as a result die within ten or more days, I decided to give these rats access to salt solution to see whether they would on their own, take enough salt to keep themselves alive and free from symptoms of insufficiency.

These tests showed that the adrenalized rats will take adequate amounts of salt to keep themselves alive indefinitely. They will take not only sodium chloride, but sodium in a number of different compounds. These studies are really still in progress, since I have not been able to satisfy myself with this simple type of explanation.

When I heard that parathyrodectomized rats lose calcium and die after a fairly short time, I decided to give them access to calcium solutions as calcium carbonate, and found that they too will take enough calcium to keep themselves alive and free from symptoms of tetany. They will show an increased intake not only for all calcium solutions, but also for solutions of compounds atomically related chemicals, such as strontium. They showed an aversion to phosphorus, which agrees with the finding that parathydectomized rats have an excess of phosphorus. Many other examples have been found.

The following list contains references to papers that we have published on adrenalectomized rats and parathyrodectomized rats.

Adrenalectomy and Pregnancy

Richter, C. P. 1936a. Increased salt appetite in adrenalectomized rats. *Am. J. Physiol.*, 115:155–61.

———. 1936b. Salt appetite of mammals: Its dependence on instinct and metabolism. Contribution to vol. "L" instinct dans le comportement des animaus et de l'homme. Paris, France.

Richter, C. P., and Barelare, B. Jr. 1938. Increased sodium chloride appetite in pregnant rats. *Am. J. Physiol.* 121:185–88.

Wilkens, L., and Richter, C. P. 1940. A great craving for salt by a child with cortico-adrenal insufficiency. *J.A.M.A.* 114:866–68.

Parathyroidectomy and Calcium Appetite

Richter, C. P., and Eckert, J. F. 1937. Increased calcium appetite of parathyroidectomized rats. *Endocrinol.* 21:50–54.

———. 1939. Mineral appetite of parathyroidectomized rats. Increased appetite for calcium solutions (lactate, acetate, glucinate and nitrate: also for strontium and magnesium solutions). *Am. J. Med. Sci.*, 198:9–16.

Richter, C. P., and Helfrick, S. 1943. Decreased phosphorus appetite of parathyroidectomized rats. *Endocrinol* 33:349.

Single-Food Choice Diets: Single Purified and Whole Foods

Some of our most interesting experiments were obtained by restricting the intake of rats to one or another of the six common sugars. We found that the amount of each sugar taken by the rats agreed almost exactly with the limits of their assimilability as determined by biochemists.

The results of these experiments have definitely thrown a new light also on the mechanisms involved in the maintenance of homeostasis of adrenalectomy or parathyroidectomy.

The following bibliography contains a few references to papers that we have published on the single-food choice sugar experiments and on self-selection experiments.

Richter, C. P. 1941. The nutritional value of some common carbohydrates, fats and proteins studied by the single food choice method. *Am. J. Physiol.* 133:29–42.

———. 1942. Increased dextrose appetite of normal rats treated with insulin. *Am. J. Physiol.* 135:781–87.

———. 1942. Total self-regulatory functions in animals and human beings. *Harvey Lecture Series* 38:63–103.

———. 1977. Six common sugars as tools for the study of appetite for sugar. In *Taste and Development: The Genesis of Sweet Preference,* ed. James M. Weiffenbach. National Institute of Dental Research, DHEW Publication No. (NIH) 77-1068, U.S. Department of Health, Education and Welfare, National Institute of Health, Maryland.

Richter, C. P., and Barelare, B., Jr. 1939. Further observations on the carbohydrate, fat, and protein appetite of vitamin B deficient rats. *Am. J. Physiol.,* 127:199–210.

Richter, C. P., and Campbell, K. H. 1940. Taste thresholds and taste preferences of rats for five common sugars. *J. Nutrit.* 20:31–46.

Richter, C. P., and Hawkes, C. D. 1941. The dependence of the carbohydrate, fat, and protein appetite of rats on the various components of the vitamin B Complex. *Am. J. Physiol.* 131:639–49.

Richter, C. P., and Rice, K. K. 1942. The effect of thiamine hydrochloride on the energy value of dextrose studied in rats by single food choice method. *Am. J. Physiol.* 137:573–81.

———. 1944. Comparison of the nutritive value of dextrose and casein and of the effects produced on their utilization by thiamine. *Am. J. Physiol.* 141:346–53.

Self-Selection of Single, Natural or Chemically Purified Foodstuffs

In an intensive set of experiments, we gave rats under all conditions access to natural foods or purified foods in separate containers and let them work out their own diet.

Results of these experiments showed that the rat is able to make beneficial choices from a great variety of natural as well as highly chemically purified foods.

It must be remembered that in all of these situations in our cages, the rat must make selections for survival. That is one of the most important uses of these cages and setups that we have. For that reason the rats make very reliable selections that make it possible for them to live and reproduce. We have tested this method on normal rats with all kinds of natural and highly purified foods. Invariably the rats make very good selections. In several instances the rats had been offered twenty to twenty-three different substances, all in purified form, with the exception of a small container of cod liver oil. The various substances were all offered in separate containers, either as fluids or as solids. The rats made very good selections as shown by the fact that one rat not only survived but mated and gave birth to babies. However, it became clear that something that the collection of substances left out was necessary for the maintenance of nursing. Although the animal gave birth to babies, it was not able to nurse them to full growth. The experiment showed that the twenty-one or twenty-two substances that were offered to the rat were almost complete, but lacked one or two important substances.

It appears from our experiences, that rats on their own will make the best selection possible for survival if they possibly can. Such studies would be absolutely impossible outside of our cages, or in the wild states.

The following list is a collection of a few of the papers that have been published on self-selection of diets in rats and in other animals.

Richter, C. P. 1927. Animal behavior and internal drives. *Q. Rev. Biol.* 2:307–43.

———. 1941a. Behavior and endocrine regulators of the internal environment. *Endocrinol* 28:193–95.

———. 1941b. Biology of drives. *Psychosomatic Medicine* 3:105–10.

———. 1942–43. Total self regulatory functions in animals and human beings. *Harvey Lecture Series* 38:63–103.

———. 1953. The self-selection of diets. *Essays in Biology.* In honor of Herbert M. Evans. Berkeley, Calif.: University of California Press.

———. 1954. Biology of drives. J. Comp. Physiol. Psychol. 40:129–34.

Richter, C. P., and Barelare, B., Jr. 1939. Further observation on the carbohydrate, fat, and protein appetite of vitamin B deficient rats. *Am. J. Physiol.* 127:199–210.

Richter, C. P., and Eckert, J. F. 1938. Mineral metabolism of adrenalectomized rats studied by the appetite method. *Endocrinol.* 22:214–24.

Richter, C. P., and Rice, K. K. 1943. Depressive effects produced on appetite and activity of rats by an exclusive diet of yellow or white corn and their correction by cod liver oil. *Amer. J. Physiol.* 139:147–54.

———. 1945. Self-selection studies on coprophagy as a source of vitamin B complex. *Am. J. Physiol.* 143:344–54.

Richter, C. P., and Schmidt, E. C. H., Jr. 1939. Behavior and anatomical changes reproduced in rats by pancreatectomy. *Endocrinol.* 25:698–706.

Poisons: Observations on Various Thiourea Compounds

I undertook a series of experiments on the ability of rats to avoid ingestion of toxic poisons substances. I did this because I found that a chemical, phenyl thiourea or carbamide that had been commonly used for taste tests in man, turned out to be highly toxic to rats. It occurred to me that this compound might be useful as a rat poison.

It turned out however, that phenyl thiourea has a definite taste and would be detected by the rats; it thus would not be useful as a rat poison.

It then occurred to me that if we could get a thiourea compound that does not have a taste, it might serve as a very good poison for rats. This idea was followed up extensively over several years and compounds were obtained that had practically no taste, but still retained their high toxicity. This compound was called Alpha-Naphthyl Thiourea, or ANTU. It served as a very good poison for rats, provided that it killed them after their first meal. When the rats did not die after the first ingestion of the poison or refused to take the poison after a second exposure, their detection of a faint odor of the ANTU on their first exposure, gave them warning for any future exposure.

For these studies we tested rats on all kinds of poisons and observed their behavioral reactions to various kinds of poisons. This gave us a considerable amount of information regarding their behavior in relation to toxic materials.

ANTU was used in a citywide campaign in the city of Baltimore over a three-year period, and killed by actual count, over a million rats. However, the fact that it has a slight odor that the rats can detect definitely limited its usefulness.

The work that was carried on in the city of Baltimore during the rat control campaign gave us an almost unlimited supply of wild rats. We worked out a fairly simple and very inexpensive trap for catching wild rats, and we were able to get men supplied by the city to trap as many rats as we needed. Some days they brought in as many as 350 rats. We had plenty of material to work on for the poisons.

The following list contain a few of the references to papers that were published on thiourea and related compounds during the period of our work on poisons.

Dieke, S. H., Allen, G. S., and Richter, C. P. 1947. The acute toxicity of thioureas and related compounds to wild and domestic Norway rats. *Journ. Pharmacol. & Exp. Therap.* 90(3):260.

Dieke, S. H., and Richter, C. P. 1946. Comparative essays of rodenticides on wild Norway rats. 1. Toxicity. *Pub. Health Rep.* 61:672–79.

Latta, H. 1947. Pulmonary edema and pleural effusion produced by acute alpha-naphthyl thiourea poisoning in rats and dogs. *Bull. Johns Hopk. Hosp.* 80:181–97.

Richter, C. P. 1945. The development and use of alph-naphthyl thiourea (ANTU) as a rat poison. *J.A.M.A.,* 129:927.

———. 1948. Physiology and endocrinology of the toxic thioureas. *Proc. Laurentian Hormone Conf.* 2:255–76.

———. 1951. The physiology and cytology of pulmonary edema and pleural effusion produced in rats by alpha-naphthyle thiourea (ANTU). *J. Thoracic Surg.* 23:66.

Richter, C. P., and Clisby, K. H. 1942. Toxic effects of the bitter-tasting phenyl-thiocarbamide. *Arch. Path.* 33:46–57.

Richter, C. P., and Emlen, J. T., Jr. 1946. Instructions for using ANTU as a poison for the common Norway rat. *Pub. Health Rep.* 61:602–07.

Domestication of the Norway Rat

The great opportunity offered to us by the availability of the large numbers of recently trapped wild rats, together with the rats from our colony, which had been used for over forty years, made it possible for us to carry on a limited study of the effects of domestication. We made comparisons of the anatomy, of the different organs, differences in toxic reactions to poisons, reactions to all kinds of situations, so that we were able to gather a considerable amount of information about the effects of domestication.

The upshot of the entire experience was the discovery that the wild rat is a really wonderful animal, very well equipped to take care of itself under all kinds of conditions. I have summed up my sentiments in an article that I published several years ago entitled, "Reluctant Rat Catcher—Is the Wild Rat a Friend or an Enemy?"

The following list contains some of the papers that we published on the effects of domestication.

Griffiths, W. J., Jr. 1944. Absence of audiogenic seizures in wild Norway and Alexandrine rats. *Science* 99:62.

Richter, C. P. 1949. Domestication of the Norway rat and its implication for the problem of stress. *Proc. Ass. Res. Nerv. & Mental Dis.* 29:19.

———. 1951. The effects of domestication and the steroids of animals and man. Symposium on Steroids and Behavior, Ciba Foundation, London.

———. 1952. Domestication of the Norway rat and its implications for the study of genetics in man. Read at the Meeting of American Society of Human Genetics. September 1952, Ithaca, New York. *Amer. J. Human Gen* 4:273–85.

———. 1953. Experimentally produced behavior reactions to food poisoning in wild and domesticated rats. New York Academy of Sciences Meeting, 22 February 1952. *Ann. N. Y. Acad. Sci.* 56:225–85.

———. 1957. Phenomenon of sudden death in animals and man. *Psychosomatic Med.* 19:191–98.

———. 1959. Rats, man and the welfare state. *Amer. Psychologist* 14:18–28.

———. 1968. Experiences of a reluctant rat-catcher. The common Norway rat-friend or enemy? Reprinted from *Proceedings of the Am. Philosophical Society* 112(6).

Rogers, P. V., and Richter, C. P. 1948. Anatomical comparison between the adrenal glands of wild Norway, wild Alexandrine and domestic Norway rats. *Endocrinol.* 42:46–55.

Book

The book, *The Psychobiology of Curt P. Richter,* edited by Dr. Elliot M. Blass, published by York Press in Baltimore in 1976, carries a fairly full list of all of my publications up to 1975. It carries all of the publications on animal behavior, grasp reflex, and the sympathetic nervous system, as well as a number of acute experiments that did not belong to any particular research project.

Acknowledgments

Over the past sixty years many medical students and members of the hospital and medical school staffs collaborated in these researches: Katherine Rice, Sally Dieke, Carl Hartman, Bruno Barelare, John Eckert, Douglous Hawkes, E. Schmidt, Emmet Holt and David Mosier. In the last years of his life, Professor E. A. Park, Head of the Department of Pediatrics for many years, dropped in the laboratory almost daily with inspiration and encouragement.

For thirty-three years Mrs. Ardis O'Connor had complete responsibility for the maintenance of standard conditions in all of my experiments and also was in charge of my colony of Norway rats.

John Paul Scott (b. 17 December 1909) in February 1963

16

Investigative Behavior: Toward a Science of Sociality

John Paul Scott

Historical Background

In writing of my career, I shall first of all place it against the background of significant world events that took place in my lifetime. I was born on December 17, 1909, and one of my early memories was seeing three-inch black headlines in the newspaper as the Germans marched into Belgium in August, 1914, and my father saying, "The Germans are just too strong." I learned to read on war news: German atrocities, pictures of ruined buildings and dead horses (no one then would publish pictures of dead men), Wilson's speeches (making the world safe for democracy), and battle after hopeless battle.

Then afterwards, the young soldiers came back from France, disturbingly stimulated by contact with a foreign culture and vastly disillusioned with war and the mud and blood of the trenches. Some of these men were my teachers in the 1920s, and as the years went on, the disillusionment began to get into literature— *What Price Glory* and *All Quiet on the Western Front.*

In 1930 to 1932, when I was a student in England, I visited the battlefields of France, where peasants were still plowing up fragments of old iron, unexploded shells, and bits of human bone. I saw the fortress of Verdun, where more than half a million men had died, and an ossuary where they were still collecting the bones of 100,000 soldiers.

In 1929 came the stock market crash that initiated the worst economic depression that the United States had ever known, not to be really ended until 1939, when World War II began. When I was a student at Oxford, I discovered that the depression had already been going on in England, France, and Germany since the mid-1920s, and the United States made it worldwide. I saw grinding poverty in Italy, and an economically dead Vienna, the ghostly capital of a dead empire. It was obvious that these events, involving economic misery for millions of people in what were supposed to be the most advanced nations of the world, were neither understood nor consciously remedied. There was no scientific knowledge that could cope with either warfare or economic disaster.

In the 1930s I read of the rise of fascism in Italy and Germany and how these two powers assisted General Franco in the bloody crushing of democratic government in Spain. The Japanese invaded China and fought there for five years. It was obvious that World War II was on the way, but no one knew how to stop it. Then in 1939 it came, to go on for six disheartening years, with the United States involved for nearly four.

Thus at the end of World War II, when I was almost thirty-six, I had experienced only sixteen years that had not been affected by world wars and world economic disaster; the five earliest years of my life, and the eleven years between 1918 and 1929, from ages nine to twenty, as I received the majority of my education, from the fourth grade through part of my undergraduate university life. And so at thirty-six I was still standing on the threshold of my professional career.

In 1945 the door opened on a new world of opportunities. Not that war was a thing of the past; the United States was to fight another war in Korea for three years from 1950 to 1953, and the longest and most disastrous war in its history in Vietnam from 1964 to 1975. But these events, while disturbing, calamitous, and often fatal for many younger people, had only secondary effects on me.

For the next thirty-five years I was able to follow what would have been considered in my youth a normal academic and research career. For twenty years after World War II, the United States and Western Europe enjoyed a prosperous and stable period of economy (again, no one quite knew why), and during ten of those years, the United States enjoyed what future generations will look back on as a Golden Age of Science. In 1957 the Russians put up the Sputnik, the first satellite, and the political leaders of the United States realized that if they were to compete with the Russians they would have to compete in science.

For the first time, the United States began to put sizable amounts of money into scientific research, and it paid off. Not that money will buy scientific discovery on a one to one basis. But it will buy the rapid development of technologies, and new techniques make new discoveries possible. For example, John Fuller and I had talked of the need for radio telemetric techniques to measure physiological reactions in unrestrained individuals, and he struggled with various crude devices for some time, finally producing one that worked. Then money was poured into space research, and technicians soon came up with reliable devices that have since revolutionized the study of free-ranging animal populations.

More importantly, money in an economically oriented culture indicates that an activity is both important and appreciated. Gifted young people flooded into scientific research, and the results soon began to be felt. Whereas previously an important basic result in a scientific field might come out once in five to ten years, major discoveries now began to appear every one or two years.

Monetary expenditures began to stabilize in 1967 and since have further tapered off, reflecting certain world events, particularly the Vietnam war. Even a well-running and expanding economic organization cannot expand payments for warfare and the pursuit of knowledge at the same time.

There are other economic factors; we are slowly coming to the end of an era when energy derived from fossil fuels seemed to be cheap and unlimited, and whether or not we find a solution to the energy problem will determine the quality of life for years to come. Like all problems, it must be faced with courage, but its solution will depend on science as well as politics.

Personal Experience

My early experiences with world events convinced me that all was not well with the world and drove me to the conclusion that I should do something about improving the quality of human life. In addition, I had two significant experiences, one bad and one good, that contributed strongly to my personal motivation.

The bad one came in the year 1919–20, just after World War I, when I was ten years old. My father, the chairman of the Department of Zoology at the University of Wyoming, decided to take a sabbatical leave and spend it in Washington, D.C., working in the Library of Congress. Unfortunately for me, he located his family in the little country village of Falls Church, Virginia, then little more than a cross-roads. He rented a rambling old house with a barn into which he installed a milk cow, I suppose with some memory of his own farm background.

My father and older brother commuted into Washington via the Washington and Old Dominion electric railway, but I was plunged, without preparation, into the fifth grade of the local public school. The full range of the white community was represented, from reasonably bright sons and daughters of commuting government workers to semiretarded louts and lasses of fourteen or fifteen years who had never gotten beyond the fifth grade. I set out to be the star student and did so with little difficulty. I always knew the answers, except on one occasion when Miss Rice, who was a young, ambitious, and well-trained teacher, got up a classroom discussion on the future growth and expansion of Falls Church.

The question was, why couldn't the town expand to the west? I supposed it was because of the rough and hilly terrain, but everyone laughed, and Miss Rice knowingly pointed out that Falls Church could not expand in that direction because of "Nigger Town." Miss Rice should see Falls Church today. The name remains, but the old village, white and black, is gone without a trace, part of the suburban sprawl around Washington.

As I saw it later, to come into a strange classroom and outshine everyone else is not a good road to social popularity, and I soon experienced the backlash on the playground. I was chasing a supposed friend in a rough-and-tumble game when he suddenly turned around and hit me in the mouth with his fist. If I had known I was supposed to fight, I suppose I could have and might have won, but I was totally unprepared and did not retaliate. I had never been in a real fight in my life, did not know the rules or the techniques, and my home teaching indicated that it was wrong. On that Virginia school ground, fighting (between boys, of course) was part of the social system.

Seeing that I did not resist, others began to pick on me, again without warning. To make a sad story short, I was persecuted on the school ground. I had a different sort of accent (I am told that I speak with an Intermountain accent), but the word got around that I was a German and therefore to be persecuted (this was just after World War I with all its anti-German propaganda). My solution was to avoid the school ground, arriving at the last minute and leaving immediately at the last bell. I only lived to get back to Wyoming and the life I had known before.

While I could in later years understand my tormentors, I hated them then and used to imagine ways in which I could kill them. But it was a deep emotional hurt, and I could not leave it entirely behind when I returned to Wyoming.

On the positive side, I had to repair the wound to my self-confidence, and for many years I had to prove that I really amounted to something and so became much more ambitious and achievement-oriented than I otherwise might have been. Also, it left me with a lifelong sympathy for the underdog. When I have had power, I have never used it to make subordinates out of others. More generally, I realized that those Virginia boys were giving me the treatment that they and their ancestors had been meting out to the blacks for generations. For a brief period, I knew what it was like to be a low-caste individual.

So much for the bad. By the time I got through Laramie High School, I had won all sorts of prizes, not only for scholarship but for creative writing and for football and track. This earned me respect and a certain amount of envy from my contemporaries, who were the same sort of cross-section of white America that I had seen in Virginia, but not infected with the same violent strain of racism, and only slightly more infected with a love of learning.

At the end of my sophomore year at the University of Wyoming, my father encouraged me to go to the Marine Biological Laboratory at Woods Hole, Massachusetts, to take a six-week summer course in invertebrate zoology. In a way, I was no stranger to the area. My father had been one of the instructors in the course years before, and I had gone down with the rest of the family when I was six months old, and the next summer at age one and one-half. But I had no conscious memories.

In June of 1929 I made the long train journey from Wyoming to Massachusetts—eighteen hours to Chicago, where I stopped to visit my Uncle Paul, and another overnight journey (by coach) to Boston and Woods Hole. As soon as I arrived, I felt at home; this was a warm and friendly place. I was met at the station by Don Charles, then in the process of getting his Ph.D. in genetics from Columbia, who was looking for a roommate. We got along well and he later introduced me to a lot of people whom I might otherwise have missed.

The day's work began with a lecture, followed by laboratory exercises, mostly dissections but often including observations of living and behaving creatures. At noon we stopped for lunch at the Mess, after which we walked around, looked at seals in the tank near the Fisheries building, or some of us played tennis. Back to work at two, to keep on till four or four-thirty, when we all went swimming in the warm salt water of Buzzards Bay. In the evenings, there were beach parties, sometimes dancing to a phonograph, and long walks in the moonlight. Occasionally there would be a lecture by some important person, which we might attend. I remember seeing Pavlov on the steps of the main building, and briefly meeting and talking with T. H. Morgan, the Nobel Prize-winning geneticist.

I was kindly received at the house of Professor Frank Lillie and his wife. He had been my father's major professor at Chicago and exercised a great deal of influence on the operation of the MBL, as the laboratory was affectionately called. Sitting across from my laboratory table was James Kendall, then a graduate student at Harvard and later to become a professor of histology at City College of New York. I was fascinated by his wit and sense of humor, and we became lifelong friends. Through him I met other graduate students at Harvard, among them George Snell of whom I was later to be a colleague at the Jackson Laboratory and who won the Nobel Prize in 1980. Another student was E. G. Stanley Baker, who was to be a colleague at Wabash College. Of the other undergraduates, very few went on to academic careers, and I lost sight of them.

To me, it was an idyllic existence. For the first time I was in the company of a group of people who were all strongly interested in common intellectual pursuits. There were no grades; hence no competition for status; all this had been left behind at our respective institutions. We met as equals, our old lives left behind us. We worked hard, but for the sake of the work itself. And we had time for the emotional and romantic side of life.

I sensed a similar atmosphere among the older scientists. Many of them brought their research and groups of graduate students down for the summer, leaving behind the committees, the responsibilities, and the clashes of their home institutions. For me it was a revelation of a good life, and I saw how an ideal professional life might be led. Wherever I have been since, I have tried to recreate this existence.

At the end of the course we had a week in which to do research. I chose to work on the response to light of brittle stars and discovered that they clumped together around a beam of light in a darkened aquarium. Unknown to me, and at about the same time, W. C. Allee was doing similar experiments from the viewpoint of the formation of aggregations and got similar results. This was my first piece of research on animal behavior, and I briefly mentioned it in my book, *Animal Behavior,* published nearly thirty years later.

Personal Characteristics

Those were two of the major experiences that shaped my character. In addition, I developed certain personal traits and capacities that I was to use in my scientific work.

One of these was an interest in ideas. Once I learned to read, I explored everything I could find on the written page, first going through everything in my parents' house, including a set of Shakespeare's plays (I couldn't make much out of them and thought they were pretty dull) and another of the works of Dickens. As soon as I was old enough, I got a card at the Laramie Public Library and regularly took out the maximum number of books—mostly fiction. In doing all this I became a very fast reader.

I became a good student early. When I went to the first grade at the University Training School, my first reaction was to rebel against the restrictions on my time, and I got into trouble because of various infractions of discipline. Then one day I decided that I would do well in school, and went home and told this to my mother. And I kept this resolve for the next twenty years, until I emerged from the University of Chicago with a Ph.D. This resolve eventually expanded into a more general one, to do anything that I tried as well as I possibly could.

Was I intelligent? My grades said I was, and during the course of the next few years I took several intelligence tests. Later, as part of a psychology course in college, I did a research project on the relationship between vocabulary and IQ on myself and siblings; and so got access to my own scores, which ranged from 106 to

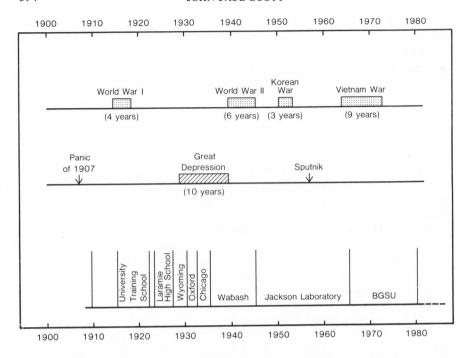

140. This wide variation first raised serious doubts in my mind regarding the meaning and reliability of such scores.

In my own case, I know exactly why the wide variation occurred. My first reaction to the tests was to try to get everything right no matter how much time it took. Later I discovered that the scores also depended on speed and so rushed through them as fast as possible. My crowning achievement was to make the highest score on the placement tests of the entering freshman class at the University of Wyoming, a point or two ahead of Norris Embree, one of Terman's thousand gifted children, who later became one of my best friends in college. It probably wasn't fair, as he had entered the university at the age of fourteen, and may not have been as highly motivated as I was.

Intellectually, I have always been something of a rebel. When I was very small, adults would ask a group of us what were our favorite colors. Mine was red, but since everyone else said red, I said blue. Today, when any scientist claims that he has discovered the secret of the universe, my first reaction is skepticism and to explore the possibility of alternate explanations.

Some of my intellectual capacities are better than others, but none are unduly low. I know that I am not as good at abstract symbolic reasoning (especially mathematics), as are many other scientists, but I can cope with this limitation, which in part comes from a lack of interest. I had early discovered that the easiest way to get a high grade in math was to learn the method of solving a problem and apply it to similar problems; no one taught me how to use mathematics creatively, and I only discovered this at the age of forty or so.

I like novelty, as I am easily bored. Consequently, I like to travel, to explore, to see new places, and to have adventures. And so I like the exploratory aspects of scientific research. I like to manipulate the English language; it is never boring as it is comparable to a chess game with half a million pieces and an infinite number of plays and combinations. I like and respect intelligent people, those who are quick on the uptake.

I like to try new things, and my general rule is to give anything a good try (i.e., do my best); if it works, fine; if not I go on to something else. I like to get things done (the mark of the born-and-made achiever) and if I want to feel good, I need to get at least one thing done every day.

If there is any theme that runs through my whole life, it is that of organization. I like to organize facts and ideas—my whole research career is based on an attempt to arrange facts so that they make sense. I like to organize people, and have helped to organize two scientific societies, the Animal Behavior Society and the International Society for Research on Aggression. And I like to think that I am a well-organized person. I can and do concentrate deeply on current projects, but I know enough to lead a well-balanced life, including occasional but not continued stress.

Some Persons Who Influenced Me

The first of these were my parents. My father was John W. Scott, a zoologist and parasitologist. He was born on a farm in Lewis County, Missouri, the oldest of nine children, all of whom attained college degrees. His undergraduate education was primarily classical, with four years of Latin and four of Greek, leading to a B.A. degree and a Phi Beta Kappa key. Sometime in the 1890s he took an M.A. in the new science of psychology, but found that field vague and unsubstantial. He got a job teaching high school in Mexico, Missouri, where he was assigned to teach a biology course. He knew nothing about the subject, and learned it by reading the textbook the night before each class.

Here he found his true vocation, one that united his intellectual interests with his farm background. As he put it to me later, he resolved to learn everything there was to know about zoology, the science of animal life. For several years he continued to teach at Westport High School in Kansas City, where I was born. He also went to the University of Chicago for advanced training and emerged with a Ph.D. in the year of the Panic of 1907. (Those were the times when the economy underwent periodic cycles of boom and bust, culminating in the stock market crash of 1929 and the ensuing ten-year Great Depression.) Consequently, there were no university-level jobs and he continued to teach in high school and to do research on embryology at Woods Hole in the summers.

In 1911, when he was forty-two years old, he got his first university position, at Kansas State University in Manhattan, Kansas. Two years later he was offered the chairmanship of the newly formed Department of Zoology at the University of Wyoming, a position that he held until his retirement some twenty-five years later.

He loved to play with young children and particularly loved to "teach" them to talk, that is, to memorize new words. Starting with me, he collected our vocabularies for the two-week period following age two. Perhaps because of his initial enthusiasm, I had the longest list of any of my brothers and sisters but, as I said, there was no obvious correlation with our IQ scores.

He was an energetic and enthusiastic person with a great deal of charm. Unlike many charmers that I have met since, his integrity was complete; he might use charm but only because he believed completely in what he was doing.

He and I always got along well, and he always showed a warm approval of my accomplishments. He loved to teach, and I suppose I must have absorbed a great deal of information unconsciously, for when I went to the university and took some zoological courses, I seemed to know the material without studying. More than anything else, he gave me a firm belief in science as a power for good.

My mother was an equally powerful person in the household. She had graduated from high school at seventeen, and immediately thereafter began to teach, continuing until she went to the University of Chicago five years later. She majored in chemistry and quit after two years to marry my father. As she saw it, there was a clear choice between a career and a family, and she chose the latter. She and my father were both oldest children, both highly articulate, and they often argued with each other.

But mostly they made a great team. My father was the creator, the entertainer, and the enthusiast; my mother was the executive, the manager, and the organizer. She also had considerable political gifts—a friendly, outgoing nature and an interest in people, their names, and relationships. Like my father, she was an ambitious person, not in a selfish way but directed toward social service. Later on she served on the local school board and helped with the Campfire Girls. Like my father, she was unfailingly supportive of my ambitions. Also like him, she put a high value on intellectual attainment. The ideas that I developed were my own, not those of my parents, but I learned to value theirs.

My parents were idealists in two senses. They believed in ideas and that an ideal world, one conforming to ideas, was achievable. They also definitely believed that the world around them was not ideal, except, in my father's case, for the world of nature as he saw it in the high plains and mountains of Wyoming, and which in his later years he worked mightily to conserve.

A third force in my early world was my older brother, four years my senior and exhibiting a classical case of sibling rivalry. He was the mischievious, misbehaving child and I was the well-behaved and conforming one. I saw no reason for getting into all the trouble that he did, and he hated me cordially. At the same time, he loved me, and later when I got into high school and college, he helped me a great deal in my dealings with the world outside the family (which my parents did not). Among other things he wanted to write, and he eventually became a newspaperman. His example encouraged me to try to write also, the beginning of my own interest in literary effort.

My first step in the direction of a career came when I was about fifteen and read a very bad book by Albert Edward Wiggam entitled *The Fruit of the Family Tree*. It was old-fashioned eugenics, based on the simple theory that all the world's ills were due to bad heredity, and if we would only apply our knowledge of animal breeding to humans, Utopia would follow. I did not really swallow this, naive as I was, but it did occur to me that if heredity had all that an important effect on behavior, someone ought to study it scientifically. And so began my interest in behavior genetics.

Education

Then there were the persons I met in the course of my education. I realize now that in many ways I had a very superior sort of education, one that would be difficult to duplicate under present conditions. I went to the University Training School in Laramie, which meant that there were small classes (never more than twenty-five), that I had the best teachers that the system could produce (I never had a poor or even mediocre one in any grade), and that the method and materials were the best known to educators at that time. In short, the system had all the best features of both public and private schools.

After completing the seventh grade, I decided to switch to the Laramie public schools, mainly because of a feeling that I needed some experience with the "real" world of my contemporaries. The eighth grade had little impact, except that I wrote and produced a melodramatic play entitled *Black Jack Pete* for the 8A (superior) section. Its only merit was that it contrasted with the usual student plays whose characters were animated flowers.

The high school was ideal in size, large enough to provide a variety of courses and activities, and small enough (around four hundred-fifty students) so that individuals were not lost. Because it was in a university town, it attracted superior teachers, and through my older brother I had leads to which were the best. I particularly remember Miss Emily Gillis, a tall young lady from Boston who had come West for adventure. She recognized my gift for writing, and in my first year I won the statewide essay contest and placed in the *Youths Companion* contest for short story writing. Then there was Mr. A. J. Conrey, the science teacher, a favorite among all, who taught me biology, physics, and chemistry.

My experience with athletics was equally important, earning me respect and giving me self-confidence. I went out for football as a freshman but did not become a regular member of the team until my senior year, as I grew rather slowly and only weighed 145 pounds at that time. Sooner than that, I discovered that I could run faster than most people, and earned a place as a sprinter and quarter miler on the track team.

When I was just under eighteen, I entered the University of Wyoming. A few years before, I had heard the Oxford University debate team and learned about the Rhodes Scholarships. I set my eye on winning one of these and organized my college career around demonstrating the qualities of scholarship, leadership, moral character, and excellence in manly sports that Rhodes had established as criteria. Scholarship was easy; I earned letters in football and track by the end of my sophomore year, and I tried for leadership in various campus activities. Moral character I let slide, not having much time to investigate the interesting and sinful side of life.

The university, like the training school, gave me a uniquely excellent education, one that would be difficult or impossible to duplicate at the present time. It was a real university, with colleges of liberal arts, agriculture, law, engineering, and education, and so offered a wide variety of courses. But there were only 750 students in all, which meant that classes were small, and the teachers had time to teach.

I met an excellent teacher of English, Professor W. O. Clough, a man with a lively creative mind, enormously well-read in a variety of fields, and very encouraging to me in my literary efforts. Besides my father, there were a handful of scientists with national reputations. Those whom I especially enjoyed were Aven Nelson, the botanist, and June E. Downey, the psychologist.

Then, at the end of my junior year, I went off to Oxford, having hastily accumulated enough course credits to graduate. That spring I took twenty-four hours of work (fifteen being the normal) and passed them with excellent marks. But I never could remember any of it afterward.

Oxford

In October 1930 I entered Lincoln College and matriculated in the Honor School of Natural Science with advanced standing so that I could immediately enter on the study of zoology, a two-year program of lectures and laboratories that culminated in a final comprehensive examination.

The Department of Zoology reflected the effects of World War I, then over for only twelve years. In 1914 the young men of the English universities had, in accordance with their training, flocked to enlist in the army, and they died in the trenches. This process, aided by conscription, went on for four years, with the result that the Oxford department in 1930 consisted of one elderly professor, E. S. Goodrich, and a group of young men under thirty. One person of intermediate age, Julian Huxley, had already moved to the University of London. The group was lively and full of ideas, but their accomplishments could not compare with those of the seasoned investigators whom I was later to meet in Chicago.

The course, which everyone concentrating in zoology took, was organized over two years—vertebrates one year and invertebrates the next. So I began with E. S. Goodrich's lectures. He was the last of the comparative anatomists, and his specialty was the evolution of the skulls of fishes. He was a meticulous scholar, and I heard about the evolution of fishes until it ran out of my ears.

Evolution was the overwhelming idea to which I was subjected at Oxford. The lecture room in the University Museum where Thomas Huxley and Bishop Wilberforce had held their historic debate was memorialized by a bronze plaque, and most of the lectures were organized around evolutionary topics. Strangely, I heard almost nothing of the genetic theory of evolution, which at the time was being developed by Fisher, Haldane, and Wright. I did hear of Wegener's theory of continental drift, used to explain similarities in fauna in continents now widely separated, but this was to pass into limbo for thirty years, to emerge as the now widely accepted geological theory of tectonic plates.

This emphasis on evolutionary history resulted in a downgrading of other aspects of zoology and particularly that of physiology. I attended one lecture by

Sherrington, given in a different department, but it was of such a specialized nature that I did not attend another. Biochemistry was introduced only in an ancillary fashion in connection with the few experimental sessions. Experimentation itself was limited, because of the strict English laws regarding vivisection; we rarely saw anything but dead animals. J. Z. Young, whose work on the physiology of the nervous system was later to become well-known, was at that time a very junior member of the department. He was very helpful to me in the summer of 1931 when I spent a few weeks at the Zoological Station at Naples.

The thinking of G. R. DeBeer, the embryologist, was dominated by evolution, but he did introduce me to the work of C. M. Child (whom I was to later meet in Chicago) on metabolic gradients.

Of the newer subsciences of zoology, ecology was represented by another young man, Charles Elton, whose little book *Animal Ecology* was a landmark in the development of ecological theory. He introduced me to the concepts of food chains, of population dynamics, and the older concept of succession developed by V. E. Shelford in the United States.

Genetics was represented by E. B. Ford, who became my tutor because of my previously announced interest in this field. He did a conscientious job, which consisted of assigning me an essay topic each week and listening (usually with closed eyes) while I read it aloud; then saying "very good." This did introduce me to the sort of topics that I would meet on the comprehensive examination, but I learned very little that was new about genetics, which, unlike the older portions of biological science, had developed largely in the United States. Ford's own research on ecological genetics, strongly related to evolution, was still in the future.

We heard little about behavior, except that Oxford was the place where I first heard of Eliot Howard's *Territory in Bird Life,* published in 1920, and the first major scientific work on the phenomena of territoriality in animals. Julian Huxley, who had discovered ritualization in bird behavior, had already gone to London, and I did not hear about his work until years later. There was at that time no department of psychology at Oxford; a few bold spirits were reading William James, but the first chair of psychology was not established till years afterward.

I spent my last six months at Oxford preparing for the comprehensive examination, and in so doing made a typewritten outline of all the information to which I had been subjected, and organized it in a systematic fashion. I had much previous experience with examinations at the University of Wyoming and had no doubts about my ability to deal with them. The result was that I passed the examination with a "First," the highest honor that the English university system bestowed on its graduates. Since I had also earned an Oxford "Blue" by running the 100-yard dash against Cambridge (I came in second), an honor that in some circles was held in greater esteem, I stood on the threshold of fame and fortune had I wished to remain in England.

Among my undergraduate acquaintances in Oxford, two were special friends. One, H. D. Springall, was a chemist who later became the vice-chancellor of the University of Keele in Staffordshire. The other, W. L. Russell, took zoology with me and later followed me to work with Sewall Wright at Chicago, where we enjoyed bouncing ideas off of each other in long discussions.

In the spring of my last year at Oxford, I came to a choice point. I still held to my ambition of becoming a writer, and while overseas had written some stories and poems that I submitted to various publications. The only ones that were published were a few poems that appeared in "slim volumes" of Blackwell's *Oxford Poetry,* annuals including various undergraduate efforts.

But I had also become engaged to be married, a goal that was not compatible with that of a solitary writer starving in a garret. So I applied for graduate assistantships at three universities: the University of California at Berkeley, Chicago, and Harvard. If I had been offered a stipend at Harvard I might have taken it, as it was closest to my future wife's home in Vermont. As it was, I had a choice between Chicago and Berkeley, and on my father's advice chose the former. So I passed a major choice point in my career, and who knows what might have happened had I chosen otherwise; I might now be a respected elder academic in California.

About this same time I came to another decision. Up to this point my life had been that of competitor and status seeker. And I had reached the pinnacle of undergraduate success, the highest academic honors from one of the oldest and most famous universities in the world. I asked myself, why am I knocking myself out to achieve status, especially as it leads me into activities that are frequently laborious and uninteresting? In part it was to demonstrate to some louts in the fifth grade in Virginia, who couldn't have cared less, that I really amounted to something. So I resolved, now that I was in the process of entering my adult occupation, that I would do what I thought was important, theoretically, ideally, and personally, and let status take care of itself. And so I set myself free.

Chicago

My choice of Chicago was a fortunate one. The zoology department was then at its zenith, with ten members of the National Academy of Science in its ranks, and my major professor, Sewall Wright, was the star of them all.

He was a first rate scholar *and* a genius. The two things are not necessarily synonymous, and this was the first and only time that I was to have prolonged contact with such a person. Such terms are meaningless except at the extremes of distributions, and Wright qualified in both respects. At the time I met him, he said that he had read every paper that had been published in the field of genetics, and it was his habit to read papers critically. He was, moreover, a mathematical genius who happened to be interested in genetics, one of those people who think naturally in mathematical symbols and use them as easily as a second language. He began early; one of the stories that I heard at his ninetieth birthday celebration was that when he entered the first grade he already knew how to extract cube roots. He also told me that he had almost no formal mathematical training; he could do all that for himself, and I suspect that it might have inhibited his creativity if he had been subjected to it.

Among his other accomplishments Wright developed the coefficient of inbreeding, a widely used mathematical tool among geneticists. He was the creator of path coefficients, a method of statistical analysis now widely used in a variety of fields. But above all, he created a genetical theory of evolution, one that went far beyond those of his contemporaries, Fisher (another statistical genius) and Haldane. At that time we used to say that there were only three people in the world who could understand each other in this field, and the understanding was by no means mutual.

Wright knew nothing about behavior. Here I was, working with one of the world's leading geneticists, and I decided that I would profit most by working in his laboratory rather than striking out on my own. So I took the problem of working out the embryology of a mutation that Wright had discovered in the guinea pig, a dominant gene for polydactyly, or extra digits. And so I lived with guinea pigs for the next three years. I discovered that the normal embryology of the guinea pig had been only casually described, and worked out that along with the development of polydactyly, which I explained as the result of a gene that stimulated growth at a critical period in development (Scott, 1937).

The idea of critical periods was nothing new, although it had been little over a decade since Stockard (1921) had published his definitive monograph. What I did was to show that speeding up an organizational process, as well as slowing it down, as had been shown in many other experiments, could explain critical-period effects. It also prepared me to recognize critical-period phenomena as I encountered them in the development of social behavior. (In the same decade, Lorenz [1937], another biologically trained scientist, called attention to critical periods in imprinting.)

The other scientist at Chicago who exerted a major influence on my work was the ecologist, W. C. Allee, who had been a student of Shelford and who had returned as a professor at the University of Chicago. In the course of his field research he had observed aggregations of the water isopod, *Asellus,* and this led him to look for other such phenomena. The sociologists of that era, the 1920s, were fond of pointing out the theoretical and hypothetical benefits of life in social groups, and it occurred to Allee that it ought to be possible to demonstrate measurable physiological benefits. This led to his *Animal Aggregations* and a line of research that he pursued through many years. It also led him into animal behavior; he was at that time the only major figure in American zoology who was working in that field. He and his students confirmed and extended the peck-order research of Schjelderup-Ebbe, and Allee was one of the first persons that I knew who recognized the importance of Lorenz's early work.

Allee did not impress me as a genius. Indeed, as I now look back on it, he was in many ways a person much like myself, a hard-working scientist who could recognize good creative ideas in others as well as generate them. He was a humane person; he was the secretary of the zoology department and looked out for the welfare of graduate students and found them jobs. And he had a strong interest in applying his research to promote human welfare generally, as befitted a person with a background in the Society of Friends.

I got some *Drosophila* stocks from Wright, who kept them for classroom experiments; Allee let me have a little space in his controlled-environment laboratory; and I began my first study of behavior genetics, based on the variations of phototropic response of these insects. To begin with, it seemed to me important to demonstrate more than that heredity had an effect on behavior, that major effects could be associated with a single gene. Wright suggested that it would be necessary to make extended backcrosses into a nonmutant stock, and I began this technique with two mutant stocks, a brown-eyed stock that appeared to be more phototropic than others, and a white-eyed stock. As it turned out, the brown-eyed gene had no effect but the white-eyed gene did (Scott, 1943), as compared with the wild-type red. The single gene technique has since been widely used by other experimenters in order to study possible pleiotropic effects of such genes as albinism.

Theodosius Dobzhansky (who collaborated with Wright in many experiments) wrote to me and wanted me to follow up the fruit fly experiment with a more sophisticated genetic technique, but I did not do this, as I was not interested in genetic theory per se, but in behavior as a primary phenomenon. Later Dobzhansky and his students (as for example, Jerry Hirsch) themselves did many experiments on behavior genetics in fruit flies. But I felt that if the work were to have human relevance, it would have to be done with mammals.

At the end of my three years at Chicago I emerged with a Ph.D. in Zoology, a qualified researcher in Developmental Genetics, and a gleam in my eye with respect to Behavior Genetics. In ordinary times, I would have expected to step into a substantial university position. But this was 1935, the middle of the Great Depression. There was one job available, at Wabash College, and I accepted it quickly.

Work

Wabash College was, aside from the salary of $2,200 per year, a good place to work. At that time there were four widely respected academic institutions in Indiana: The University of Indiana, Purdue University, DePauw University, and Wabash College. The last was an undergraduate institution for males only; it had high standards and, like all academic institutions at the time, had barely enough money to get along.

I was appointed associate professor and chairman of the Department of Zoology. Within a few years I became a full professor; with respect to status I have often said that my academic career began at the top. The other member of the department was E. G. Stanley Baker whom I had met while at Woods Hole; we got along well and arranged the courses so that there was a decent teaching schedule and enough time for research. We had half a building to ourselves, and I soon acquired a couple of rooms in the basement for a mouse and rat colony. The college placed no emphasis on research; as long as one did a good job of teaching, anything else was fine. Later, Baker went off to get a Ph.D., and was replaced by Howard Vogel, who became a lifetime friend.

More than this, Wabash brought me two important experiences. The first of these came through my wife, Sally Fisher Scott. She was the daughter of Dorothy Canfield Fisher, a popular novelist and feminist who reached her zenith in the 1920s, and much of Sally's early life had been a struggle to escape from her mother's all-pervading aura. When Sally and I first met in Oxford, I had read only some pieces by Dorothy Canfield in the *Good Housekeeping Magazine* entitled *Home Fires in France.* Their literary merit was slight, so I thought, and this low opinion of her mother, contrasting with the usual adulation of her fans, got Sally and me off to a good start.

Unlike me, Sally was not strongly ambitious, but she breezed through the Oxford examinations in English Literature for an easy Second Class, and she went home to teach for a year in the English Department at the University of Vermont, while I spent an energetic first year at the University of Chicago, culminating with my passing the preliminary Ph.D. exam while the temperature was 105°.

We were married a week or so later, in June 1933. A year after that, our first child, Jean, was born and after another two years our second, Vivian. Sally loved being a mother and applied all of her brilliant mind to the task, for here was one area in which she felt she could easily do a better job than her mother. But this was not enough to keep her busy. She wanted to share my working day as well as life at home. Her own training had included a minimum of science, but she observed that I, in common with all the other faculty members at Wabash College, had no secretarial help (the president had a secretary, and there were one or two others in the university office).

So she hired a competent neighbor lady to be a baby-sitter (experienced house-keepers were then working for thirty-five cents an hour and glad to get it) and came up to the office for an hour or two every morning. She taught herself shorthand in a few weeks, and we were off to a collaboration that lasted nearly forty years. I provided the ideas and dictated them to a sympathetic and appreciative audience of one, Sally typed them, and they always came out better than I had first said them. Besides that, she raised questions and suggested new ideas that were often incorporated in the next draft. As we got better at it, this enormously increased both the volume and quality of our joint scientific literary output compared to my earlier solitary efforts. In effect, she was an editor, collaborator, and secretary, with talents at least as great as my own. Later on, when we became more affluent, I relegated some of the secretarial functions to others, but no one else could perform the other roles. Much of my work could well have had joint authorship, and I am sorry now that I did not put her name on it, but it seemed important then to maintain separate public identities, as she was developing one as a writer of children's books (she published some twenty-two of them).

The second significant experience came from my teaching. The one unique feature of Wabash College's educational plan was a comprehensive examination for seniors, modeled in part after the Oxford examinations and initiated largely by Insley Osborne, a former Rhodes Scholar. Naturally, the seniors resented having to do this extra work, so different from the lecture, quiz, and forget routine to which they were accustomed. So the college devised a Senior Study Camp held during the spring of the senior year at The Shades, then a private summer resort. The boys (Wabash was all male) studied and exercised during the day, and at night groups of professors came down and organized discussions in which all took part.

One night it was my turn. As a ploy, I attacked the social sciences, saying how unscientific they were compared to biology, my own subject. (Is this faintly reminiscent of E. O. Wilson?) The historians and social scientists responded like bulls to a toreador, and a grand and reasonably friendly argument resulted.

I tried this several times, with equally good results. Then one night as I was driving home, it occurred to me that perhaps this had more possibilities than mere intellectual entertainment. What if one *were* to apply biological methods and concepts to social phenomena? The more I thought about it, the more possibilities unrolled before me.

The first step, obviously, was, to describe and classify social behavior and derived phenomena in every animal species. This meant looking at the entire animal kingdom in a new light and doing the job that Linnaeus had long ago done for animal and plant anatomy. Classification would in turn lead to new ideas and theories, just as Linnaeus' orderly arrangement of similar species led to Darwin's theories of evolution. Theories in turn should lead to new experiments.

But evolution is only one of many biological concepts. Among the major ones are *variation, function and adaption, organization, systems, levels of organization, emergence and creativity, development,* and *process.* Would these work if applied to social phenomena? It seemed to me that they would, and for the first time social science began to make sense to me.

I began to feel a little like the apostle Paul on the road to Damascus. I had seen the light; for the first time I could integrate the training I had with my talents and ambitions, and above all with the understanding and eventual solution of the major social problems that I had experienced. I could see a clear road ahead toward a new interdisciplinary science, and I debated as to whether it should be called biosociology or sociobiology. I came down in favor of the latter, on the analogy with biochemistry; the more basic science coming last.

I was tempted to become a Messiah and announce that I had the secret of the universe. But when I waxed enthusiastic, my friend Bill Russell asked, "Are you saying that everyone should study social behavior?" There was only one answer, "No." Besides, even before I became a scientist, I was always as much a doubter as a believer, and how could I ask anyone else to become the sort of true believer that I was not? Then there was the sheer enormity of the task that I had laid out. No one person could accomplish more than a fraction of it in a lifetime; it would take the cooperation of hundreds and even thousands of scientists.

From a personal standpoint, I gained an eminently satisfactory framework of ideas. When a person in my field makes a brilliant discovery, I can express honest appreciation rather than jealousy, seeing it as one more brick in the building of a vast structure.

As I saw it, the next thing to do was to learn more about the social sciences; I could never hope to be taken seriously in other fields unless I could speak their languages and know their concepts. Second, I ought to write a general book on the subject (this is still to be done in final form). Third, I ought to put my money where my mouth was and make a study of species that would test my ideas and serve as a model for other scientists.

I had now been at Wabash College for three years, and was beginning to feel restless; up until now there had been a change in my career every three to four years, but life at Wabash went on just as usual. At this point Sally received a small inheritance of $3,600. We decided to use this to spend a year with each of us doing exactly what we wanted. So we moved to Boston. Sally took nursery school training, with direct application to our two- and four-year-old daughters. I studied social science in the various libraries around the city, mostly at Harvard, and began to write a book on social organization in animals and man. I wrote in the morning and read in the afternoon. Each new set of data produced new and exciting ideas, and I also began to get all sorts of insights into human behavior. From the practical standpoint, I began to understand myself, my children, my students and my scientific colleagues.

E. O. Wilson must have had similar feelings when he began to work on his *Sociobiology*. Indeed, when his book came out, I wondered if he had not done the job I meant to do in 1938–39. Actually, Wilson's book is only one of the sections in my project, that dealing with evolution. Unfortunately, he and I had very little contact prior to the publication of his 1975 volume. People kept telling me about a bright young man working with the social insects. But entomologists are a breed apart from other biologists; the insect world is their universe, and Wilson never came to the Animal Behavior Society meetings. To give him due credit, Wilson did read much of the scientific literature on vertebrates and other animals, and it is the scholarly nature of his work that is responsible for much of its impact. So I could applaud; his was more than a brick—it was a room in the house I had hoped to see built.

As for my 1938–39 book, it was obvious that one year was not long enough. I showed a draft to two people: W. C. Allee, who was positive and encouraging, and Anna Freud, whose lectures at the Harvard Medical School I was attending. She had little to say except that the emotional tone seemed cold (this was before I began to collaborate with Sally).

So I came back to Wabash in September of 1939. Our rented house was not ready, so we camped out at Turkey Run State Park on a crowded Labor Day weekend. The next morning we woke up to a neighboring camper's radio. The voice was that of King George of England announcing that the Germans had invaded Poland and that Great Britain had, as an ally, declared war on Germany. World War II had begun and would occupy everyone's attention for the next six years.

The War Years

December 7, 1941, was just ten days before my thirty-second birthday. That afternoon the administration and staff of Wabash College met in a miniature council of war. It was obvious that this all-male college was going to suffer. I was against war and participating in all its horrors; yet I could see that there were some situations in which war was preferable to nonresistance.

On the one hand, as an old team player, I was impelled to join up and fight. Many of my contemporaries did, receiving commissions to work in safe jobs behind the lines and to train younger men to go out and die. On the other hand, to take part in such destructive behavior, even indirectly, was emotionally and ethically impossible for me.

So I stayed home, safe from the draft because of my age and parental status but with a perpetual stomach ache. If I had known what the Germans were doing to the Jews, I might have acted otherwise, but for all we knew, the meager reports might have been Allied propaganda. The full horror of the Holocaust was not revealed until 1945.

My personal solution was that if we were fighting to save civilization, then someone had to make sure that civilization survived, and I did my best to keep up my scientific work (Scott, 1942b). It was not easy, and I marvel now that I got anything done at all. As I and most other scientists found, five years of our professional careers were blanked out. I was all ready for my life work and found myself able only to inch along.

One of the things that I did was to write a short book applying what I had learned about fighting in animals to human affairs. I called it *The New American Destiny*—I was always good at titles—but when I tried it out on the University of Chicago Press their reader said it was just another book drawing analogies from animal to human behavior. This later became the foundation of my book *Aggression,* published by the same press in 1958, but now suitably revised to meet professional rather than popular standards.

It was obvious that even if I could get my book published (not many books were published in the 1930s) no one would pay any attention to it. Still following my dream of becoming a literary author, I had written it in a semipopular fashion, which was inappropriate for the times (S. J. Gould was still far in the future). And, from the scientific viewpoint, it was obvious that adequate studies of animal societies were meager.

So I began to do field and laboratory research on animal behavior. We rented a six-acre farm on which I raised sheep and so produced a naturally formed flock. My study of its social behavior and organization resulted in my 1945 monograph. I began a parallel field study of mountain sheep in Yellowstone Park, but this ended in the summer of 1942 when all travel was curtailed; I got little more than some movies of various flocks of rams in the month of June.

I intended the sheep monograph to be a model study of an animal society, employing biological techniques and concepts. Technically, I followed the basic outline of Observation, Classification, Hypothesis Formation, and Experimentation. Observation led to the concept of the "behavior pattern," a segment of behavior having a definite function and serving as the unit of description. Each species should have a characteristic and finite set of behavior patterns whose description formed a basic prerequisite for behavioral study. Unknown to me at the time, Tinbergen was developing a similar concept and technique, which he called the *ethogram.*

Animal Behavior and Sociobiology

After my experiences in the 1940s I had decided not to write any more books unless someone asked me; I was tired of writing them on speculation and being turned down. Then, about 1955, Ralph Buchsbaum, who had taught at the University of Chicago and later went to Pittsburgh, asked me to write a short book on animal behavior, as part of a series that he was editing for the University of

Chicago Press. The general idea was to produce a monograph series on a level suitable for graduate students. I would have preferred to do something on sociobiology, but Ralph wanted it more general.

And so Sally and I wrote the first edition of *Animal Behavior.* As usual, I dictated it to her, a warm and sympathetic audience of one, and this is the tone that pervades the book. As soon as he saw the manuscript, Buchsbaum became enthusiastic and encouraged me to make it readable and nontechnical, using lots of pictures, following the model of his successful *Animals Without Backbones.* I myself had always believed that good writing could make even the most complex ideas interesting and intelligible, and was happy to respond. I was also fortunate in having Catherine Nissen as an editor at the University of Chicago Press. An accomplished primatologist who had home-reared the chimpanzee, "Vicki," she also caught fire with enthusiasm and contributed much to the book.

Animal Behavior, published in 1958, was the first book to appear in a growing field. Anyone who reads it now can see that it includes all the important theories relating to animal behavior, and that its basic organization relates to levels of organization. I did not actually state my ideas of behavioral systems, as I felt there was not yet enough evidence to justify the concept, but included them only as behavioral categories.

The book was an instantaneous success, but not in the way that I had originally planned. Instead of stimulating my scientific colleagues to new ways of thinking, it was sent out by a book club, appeared in paperback, and was read even by high school students. I have had the gratifying experience in later years of having young scientists approach me with appreciation and say that reading this book first got them interested in animal behavior.

It was not ethology per se, and it was not sociobiology, except for one chapter. Ethology had been defined as the science of behavior, but it was at that time dominated by the concept of instinct. As a geneticist I found little use for it, as it assumed genetic determination without experimental test and therefore often became an arbitrary label.

More useful to me was the concept of the behavior pattern, an organized unit of behavior. How, when, and by what it was organized was a problem for experimental analysis. Classifying behavior patterns according to similar functions produced nine major groups, and I hypothesized that each of these comprised a behavioral system with underlying emotional, motivational and physiological organization. Where clear-cut popular terminology did not exist, I used the biological technique of inventing new terms based on Greek and Latin, and did so with the help of John Charles, the classics professor at Wabash College. And so such terms as *allelomimetic* (doing what the other animals are doing with some degree of mutual imitation), and *epimeletic* (giving care or attention; I think this is by far a superior term to *altruism* as used by later sociobiologists, with its burden of surplus meaning) came into being.

Research on Fighting and Violence

Still another term was *agonistic behavior,* defined as behavior adaptive in situations of conflict between members of the same species. The ever-recurring threat of war and my own past experience gave this term impelling urgency, and it has been the most widely adopted of these terms.

One of the things I had done in the summers of 1938 and '39 was to go to the Jackson Laboratory as a summer investigator at the invitation of Bill and Elizabeth Russell, who had gone there after obtaining doctoral degrees with Wright at Chicago. So I looked at differences in fighting behavior between males in various inbred strains of house mice. One of the people who helped was George Snell, who provided some of his surplus mice.

I discovered that there were striking strain differences, which I described in a qualitative fashion in a paper in the *Journal of Heredity* (1942a). I told this to W. C. Allee, one of whose students, Uhrich (1938), had previously done a study of fighting albino mice. So Allee and his then research assistant Benson Ginsburg did a systematic round-robin study of fighting between these strains. When I saw the results, I immediately saw that the most striking of these was not so much that strain differences were confirmed, but that the experience of defeat produced very long-lasting results that could be interpreted in terms of Pavlovian conditioning. This in turn set the stage for a series of experiments of my own on the effects of training in victory and defeat and the genesis of maladaptive behavior (Scott & Marston, 1953). It also presaged a long and productive collegial relationship with Ginsburg as a summer investigator at the Jackson Laboratory.

Although my biological training included the concept of systems, I first approached the problem of analyzing destructive violence from the mechanistic viewpoint; that is, of one-way causation. From this I progressed to the concept of multifactorial causation, a commonplace concept in genetics where many genes affect the same characteristic. Then I classified causes into groups according to levels of organization: genetic, physiological, organismic, social, and ecological. My research on the dominance-subordination relationship in goats led me to the conclusion that a major determinant of agonistic behavior is the nature of the social relationship in which it is expressed. But I was still thinking of the social relationship as a one-way determinant. It was not until many years later that I synthesized these findings with those of others to develop a *polysystemic theory* of agonistic behavior. At the Center for Research on Social Behavior, which I founded at Bowling Green State University, one of our projects was a book on violence (Neal, 1976). One of the authors, Don K. Rowney, a brilliant researcher of Russian history, waxed enthusiastic over systems theory, which was new to him, and I was stimulated to new insights.

Briefly put, polysystemic theory states that agonistic behavior becomes organized on every level of systems organization, from genetic to ecological and that it is affected by every process of organization, developmental and social interactional in the individual, and by cultural and biological evolution and even ecosystem change in populations (Scott, 1975).

Agonistic behavior can be either useful and adaptive, or destructive and harmful, with all gradations between. Those experiments that produce destructively violent fighting in otherwise normal individuals overwhelmingly support the conclusion that destructive violence is strongly associated with either social disorganization, (i.e., organization that has been destroyed or weakened), or lack of social organization.

From this it is but a step to the theory that destructive agonistic behavior can be brought about by disorganization at any systems level; harmful mutations in genetic systems, brain damage and endocrine disorders in physiological systems, mental disorders on the organismic level, and disorganized or unorganized social relationships. The effect of ecosystem disorganization is more hypothetical; it probably works through upsetting social organization.

The above theory should be part of the armamentarium of any clinician. But unlike the usual rule with respect to matters of health, which is to look for physical causes first, the first thing to examine in a problem of aggression is the social relationship within which the undesirable behavior occurs. Agonistic behavior *is* social behavior. Further, the nature of the physiological, emotional, and motivational basis of agonistic behavior in vertebrates suggests that one should first look for predisposing circumstances within the past few hours or days rather than the remote past.

What of warfare, or group violence? Warfare is not disorganized behavior but behavior highly organized for the purpose of destruction. But even here polysystemic theory sheds some light. Warfare can be considered disorganized behavior on a higher level, that between nations. The obvious remedy is international organization of a satisfactory sort. This can be achieved even by warlike nations, as for example in the peaceful relations between the Scandinavian nations. The weaknesses and failures of past and present attempts at global international organization should not blind us to the fact that this is the road to follow.

Within nations, how do we organize satisfactory social relationships within which agonistic behavior appears only in a useful context? The answer, based on the Pavlovian principle of passive inhibition, is that being peaceful in a particular relationship forms a habit of being peaceful, which grows stronger with time. The best way to induce peaceful behavior is to stimulate constructive and positive behavior. Skilled teachers are adept at doing this, beginning in nursery schools, and are generally successful. Among adults outside the educational institutions, the best method is employment in a constructive, satisfying, socially respected, and reasonably paid job. It follows that a major and highly effective way to reduce violence would be to insure full employment. But it may take a future political Messiah to bring this politico-economic change about (Scott, 1977).

With respect to research itself, the work that I and others initiated on the phenomenon of fighting in mice has expanded into an area involving dozens of researchers all over the world and resulting in the publications of hundreds of papers and books. It is highly gratifying to me that as the field matures, the phenomena that I originally described are now generally accepted and verified, and that when drug tests are made, there is a tendency to look at effects on the whole range of mouse behavior and social organization.

The Jackson Laboratory and the Rockefeller Project

In 1945 I had been at Wabash College for ten years and was thirty-five years old. My elder son, John Paul, had been born in 1943, and a second, David, was to appear in 1946, completing my family. I had discovered for myself a new field of research, but there were obvious limitations to what I could accomplish at a small liberal-arts college with no graduate program. Under what I would have then called normal economic and political conditions, I would have long since moved to a different sort of institution, but everything was frozen. If I had showed enthusiasm, I might have gone to McGill University in 1940, but did not want to involve my children in the foreign and bitterly divided culture of Quebec.

Then in the spring of 1945, my first real opportunity came. Alan Gregg, the director of Medical Sciences for the Rockefeller Foundation, had long believed that psychiatrists and psychologists were paying too little attention to heredity. His section had previously financed a large project on dog behavior at Cornell for C. R. Stockard, the brilliant embryologist whose work had first established the existence of critical periods. Then Stockard died suddenly, leaving his work unfinished. Even worse, the program had been set up to test the hypothesis that the dog breeds represented "endocrine types," and the breeding results showed that physical form and endocrine function segregated independently. So the whole project folded, leaving behind only a book that was essentially a monument over the grave of a dead theory.

Gregg was a Harvard classmate of C. C. Little, the director of the Jackson Laboratory at Bar Harbor, Maine. Little was a very unusual administrator in that he had been a university president with an earned Ph.D. in mammalian genetics and had started the first inbred mouse strain in 1909 (this strain, the dba, is still going, more than seventy years later). Gregg decided that Little was the obvious person to undertake the project in dog behavior genetics, especially since everyone knew that he was intellectually brilliant, and this might get him back into research.

Little responded with enthusiasm. He had been keeping the Jackson Laboratory going on a shoestring since 1929, and Gregg was offering him $50,000 a year for ten years plus an extra $50,000 to set up a laboratory. This was in 1945 dollars that had a buying power of at least five times those of today.

Then there was the question of who was to do the work. They did not have far to look, as I was at that time the only person in the country with formal training in genetics who was interested in the genetics of behavior. Cultural inheritance was also in my favor. Little's mentor at Harvard had been W. E. Castle, the first mammalian geneticist. He was not a brilliant man, but he trained an extraordinary crew of graduate students, including Little himself, Sewall Wright, and George Snell. In 1945, the small research staff at the Jackson Laboratory included Snell, Bill and Elizabeth Russell, who were students of Wright, and Lloyd Law, the youngest and last of Castle's students. In this group, my credentials were ready-made.

I decided to take the position and so set the stage for my next twenty years of work. Little kept his promise; I really did have a free hand to do all the things I had dreamed about at Wabash College and during my year in Boston. I invented a title for myself, the chairman of the Division of Behavior Studies, and set out to find a staff. Learning from Stockard's example, I decided that I must have a coworker of roughly equal capacities who could take over in case I became incapacitated. I also saw that one of the assets of the Jackson Laboratory was its location; people would like to come there in summer for work, so I formally organized the Summer Investigators Program. In future years we were to benefit by the presence of dozens of visiting investigators: biologists interested in animal behavior, plus comparative, physiological, experimental, and developmental psychologists. The story of their work would take at least another chapter. During this first year of the program I set aside $1,000 to pay a group of five potential permanent staff members to come for the summer of 1946. The group included C. S. Hall, then a professor of psychology at Western Reserve University, a former student of Tryon, who was later to christen a new field of behavior genetics (which he called *psychogenetics*) with a chapter in Stevens's *Handbook of Experimental Psychology* (1951). Then there were Benson Ginsburg and Elizabeth Beeman with relatively new Ph.D.'s from Chicago obtained under Wright and Allee, respectively. Still others were John Fuller, a professor at the University of Maine with a Ph.D. in physiological ecology, and Howard Vogel, my former colleague at Wabash.

The payoff was immediate. Hall only stayed for six weeks but made a major discovery. He decided to test two strains of inbred mice for audiogenic seizures, following up similar work on rats. The apparatus was simple and inexpensive—a wash tub, an electric doorbell, and a stopwatch. The procedure was equally simple. The experimenter placed a mouse in the tub and rang the doorbell for two minutes. The results were dramatic. Of the first two strains that Hall tested, the dba 2 strain mouse would run wildly around the tub, go into a convulsion, and die. If there was ever an example of maladaptive behavior, this was it. A C57BL6 mouse on the other hand, would show an initial startle reaction to the noise and emerge from the experience unharmed. Hall thus had the good luck to select two strains that showed a clear-cut genetic difference.

All of us were excited by the discovery. Ginsburg and Fuller and Emilia Vicari (at that time a member of our current staff) went on to do a number of analytic experiments with seizures in subsequent years. I myself thought of it as an interesting phenomenon, but one that seemed to be confined to rodents, with no application to humans except that it definitely was related to the problem of mental health.

From the very outset of my interest in animal behavior, I saw that if I and other researchers expected substantial financial support it would have to be associated with some major practical problem or problems. An obvious application was that of mental health, and I pursued this for some years.

My first idea arose out of my research with fighting mice, and I employed the then common theory that maladaptive behavior arose out of mental conflict. In a fight, a mouse is always confronted with the decision of whether to fight or run away, implying an emotional conflict between anger and fear.

My assistant Mary-'Vesta Marston and I arranged such a conflict by first giving male mice an experience of severe defeat, which had the effect of inhibiting fighting. Then we gave the same mice daily training in fighting, the technique being giving each mouse an opportunity to attack a helpless mouse that we bumped against him. After some weeks, a mouse trained in this fashion did indeed begin to do peculiar things, waving a hind foot in the air, and jumping up in the air and shaking. These could be interpreted as inhibited patterns of attack. Had I investigated the physiological reactions of these mice, I could probably have found even more interesting things, but it is hard to do a physiological analysis on a mouse without killing it.

It was not until some years later that I saw that audiogenic seizures provided a general model for all the cases of experimental maladaptive behavior that had been reported in animals. Seizures occur only as a result of a combination of four conditions: (1) overwhelmingly strong motivation, produced in this case by auditory stimulation, (2) a lack of opportunity to escape, (3) an inability to adapt to the situation (it never occurs to a mouse to put its paws over its ears), and (4) genetic susceptibility.

From the viewpoint of this four-factor theory, my experiment with conflict took on a different meaning. The experimental mice were unable to escape from the situation (a fact that usually escapes the notice of the experimenters). They were unable to adapt because they had been so strongly trained in defeat. The motivation to fight, according to the laws of learning relating to repeated reinforcement, should have increased with each training session. The fourth factor, genetic susceptibility, was not measured, because only one strain was used, but these mice obviously were susceptible.

All the numerous experiments on experimental neurosis by other experimenters can be interpreted in a similar fashion.

In my own work with dogs, I was to encounter repeatedly a major sort of maladaptive behavior: the kennel dog or separation syndrome. The principle diagnostic symptoms are persistent fear of anything strange, and a tendency to be untrainable. A dog that is raised in a kennel or any other limited environment for as long as six months and then suddenly placed in a different environment has a very high probability of developing the syndrome. At Bowling Green, I and my graduate students made an extensive experimental study of this phenomenon, varying both breeds and the amount of early separation experience.

Those animals that had no experience outside the kennel before five months of age were not only fearful in strange situations but poor learners. In a few cases we were able to place such dogs in a home where those belonging to one breed exhibited the typical kennel dog syndrome. In another more aggressive breed, the dogs did not flee but bit people. Thus, the genetic factor determined the symptoms.

The separation syndrome therefore involves a high degree of motivation (separation distress becomes more intense with prolonged separation), an inability to escape from the situation (if allowed, such a dog would try to go home), and an inability to adapt (related to high emotional arousal). When one applies therapy in the form of desensitization techniques, one can get the dog used to a particular strange object but when a new one appears, the dog reverts to its old symptoms.

Our experiment with early separation showed that without such experience, separated older animals were difficult to train in a variety of tasks. Sam Corson of Ohio State University discovered a more dramatic instance that has human implications.

I had sent him some surplus adult dogs from the Bowling Green laboratory for use in his Pavolvian conditioning experiments. Some of them, particularly the

Beagle-Telomian hybrids, could not be trained to stand still on the Pavlovian stand. Knowing that amphetamine had been used to control hyperactivity in human children, Corson tried it on the dogs, and it worked. Ben Ginsburg at the University of Connecticut obtained similar results with some dogs that we sent him.

To make a long story short, we did an extensive series of experiments at Bowling Green, but found that amphetamine worked *only* if the dogs were sent away and developed the separation syndrome. This implies that amphetamines and similar drugs should be effective on only those hyperactive children who are experiencing separation distress. It would also explain some of the addictive properties of amphetamines. The experiment further implies that severe cases of separation distress may account for some school and college failures.

More generally, functional maladaptive behavior is rooted in major sources of emotional motivation. Since Freud's day, psychotherapists have been principally interested in three of these sources of motivation, sex, anger, and fear. The discovery and analysis of separation distress adds a fourth sort of emotional motivation that may lead to maladaptive behavior, given the necessary combination of the other three antecedent conditions.

My work on the development of social behavior and the process of attachment also has implications for problems of mental health and maladaptive behavior.

Conference on Genetics and Social Behavior

During the first six years of the Rockefeller Project I organized three research conferences, each of which had far-reaching effects. The first, on "Genetics and Social Behavior," was held at the end of the summer of 1946, with the specific purpose of obtaining the best possible advice concerning the conduct of our research. To this we invited (1) anyone who had published research on the genetics of behavior (H. J. Bagg, B. E. Ginsburg, C. E. Keeler, D. M. Levy, C. S. Hall, W. M. Dawson, L. V. Searle, H. H. Strandskov, E. M. Vicari); (2) the leading comparative psychologists in the United States (R. M. Yerkes, T. C. Schneirla, O. H. Mowrer, C. P. Stone, C. R. Carpenter—Stone, from Stanford, had been another successful trainer of graduate students; among them were the future kingpins of primate research, Carpenter, Nissen, and Harlow); (3) various scientists who had connections with Little and the Jackson Laboratory (W. L. Russell, G. W. Woolley, D. T. Allen, M. Kennard); (4) a group of leading science writers (W. Kaempffert, R. D. Cook, D. Dietz, G. Lal, A. Scheinfeld); (5) a group of leading physiological psychologists and psychobiologists (C. T. Morgan, F. A. Beach, N. E. Miller, E. W. Dempsey, H. S. Liddell, W. C. Young); (6) a pair of prominent social psychologists (G. and L. Murphy); (7) A group of brilliant young animal behaviorists (N. E. Collias, E. Beeman, J. L. Fuller). Invited because of their official positions were Alan Gregg and Robert Morison of the Rockefeller Society, and Fairfield Osborn of the New York Zoological Society.

It was a star-studded gathering, and the conference was enormously successful. Yerkes, as dean of comparative psychologists, was chairman (few will now remember that his Ph.D. dissertation, *The Dancing Mouse,* was a study in behavior genetics), and the secretary was Frank Beach, my contemporary from the American Museum. Yerkes was a forceful and impressive person, and I have always remembered one of his remarks: "I have never seen a disembodied mind." He also told me that Beach was slated to succeed him at Yale, as "Beach is the only young psychologist in the country who is seriously interested in comparative psychology."

Beach was to be enormously and unselfishly helpful during the next several years, at first informally, and later as a member of our scientific advisory board. He helped us with outside contacts and finding sources of supplementary income, as well as by using his lively and inquiring mind as a sounding board for our ideas. On our part, we furnished him with some dogs on which he could do sex research, and he pointed out to me my first example of the separation syndrome. The dogs that had been normal well-adjusted puppies at the Jackson Laboratory, became permanently timid and fearful when sent to Yale. I made a mental note of the occurrence, but did not appreciate its full significance until several years later.

C. R. Carpenter had served as a captain in the army and appeared at the conference in uniform. He had an authoritarian manner and to some he seemed to be acting like a jack-booted Prussian. But I had enormous respect for his work on the howling monkeys, the first major study of a mammalian society under natural conditions.

Carpenter had withdrawn from active primate research; he was a proud and ambitious man, and he knew he had done important work, but the applause (among psychologists at least) was muted or nonexistent. He decided that if he was to advance professionally, he would have to do something else, and he turned his talents to the production of scientific movies, of which several are still available.

But he also firmly believed that the remaining species of wild primates *must* be studied. He enlisted the aid of Fairfield Osborn, then the president of the New York Zoological Society, and got him to set up a program for behavioral research at the Bronx Zoo. Osborn was helpful in other ways, particulary in financing another conference in New York, which I shall describe below.

For the rest, the most important aspect of the conference was that it was one of the first postwar meetings. The war was over, right had triumphed, and scientists could once more live in the fresh air and sunshine of discovery.

The Summer Investigator's Program at the Jackson Laboratory was an ideal way to recruit new staff. They saw the Laboratory under ideal conditions, and stayed long enough so that I could get to know them. Our choice for a permanent staff member eventually narrowed down to Fuller, and he agreed to come permanently in 1947. It was a fortunate choice. He and I were about the same age, personally congenial, and mutually stimulating intellectually. We agreed to divide the work of the Rockefeller Project in a nonoverlapping fashion, he to take the older puppies and to concentrate on physiological measures, while I took the younger ones and specialized in development and measures of social behavior.

It was a fortunate choice in another way. By now I had gone beyond my initial interest in genetics and behavior and was much more interested in social behavior, but to Fuller behavior genetics offered a new and fascinating challenge. He, together with Robert Thompson, was to write the first major book on behavior genetics (Fuller & Thompson, 1960).

Results of the Rockefeller Project

Fuller and I collaborated for some thirteen years before the data-collecting phase of the Rockefeller Project finally came to an end. We had developed our tests on five different breeds and selected two of them, the African basenji and the American cocker spaniel, for cross breeding. Our experimental design was a classical Mendelian one that permitted analysis by both factorial analysis and analysis of variance. We made reciprocal crosses and then backcrossed the F_1 males to their purebred mothers, so that the backcross and F_1 generations had the same mothers. With two litters from each mating, both litter effects and genetic effects could be analyzed against the same maternal environment. We also crossed the same F_1 males to their sisters to produce an F_2.

Midway in the experiment, the technique of analyzing F_1's from all possible crosses between several different strains was developed by Mather (1949). In some ways this technique would have given us more information, since we could have involved four or five breeds instead of two, but it was too late to change. Also, I have since seen the results of this technique applied to mice, and it also has its difficulties and limitations. It would be very difficult, for example, to get the same female to successfully mate with five different males.

How to analyze the data? We had tested some 300 puppies in each of thirty odd tests, and I calculated that we had at least 8,000 separate pieces of data, and many of the tests included multiple measures. Early in the enterprise, Joe Royce, then a graduate student of Thurstone, came to do a factorial analysis of the data, in the hope that the factors might represent genes with multiple effects. Later, Anne Anastasi and a student of hers performed another factorial analysis on some of the problem-solving tests. These calculations were performed on electric adding machines and required weeks of work to calculate the necessary correlation coefficients.

Finally, when the data were complete, we enlisted the aid of Loring Brace, a student in physical anthropology at Harvard, to make a grand final analysis, including all sorts of different measures. The basis of his work was a 50 × 50 correlation table, a task that would have been impractical a few years before. When we started the project, IBM cards were available, but the only machine we had was a mechanical card sorter. But Loring Brace could work with electronic computers. Even so, he had to devise all the programs, and he finally got the work done by arranging to use a life insurance company's computers at night (these arrangements somehow involved a case of scotch). It took him all of one night, after weeks and months of preliminary processing of the data, to accomplish what today's computers could do in a few minutes.

The results were interesting from a negative as well as from positive viewpoints. In the first place, correlations between different behavioral tests were low, averaging about .30. The dogs reacted to each test as if it were a new situation, with little carryover. There were breed differences, but one breed would do well in one test and poorly in another. Consequently, there was no evidence in favor of a general factor of intelligence, or *g*.

Nor was there any good evidence of general temperamental factors. The same breed might be fearful in one situation and confident in another. Basenjis were fearful of strange situations but not at all fearful of other dogs.

The data provided a clear test of Sheldon's somatotype theory of behavior. A variety of physical measurements were made, and these correlated highly with each other, yielding a factor of physical size, but no indication of separate physical "types." Furthermore, measurements of physical size did not appear in the factors that included behavioral tests; that is, physique had little effect on behavior. This does not mean that physique cannot affect behavior; greyhounds run faster than dachshunds, and large dogs become dominant over small ones. What it does mean is that physique affects behavior only in direct and obvious ways, directly contrary to the racist notion that physical type is correlated with intelligence, temperament, or what have you.

On the positive side, the data showed that emotional and motivational factors do have important effects on performance in tests involving problem solving, a finding that needs to be pursued in human experiments. It is possible that human differences in "intelligence" reflect only differences in motivation rather than cognitive capacities, and this research should be followed up.

The analysis of variance produced a different set of findings. All the traits that we had tested showed some degree of heritability (the proportion of variance attributable to hereditary variation). This was not surprising, since we had chosen many traits because they appeared to be variable, and had deliberately selected breeds that were widely different. Nevertheless, it supports the general hypothesis that in any population in which wide genetic variation is permitted, any trait that is studied will show some degree of genetically determined variation.

It does not follow that behavior is genetically determined; only that some of the *variation* in behavior is genetically determined.

I had started with the notion that it would be important to study behavior developmentally, as in a young animal one should see the effect of genetic variation before it became contaminated with the effects of experience. Actually, I found that the behavior of young animals was extremely variable; it was only later that behavior became more fixed as the result of habit formation. Genetic variation had the effect of making it more probable that certain alternative behavior patterns would be chosen by the animal in preference to others in the repertoire. Genetics does not put behavior in a straitjacket.

Analysis of variance showed that the average amount of variance that could be attributed to breed differences was 27 percent of the whole. A certain amount could be attributed to within-breed differences, perhaps 12 percent, making a total of approximately 40 percent. This was under conditions in which we had made every possible effort to eliminate environmentally caused variance. The heritability of different traits based on breeds alone varied from nearly zero to around 66 percent, the highest being physiological traits such as heart rates. Perhaps because we had chosen breeds of approximately equal size, the heritability figures for physical measures averaged almost the same as the behavioral ones.

Such results have many implications for human behavior genetics. For example, we found that part of the difference between purebred puppies of different breeds was caused by growing up with similar animals; if puppies of two breeds grew up together, the differences might be lessened. Or, especially in measures of social interaction, their differences might be magnified. The whole question of interaction between similar and dissimilar genotypes is still to be adequately explored. This is especially pertinent to the behavior of identical twins (similar genotypes) and fraternal twins (different genotypes).

Compared with similar heritability figures in human research, especially those for intelligence tests, the human figures appear to be far too high. It should be pointed out that human intelligence testing involves the use of language, whereas problem-solving tests in other animals seldom if ever involve symbolic reasoning. It is possible that, since intelligence tests involve the use of language, that there may be highly heritable variations in certain kinds of language ability and symbolic reasoning capabilities. If so, the evidence will never be convincing until these capacities are discovered and their physiological basis demonstrated. Also, since language acquisition involves so much practice, it is possible that initially small genetic variations are magnified by use. The fundamental question is: does language and symbolic reasoning involve an entirely new genetic system? Alternatively, it is possible that cultural inheritance and genetic inheritance are inextricably intertwined and confused.

Origin of the Animal Behavior Society

At the 1946 meeting, Carpenter and I, together with Schneirla and Beach, laid the groundwork for the formation of what I called "The Committee for the Study of Animal Societies Under Natural Conditions." Bound by my informal contract with the Jackson Laboratory, I personally was to study an animal society (canine) under very artificial conditions indeed, and Carpenter was also out of field work. But we could influence others.

I had been regularly attending meetings of the American Association for the Advancement of Science ever since my Ph.D. had been awarded. At that time the biological societies all met together under the aegis of the AAAS; the two in which I was particularly interested were the American Society of Zoologists and the Ecological Society of America. I regularly presented a paper on behavior each year, but it usually got thrown in with a group of unrelated topics, as did the other five to ten papers on behavior.

So I arranged a joint session each year, calling it "Animal Behavior and Sociobiology," and invited researchers to submit papers. I usually acted as the session chairman, and after the papers, we would have a luncheon meeting with informal discussion. The first of these meetings took place in 1946.

As the years went by, these sessions became larger and more successful; we always had bigger audiences than any other session. Finally, Lee R. Dice, then president of the Ecological Society of America, suggested that we organize more formally, as sections of Animal Behavior and Sociobiology in each of the two parent societies. He predicted that we would eventually wish to become a separate society, and this finally happened in 1964 at the Montreal meeting. By this time Martin Schein was acting as secretary of the Sections: his enthusiasm and hard work were largely responsible for the success of the meetings, and he became the first president of the Animal Behavior Society.

Conference on Methodology and Techniques

The original Committee for the Study of Animal Societies under Natural Conditions had no formal organization. I was its secretary, and I kept it alive by dint of having adequate secretarial service of my own, and time for correspondence. Membership was by invitation, and membership was offered to all interested scientists. We agreed that there was a strong need for developing adequate techniques for field studies, and we invited the leading zoologists and psychologists in the country who had done such studies to present papers at a conference sponsored by the New York Zoological Society. Among others I could with good conscience invite my father, J. W. Scott, to present his pioneering study on the social organization of sage grouse; there were other reasons for inviting him because the New York Zoological Society was setting up a field research station in Jackson Hole, Wyoming, and it was important to have good will on both sides.

My own paper I subtitled "A Study in Social Systematics." I intended it to be a cogent example of how to study social systems, but perhaps because it involved dogs and wolves, and almost no one else was studying these species at the time, it attracted little attention in this country.

The principle technique emphasized by the conference was that of individually identifying the animals studied, and several devices for doing this were suggested. The monograph entitled "Methodology and Techniques for Studying Animal Societies Under Natural Conditions" is still useful reading for anyone entering this field.

An important new face in this group was John T. Emlen, an ecologist who had done field research on both birds and mammals. He later inspired some of the major research in mammalian societies. For example, for years Carpenter and others had talked of the importance of studying the disappearing wild primates, but it was not until Emlen and his student George Schaller went to Africa to study the mountain gorilla, that this branch of primatological research got under way.

Conference on Early Experience and Mental Health

I had originally planned to hold a major research conference at the Jackson Laboratory every five years, and as the time approached it was obvious from our developmental studies with the dog that this species exhibited a major critical period beginning at about three weeks of age. I decided to hold a conference around this central theme and to place it within the general framework of research on early experience. At the same time there was a high degree of interest among experimental and clinical psychologists in testing some of the theories concerning early experience proposed by Freud. The conference therefore struck a responsive chord that was felt not only by the conference members, but also by others who heard about it.

The next ten years were to see a burst of research in this area along many different lines, including the isolation experiments by D. O. Hebb and the research on traumatic perinatal experience in rats by Seymour Levine, Victor Denenberg, and others.

Research on Early Experience

One day there came to Bar Harbor two young scientists from the newly formed National Institute of Mental Health. In effect they said, "Look here, Congress has appropriated some money for research; why don't you apply?" I asked about the requirements; there was a form of one or two pages on which one filled out one's scientific life history and briefly outlined an idea for research. If the applicant had a good scientific reputation, and the budget seemed reasonable, he got the money with no strings attached. (Compared to what goes on today, with endless forms and a demand that one state exactly what he will do throughout the life of the grant, it all seems like a wonderful dream.)

So was born MH123, a grant to support research on the effects of early experience on fighting behavior. I used it to pay for my research on mouse fighting and to hire a new staff member, Emil Fredericson, a new Ph.D. who had worked under Calvin Hall as a clinical psychologist.

His mentors agreed that he was too sensitive to be a good clinician, but he was a gifted researcher. He was the only person I ever knew who could think like a mouse; that is, he could design experiments that fitted murine capacities and hence were meaningful to both mouse and man. He developed a more humane measure of mouse fighting—time to the first attack, instead of letting them fight it out in a bloody battle—and he first called my attention to the phenomenon of distress vocalization resulting from separation in young puppies, which later became the subject of much of my research on attachment and separation. But Fredericson, in spite of his success, was still bound to be a clinician, and he left after three years.

Meanwhile the laboratory was graced with a brilliant young postdoctoral fellow, sent to us by Lee Dice of Michigan. This was John King, who had recently completed a beautiful field study of social organization in the prairie dog (really a prairie ground squirrel). In my opinion, this is one of the best field studies of a mammalian society that has ever been done.

When Fredericson left, the obvious thing was to offer Jack King the position in MH123 and, to our good fortune, he accepted. He attended the Conference on Early Experience, and from this he developed his own line of developmental research, but that is his story. I sometimes wonder if I did not help to divert a wonderful field sociobiologist into another line of research, but he was also a family man, and family life does not fit with excursions into the wild, as I had found myself.

Social Development and the Critical Period Phenomenon

As with any other field of research, the primary task is description, followed by classification. When I and my research assistant Mary-'Vesta Marston described the postnatal development of puppies and classified our results according to the times at which particular processes were active, we found that early development could be divided into three periods: (1) *Neonatal,* marked by establishment of the neonatal process of nutrition, nursing; (2) *Transition,* including a changeover from neonatal to the adult forms of nutrition, sensory capacities, locomotion, and learning; and (3) *Socialization,* including the establishment of primary social relationships and attachment.

At first I assumed that similar periods, existing in the same order, would be found in other mammals. This turned out not to be the case. The same processes could be recognized but not in the same order. For example, in the puppy the eyes open first, then the ears. But in the mouse the developmental changes occurred in the opposite sequence. Even more striking, a comparison of human development indicates that the three periods, with their accompanying processes, occur in a different order: neonatal first, socialization second, and transition third. This results in humans forming their primary relationships with their caretakers rather than with their peers, and this in turn is associated with a different form of social organization: a group including all ages and both sexes, as contrasted with the pack organization of dogs and wolves.

One of the major outcomes of this research was to generate hypotheses that could be tested by students of human development. My feeling was, and still is, that a strongly stated, testable hypothesis will produce useful results, irrespective of whether it turns out to be correct or not. On the basis of our descriptive results plus some of Fuller's developmental studies, I stated that the neonatal puppy did not seem to form associations, and that the capacity to learn appeared rather suddenly toward the end of the transition period. My colleague Walter Stanley challenged this hypothesis and embarked on several years of work and some elegant experiments showing that the neonatal puppy actually could make associations, although they were mostly limited to sucking behavior and took place much more slowly than the one- or two-trial learning of older puppies.

Such results stimulated the researchers in human development to look again at the learning capacities of neonatal babies, and they have discovered that the human infant also organizes its behavior as a result of experience.

However, both these human experiments and those of Stanley have concentrated on demonstrating that some form of learning takes place and have neglected the more important problems of developmental changes in learning capacities and whether or not any of this learning is retained. Some exceptions are B. A. Campbell and Z. M. Nagy who are researching the problem of infantile amnesia. There is evidence that while some neonatal behavior is modified by learning, it is never recalled or used in later life.

Two experiences impressed me with the concept of critical periods in behavioral development. If there is anything that is characteristic of sheep, it is to follow the flock. Yet when I took a female lamb and raised it on the bottle for the first ten days of life, it followed its human caretakers everywhere and never followed the flock, even though it lived in the same field for the next several years.

The second was observing the process of primary socialization in the puppy, either to dogs or to humans. At the proper age, a puppy taken from the litter and exposed to humans will form strong attachment within twenty-four hours. When I first reported this to the 1951 Conference, we had only observational data to support it, and this raised some doubts in the mind of the audience. Obtaining experimental proof was difficult, mainly because puppies formed attachments so easily to people that it was difficult to find a puppy that was not socialized. Finally, Freedman, King, and Elliot (1961), my associates at the Jackson Laboratory, ran an extensive experiment in which the untreated controls were those reared in our large outside pens whose high board fences cut off visual contact with people. The results definitely established the dimensions of the critical period.

I was content at the time to leave it at that, a major phenomenon definitely established. Meanwhile, excitement over critical periods had spread to several other areas. One of these was a postnatal critical period for handling effects in rats. Ted Schaefer, then a graduate student at the University of Chicago, was the first to examine it systematically and discover that the quality of handling made no difference. Victor Denenberg and Seymour Levine, both of whom were at various times summer investigators at the Jackson Laboratory, then took up the problem, and discovered that the endocrine changes associated with the stress of handling permanently modify the physiological and emotional organization of the young animals. This caused Denenberg to challenge some of the statements I had made, but it was not until I attended the 1967 Conference at Prague, in Czechoslovakia, where the theme was critical periods, that I was stimulated to develop a general theory of critical periods (Scott, Stewart, & DeGhett, 1974).

Essentially, this theory states that organizational processes are most easily modified at the time when they are proceeding most rapidly; either before the process has begun or after it has ceased, a process cannot be modified, except that the capacity for it may be destroyed. I also concluded (Scott, 1979) that there are two kinds of organizational processes: developmental processes and maintenance processes. The latter go on almost continuously and hence do not show critical periods, while the former, depending on the time periods in which they proceed, almost inevitably show one or more critical periods.

My first idea, based on the situation in the dog, was that there was a single critical period of behavioral development, because in that species so many organizational processes are crowded into a brief period of life. But comparison with other species, especially humans, with their very long periods of development, showed that there could be as many critical periods as there are developmental processes, and these might be either quite short, as in the period of primary socialization, or very long as in language development.

This general theory can be applied to any developmental process in any sort of system. It is particularly applicable to social systems. Indeed, I have found that social psychology has produced only two reliable phenomena: critical periods and social facilitation.

Aside from its practical consequences, a theory is important only if it generates research; that is, it is interesting enough so that people want to test it. Critical-period theory raises the question, what is the nature of the organizational process? I found this question particularly interesting with respect to the process of socialization, and while at Bowling Green entered on a long series of experiments regarding it.

One of the best measures of attachment is the emotional distress that results from separation. My colleague Emil Fredericson first called my attention to this phenomenon, when he took home a four-week old puppy and found that it vocalized continuously for twenty-four hours. This led to a somewhat low-pressure research program at the Jackson Laboratory that developed into a major project after I came to Bowling Green. I still do not have all the answers.

The first key to understanding the process was to analyze the external circumstances. Cairns and Werboff (1967) found that all that is necessary for a puppy to become socialized to a rabbit is visual contact, as we had found with canine attachment to humans. They also found that active interaction would facilitate attachment. I theorized that repeated separation should strengthen attachments on the grounds that there was an internal reinforcing mechanism that punished the puppy for separation (separation distress) and rewarded it for reunion (relief of distress and perhaps some positive emotion as well). This hypothesis proved difficult to test because puppies so rapidly reached the maximum degree of reaction to separation.

I have now come to believe that any form of emotional arousal, pleasant or unpleasant, will facilitate the attachment process. This thesis is still to be tested adequately, but has obviously important implications. Sexual behavior seems to fit the hypothesis, but what if physical punishment also brings about attachment?

To sum it up, we have learned a good deal, through pharmacological experiments and studying interaction with other emotions, about the physiological basis of separation distress. My colleague at Bowling Green, Jaak Panksepp, has gathered evidence that this painful emotion is related to the brain opioid system. But we still know almost nothing about the physiological basis of attachment itself.

Another unsolved developmental problem concerns the effect of sexual maturity on the development of cognitive capacities. In humans, there is good evidence from recovery from brain lesions that the brain becomes "frozen" with respect to language learning shortly before puberty. In dogs, there is a period from approximately eight to twelve weeks when a puppy learns everything with ease, although not very skillfully, whereas the same things can be taught only with great difficulty later. What would happen to cognitive development if sexual maturity were artificially advanced by hormone treatments? We know that castration, which prevents sexual maturity, has an effect on physical growth; what does it do to the nervous system?

Most thinking along these lines has been in terms of neotony, or early sexual maturity of a larval or immature form, such as in the axolotl, a salamander. But the human case, with its late sexual maturity, suggests the contrary: that sexual maturity has been delayed, prolonging the early developmental period. One also wonders what effect the current tendency toward earlier and earlier sexual maturity will have on human language and cognitive development.

Center for Advanced Study

In 1963–64 I was invited to spend the year as a fellow in the Center For Advanced Study In the Social Sciences, at Stanford. This institution had been organized by a group from the University of Chicago with the aid of a $25 million grant from the Ford Foundation. The plan was to invite fifty fellows of different ages and nationalities, and representing various basic and applied social sciences, to come and do research and writing for a year under conditions as ideal as the Center could make them. Benson Ginsburg, then at the University of Chicago and a former fellow, was instrumental in my appointment.

There were three of us interested in behavior genetics; myself, Bob Thompson, and Gardner Lindzey. Lindzey, a student of C. S. Hall, had been an early summer investigator at the Jackson Laboratory, at first interested in early experience, but later becoming involved in behavior genetics. Bob Thompson, originally from McGill, had worked with John Fuller at the Jackson Laboratory as a fellow. We three already knew each other so well that we exchanged few new ideas. Instead, Sally and I worked on the final version of *Genetics and the Social Behavior of the Dog,* plus the first draft of *Early Experience.*

I decided that if I were ever to leave the Jackson Laboratory (I was then fifty-four years old) now was the time. I looked at openings in California. There was a new university at Santa Cruz, with an interesting idea behind it, that is, residential colleges in the Oxford model and each devoted to a specialty, such as biological science, but I realized that this would be living Wabash College over again. The University of California at Davis had an opening in the psychology department, but psychology was at that time at the bottom of the pecking order in that predominantly agricultural college, and Davis was no Garden of Eden as far as living conditions went.

A choice finally narrowed down to two state universities in the Middle West, each with its advantages and disadvantages. The University of Wisconsin at Milwaukee was a well-established and well-supported institution, but the job was in the Department of Zoology, and I would have had to teach some of the old standard courses in which I had less than a burning interest. John Emlen, who had located the job for me, was a little annoyed when I turned it down.

Bowling Green State University, in Ohio, was a unique institution. Originally one of the five state universities in Ohio, and one of the youngest and weakest of the lot, it was just on the point of converting to a real university with extensive graduate programs.

I was invited there by John Exner, the chairman of the psychology department. He told me that if I came, I could have anything I wanted, including his own job as chairman. In him I recognized a first-rate promoter, and I sensed that the department was going places. The department was very small, with ten or twelve members. Four of these were young men who were headed toward national reputations: Exner, Bob Guion, Ben Rosenberg, and Brian Sutton-Smith. But they needed someone with an established reputation, such as myself. More importantly, from my viewpoint, the university had just gone through a revolution. A previous president, McDonald, had been an autocrat with delusions of imperial power (one of his creations had been a presidential suite in the Student Union, deluxe living quarters where he could entertain important guests). He had the university organized like an army, with himself as general and chains of command extending down to the lowliest employees. No caretaker could lift a broom without his say-so. Among the lowliest members of his hierarchy were the students and faculty, and the result was, in the early 1960s, a combined faculty-student revolt—the whole bit, with rallies, marches, and throwing rocks at the president's house. A few years later, such actions would be nationwide, but no one heard of this one outside northwest Ohio.

The leaders of the revolt got the ear of the board of trustees, and the outcome was a faculty charter, essentially a grant of power to the faculty, defining their rights, duties, and privileges. It established what was essentially a democratic form of governance, with the very important right of review of administrative officers. Department chairmen got reviewed every four years, deans every five, the provost every six. About half the deans and department chairmen had been fired, and when I arrived, the atmosphere was clean and sweet. I interviewed the president, Bill Jerome, an unusual administrator who respected the limitations under which he worked. As he remarked, I was looking at his qualifications as much as he was looking at mine; we found each other mutually satisfactory.

The result was that I took the job, as a research professor, with no teaching duties except to offer one course per semester of my own choosing, and at a substantial raise in salary. I also volunteered to help set up a new Ph.D. program in psychology.

In practice I have found that there are two principles of social psychology that always work: social facilitation and critical periods of organization. Obviously, Bowling Green State University was in a critical period: rapid growth and expansion of resources, and setting up newly organized functions. This was a time, therefore, when a little effort would produce large effects. As I look back on it now, the Jackson Laboratory also had been in a critical period when I went there in 1945; the situation with the psychology department was very similar in that it was relatively small and was to undergo a period of rapid expansion and change of function.

And so I tried once more to create a scientific Garden of Eden similar to that of the MBL. Not being the chairman, I could not strongly control the department as a whole, but I did try to introduce humanity, cooperation, and mutual appreciation into its dealings. Bob Guion, who was to be the next chairman of the department, saw in me an example of how to produce a cohesive organization, and in his appointments to new positions, he always tried to pick persons who had interests in more than one field. The result is that the department has never suffered from the bitter factional disputes (say between experimentalists and clinicians) that have plagued the relationships within other departments of psychology.

In some ways, my return to academia was a shocking experience. In the 1930s a remark attributed to a Princeton professor, "publish or perish," was repeated as a funny story, as were tales of deans who counted the numbers of publications or stacked them up and measured the height with a ruler; everyone knew that what really counted was the quality of research, which was usually expressed in a relatively small number of publications.

But in the 1960s, academics were taking these jokes seriously. Everyone *had* to publish, whether the material was good or bad, and professors were actually counting each other's publications as if this meant something.

The result was a vast outpouring of short, sloppy, and uninformative papers. The response to this was to try to judge quality by counting the number of papers published in journals with stringent refereeing policies. It soon became known that the easiest way to publish a paper in such a journal was to submit one in a well-known field. The result was that papers involving slight changes of well-known themes proliferated. Further, the refereeing procedures themselves, involving comments by unpaid and anonymous scientists, had the effect of slowing down publication for as long as two years, partly because the referees were slow, and partly because of time wasted by authors in responding to suggestions that were frequently trivial and might vary in quality from excellent to awful.

It was a rat race from which scientists still have to extricate themselves. I found that if I wish to know what is going on in a field, the only way is to go to a meeting, and preferably one in which there is no selection of papers. Only there does one find genuinely important discoveries being reported.

One of the things that I had hoped to do at Bowling Green University was to shift gears, to deemphasize the discovery aspect of my scientific work and to concentrate my efforts on the practical application of the results.

With this in mind, I set up the Center for the Study of Social Behavior. Originally, I meant it to be a center for the study of social control, which I considered to be the major problem of practical application, but I received so much negative feedback to the work "control" that I changed it to a more general title.

The heart of the center operations was a weekly luncheon meeting of persons who were considered most gifted in research in the social sciences in the university. We met and discussed our ideas, and from these emerged projects. The first of these was a volume entitled *Social Control and Social Change,* which Sally and I edited (Scott & Scott, 1971). This book contains most of my ideas about the application to human affairs of the basic research that I and my colleagues have done.

Other projects of the center were a study on violence edited by Arthur Neal (1976), and a study on university organization and governance, headed appropriately by D. K. Rowney, whose specialty was the analysis of bureaucracy in the Russian Empire. The latter project never produced anything tangible, partly because most of the participants were so immersed in the educational system that they could see nothing but the inner details. Certainly they were not revolutionaries.

Rowney, on the other hand, was exploring systems theory with fascination and enthusiasm, and one of our later projects was to explore systems theory as it affected various academic disciplines. This led me to rethink much of the research that I had done and to use the systems concept as an integrative device.

Looking back on my stay at Bowling Green State University, it was an enjoyable experience to take part in boom times in the academic community, and I hope that I helped to build and maintain a better university.

But these times were already coming to an end, even as I arrived on the scene. In 1965 Lyndon Johnson, almost single-handedly, escalated our Vietnam venture into a full-scale war, a war that was to poison every aspect of American life and was to bring death and disaster to a whole generation of young people, including one of my sons.

Johnson thought he could pay for a war and social programs at the same time; this was what began the serious elevation of inflation rate, and along with it erosion of the support for scientific research. The scientific community became gradually poorer, and the golden age came to an end.

Balance Sheet

Now, some forty-odd years after I began the scientific study of major social problems, it is perhaps time to draw a balance sheet. Is anything really different as a result of the efforts of myself and hundreds of other scientists in those years?

The answer is, not all that I had hoped, but neither is the situation hopeless. If I have not changed the world, I have at least changed myself. I know how to make social changes, I know the sorts of changes that need to be made, and I understand why some kinds of change can only be accomplished over long time periods.

Of the major social problems, including violence, poverty, mental health, and environmental degradation, we have made most scientific progress in the area of destructive violence. We have the basic information necessary to create constructive and peaceful societies. The basic research has been done; we now need applied research and perhaps some scientific Messiah to get people to use it.

To a lesser degree, the same might be said of mental health. I have developed a theory of social needs, by which it is possible to analyze any given way of life, such as that provided by a university or other institution, and determine whether existence within it is generally satisfactory, or whether some needs must be met outside.

From my own experiments and those of others I have concluded that there are three essential components to situations in which adaptation fails: a high degree of motivation (whatever its source), inability to adapt to this situation, and lack of opportunity to escape. The behavior of any individual exposed to this combination of circumstances will eventually become disorganized. Fourthly, there is the factor of genetic variation; some individuals break down sooner than others, and genetics may determine which set of symptoms appear.

There are many unsolved problems, of course. The above formulation provides a guide to living the kind of life that does not lead to breakdowns, but it says little about repairing those that have occurred. Perhaps a mental breakdown is an example of disorganization that makes reorganization possible, and in this way is an adaptive response. But our treatments of serious cases of disorganization, such as schizophrenia, are still largely palliative.

Environmental degradation is largely a politico-economic problem, an area in which we have made little or no scientific progress. Man is not the only species that renders the environment unfit to live in; every species, left to itself, tends to make life unlivable through the process of living, by using up all available resources or creating unusable wastes. Man *is* the only species with the capacity to constructively modify ecosystems. Much of the basic scientific information is there, but not the knowledge of how to get our politico-economic systems to use it.

As for poverty, this again is not one of the areas for which my research has relevance. Overall poverty, from a biological viewpoint, should result from a lack of available energy, whatever its source, which might be the sun directly, other plants and animals as utilized by agriculture, fishing and forestry, fossil fuels, or nuclear energy. Fossil fuels are being used up at an increasingly rapid rate, and as this happens our nation inevitably becomes poorer. Nuclear energy has not yet become the boundless supply of inexhaustible energy envisioned by optimistic physicists. It requires enormous outlays of capital to build plants and dispose of wastes, and this uses energy.

The more urgent aspect of the problem is *relative* poverty, masses of very poor living alongside the very rich. Our society has made some progress by buffering the politico-economic system with devices such as unemployment insurance, welfare programs, social security, medical insurance, and the like. As a result, we seldom saw beggars on the streets of cities as we used to do. Unfortunately, as I write, there is a reactionary faction in power that is intent on restoring relative inequality, making the rich richer and poor poorer; the beggars and homeless are back.

If carried out, such a policy can only lead to a crisis, but hopefully there will be a reaction in the opposite direction. I stress again that neither the economic policies advocated by the present United States government, nor the reactions against them, are guided by scientific information and principles. In this area, we still have a long way to go.

In these pages I have classified the scientists I have met into four types that are not mutually exclusive: Messiahs, Geniuses, Scholars, and Promoters (a.k.a. hustlers). Like all typologies, this one picks out individuals that show excessive development of one characteristic or another that is present to some degree in all individuals.

So it is only fair that I try to classify myself. In the first place I am not a promoter; I abandoned this dimension when I gave up the pursuit of status. Nor am I a messiah. I have influenced other scientists, but it has been through example, appreciation, help, and organization. Nor am I a zealous scholar. I believe in sound scholarship, but as a means to an end, not an end in itself. That leaves the genius category. In all honesty, I do not believe that I have the combination of great ability and enormous motivation that has characterized the geniuses I have met.

So what am I? A person who does not fit conventional stereotypes. If I have an outstanding characteristic, it lies in a combination of qualities, rooted in the desire to organize the world in a meaningful and understandable combination of symbols. Much of what I have accomplished has resulted from hard work, but many people do this. I am fundamentally lazy, and I work hard only where it gives promise of being effective. And I have a strong interest in the unknown.

This suggests that I have omitted another kind of scientist, the Explorer. I like to try new things, and this may be my strongest point. I have started and helped to start many new things: new fields of research, and new scientific organizations. I have very little interest in working out the last final details of an older discovery.

Why have some of my new ideas caught on and others not? The answer seems to lie in timing. All that I have described in these pages were sound (I believe); for some of them the time was right. The disadvantage of being an explorer is that one is, by definition, always ahead of one's time. But it is enormously exciting and never dull.

As I reread these pages, I find that I have omitted much of the personal side of my later career. There was simply not enough space, and I apologize to my colleagues and graduate students whose names I have not mentioned. Science is more than ideas and symbols; it does not exist apart from human beings. Where I once conceived of science as an abstract body of knowledge, I now see it as a continuing stream of intercommunicating people, each as dependent upon each other as the drops of water in a river, and none without effect.

References

Cairns, R. B., and Werboff, J. 1967. Behavior development in the dog: an interspecific analysis. *Science* 158:1070–72.

Freedman, D. G., King, J. A., and Elliot, O. 1961. Critical period in the social development of dogs. *Science* 133:1016–17.

Fuller, J. L., and Thompson, W. 1960. *Behavior Genetics.* New York: Wiley.

Hall, C. S. 1951. The genetics of behavior. In *Handbook of Experimental Psychology,* ed. S. S. Stevens, New York: Wiley, pp. 304–29.

Lorenz, K. 1937. The companion in the birds' world. *Auk* 54:245–73.

Mather, K. 1949. *Biometrical Genetics.* London: Methuen.

Neal, A., ed. 1976. *Violence in Animal and Human Societies.* Chicago: Nelson Hall.

———. 1937. The embryology of the guinea pig. III. Development of the polydactylous monster. A case of growth accelerated at a particular period by a semi-dominant lethal gene. *J. Exp. Zool.* 77:123–57.

Scott, J. P. 1942a. Genetic differences in the social behavior of inbred mice. *J. Hered.* 33:11–15.

———. 1942b. Science and social action. *Science* 96:39–40.

———. 1943. Effects of single genes on the behavior of *Drosophila. Am. Nat.* 77:184–90.

———. 1975. Violence and the disaggregated society. *Aggr. Behav.* 1:235–60.

————. 1977. Agonistic behavior: function and dysfunction in social conflict. *J. Soc. Issues* 33:9–21.

————. 1979. Critical periods in organizational processes. In *Human Growth,* F. Falkner and J. M. Tanner, vol. 3., pp. 223–41. New York: Plenum.

Scott, J. P., and Marston, M. 1953. Non-adaptive behavior resulting from a series of defeats in fighting mice. *J. Abnorm. Soc. Psychol.* 48:417–28.

Scott, J. P. and Scott, S., eds. 1971. *Social Control and Social Change.* Chicago: Univ. of Chicago Press.

Scott, J. P., Stewart, J. M., and DeGhett, V. E. 1974. Critical periods in organizational processes. *Devel. Psychobiol.* 7:487–513.

Stockard, C. R. 1921. Developmental rate and structural expression: an experimental study of twins, "double monsters" and single deformities, and the interaction among embryonic organs during their origin and development. *Am. J. Anat.* 28:115–267.

Uhrich, J. 1938. The social hierarchy in albino mice. *J. Comp. Psychol.,* 25:373–413.

Niko Tinbergen (15 April 1907–21 December 1988). Photograph by Elisabeth Walker, 1980

17

Watching and Wondering

Niko Tinbergen

At some moments I feel that autobiographies are written only by very great or very vain persons, but at other times I think that as "inside stories," however subjective, written by people who were there when exciting things happened (and the birth, or rather the rebirth of ethology *has* been exciting) they can be interesting. What follows here was written at such an "other time," and in response to Dr. Dewsbury's enthusiastic prodding. It also seemed to me, when I saw the list of contributors to this volume, that the accounts given by a number of people who, each in his way, have played a part in the shaping of modern ethology, would be bound to reveal how different the nature, the formative experiences, the challenges and opportunities, and the responses to them, have been for each of them; in other words how kaleidoscopic the network of circumstances and events have been that resulted in "ethology" as it is now. For I would be surprised indeed if my colleagues had not found, as I have, that not only was it impossible to give a smoothly flowing account of what has happened—I myself, at least, became overwhelmed by the jumble of memories that came back to me once I started to draft this retrospect—but also that few if any of them could honestly describe their lives as deliberately planned, consistently pursued, and, throughout, goal-directed exercises. Frankly, I rather distrust autobiographies whose authors claim to have had their plans ready-formed when they were still in their cradles—which might anyway be true only of some power-lusting politicians "who always knew they would become Prime Minister."

Although I have tried to keep my story, on the whole, factual, I have not been able to stop myself from expressing some value judgments; they too are part of my story.

Formative Years

When I look back over the three score and fifteen years of my life, I remember having shown unmistakable signs of an interest in the outdoors and in living things from when I was five or six years old. Whatever the influence of "nurture" has been (and it has been considerable), that of an innate deviation from the norm must have contributed a great deal to determining the course of my life. Neither my four-years-older brother, nor the sister who preceded me by two years, nor the brother who was born two years after me, though all exposed in their early years to very much the same family and physical environment as I, have become even amateur naturalists. They became, respectively, a theoretical physicist (who later became an econometrist); a language student-cum-teacher; and a director of the Hague's municipal energy works. Only my much younger brother Lukas (September 7, 1915–September 1, 1955) became a professional zoologist, but his interest appeared a little later in his life. I still remember vividly how, home after a family summer holiday in one of the wilder stretches of our native Holland, he ripped down in one evening all the pictures of cars, railway engines, and other wonders of modern technology from the walls of his little room and replaced them with wildlife posters—much to the amusement of our father who, himself a grammar-school master, was a great believer in educational freedom.

In my more mature years, I have often wondered about the origin of my leanings toward the study not just of living creatures but of living things in the wild—zoos and natural history museums have always rather bored me. I certainly cannot claim to have "always" had a scientific interest; indeed, many aspects of science were and are not to my liking (and now even frighten me). I am neither exceptionally bright nor totally committed even to my science; at least I know many colleagues who are in both respects very much my superiors. I have, however, a strong and a deep-rooted love of natural beauty that I have come to believe to be (in my case) the largely innate, typically masculine love of the hunting range (on which imprinting on specific habitats has of course been superimposed).

Although circumstances have never allowed me to actually be a hunter (except during a year's stay in the Arctic, where a certain amount of hunting and fishing was as necessary as shopping is in our urban world—though more enjoyable) it is in the spending of long days in uncultivated wild countryside (the wilder the better) and in the outwitting of elusive animals—whether to enjoy their beauty, or to see them behave naturally, or to fool them by my experiments—that I have always found my deepest fulfillment. Knowing from personal experience how it feels to have killed, cleanly and without cruelty, one of those extremely alert Arctic seals after a long stalk over the fjord ice, I can testify that the experience of the genuine hunt (as distinct from, say, the English way of fox "hunting" or the massacring of grouse) is indistinguishable from that of watching, unseen, from a well-built hide, the natural behavior of, say, a family of shy hawks, or that of succeeding, after long preparations, in filming an oystercatcher expertly opening a mussel on the low-tide seashore without him being aware of me.

From early childhood I was taken, and soon went alone or with a few friends, on long walks in the countryside around my birthplace, The Hague in the Netherlands. This was at that time so little industrialized and the town was so small that the sandy seashore, the very rich coastal sand dunes behind it, the meadows, hayfields, and the inland waters of the polders—then still largely unpolluted and extremely rich in wildlife—were all so to speak on our doorstep; we had an area of some two hundred square kilometers within easy walking, bicycling, or skating distance. This country, part of the fertile delta of Rhine, Meuse, and Scheldt, offered habitats to a great variety of plants and animals. It was far less spectacular than, say, mountain ranges such as the Alps or the Rockies, and poor in species compared with tropical forests and savannas, yet Holland had certain unsurpassed qualities of grandeur, mainly due to its very flatness and openness under its unique, wide skies with their towering Atlantic clouds, so well known from the paintings of Ruysdael, Van Ostade, and—finest of all—from Vermeer's *View of Delft.*

All this country was our "hunting range," ever-changing with the seasons, with the time of day, with the weather, with every new encounter with its natural inhabitants. My memory is packed with experiences that have made that little part of the world as alive to me as his hunting range is to every hunter. Fit as I was, my body rarely let me down, and it was not until some ten years ago that I began to feel that my legs sometimes betrayed me; and nowadays a two-hour, slow walk through the easy dunes of the Ravenglass Nature Reserve, in the enthusiastic and congenial company of my good friend and fellow naturalist Jimmy Rose, is about my limit, and makes me long for a rest and a snooze. Yet even now, literally *every* walk provides *some* new experience.

My attachment to coastal flatlands is to me a clear case of habitat imprinting. However much delight I have taken, later, in the year spent together with my wife among the Eskimos of East Greenland, or in traveling with Jan and Alekijn Strijbos through the semidesert of the Karroo, or in joining Hans Kruuk and his colleagues in their studies of hyaenas and lions on the Serengeti Plains, or Iain Douglas-Hamilton in watching the magnificent elephants of Manyara, it is the sandy shores of Holland that are my real home range, where I feel at home in all seasons and all weathers, at all times of day and night. I do not think that I have ever been bored by coastal scenery, the open sea, deserts, ice and snow, a sunrise (so much more a time of hope than a sunset), severe gales, beautiful flowers or animals, and certainly never by seeing wild animals go about their business in their natural habitat. The first time I realized how strong my habitat attachment was, was when, seventeen years old, after a camping walk of several weeks through the magnificent Swiss Alps with their beautiful wild flowers, I found myself heaving a sigh of relief when I returned to the far less spectacular and far less romantic plains of Holland. In the Alps I had felt hemmed in.

I find it difficult to characterize in general terms what were the kind of phenomena that fascinated me in my youth. Instead I mention, for what they are worth, a sample of specific experiences that come to mind. For instance, I feel privileged to have seen, and still recall as if they happened yesterday, the foaming North Sea during severe onshore gales—exhilarating to the landlubber, but on some occasions horrifying, as when, as a boy of four or five, I saw one of those magnificent Baltic square riggers wrecked on the shallow sands and battered to pieces (many years later I was delighted to sail on a similar ship through the East Greenland icepack). I remember the same Dutch seashore in the grip of a severe winter, covered with ice, the open water smoking, while for days the massive emergency migration of countless shorebirds streamed past, leaving these suddenly inhospitable shores covered with dead and dying birds—those who had responded too late to this rare lick of the "Siberian tongue."

Another vivid memory is that of seeing, from under a washed-up basket on the shellbanks of the Hook of Holland, a little tern feeding its downy young not a foot from my eye. Yet another is that of watching at dusk on a warm summer night the newly emerged pine hawk moths pay the first visits of their lives to sweet-scented honeysuckle flowers, hovering lightly, ghostlike, from one flower head to the next. I shall not forget either the frightening yet exhilarating moment when, in East Prussia between the wars, I was chased up into a slippery birch tree by an angry bull moose, whose mating call I had just been taught to imitate. And later, seeing and hearing the mighty East Greenland icebergs "calve," throwing off blocks of ice the size of major apartment houses, sinking slowly down into the green grey Arctic Ocean, and wallowing lazily for perhaps half an hour, causing "surface seaquakes" that sent ever-widening circles of major waves further and further away until they reached us in our humble but seaworthy canoe—such moments of majestic natural but impersonal violence were fascinating, and just as exciting as a major tropical thunderstorm.

Yet another memory, of a totally different character but equally interesting in its own way, is that of seeing a male grayling butterfly pursue not only passing females of his own species, but also flying songbirds ten times his size, or falling leaves, or even his own shadow (when he ended up, of course, in a ridiculous collision with the ground). Other moments that I still relive in my dreams are dead-calm June dawns in the sand dunes near the Hague and later, on countless occasions, near Ravenglass in Cumbria, when, ambling along on my own or with at most one or two friends, I found on the dew-moistened, rippled sand the most exquisite tracks left by a fox perhaps an hour earlier on one of his nightly visits to the gulleries or by a running hedgehog, and tried to decipher what exactly they had been doing.

The variety of these "mental engravings" is near endless, and they range from the beautiful to the eye-opening or intellectually challenging kind; vision plays the major part, but sounds, the feel of the sand, of the air, and almost equally important, the smell of the coast, of moisture-laden air with its multiple scents, all enrich the images. I know I am expressing myself poorly, and even now I cannot really say what exactly made me observe these things so intently, what drove me to watch all this, or to sit for so many hours in front of the little aquarium in our garden in which a brilliantly green-, blue-, and red-colored stickleback built his nest and guarded his brood, or why I stopped and looked on, fascinated and amused, when barn swallows dived like stones into a pine forest just after, hundreds of meters higher in the blue sky, two young hobby falcons began to chase each other in harmless play.

Sometimes I even imagined that I could *feel* what a wild animal must feel, for instance when I found I could walk up to two furiously fighting little wrens and hold them in my cupped hands, a writhing little tangle of brown feathers, legs and beaks until, suddenly coming to their senses, they shot away to safety; or when, many years later, I was charged by a fast-moving lioness who thought I was a threat to her cubs. I knew of course I was safe enough in my friend's Land Rover, but I remember how, realizing I was the target of her piercing yellow eyes, I was momentarily scared yet thrilled that "this was how a little songbird must feel when a hawk swoops down on it."

At other moments I relive highlights of my research, from the moments when, as a schoolboy, I watched a hooded crow "cache" freshly stranded bivalves at the foot of the sand dunes, and realized why he had to cover the cache with sand, or when I experienced what a powerful deterrent for a fox or a stoat the stinking feces must be that a disturbed mallard duck sprays over them, as she did over me when I flushed her from her nest. Such "flashes of insight" felt the same as when, much later, I suddenly realized that many signaling movements of animals were due to dual or even multiple motivation; or when I suddenly "saw through" the secondary adaptations to the signaling function that obscure the origin of so many "ritualized" displays.

Equally exciting, though of a quite different nature, were the times when a long-term experiment began to yield, step by step, "statistically significant" results that made sense in terms of what one had supposed to be the truth. (Later, one also learned to tolerate, and even to take positive delight in results that refuted one's pet hypothesis; this can overcome—though only very slowly—one's initial annoyance after one has experimented a number of times, and make one realize that these are the kind of data that make one rethink and, with luck and perseverance, find a new, and *always much more interesting* interpretation, with its sequel of new experiments.)

But, to return to my main story, leaving the reader to ponder how to characterize a "curious naturalist" of my type, I am afraid that from the start I have considered school—indeed, until I was about twenty years old, all institutionalized and regimented instruction—rather a nuisance and, even though it was at school that I felt at home with fellow sufferers of my own age, a frustrating restriction of my freedom. The sheer *boredom* of having to learn the multiplication tables, the names of the railway stations between The Hague and Utrecht or Groningen, the produce exported by Brazil, or the year of the battle of Nieuwpoort (1600—now the only one I remember)! Even now one of my memories of the schoolroom is looking out longingly through the nonfrosted higher part of the window at a bit of bleak November sky and seeing flight after flight of hooded crows migrate past; or (even more cruel!) on sunny April mornings, the blue spring sky with hundreds of black-and-white patterned lapwings hurrying back North, which made me look forward to finding once more their beautifully blotched eggs on the dry mossy dune ridges that lapwings often prefer. I also remember my father's amused chuckle when, on a walk together with him and one of my friends, I exclaimed angrily: "If only that rotten Count Floris had not invented schools, we would be free tomorrow!" (It *was* Count Floris the Fifth of Holland who was the culprit—it said so in our history books.) I did not at all understand why my father was amused; it was such an obvious deduction!

Later of course I realized that, while by my "talent" for playing truant [as Dick Hillenius once wrote], I had protected my most valuable gift, that of imaginative and at any rate independent thinking, I had cut myself off from much that might have aroused my interest, and might indeed have been useful to me.

My introduction to biology came through my parents, even though neither of them had more than the natural delight of healthy, happy middle-class city dwellers in the world of nature round the town, and in its city gardens. We *were* a truly happy family; our father a liberal-minded, very hard-working man with many intellectual and social interests and a fine, if at times somewhat prudish, sense of humor; a man of total honesty; our mother the ever-cheerful, understanding, caring center of "hearth and home." Both had to work extremely hard—all five of us had a university education and were given one opportunity halfway through for a period of study abroad—and the atmosphere at home was one of having to work as a matter of course, yet never at the cost of happiness and freedom and play. Of course having to wipe your feet on the doormat, having to wash your hands now and then, and having to tidy up or to wash the dishes were irritants but they were parts of normal life. The standards set by example were high, but I often marvel at the freedom I was given to indulge in my many sports and games, in my many long skating and camping trips, in my choice of friends—especially when, once I was a father myself, I began to realize how often my parents must have worried especially about me, in so many respects the odd one out.

While my brothers and my sister did not share my obsession with wildlife, they had strong social and political interests on issues that left me rather cold. Although my father, as he told me later, had often wondered whether a boy obsessed with sports, with camping and skating, and with nature photography would ever be able to earn a living, he did encourage me in my biological and camping activities. And although he was a schoolmaster himself, he did not attach undue value to sweating for good marks; my school reports usually had monotonous just-pass marks for all subjects except for sports and for drawing—two "tens" peaking above a row of sixes and fives. If, through carelessness on my side, a four appeared, his asking, in his mildly admonishing way, whether "this was really necessary" was enough to make me squirm—for a while.

The interest in our native flora and fauna was at that time greatly stimulated in the Dutch middle classes by the work of two quite exceptional, uniquely gifted men, E. Heimans and Jac. P. Thijsse, both schoolmasters who, at the turn of the century, began to publish their simple, but wonderfully refreshing, indeed never-equalled books on our native flora and fauna; founded a monthly (*De Levende Natuur*, in which I have published popular accounts of much of our work done between 1925 and now); and organized national societies for the conservation of "areas of outstanding natural, recreational and biological value" (as we would call them now), then still to be found in the Netherlands; and did other things that made the many millions crowded together in our little country "nature conscious."

A yearly event was the publication, by the biscuit factory owned by the Verkade family, of a book in album format on some aspect of the geography or the natural history of the Dutch countryside, for which Thijsse usually wrote the text and well-known Dutch artists painted the water-color illustrations. These, always 144 per album, of the format of visiting cards, were wrapped with the firm's products. For many years, the nature library of many families was enriched every "Sinterklaas Eve" (December 5) by the purchase of an empty album, in which, in the course of the next year, the gradually acquired pictures were glued in. There was a special Verkade service for the exchange of duplicate pictures. I (and many of my contemporaries) still have most of these Verkade albums: "Winter,"

"Spring," "Summer," "Autumn," "Our Blond Dunes," "Our Colourful Meadows," "Heath and Pines," "Our Rivers," "Our Flowers and their Insect Friends," and many others. Some of them have even recently been reissued.

My school career remained unglorious through secondary school; my interests centered on drawing, sports, and natural history. I owe a great deal to my biology master, Dr. A. Schierbeek, who accepted my lukewarm interest in morphology and taxonomy and encouraged me in my "studies" of wildlife in its natural surroundings. He also supported and encouraged in an unobtrusive way the activities of our natural history club.

When I left school I was not at first inclined to accept my parents' offer—which all my brothers and my sister hardly hesitated to accept—to let me go to university. I had vaguely thought of emigrating to Canada, of becoming a professional photographer, and, had professional hockey then existed, I might even have chosen a sports career. But my parents were advised by two friends of the family (Dr.Schierbeek and my elder brother's professor of theoretical physics Paul Ehrenfest, who were both convinced that I had the makings of a professional biologist) to send me for a couple of months to a place where I could see biological fieldwork, of a kind that would appeal to me, in action. It was the resulting stay, during three exciting autumn months, in 1925 at the "Vogelwarte Rossitten" (the first bird observatory, founded by the German student of bird migration and pioneer of bird banding Professor Johannes Thienemann) that made me decide to enroll, a term late, as a biology student at Leiden University. Seeing the spectacular autumn bird migration along the Kurische Nehrung (now part of the USSR), the curious "moving dunes" on this narrow strip of sand with the fascinating indigenous moose had reinforced my naturalist's motivation, and although I was not at all attracted by the classical comparative anatomy and taxomony that made up the bulk of the curriculum at Leiden, I realized that I simply had to take this hurdle and grind through the stuff that, until I knew better, I considered dull and boring.

When I compare what I remember about my first twenty-five years or so with those of many of my fellow biologists, it seems to me that I was on the one hand a fairly poor student, rather lazy, self-centered, and narrowly interested, yet on the other hand one who did have an inquiring and independently thinking mind. I still remember on what issues I was intuitively dissatisfied with the lines of thought explained by my biology master, and have found that those were exactly the issues on which I later developed explicit ideas of my own that led to my best research. I was extremely fortunate in having been raised in a family that I have gradually learned to see as in so many ways exceptional, and in some of my teachers. But there is no doubt that I also was taught by some exceptionally bad and dull teachers, who managed to kill whatever budding or potential interest I had. For instance it was not until my early forties—for so many people a time of reorientation or at least stock taking—that I regained my interest in history, in archaeology, in literature and in other forms of art. Yet I had always greatly enjoyed our days of outdoor sketching, which my father loved and in which his sons often joined him, but I usually became soon bored when he took me to some of the (very fine) museums to be found in The Hague, Amsterdam and Haarlem, the very museums where, in my later years I spent many an hour looking, again and again, at the Holbeins, the Rembrandts, the Vermeers, and the Frans Halses.

Curiously enough, although I always went my own way and paid little attention to the political and social issues that were to determine the course of our lives and that preoccupied my brothers, I must have been influenced by their social concern for the terrible unemployment of the twenties and the cruel social inequalities in general that turned two of them into active socialists, for it was with a certain sense of atonement that, in the last years of my tenure in Oxford, I began to take part in teaching in the newly founded "Human Sciences" honors course, and that after my retirement, I spent so much time and attention on the study, done jointly with my wife, of certain mental disorders in children.

Within biology itself it was not until, as a young instructor, I had to teach vertebrate comparative anatomy and, later, the biology of certain taxonomic groups that I became genuinely interested in this kind of subject, in which I learned to give attention to functional anatomy, adaptive radiation, and evolutionary aspects in general. That I had not widened my horizons earlier was partly due to my own intellectual limitations, partly to the fact that I started my studies in Leiden at the tail end of a period of the most narrow-minded, purely "homology-hunting" phase of comparative anatomy, taught by old professors just before they were succeeded by the younger generation.

At an early stage, my special interest in the *visible* natural world made me turn to the camera as well as to sketching, rather than to, for instance, music or the theater; and still photography and, later, cinefilming of animals in their natural setting and of the way they lived, and of nature in general (including landscapes and cloudscapes) became one of my principal hobbies and soon proved to be very useful in my work as a teacher and as a student of animal behavior. The explosive growth of the market for books on behavior, as well as the television industry provided the funds for this. A certain flair for simple but clear, unaffected speaking and writing, in no way outstanding but, judged by the responses of my audience, appealing to them, has certainly contributed to what influence I have had on my students and on nonprofessional audiences. (My experiences in this field have convinced me that the recognition a scientist receives in the form of promotion, honors and prizes depends not merely on the qualities of his research, but on his "salesmanship" as well—on whether or not he can communicate about his work in a way that others find interesting, and whether while he speaks, he pays attention to, and acts upon, the feedback he receives from his public.)

It is not only in having been raised in a happy and stimulating family, having been allowed to indulge freely in my hobbies as a budding naturalist and a sportsman, and in having grown up in a naturalist's paradise that I have been exceptionally fortunate, but also in having been befriended by a number of remarkable friends many years my senior and, later, by contemporaries, and, still later, by my pupils.

The late Gerard Tymstra, ex-artillery officer, mathematics teacher, and gifted naturalist, inspired and encouraged me not only in my early years as a naturalist (it was his example that made me turn to the herring gulls as objects for serious

studies on bird sociology), but also in my budding interest in human affairs generally; in particular he taught me by his own "manwatching" (as Desmond Morris
was to call it later) what an interesting creature *Homo sapiens* is.

My biology master in The Hague, Dr. A. Schierbeek, already mentioned, encouraged every interest in natural history that I showed and taught me much about
how to keep aquaria, how to collect seashells and other tideline treasures, how to
make a herbarium, where to find books and other reading matter, how to keep
records of my observations, and so on.

From the moment I arrived in Leiden, without as yet having any specific plans
for what I wanted to do with my life, I was befriended by two other gifted men
who were to have a decisive influence on the nature of my future scientific work:
Dr. h.c. A. F. J. Portielje, who was in charge of the animals in the famous Amsterdam Zoo "Natura Artis Magistra," ("Artis" for short); and Dr Jan Verwey, who
was to become the outstanding marine ecologist of the Netherlands but was then
still a young assistant in the zoology department in Leiden and still very much a
student of bird behavior. (He had then already done the remarkable fieldwork on
the breeding behavior of the gray heron on which he published, in 1930, his fine
monograph "Die Paarungsbiologie des Fischreihers", now almost, though not
totally, forgotten).

Portielje had an immense influence on many people, young and old, by his
enthusiastic, perceptive and sensitive, empathizing stories on animal behavior, of
which he had as much firsthand knowledge as Oskar Heinroth. But his passionate
belief in the possibility of knowing about the subjective feelings of animals, which
I rejected with all the arrogant certainty of my age, made us cross swords time and
again. I owe much to his generous and truly friendly willingness to enter so often
and so spiritedly into private debates with the stubborn, opinionated, and rather
destructively arguing youngster I then was. Dr. J. A. Bierens de Haan, at that time
Holland's leading animal psychologist, was equally kind and courteous to me,
even though he used to say that by denying ourselves the ability to know an
animal's subjective feelings we were "like people who put on monochromatic
spectacles when looking at colorful paintings."

When, shortly after I had met him, Jan Verwey left for the (then) Dutch East
Indies to take up a post at the Marine Biological Laboratory near Batavia (now
Jakarta) he gave of his valuable time to keep up a lively correspondence with me. I
have never forgotten one particular sentence in one of his letters to the effect that
he believed that the most challenging aspect of living animals would turn out to be
their social behavior. Having already been alerted by Tymstra to the fascinating
structure of life in a herring gullery, I turned my attention to work that was later to
be described in my books *The Herring Gull's World* and *Social Behaviour in
Animals* and, much later, in the television film, made with Hugh Falkus: "Signals
for Survival," which, I am told, is still used regularly in university courses on
animal behavior and communication.

I relate all this in some detail because it does show that my early work did not at
all start in a vacuum or from scratch; on the contrary, I had these personal friends,
guides, and mentors, and also exceptionally fine opportunities. And I had read,
with total absorption, the publications of such different men as William Long,

Edmund Selous, Julian Huxley, Eliot Howard, and, equally decisive for my later choice of subjects: J. H. Fabre. Of Charles Darwin's work, I read at that stage only *The Voyage of the Beagle* and, without being much influenced by them, the three volumes on Darwin's life and letters. And from Thijsse (who had a soft spot for digger wasps) and Fabre it was a small step to Ferton (whose little experiments I admired) and from there to Karl von Frisch.

I was at that time dividing my attention (when I did not "have to" go bird watching or to play grass hockey, which were both passions with me in my undergraduate years) between, on the one hand, observing closely how the behavior of individual herring gulls fitted them for membership of their "bird city," as yet without really experimenting, although I did color-ring a number of breeding birds and so joined the (then) small band of people who followed recognizable individual birds through the years; and, on the other hand, reading the papers and repeating some of the experiments of von Frisch, which seemed (and still seem) to me methodologically faultless and beautifully elegant in their sophisticated simplicity. I was as fascinated by his work on the senses of honey bees as by that on the sense organs of fishes. I had a tremendous admiration not only for his style of research, but also for the way he described it (and the civilized way in which he, the younger man, rejected the boorishly expressed criticisms by men more than a generation his seniors—something that at that time in Germany took considerable courage). Later I also had the privilege to hear him lecture and was likewise impressed and charmed by the elegant lucidity of his speaking style.

Early Research and Teaching

Still, toward the end of my academic studies, in the early thirties, I had no clear idea at all of what I wanted to do, except to study live, intact animals in their natural environment. I certainly did not dream of helping to get a new branch of science, "ethology," off the ground! Understandably, several of my colleagues and also younger ethologists have, with hindsight, given their impressions of what we had done in their various outlines of the history of ethology—impressions that ascribed a little too much planned foresight to us, or at least to me. These, and comments on the occasion of the awarding, in 1973, of the prestigious Nobel Prize for "Physiology or Medicine" to Karl von Frisch, Konrad Lorenz, and myself for our "work on the elicitation and organization of behavior patterns," did not truly reflect the haphazard, kaleidoscopic attempts at understanding animal behavior done by the future ethologists.

Rather than being farsighted ventures into the unknown, with the intention to "map" that unknown territory systematically, they were in fact no more than tentative, groping attempts at seeing some sense in the variety of animal behavior systems that fascinated, yet bewildered, us, and the understanding of which had in many ways been made difficult rather than facilitated by the many early brands of psychology to which we turned for enlightenment, but which had disappointed us so bitterly.

As I have described much later, in my book *Curious Naturalists,* it was the combination of my inherent curiosity about the ways of wild animals, of my admiration for von Frisch's work, and of a sudden, totally unexpected opportunity that started me on my first systematic field experiments: those on the homing abilities and the bee-hunting behavior of the digger wasp *Philanthus triangulum.*

At the same time I had also, under the influence of, mainly, Jan Verwey and of Konrad Lorenz's first papers on the behavior of jackdaws, a strong inclination to concentrate on "plain observation." This was the time when several of us, among them Frans Makkink and Friedrich Goethe, were aiming at making "ethograms"— studies-in-depth (for those times) of the "behavior repertoires" of individual species.

A fine opportunity to turn from the herring gull to a totally different type of bird came in 1932. After having been awarded my degree of Ph.D. on a short thesis on the homing of *Philanthus,* I married my girl friend Elisabeth Rutten, and together we considered plans for the future. By an exceptional stroke of good luck we were allowed to join the small Dutch contingent for the "International Polar Year 1932– 33" (the forerunner of the International Geophysical Year), which had been allocated the district of Angmagssaalik on the east coast of Greenland, the homeland of the 800-strong tribe of the as yet practically un-Westernized Angmagssalingmiut Eskimos. Throwing the prospect of a secure job to the winds, we jumped at this splendid chance to spend a year doing fieldwork studies on some birds of this mountainous, truly arctic strip of land between Greenland's inland icecap and the broad ribbon of pack ice and icebergs of the Denmark Strait between Greenland and Iceland.

We never regretted that decision, even though for several years afterwards we had to live on a shoestring. In that unforgettable year, our main efforts went into a step-by-step, full-time study of the breeding behavior of the snow bunting, from the moment of their arrival from the south in March till their departure in late summer. We did interrupt this for a couple of weeks (which did not harm our work since the snow bunting pairs along the fjords varied a great deal in their seasonal rhythms, dependent on locality and altitude) to cash in on a rare and totally unexpected opportunity to study territory establishment and pair formation in the red-necked or northern phalarope, one of those few bird species in which the males are cryptically colored and the females have a brightly colored plumage.

During the rest of that Greenland year, we lived in intimate contact with the Eskimos, whose guests we were for the winter and spring, and we studied, in an amateurish way, the life-style of this fascinating tribe, still living their age-old lives of hunters (of various seals and polar bears mainly) and fishermen (of cod, halibut, red perch and char), with a little gathering of berries, roots, and shellfish by the women and children. (Not long after our stay, the Westernization of this last untouched tribe was carried out with great, deplorable, yet of course inevitable, thoroughness.)

Finally, we could spend quite some time on a study of the social behavior of the many huskies, who, with their wolflike pack-living habits, fitted so well in a society where every hunter had a team of sledge dogs of his own, whose members lived around the hunters' houses, and who had themselves divided the space in each settlement into group-territories located around their owners' homes.

Our reason for selecting the snow bunting was the fact that Eliot Howard had made a detailed study of the highly territorial British buntings, while E. M. Nicholson had, after a summer visit to Greenland, made a brief, though explicit, remark suggesting that at least in West Greenland the snow buntings showed at best only very slack territorial behavior. I hoped that a closer study might perhaps reveal an ecological reason for such a discrepancy between closely related species. In fact we found that the snow buntings were as fiercely territorial as their British relatives and merely relaxed the system late in the season, when they roamed widely in search of food for their growing young. In spite of this we found our study of the snow buntings richly rewarding.

Our reason for having a quick look at the phalaropes was that we wanted to check whether, as was generally expected, the reversal in sexual color dimorphism would be matched by a reversal of territorial, mating, and breeding behavior, and this we found to be the fact, with some nice unforeseen but understandable touches such as the performance, by each female when about to lay an egg, of a special ceremonial song flight, which served to lure her mate to the clutch which he, as the cryptically colored partner, would have to incubate—no female phalarope incubates or cares for the young.

In both these studies we did a great deal of purely descriptive work, but we also had our first taste of using our observations for interpretations, sometimes based on "natural experiments," concerning a number of general problems such as the significance of territory, of plumage coloration, and of specific, specialized behaviors. Examples of the last were the nest-showing ceremony of the phalaropes and the displacement-pecking of male snow buntings during boundary fights, which made us search for suspected, but never found, tiny food items on the bare snow where they had drawn their territorial boundaries (we called it "substitute feeding," not having thought of a better name yet).

Our dog studies led us to speculate on the pack structure, dominance, and group-territorial behavior of their ancestors the wolves (of which such behavior in the wild was to be described later); and, although we did not realize it at the time, our living with the Eskimos taught us a great deal about the life-style of primitive hunter-gatherers that biologically oriented anthropologists studied later in such great detail. Living with such people, adopting their ways, and working hard on learning their language so that we could communicate satisfactorily with them gave us above all a feeling for their system of values (and for many far from admirable aspects of our Western set of values) and this experience changed our outlook on life permanently. We were indeed lucky to have this great experience so early in our lives, and we later realized also how valuable had been my wife's years spent as a child among the quite different inhabitants of Indonesia, where her father's work as a field geologist took the family from one outpost to the other in a long series of stays in many different parts of that vast archipelago.

Back in Leiden after our "Arctic interlude," my Professor Hilbrand Boschma, himself a field naturalist specializing in invertebrate taxonomy, generously allowed me, together with a few undergraduates, to resume my research on the homing of digger wasps by stretching my official holidays of a fortnight yearly into fieldwork periods of months, so combining research with teaching and pleasure. It was the students, soon to be joined by postgraduates working for their doctorates, who helped me build up the kind of research group that, despite its continuous turnover in membership, was my "club of coworkers" from then until my retirement. As a way of teaching, but also as a style of life, this working together on problems that we began only gradually to see as worth investigating, and of which I knew just as little as my students—such collaborating-in-research has been both a source of intense delight and a continuous challenge to us all. Looking back, it is as impossible for me as it is for many of them to know with certainty which of the many ideas that have in my lifetime contributed to the growth of "naturalistic" ethology, and later to a primitive kind of eco-ethology, originated in whose mind(s). Much has undoubtedly been the result of group work, of communal, often at first vague "wondering."

In those days, the prewar thirties, ethology slowly began to take more definite shape, or rather shapes. Our aim of producing ethograms of *species* or small taxa began to move more and more toward attention to more or less general *problems*. I find it difficult now to say how exactly this happened, but it was of course the same kind of process that has occurred in young sciences before, for example, in taxonomy, in physiology, and for that matter in history, in meteorology, in oceanography, in geology, and in many other disciplines. It happened with Konrad Lorenz and with myself independently, each with his early pupils, and before we knew each other personally, although we each had special interest in the other's work. However much we had in common as naturalists, our approaches did differ; Konrad's way being more observational and interpretative, mine more experimental, verifying, and also more ecologically oriented. But even so we had much in common; thus Konrad's refreshing jackdaw studies and his "Companion" paper encouraged me in my largely observational herring gull studies.

But I had independently begun to organize practical work for students on a fairly large scale. This job was simply given me as my main task by the new head of our department, Professor C. J. van der Klaauw, who saw something of potential value in our earlier fumbling attempts in this direction (and who knew how to "stretch" his young staff). It was for these practicals in animal behavior (in which third-year biology students were given solid "blocks" of six weeks without any other lectures or practicals, so that they could devote day after day to little research projects on animal behavior, which they could select for themselves under my guidance and on which they had to report to their fellow students at the end of the course) that I began to look out for suitable objects.

I wanted wild, not domesticated species and those that could be kept in the laboratory under reasonably natural, richly "patterned" conditions. Definitely *not* white rats, to which behaviorism had given me a mental allergy from which I have never fully recovered. This made me return to my boyhood loves: the ubiquitous and beautiful, hardy and tame three-spined stickleback and its relative, the smaller and quite different ten-spined species; to our three most common native species of newt, to freshwater beetles and bugs such as *Dytiscus* and its larva, to *Notonecta, Nepa* and their relatives, and to the herring gulls, which we went out to study in their breeding colonies.

It was an exciting time, when each student could, quite literally, by patient watching, wondering, and doing some guided, sometimes experimental "research," discover a quite respectable chunk of truly new facts about the behavior of these species. We also put our *Philanthus* camp on a permanent basis of two summer months' camping in the pinewoods near Hulshorst, some eighty miles away from Leiden. There, exploratory studies of local species of insects and birds grew out into doctoral theses and postdoctoral work. It was in Hulshorst, for instance, that Gerard Baerends's fascinating and highly original monograph on the Sand wasp *Ammophila,* Jan van Iersel's penetrating analysis of the homing behavior of the large, fly-hunting digger wasp *Bembix,* and our more modest, but at the time exciting observations on hobbies, and also our work on the mating behavior of the grayling butterfly were done. Later, my younger brother, originally one of my own students, chose Hulshorst as the basis for his very detailed, ecologically seminal work on the kestrel, the sparrow hawk and its feeding ecology, and later his pioneering quantitative work on the ecological relationships between the insects of the pinewoods and their avian predators, published, largely posthumously, by the devoted efforts of some of his pupils and colleagues.

Those prewar years were times of intense activity, of collaborative reconnaissance, and of discovery. We began to realize that we were trying to find our way into a quite new field. How simple yet how packed with tasks was our yearly work schedule in those times. Throughout the academic year, I had to teach animal behavior and comparative anatomy of vertebrates (which gradually began to fascinate me when the study of function and of analogies or convergencies was added to that of homology, a revival of truly comparative study of structure without forgetting about function, which had been spearheaded in the late twenties by the German schools of Erwin Stresemann and Hans Böker); in the spring my early mornings, from before sunrise to approximately 8:00 A.M. were devoted to gull watching (I had to start my normal lab day at nine); the summer was spent in our "insect camp" in Hulshorst; and the autumn and winter were for teaching, for writing up our results, for reading, and for preparations for the fieldwork of the next season.

Contact with Lorenz

This period of, as yet, tentative exploration of isolated problems in a great variety of animal species came to an end (at least for the time being) through my suddenly intensified personal contact with Konrad Lorenz. His seminal paper on "The Companion in the Bird's World" (of which the greatly abridged American version gave but a pale image) was published after we in Leiden had concluded our first series of dummy experiments in which we had analyzed, in far more detail than had hitherto been done in any animal, the complex, under optimal conditions fairly schematical, mating sequence of the three-spined stickleback. Together with my student, the late Joost ter Pelkwijk, I had published a first, brief account of this, in which we could show that in *each* of the two sex partners *each* step in the sequence that began with the first encounter and ended with spawning, fertilization, and the departure of the female, was triggered by relatively simple stimuli from the opposite "actor"—stimuli, moreover, that, though provided, on each step in the sequence by the same male and the same female, differed for both of them for each separate step. (It is often not realized that, in our well-known—and frequently reproduced—diagram of the mating sequence of sticklebacks that arrows that zig-zag back and forth between the male and female, indicate *true cause-effect relationships*—not just sequences—and that each one of these relationships was demonstrated by our experiments with dummies.)

Just at this time Konrad Lorenz came to Leiden at the invitation of Van der Klaauw to take part, together with Portielje and Bierens de Haan, the Finnish eco-ethologist Pontus Palmgren, the Belgians Dr. and Mrs. Verlaine, and myself, in a symposium on "Instinct." I was fascinated both by Lorenz's paper and by what he said during those few days' discussions; from his side he immoderately enthused about our stickleback work: "That is *just* what we need!" was his often repeated comment. I was frankly astonished at his undue admiration for our work; in my view the experiments were so commonsensical and so much dictated by the relative rigidity of the hundreds and hundreds of mating sequences we had seen (both in our aquaria and in the easily observable wild fishes in their natural habitat: the meres and draining ditches of our Dutch polders), that, although I felt of course greatly flattered, I honestly could not see what all the fuss was about!

It was during that symposium that the Lorenzes invited us—myself, my wife, and our toddler son—to spend the spring in their home in Altenberg, not far from Vienna. Van der Klaauw generously arranged for the leave necessary for this prolonged absence—my official annual holiday was still a fortnight, and "sabbaticals" were unheard of for youngsters of my humble status.

Those delightful spring months at the foot of the Vienna Woods in early 1937 laid the foundation for our lifelong friendship and collaboration, even though we each chose our own course and in some respects drifted apart scientifically (with Konrad always in the lead in theoretical and philosophical matters). Lorenz was at that time deeply involved in his studies of greylag geese, which he hand raised and also tried (in vain until he fed them germinating, vitamin E-containing wheat!) to get to breed in the ponds we dug in his parents' large garden; and in the comparative study of the "courtship antics" of as many species of dabbling ducks as he could lay his hands on. But we also did a great deal of sitting and watching other animals and quite a lot of experiments on "effective stimulus situations" that either elicited or oriented behavior. (The distinction between these two governing principles, made conceptually by Lorenz, was worked out at that time, notably by our joint paper on egg retrieval by greylag geese, and in the paper that I wrote together with my pupil D. J. Kuenen on our work on the stimuli that elicit and orient the gaping of nestling thrushes, for which I was then doing the last few experiments and of which I was completing the text in Altenberg.)

It was in Altenberg too that Lorenz and I did the series of experiments with cardboard dummies of flying predators and other objects, sailing them along a string running from one treetop to another high over the heads of hand-raised birds of varous species of fowl. To some of these models the birds responded by crouching as they did to real hawks overhead. We did this with very young and "inexperienced" animals, knowing that the gist of such "species-specific" behavior could not have been learned either from us or by bitter experience, and thinking (in our simplistic overreaction to American behaviorism) that it therefore must be "innate." Later of course we learned from the work of Schleidt, Melzack, Hailman, and others that in our schematizing classification of behavior *components* into "innate" and "acquired" we had grossly neglected the very real, in fact central problem of the *processes* that control the development of behavior. (But we were not quite as naive as our good friend the late Danny Lehrman made out later in his famous "Critique" paper; for instance in our thrush work, Kuenen and I *had* deliberately raised a whole brood of blackbirds in an almost totally white environment, feeding them from white dummies only, and so on, and testing them after a week of this exposure treatment; and we had found and reported that, when given the choice between white and black dummies, these birds all gaped best in response to and toward the black models.)

Also, critics of Lorenz's and my own overemphasis on the "innateness" of much behavior (so useful in those "Kuo and Schneirla times") have been far too schematic in their almost total ignoring of the fact that it was, after all, Lorenz who emphasized and reconfirmed Heinroth's discovery of "imprinting" in young goslings; that it was he and I myself who had described a variety of hitherto hardly known, quite subtle, and quite vital learning processes such as the individual "recognition" and bonding between the members of a pair of geese or of a pair of gulls; the work by our "digger wasp group" on the extraordinary achievements of *Philanthus* when learning, during extremely brief "locality studies," the layout and (selectively) certain aspects of the landmarks lying around their burrow entrances.

I still think, to this day, that the demonstration by W. Kruyt and myself, that under certain circumstances a wasp can become conditioned in a mere *six seconds* to a new, very detailed complex of landmarks, which she remembers throughout her flight to the hunting area, through her long hunting activities, and through her way back home, was one of the neatest experimental demonstrations of the function of an extremely specialized form of exploratory behavior—of behavior of which the *only* function is to explore, to register, and to remember many of the topographical details of a very complex situation.

Nor has sufficient attention been given by learning psychologists to the beautiful and convincing experimental demonstration by Baerends that *Ammophila* is capable of a "delayed response" up to sixteen hours after an "inspection visit" to her burrow, when she learns whether or not the larva need a new supply of food and if so, how much. Finally, our early emphasis on "innate predispositions to learn" proved later to be crucially important.

Looking back at all this, I feel that where we went wrong was not in denying or failing to study the functions and forms of learning, but in seeing the classifying (into innate and learned) *behavior,* of genetically controlled and experientially acquired *properties*) as our *goals* rather than as the first approximating steps in the unraveling of internal and external conditions and events in the ontogenetic *programming process* that results in proper functioning behavior machinery. I have over the years come to suspect that some people, perhaps some cultures, incline more to being satisfied with classifying, with *pigeonholing as an aim in itself,* than with *distinguishing* between different classes of phenomena *as a first step toward their scientific understanding,* and I believe that it is no accident that the pragmatically thinking Anglo-Americans tend to take the latter approach to the nature-nurture problem.

On my way home from Austria after those marvelous four months with the Lorenzes, I stopped over for a brief visit to the great von Frisch, who had kindly and encouragingly replied to my request for a personal meeting. As I have written elsewhere: "My recollection of this visit is a mixture of delight with the man von Frisch, and an anxiety on his behalf when I saw that he refused to reply to a student's aggressive *Heil Hitler* by anything but a quiet *Grüss Gott.*"

A Sniff at Behaviorism

Back in Leiden, I found that my wanderings in the wonderland of animal behavior had begun to become a little less aimless, but I was still very far from even attempting to see the study of behavior as a more or less autonomous, potentially "respectable" branch of biology. Perhaps unduly impressed by the sheer numbers of American psychologists, by the masses of behaviorist textbooks (but also puzzled by their bewildering jargon and by the voluminous literature poured forth by colleagues deeply involved in working out *general* theories about behavior by studying almost exclusively *one* domesticated animal species in what seemed to me extremely impoverished and curiously contrived conditions), I applied for, and once more was granted, leave and funds for a three month visit to the Eastern United States in the late summer and autumn of 1938. The Netherlands-America Foundation generously gave me the cost of a return fare by freighter from Antwerp to New York, and I eked out a modest but extremely satisfying three months, partly by sponging on friends old and new ("baching it" as a guest of Ernst Mayr, whose family were on a visit to Germany), partly by giving lectures in broken English to patient and indeed keen audiences interested in either Eskimos or ethology. Top floors of YMCAs, the New York subway (one nickel), streetcorner or roadside diners, and Greyhound buses were my props in this first visit to fabulous America. Hooked on wild natural habitats as I was, I yet fell deeply under the spell of Manhattan's skyscraper skyline. (I also became preadapted to the *New Yorker* and, later, to Alistair Cooke.)

I find it difficult to assess what this visit did to my scientific development. Robert Yerkes had welcomed me very kindly to his laboratories in New Haven, Connecticut, and in Orange Park, Florida, and I was also allowed to spend some time in the American Museum of Natural History. What I saw of work done with rats and chimpanzees frankly bewildered me, and I felt more at home in the stimulating company of Ernst Mayr, in whose mind the general framework of *Systematics and the Origin of Species* must, as I realized later, already have been taking shape; and in Kingsley Noble's Department of Animal Behavior, where I met people who were later to become good friends and fellow bridge builders between psychology, general biology and ethology—among others the shy, keen, and thin-as-a-beanstalk Danny Lehrman, still a young student. Yet at this time, I showed little overt mental growth; there was too much to cope with, and the threatening war, the Munich crisis (solved while I listened, tense, to a radio far out on a flat Florida beach, and which as a European I recognized of course straightaway as a maneuver ensuring us merely a brief respite from a European war). Having lived for some time in Germany and in Austria I had no illusions about the aims of Nazi-dominated, resentfully aggressive Germany; and worrying about my family back in Holland prevented me from concentrating fully on animal ethology. Yet I learned something about human behavior.

On the whole, the years from 1930 to 1939 were very productive ones. Working, and teaching about animals, both in the field and in the laboratory, appealed not only to myself but to many of our students, and they flocked in growing numbers to our annual six-week "block practicals." Nor had I difficulty finding young coworkers at various stages in their studies, and I remember with great pleasure the weekly seminars we had at our home. A long series of our papers about many different animals appeared in those years, and these were later to provide much of the evidence—now seen so clearly as being "merely qualitative" and in many respects admittedly primitive, but firsthand and factual—that I was to use later to illustrate our approach to animal behavior in my book *The Study of Instinct.*

It was a time of enthusiastic exploration, when new discoveries about the way animals live were made literally every day. It is now difficult to realize that this was done in a time when ethology was still represented by no more than two young university teachers in the whole of Europe, both working on a pittance. For years my salary was the equivalent (then) of £150 a year (compared with a professor's salary of some £700), later to be raised by another £100. The part of our zoo lab's annual budget set aside for ethology was £40; each of us traveled the eighty miles to and from our summer camp on bicycle; we had our own little tents, and each student paid sixpence a week toward a fund from which we bought our saucepans, mugs, and so on.

Once I applied for, and to my delight received, a grant of some £5 to buy a large tent to accommodate our growing family (the argument I used was that my wife did the cooking, and other chores for all of us). We had to buy and mail our reprints at our own expense; yet it was through exchange of these that we had to and did build up our own vital collection of reprints received in return. For several years after our return from Greenland I had to earn the money for this, and even for the stamps for letters to our parents, by giving illustrated lectures on "Our Year Among the Eskimos." Yet those years were very happy ones. Is it amazing that I often feel a little out of sympathy with present-day students who complain about lack of funds and about the insufficiency of their personal grants? Certainly no one can deny that, for the money spent on us budding ethologists, our "productivity" was prodigious by modern standards.

The War Years

With the outbreak of World War II, our work, and the collaboration with our German colleagues were badly affected. Although my country was overrun after only five days unequal, but in places fierce, fighting, and I thus came to live on the same side of the front as Lorenz and my other German friends, our contact, as long as it lasted, was restricted to correspondence on scientific matters. Fortunately, none of our German colleagues came to call in person; our decision to tell them, if ever they were to present themselves in the hated German uniform, that they would be welcome only after the war and in civilian clothes, would have created needless alienation. Yet, feeling that science should remain international and cooperative, I continued publishing in the German *Zeitschrift für Tierpsychologie* (in which our grayling butterfly paper appeared in 1942) and I even accepted, early in 1940, a well-intentioned distinction from the German Ornithological Society (communicated to me by a letter jointly signed by my revered senior colleagues Oskar Heinroth and Erwin Stresemann). I do not apologize for this even though, much later, people who were born after the war was over (a group of Canadian "Maoists" who heckled me through two lectures given at Vancouver's University of British Columbia and Simon Fraser University) have accused me of "political collaboration" with the Nazis. The reality was somewhat different: at the same time as I kept up contact with my German friends in the pursuit of our scientific aims, I was imprisoned as a hostage by the Nazi occupying authorities because of my open resistance against their policies, in particular the dismissal of Jewish colleagues and the interference with our freedom of thought. All of us hostages considered our status as an honorable one—toward the end of the war a standard joke circulated in Holland saying that upon the return of our government-in-exile a questionnaire would be sent to all citizens asking: "Have you ever been in prison? If no, why not?" To make the distinction in everyday life between political resistance and scientific and human "antarchy" was of course difficult, and in the beginning we made many naive mistakes.

Life in the hostage camp, where we were herded together—some 1500, on the whole interesting, people of all social strata—was a curious mixture of humiliation (such as when the Dutch SS guards, simply to intimidate us, would shout at each other: "Time for another little firing squad!") and learning to cope with the reality of having to decide about one's priorities in such an unprecedented personal situation. Because we had every reason to believe the German threat that "for every one of us who would try to escape, ten Dutch *Jews* would be executed," we agreed on a resigned common policy of accepting our position and of making the best of it by preparing ourselves for our postwar tasks. For this, the cultural hotchpotch of such a community, a second period of living in an intellectual melting pot not unlike one's student years, offered a grand opportunity. Most of us did a lot of listening to others, of rethinking, and of planning.

For my future role as a cofounder of the emerging science of modern ethology it was, I felt, essential that I began to think of myself as a potential exporter of "Lorenzian"—in essence Austrian, Dutch, and Swiss—ideas to the English-speaking world, in which promising starts (Watson, Lashley, and others in America; Selous, Huxley, and Howard in Britain) had, in very different ways, failed to grow into a biological science of behavior. I intended to carry out this overconfident plan of internationalizing our young science by repeated visits and lecture tours in Britain and the United States and by attending international meetings (all at the time still no more than hopeful daydreams). I thought that, as the senior ethologist of a small, border country between Central Europe and the Anglo-American world, I might well be a suitable "missionary" for ethology (as Lorenz and I had already in 1937 decided, following Ernst Haeckel, to call our branch of zoology).

The Move to Oxford

After the war, this ambitious plan was in fact realized, though only step-by-step. Already, before the war, when I published mainly in German and Dutch, I had started to write some ethological papers in (a sort of) English. Postwar visits to Britain and the United States did make a certain impact, but it was not really lasting, and in 1949, after a few years of peacetime reconstruction at Leiden University, I accepted Professor (later Sir) Alister Hardy's offer of a "University lectureship" in animal behavior in the Oxford Department of Zoology and Comparative Anatomy. The offer of the position, which he had created by "juggling with two half-vacancies," and which was less prestigious and certainly less well paid than my full professorship in Leiden, was conveyed to me personally by David Lack on a visit, with his then fiancée, later his wife, Elizabeth, to our reestablished Hulshorst "insect camp."

For myself and my family this was burning our bridges behind us, with all that this implied. There is no need here to say more than that many in Holland, not only in the administration of the Dutch (state) universities in The Hague, but also among some of my colleagues, viewed our move from Holland to England with disapproval.

For my work, our settling in Oxford added urgency to the publication of the lecture series I had been invited to give in the winter of 1946–47 in New York under the auspices of the American Museum of Natural History and Columbia University, and that I had written up shortly afterwards in the form of a book to be called, not quite aptly, *The Study of Instinct.* (The book was a worked-out version of my mid-war paper "An Objectivistic Study of Animal Behavior.") I also had to start teaching—within the admirable but in many respects (to a foreigner) alien, mysterious, and unfathomable Oxford framework—"animal behavior" to zoology students and to whoever wanted to come and listen. (However happy I have been in Oxford and as a naturalized Briton, I still do not feel *quite* at home there even now, but then neither am I at home any more in my native Holland—like many emigrants we have seated ourselves between two stools. For our children, however, the move has had many advantages.)

Being a member of the Oxford setup gave me the unique chance to absorb, through daily personal contacts, the typically ecology and evolution study-oriented atmosphere of Oxford zoology. Life in this academic community, in which, under Alister Hardy, such pioneers as Charles Elton, Edmund ("Henry") Ford, John Baker, Bernard Tucker, George Varley, and Arthur Cain were all building up their research and teaching centers, influenced my entire outlook profoundly, and the group I now began to build up, from very modest beginnings indeed, began to produce work with a distinctly Oxonian flavor.

Interestingly, it became clear to me only then how much I had learned, during the war years, from a fellow hostage, the portrait painter Karel van Veen, who generously taught several of us for two years in daily sessions of several hours each, how to draw portraits in pencil. His basic procedure was: sketch first, in no more than pale, very vague forms, the pattern of dark and lighter *areas* (no lines) of the head. If you get their shapes and their relative positions right, you will straightaway have a perfect likeness, even an impression of the mood you want to capture your sitter in. Then begin, all over the face, a series of gradual "detailing" elaborations, ending, as a kind of "punchline," as one would say in writing a film commentary, with rubbering-in (through a little hole out in a piece of paper) the highlights of the eyes.

I have only gradually come to realize that this procedure of "descending" from a general sketch to an ever more detailed picture is also typical of the way in which I tackle a scientific problem. In my teaching, I liked to contrast this to the procedure, so common in for instance neurophysiology, of "ascending" from the study of simple components, such as, say, the achievements of sensory cells, nerve cells, and motor units in the muscles (each first analyzed themselves in the descending way) toward the investigation of ever more complex systems in which the senses, the nervous system, and the muscle system cooperate in an integrated, higher-level system. (Of course we too "ascend" when we move from a study of individual animals to that of communities and ecosystems.)

The Study of Instinct, written largely in 1948 but, because of the necessary editing and the transfer from the New York branch of the Oxford University Press (where the text had been duly Americanized by my good friend Danny Lehrman) to the Clarendon Press in Oxford (for whom it was turned into something more like Oxford English by another friend—who was to become a powerful ally and protector—Professor Sir Peter Medawar), did not appear in print until 1951. It was my first, soon-to-be-outdated attempt at writing a more or less balanced, comprehensive yet short and, I hoped, readable outline of ethology as I saw it then. It was a brave—somewhat premature, even cheeky—venture, and I look back on it with very mixed feelings. The provocative title reflected our still unsophisticated way of looking at problems of ontogeny, a way incidentally that was not really excusable at a time when studies of the ontogeny of form and function in other areas of biology were already far ahead, though still themselves groping for the most useful conceptual framework.

But my book did at least stir up things and made many psychologists take notice of the growing interest of zoologists in a field that had until then been almost exclusively their "hunting ground" and phenomena that had been their "fair game." It now seems to me that the main contribution of this book (and a later, shorter, and more popular but conceptually face-lifted text in the *Life Nature Library: Animal Behavior*) is to be found, not only in its incorporation of material on many different species of animals and its trend away from artificial, laboratory-based, experimental arrangements, but also in its logical/semantic analysis of the basic scientific question "why?" (To this day it is difficult to make young scientists see that their work, and their publications, always have to begin with watching, wondering, and the clear formulation of a *question*.) I still think that, with some modifications and refinements, my distinction of "the four why's" (to which I returned in my later paper "On Aims and Methods of Ethology") has helped in the clarification of our scientific thinking about behavior (and indeed about life processes in general).

Later I used to distinguish between *two* "why's," namely, (1) the question of how the *effects* of behavior, or any life process, influence *survival,* or, as we say now, a little pedantically, "inclusive fitness," and (2) that of *"what makes behavior happen";* and to *subdivide* this second question (that of "causation") into those of "moment-to-moment control" or physiology in its widest sense; the development of the behavior machinery in the individual; and the evolution, or succession of developments, that result in evolutionary change.

Each of these three questions about "what causes behavior" requires its own approach; in addition, the last one, that about past evolution, has a fundamentally different character because it deals with nonrepeatable, strictly speaking, unique events, *each of which has happened once in the past.* This makes the study of past evolution a branch of the study of history, and as such—because you cannot really experiment on such processes but have to rely on extrapolation of experiments of *possible future evolution*—distinct from the study of moment-to-moment control and from ontogenic developments, which are phenomena that happen time and again before our very eyes.

There are other worthwhile theoretical aspects of *The Study of Instinct* (such as the application of the hierarchy principle and my linking of this with Paul Weiss's work on the next lower levels of integration of the nervous system; some discussions on the interaction of such major systems and in particular on "ambivalent" behavior; useful remarks on experiments at the highly integrated level of observable behavior), but also definite flaws, such as a fairly poor treatment of ontogeny, and regrettably a total disregard for questions of feedback, especially on what makes behavior, or bits of behavior, *stop,* slow down, or change.

I do not think that I have made any worthwhile contributions to theoretical ethology beyond those made up to and in my "Aims and Methods" paper. But in the work done by the successive generations of graduates who came to do their "D.Phil" or their postdoctoral research with us, I see a distinct, gradual development in the direction of ecology- and evolution-oriented approaches. The two grade, of course, into each other, for we have been preoccupied not only with questions of "survival value" and of the relations between the requirements imposed on animals by their environment (*"Umwelt,"* von Uexküll) and the way they actually meet these requirements (never "perfectly" but always *sufficiently* well for the animals and, through them, their species, to survive), but also with applying our insights to the understanding of natural selection as a (demonstrable) *stabilizing* process and (by implication and extrapolation) a *molding* process.

It was also in those early Oxford years that I felt the need to write two other books: a brief general one on *Social Behaviour in Animals,* in which naturalistic, observational methods and the distinction of the "four why's" were applied to social systems rather than, as in *The Study of Instinct,* to individuals; and *The Herring Gull's World,* in which I wanted to show *in concreto* what I had done so far in terms of a more or less monographic "ethogram" of one of my favorite species. I wanted to publish this at that stage because I hoped to interest my future students in branching out into comparative work on as many species of gulls as possible. I like to think that both these books, however outdated they may be now, have served their purpose.

In Oxford I had from the start plenty of moral support, but at first, apart from an initial tiding-over grant of £500, which Hardy had wheedled out of the Agricultural Research Council, hardly any funds. But, although Oxford University itself did not, until some fifteen years later, create a second post for animal behavior—a lectureship that was soon taken up by Richard Dawkins, who has so far remained faithful to the place—funds were soon coming our way, first by a generous grant of £2,000 per year for all in all ten years, provided, without any strings attached, by the Nuffield Foundation at the prodding of Sir Peter Medawar, and afterwards, until my retirement in 1974, by even more generous grants given by the Nature Conservancy, carrying out Max Nicholson's forward-looking policy. Typical of the spirit in which this money was provided was Nicholson's answer to my question about what the Nature Conservancy wanted me to work on: "The same kind of work that you have been doing so far." (But later the Natural Environment Research Council—the name for the Civil Service-like organization that swallowed up the Nature Conservancy—drew the line when I suggested that one of my "research officers" be allowed to study the behavior of children. Taken aback by this refusal, I wondered at the time whether the new, to my taste now rather too rigidified body, considered that children were neither "nature" nor worthy of conserving?)

One of the attractions of Oxford as it then was (and largely still is in spite of its growing complexity) was the very great freedom it allowed its scientific staff. In accepting Hardy's offer of a humble lectureship instead of the full professorship I had held in Leiden, I had been in a position to bargain for some concessions, the main one being that in the spring I would always be free to do fieldwork with my graduates. I was also allowed time off for frequent visits abroad, to attend congresses and conferences, and to accept occasional one-term visiting professorships in other countries. Because my main aim was "to spread the ethological gospel" in the English-speaking world, most of these visits were to Canada and the United States, although I traveled frequently to Europe, spent one term in South Africa, and had repeated brief commitments in East Africa.

Already, before leaving for Oxford, I had approached my German, Dutch, Swiss, English, and American colleagues with the suggestion that, in order to fill the niche left by the deceased German *Zeitschrift für Tierpsychologie,* we found a truly international journal. Together with Gerard Baerends of Groningen, Netherlands and Bill Thorpe of Cambridge, England, I approached the publishing house of E. J. Brill in Leiden. They knew a good thing when they saw one (although they probably did not realize just *how* good a thing it was going to be for them) and thus the journal *Behaviour* was founded. Little did we anticipate in those years that there would soon be ample room for not only the revived *Zeitschrift für Tier-*

psychologie but also for the British journal *Animal Behaviour,* and that, even so many behavior studies would find their way to the pages of journals in the fields of zoology, ornithology, and many other scientific areas.

With all this and many other commitments, the 1950s and 1960s were very busy years, and they passed quickly. I like to believe that my personal ties of friendship with so many colleagues abroad have helped consolidate the remarkable spirit of cooperativeness that is even now such a striking characteristic of the international ethological scene.

Never a very steady worker, nor a really good organizer, I had soon to shed parts of my self-inflicted work load, and I have to admit that by the late sixties, some five years before my formal retirement, I began to lose my grasp of the new ethology that was being build up by able youngers all around me, and I was forced to leave more and more of my original work to my younger colleagues, of whom Mike Cullen, now Professor at Monash University in Melbourne, and Richard and Marian Dawkins were rapidly becoming the new leaders.

The work done in these early Oxford years, which I see as preparatory to what has now become behavioral ecology and related fields, was a combination of two of my favorite themes.

Having since before the war begun to do and to supervise work on the many different antipredator functions of external coloration in a variety of animal species (in which we studied the predators' behavior as well as the prey's coloration and, later, its correlated defensive behaviors) we began to turn more and more to field experiments on the survival value of such adapted structural-cum-behavioral features. The best known of these projects is that on the function of "eggshell removal" in the black-headed gull, which, however primitive it still was, did show nicely how enlightening it was could be to study the *effects* of behavior together with its *controlling mechanisms* and its *ontogeny.*

At the same time we inserted our interest in survival value and selection pressures into our comparative behavior studies, especially into those of the various species of gulls that we began to compare with the herring gull and with each other. I had only gradually begun to realize that it was in Lorenz's otherwise fascinating pioneering studies in this field that never fully satisfied me: his overemphasis on finding homologies, especially in signal movements; and his insistence that their adaptive radiation (expressed in interspecies differences) was governed entirely by the demand of this one function.

Our functionally oriented studies (inspired very much by "functional anatomy" as I had seen it practiced for instance by the school of Erwin Stresemann of Berlin) showed us that the gull species that we studied differed in *many* respects, which moreover were functionally interrelated, and that the evolution even of signal movements was very much influenced by the ecological niche occupied by each species.

Esther Cullen's classic study of the "Adaptations to Cliff Nesting in the Kittiwake" showed, as convincingly as a largely nonexperimental study could do, not only that this species differed ethologically and morphologically from other gulls in what could only be described as a *system of interrelated traits* but that its adaptation to a special ecological niche (that of a largely pelagic feeder who came on land only to nest, and chose the least predator-exposed nesting habitat: sheer cliff sides) had carried with it, as necessary corollaries, an extremely complex behavioral switch, of which the signal movements formed only a part.

To me, comparative studies of closely related but diverged (as well as convergently similar) species are at their most fascinating, indeed make sense only, when both homologies and functional aspects of adaptive radiation—of divergence of homologous characteristics and of convergence of less closely allied species—are considered. Of course this way of looking at behavior was not new; as in so many modern developments, Charles Darwin had already shown us the way, but his behavior work had long remained underestimated, almost ignored.

This characteristic of the work of our group has been emphasized aptly by Marler's and Griffin's (1973) comments in *Science* apropos the awarding of the 1973 Nobel Prize in Physiology and Medicine to Ethology: "[they] showed how an ecological decision made in phylogeny can reverberate through many aspects of the Biology of a species." With hindsight, I am amazed (and also slightly embarrassed) to realize how long it had taken us to develop this approach, but at the same time I feel that, however far modern "behavioral ecology" has evolved since Esther Cullen's rightly famous kittiwake work, we have laid at least one of the foundation stones of this fascinating and flourishing "growing point" of ethology.

But I have to confess that, just as I have been unable to follow the new developments in cybernetics and in the neurophysiological approach to behavior mechanisms, I have been left far behind and cannot possibly hope to judge critically the value of the many brands of behavioral ecology, and of modern evolutionary studies. I just "feel in my bones" that much of it is important—and also that it is a pity that the younger generation has already forgotten the early work that paved the way for their present ideas, as well as for their bread and butter! This is nothing unusual—young people prefer to say "but" rather than "and"—but it is sad that so few young colleagues see the continuity in the history of ethology. That continuity *is* there, even though at every stage there may have been many dead-end attempts, and there certainly were many red herrings drawn across our paths.

My own participation in the evolution of ethology had practically come to an end when, a year before my retirement as professor of animal behavior in Oxford, ethology was awarded the most highly regarded acknowledgement that a young— or for that matter an old—branch of science can be given: a Nobel Prize. It must have been a difficult decision for the electors responsible for selecting the laureates for 1973 to justify the awarding of the prize for "Physiology or Medicine" (as the curious formulation happens to be) to Karl von Frisch, Konrad Lorenz, and myself.

For myself this distinction came totally out of the blue (so much so that, in my confusion when a spokesman of the Swedish Embassy telephoned me the news, I could do no better to respond with a silly "Are you sure?") but I had sometimes wondered whether Lorenz might not one day qualify for a Nobel Prize for the potential value of his work to the medical field of psychiatry.

This is not the place to comment on the institution of the Nobel prizes, nor the question whether or not we were the right individuals to be so honored (since the prize can only be given to living persons, Erich von Holst, regrettably, was not one of the three—which is the maximum allowed number of laureates on any one occasion). My own reaction was a mixture of personal pride, of embarrassment at having been placed on this kind of pedestal together with two men whose achievements I had so long admired, and of a not wholly pleasant awareness that too many people suddenly began to look up to me as if I were a kind of superior know-all, understand-all, and in general a *wise* man! This has, in my post-Nobel years, led time and again to feelings of inferiority, of my not being able to live up to expectations. If *only* people realize that a Nobel Prize, just as any other distinction, is no more than a pat-on-the-back, a "well done" from other, far from infallible, human beings! Of course the gesture of the Nobel Foundation has had great value for the public image of ethology, but, the human species being a competitive creature, it has also caused a great deal of ill-feeling in certain, related sciences.

Early Childhood Autism

The end of my scientific career proper did not of course lead to a complete cessation of all my scientific activity. This was due to two of those accidental events that intervene in the lives of most of us. The first one was the initiative, taken in the early 1960s by Hardy's successor Professor John W. S. Pringle, to establish in the University of Oxford a new honors course in "The Human Sciences." I had by that time begun to have my own ideas about the value of ethology and ecology for a better understanding of (as I am used to call it) "The Human Predicament," and was only too glad to be asked to contribute to this new course. This meant a great deal of reading, and a slackening of my own research on animals, which in the last decade of my tenure had anyway become more and more confined to encouraging and supervizing my research students, and to the time-consuming struggle for funds and other matters that demand the attention of senior, yet still office-holding scientists. Apart, however, from occasional general papers on single aspects of the human predicament, and on child behavior, I did not make any specific research contributions to the field. I did enjoy the seminars and personal discussions I had with the enthusiastic and committed students who entered for the course. With some of them I am still in contact.

The second accident was the fact that Dr. Bill M. S. Russell, one of the first members of our research group, had shown me the way toward human ethology, and that my colleagues Dr. (now Professor) S. John Hutt and his wife, the late Dr. Corinne Hutt, then working in the Park Hospital for Children in Oxford, asked for my help in their attempts to understand the worrying, newly recognized mental disorder now widely known as early childhood autism. At the time, these children puzzled me, as did so much human behavior I saw around me (and in myself). But one particular sentence in their book *Direct Observation and Measurement of Behaviour*, stating that apart from "gaze aversion," all the behaviors seen in autistic children could also, on certain occasions, be seen in normal children, drew the attention of my wife (who has been a gifted child watcher as long as I have known her). She said to me: "But normal children *do* do a lot of gaze averting when they are not at ease," and it was this remark of hers that suddenly made me sit up; I had the interesting and stimulating "Aha!-experience" of seeing in a flash that she was right and that therefore autism was a deviation from normal behavior development that was quite clearly in essence an exaggeration of normal, anxious, apprehensive "social avoidance behavior," which is such an easily recognizable (and very useful) component of normal child behavior. From then on, my wife and I collaborated on what was to become a ten-year study of autism; a study that, while doing no more than applying some old, well-proven, typically ethological methods to child behavior, led to a new theory about the nature of autism, about its ontogeny, and ultimately to the therapy that we have been fortunate enough to apply and have seen applied with a gratifying degree of success to a growing number of such tragically afflicted, but (contrary to professional dogma) quite curable children.

Although I sometimes feel disappointed at having felt my scientific alertness and drive wane relatively early—partly as a consequence of less than optimal health, partly because I know and accept that I am a kind of hybrid person, neither artist nor scientist but a bit of both—it does give me great satisfaction to have, as I often feel it, "atoned" for a life in which I have frankly done little more than enjoy myself, by now having done something that begins to prove to be of social value. The therapy for autism that, developed more or less empirically by the psychiatrist Dr. Martha Welch of New York and beautifully consistent with our ethological interpretations, is now beginning to bring happiness to a growing number of severely afflicted children and their families. The whole exercise in this little corner of psychiatry has, in addition, been worthwhile because we have been able to convince at least some psychiatrists and psychotherapists of the very great potential value of ethology to psychiatry in general, which in our opinion is in many respects still in a pre- or even pseudoscientific phase.

At the moment of writing, most psychiatrists are still ignoring, denying the value of, or even actively resisting the application of ethological methods of study that, accepted though they are throughout biology, are new and incomprehensible to them, but it seems clear to us that this is a futile, defensive, rearguard action. It is necessary to remember that psychiatry, and medicine in general, have always been known for their conservatism and have only too often entrenched themselves in untenable positions, usually with unhappy consequences. There is no doubt though that ethology will become accepted even by them if we show that we can "deliver the goods." Even so there is no doubt in my mind either that the uphill struggle that we face will be much tougher than the one we have faced (and concluded successfully in the 1930s) when we began to make ethology a "respectable" branch of zoology. To have been "in the fray" in both these battles has at times been, and still is, taxing, but it has its rewards.

As I have said already, Sir Alister Hardy, who retired in 1961, was succeeded as head of the Oxford zoology department by Dr John W. S. Pringle, a nerve and muscle physiologist from the school of Sir James Gray in Cambridge. He set himself the task of continuing Hardy's admirable work of greatly expanding the scope of Oxford zoology, of making the university honor their promise, perhaps somewhat rashly given, to build a grand new laboratory, and to give zoology much more generous funds and facilities. In all this he has had remarkable success. A number of new appointments were made to our staff, and the scope of the teaching and research programme was considerably widened. The new lab was eventually built, a monument to his vision and perseverance. In teaching as in research, physiology (including its biochemical and biophysical subsciences) was soon given the same weight as "whole animal studies" such as ecology and ethology.

East Africa

When more and more gifted young field biologists came to do graduate work with us, I often wondered whether society would keep wanting to employ their expertise. I need not have worried, for throughout my period of office, the demand for "eco-ethologists" grew much faster than the supply, even though in the early sixties other centers of the same kind had sprung up, mainly in Holland, Britain, and in the United States. For us in Oxford it was an important event when John S. Owen, director of the Tanzania National Parks, visited us to lecture on his intention to establish, in the great game reserve of the Serengeti (made famous by Bernhard and Michael Grzimek of Frankfurt) a research institute for just the kind of work in which the approaches of Charles Elton, David Lack, and myself could be combined, with the grand, ultimate aim of understanding "how the ecosystem ticks" and of providing sound guidelines for the management of the reserves for the native African flora and fauna.

It was indeed a grand conception, and one that would need many years of work by a team of workers of world class. It was therefore a stroke of exceptional good luck for some of our best young eco-ethologists that through the untiring efforts of John Owen, the Serengeti, the Ngorongoro Crater, and Lake Manyara National Park became for them, at a modern, more sophisticated level, what the Dutch sand dunes and Greenland had been for myself: a splendid opportunity to learn their trade by being left, scientifically speaking, to sink or swim. They all swam. It was a great day for all of us when Oxford University acknowledged the importance of John Owen's pioneering work by awarding him the title of D.Sc. *honoris causa.*

I was exceptionally fortunate too in being invited to pay a number of winter visits to my young colleagues out in the African "veldt" (there was a time when the Serengeti team counted no fewer than six Oxford graduates) and to help in a minor way in putting the Serengeti Research Institute on its feet. My repeated stays in the hospitable homes of Hans and Jane Kruuk in Seronera and of Iain Douglas-Hamilton in Manyara, where his rondavel overlooked the beautiful Ndala River, were like visits to Paradise, of which the impact on my life is impossible to describe.

Epilogue

Looking back over all those years that have, after all, passed so quickly, I often find myself asking whether I have done the right things, and done enough. Of course there are a lot one regrets—missed opportunities, wrong handling of problems and people, neglect of duties, and, if one insists on finding fault with oneself, many more things that annoy one or make one blush. But I would be a hypocrite if

I dismissed what I have done as insignificant, and I would also be ungrateful if I did not acknowledge that I have been extremely fortunate in having been born a healthy and reasonably bright boy, and in having been given such fine opportunities, provided first of all by my quite exceptional parents, second by the naturalist's paradise in which I grew up, and third, to no small degree, by having been exposed to the influence of such men as Jac. P. Thijsse, Jan Verwey, Julian Huxley, Erwin Stresemann, Ernst Mayr, Karl von Frisch, and Konrad Lorenz, to mention only a few of the many persons who have affected my interests and my work.

Admittedly, I have been, compared with them, rather like a butterfly flitting cheerfully from one flower to the next, rather than like a steadily working, "flower-constant" worker honey bee. But such has always been my nature, and if, in the words of Dick Hillenius, I have "always been a boy with a talent for playing truant" and as a consequence have both missed much and gained much, I have at least been true to my nature.

My most "creative" work was done before I was forty; my achievements in communication, my writing, speaking, photography and filming, peaked somewhat later; and I was lucky in finding, toward the end of my life, an intellectually less-demanding but socially extremely rewarding task in the study, done jointly with my wife, of childhood autism, a study that taught me, among other things, to appreciate at its true value her exceptional gifts as a childwatcher and childminder.

Finally, I am sure that I share with many of my colleagues the awareness of having, so far, left too many tasks undone, most of which I will never complete. Apart from fields from which I have either shied away (for instance, cybernetics and philosophy) or that I have been at times keen on but unable to enter (such as skin diving and sky diving), I will never get down to publishing the once-planned book on sticklebacks, nor that on the adaptive radiation of behavior in the gull family, nor that, once almost finished, on "The Predicament of Man." A few more limited tasks may still get done, but in most scientific fields, the literature is simply gaining on me, and the problem of "information input overload," real enough for people in their thirties, is now unsurmountable for me. Ironically, this overload is in part the outcome of what we ourselves set in motion, and hoped for, when we began to develop and teach ethology! But what men in my position can truly say is that they have been privileged to have lived in such an interesting time and to have witnessed and assisted in both the birth, or rather the rebirth, and the coming-of-age of a fascinating new branch of biology.

If I were to characterize what my scientific life has been about, I would quote the lighthouse keeper of Goethe's *Faust II:*

> Ihr glücklichen Augen!
> Was je ihr gesehn,
> Es sey wie es wolle,
> Es war doch so schön!

Untranslatable (many scholars have tried it), but meaning roughly: "You fortunate eyes! How much you have seen! How generously you have shown me a world of true beauty!"

And here I could have ended this sketch of my scientific career as I see it. But had I done this, my story would have been incomplete. By having become increasingly interested in the ecological and evolutionary aspects of behavior, and in particular in the eco-ethology and the history of our own species, I could not avoid using what information has come my way for an assessment of the human condition in our time. Because that assessment turns out to be far from optimistic, in fact leads me to the firm belief that *Homo sapiens* is losing fitness—is on a course of accelerating *dis*adaptation—I feel compelled to state this once more, however briefly, publicly, and in print.

In a nutshell, I (and not I alone) see the present condition of humanity as follows:

1. The human population of the world is still growing at a prodigious rate.

2. In order to feed this population, however suboptimally both in quantity and in quality, our food production is not sufficient even now, and, in spite of Green Revolutions and other technological achievements, the carrying capacity of our planet is being reduced rather than increased.

3. We also continue, at an ever-accelerating pace, to use up our nonrenewable and even many of our potentially renewable resources, and are not making serious, at least not nearly sufficient attempts to switch to renewable ones.

4. Our industries, likewise growing at an unprecedented and increasing rate, produce waste that pollutes water, air, and soil, and, with that, our food.

5. Finally, the world's population, divided into autonomous nations, who all compete for the vital resources, are arming themselves and either prepare for or are actively engaged in warfare: neither the late League of Nations nor the United Nations have so far been able to stop these developments.

Although, unlike Konrad Lorenz, I do not so far accept that we possess an irrepressible *endogenous urge* toward violence, we do have the innate propensity to behave violently and even to kill our fellowmen *in response* to provoking stimulation; this behavior system, even if it were purely reactive, is so deeply ingrained that, up until now, it has, during our known history, led to ever more destructive wars. It is deeply ingrained because natural selection has, in our hunter-gatherer ancestors as in other pack-hunting animals, produced intergroup territoriality. In humans, this group territoriality, which in animals does not lead to mass killing (except in some ants where mass killing is however part of specialized *feeding* of the brood when it is need of protein), has evolved from threatening exchanges into real killing ever since our ancestors learned that a dead man does not return to fight another time. The geologically recent development of agriculture and of industrial production has not *caused* war, but it has increased the likelihood of its occurrence and the possible scale of destruction.

These five points state no mere opinions—they state *facts*—as everyone can testify who has read the daily press and, for instance, the publications of William Vogt, René Dubos, Barbara Ward, and the Worldwatch papers and books by Lester Brown and his colleagues. (There are many other factually reliable sources of information.)

The inescapable conclusion is that the outlook for the survival of the human species is very gloomy indeed. I do not want to be misunderstood or misrepresented: I do *not* say that we *are* doomed. But what I must say, and say most emphatically, is that, *if we do not change* our life-style and reverse the trends I have listed, *Homo sapiens,* and with him many higher forms of life on Earth, will soon, at best experience an unprecedented population crash or, at worst, and in my opinion more likely, become extinct.

It is not beyond the rational abilities of our species to apply our powers of thinking, of foresight, and of acting to the task of our own survival. After all, we do act with foresight in order to prevent or cope with possible disaster to our families whenever we pay heavy life and other insurance premiums; that is, when on the basis of rational considerations, we willingly make short-term sacrifices in order to gain a long-term advantage.

What is lacking in most people is the knowledge, and in many knowledgeable people even the will, to abandon our present lethal course in order to ensure our collective survival and that of the generations to come. So far we have failed to will a collective change of course in the interest of the survival of our species. But this does not mean that, if it comes really to the crunch we shall not be able to will it.

However, this crunch seems to approach with frightening inevitability, and it is even arguable that it has already started. Therefore, even though many have said it before, I *have* to say it once more: if we are to survive and, we hope, to return to a happier state, we shall have to halt and then reverse the trends that I have listed above. There is no choice, and time is running out.

References

Baerends, G. P. 1941. Fortpflanzungsverhalten und Orientierung der Grabwespe *Ammophila campestris* Jur. Tijdschr. *Entomol.* 84:68–275.

Cullen, E. 1957. Adaptations in the Kittiwake to Cliff-nesting. *Ibis.* 99:275–303.

Dubos, R. 1968. *So Human an Animal.* New York: Scribner.

Fabre, J.-H. 1923. *Souvenirs Entomologiques,* 80th ed. Paris: Delagrave.

Ferton, Ch. 1923. *La Vie des Abeilles et des Guêpes.* Paris: Payot.

Frisch, K. von. 1914. Der Farbensinn und Formensinn der Biene. *Zool. Jahrb. Allg. Zool. Physiol.* 35:1–188.

―――. 1967. *The Dance, Language and Orientation of Bees.* Cambridge: The Belknap Press.

Hailman, J. P. 1967. *The Ontogeny of an Instinct.* (Behav. Suppl. 15): 1–159.

Howard, H. E. 1920. *Territory in Bird Life.* London, Collins.

Hutt, S. J., and Hutt, C. 1970. *Direct Observation and Measurement of Behavior.* Springfield, Ill.: Thomas.

Huxley, J. S. 1914. The courtship habits of the Great Crested Grebe *(Podiceps cristatus). Proc. Zool. Soc.* (London), 1914, pp. 491–562.

Iersel, J. J. A. van. 1975. The extension of the orientation system of *Bembix rostrata* as used in the vicinity of the nest. In Baerends, G. P., C. G. Beer, and A. Manning, eds., pp. 142–69. *Function and Evolution in Behaviour.* Oxford: Clarendon Press.

Lehrman, D. 1953. A critique of Konrad Lorenz's theory of instinctive behavior. *Quart. Rev. Biol.* 28:337–63.

Lorenz, K. 1927. Beobachtungen an Dohlen. *J.f.Ornithol.* 75:511–19.

———. 1931. Beiträge zur Ethologie sozialer Corviden. *J.f.Ornithol.* 79:67–120.

———. 1935. Der Kumpan in der Umwelt des Vogels. *J.f.Ornithol.* 83:137–213, 289–413.

———. 1937. Ueber die Bildung des Instinktbegriffs. *Naturwiss.* 25:289–300, 307–18, 324–31.

Makkink, G. F. 1936. An attempt at an ethogram of the European Avocet (*Recurvirostra avosetta* L.) *Ardea* 25:1–60.

Marler, P., and Griffin, D. R. 1973. The 1973 Nobel prize for Physiology or Medicine. *Science* 182:464–66.

Mayr, E. 1942. *Systematics and the Origin of Species.* New York: Columbia Univ. Press.

Pelkwijk, J. ter, and Tinbergen, N. 1937. Eine reizbiologische Analyse einiger Verhaltensweisen von *Gasterosteus aculeatus* L. *Z. Tierpsychol.* 1:193–204.

Thorpe, W. H., and Hinde, R. A. 1973. Nobel recognition for ethology. *Nature* 245:346.

*Tinbergen, N. 1932. Ueber die Orientierung des Bienenwolfes, *Philanthus triangulum. Z.vergl.Physiol.* 16:305–35.

———. 1935a. *Eskimoland.* Rotterdam: Van Sijn.

———. 1935b. Field observations of East Greenland birds I. The behaviour of the Red-necked Phalarope in spring. *Ardea* 24:1–42.

*———. 1935c. Ueber die Orientierung des Bienenwolfes II. Die Bienenjagd. *Z. vergl. Physiol.* 21:699–716.

———. 1939. Field observations of East Greenland birds II. The behaviour of the Snow Bunting in spring. *Trans. Linn. Soc. N.Y.* 5:1–94.

———. 1942. The objectivistic study of the innate behaviour of animals. *Bibl. Biother.* 1:39–98.

———. 1951. *The Study of Instinct.* Oxford: The Clarendon Press.

———. 1953a. *The Herring Gull's World.* London: Collins.

———. 1953b. *Social Behaviour of Animals.* London: Methuen.

———. 1958. *Curious Naturalists.* London: Country Life.

*———. 1959. Comparative studies of the behaviour of gulls (Laridae): a progress report. *Behaviour* 15:1–70.

———. 1963. On aims and methods of ethology. *Z. Tierpsychol.* 20:410–33.

———. 1965a. *Animal Behavior.* New York: Life Nature Library.

*———. 1965b. Von den Vorratskammern des Rotfuchses (*Vulpes vulpes* L.) *Z. Tierpsychol.* 22:119–49.

———. 1972. *The Animal in its World.* I. London: Allen & Unwin.

———. 1973. *The Animal in its World.* II. London: Allen & Unwin. (This volume and the first volume listed above, contain reprints, in English, of selected papers published between 1932 and 1972 and marked with an asterisk in this bibliography.)

*Tinbergen, N., Broekhuysen, G. J., Feekes, F., Houghton, J. C. W., Kruuk, H., and Szulc, E. 1962. Eggshell removal by the Black-headed Gull *Larus ridibundus* L.; a behaviour component of camouflage. *Behaviour* 29:74–117.

Tinbergen, N., H. Falkus, and Ennion, E. A. R. 1970. *Signals for Survival.* Oxford: The Clarendon Press. This is a picture book based on the H. Falkus and N. Tinbergen film of the same title, which has been published by McGraw-Hill's film division.

*Tinbergen, N., and Kruyt, W. 1938. Ueber die Orientierung des Bienenwolfes III. Die Bevorzügung bestimmter Wegmarken. *Z. Tierpsychol.* 25:292–334.

*Tinbergen, N., and Kuenen, D. J. 1939. Ueber die auslösenden und die richtunggebenden Reizsituationen der Sperrbewegung von jungen Drosseln. *Z. Tierpsychol.* 3:37–60.

*Tinbergen, N., Meeuse, B. J. D., Boerema, L. K., and Varossieau, W. W. 1942. Die Balz des Samtfalters. *Z. Tierpsychol.* 5:182–226.

Tinbergen, N., and Tinbergen, E. A. 1972. Early Childhood Autism–an Ethological Approach *Z. Tierpsychol.* (Suppl. 10): 1–53.

———. 1982. *"Autistic" Children—the Possibility of a Cure.* London: Allen & Unwin.

Verwey, J. 1930. Die Paarungsbiologie des Fischreihers. *Zool. Jahrb. Allg. Zool. Physiol.* 48:1–120.

Vogt, W. 1948. *Road to Survival.* New York: Sloane & Assoc.

Ward, B., and Dubos, R. 1972. *Only one Earth.* London: Deutsch.

Watson, J. B., and Lashley, K. S. 1915. *Homing and related activities of Birds.* Wash. D.C.: Carnegie Inst. Wash. Publ. 211.

Welch, M. 1982. Appendix 1. In Tinbergen, N., and Tinbergen, E. A. *"Autistic" Children–the Possibility of a Cure.* London: Allen & Unwin.

Worldwatch Institute, Wasington D.C. A series of papers and books have been published since the early seventies by W. W. Norton (New York and London) The latest book is Brown, Lester R. 1982. *Building a Sustainable Society.*

---

*Reprinted in Tinbergen, 1972 and 1973.

Edward O. Wilson (b. 10 June 1929). Photograph by Lilian Kemp Photography

18

In the Queendom of the Ants:
A Brief Autobiography

Edward O. Wilson

I have been often asked whether I arrived at a synthesis of sociobiology and the evolutionary study of human nature after a lifetime of planning. Did I start with the biology of ants (myrmecology) as a first step toward the distantly planned and ultimate coverage of all aspects of social behavior? The answer is no. My scientific career was far less grandly conceived. I began to work on ants in my teens because they fascinated me. I wanted only to be an entomologist, to ride around in one of those green pickup trucks used by the U.S. Department of Agriculture's extension service to visit rural areas. Beyond that boyhood dream, I proceeded mostly one step at a time. In a sense, the ants gave me everything, and to them I will always return, like a shaman reconsecrating the tribal totem.

So what does it take to make a myrmecologist? The following ingredients were crucial in my case. My childhood was solitary—but not lonely. I was the only child of a couple who divorced when I was seven. Although I have remained in close and affectionate contact with my mother since then, financial conditions in the 1930s made it necessary for me to live with my father, and I was raised primarily by him and my stepmother, whom he married in 1937.

My father, Edward Sr., was a typical federal employee of the Roosevelt era. He was also exceptionally peripatetic, working for one acronymic agency after another through the South, never remaining at the same address for more than two years. As a consequence I attended sixteen schools in the eleven years from first to twelfth grades, skipping the third grade because of rapid early progress. Because of the difficulty in social adjustment that resulted from being a perpetual newcomer, without siblings, and younger than most of my classmates, I took to the woods and fields. Natural history came like salvation at a very early age. It absorbed my energies and provided unlimited adventure. In time, it came to offer deeper emotional and aesthetic pleasure. I found a surrogate companionship in the organisms whose qualities I studied as intently as the faces and personalities of boyhood friends. I also had a real friend in each one- to two-year period, a close chum my own age I somehow acquired and turned into a part-time zoologist. But the full balance was attained, to a degree much greater than is ordinary even for professional naturalists, by the inexhaustible resources of the outdoors.

Add to that, a drive acquired early, and which frankly I do not understand to this day. I have always been a workaholic. When I became a Boy Scout, I kept on achieving until I had acquired the rank of Eagle Scout and beyond, accumulating a large percentage of the merit badges that could be earned.

To this not entirely flattering revelation can be added the disclosure of certain physical infirmities that predisposed me to work on insects. I had use of only my left eye; the right was mostly blinded by a traumatic cataract when I carelessly jerked a fish fin into it at the age of seven. As a result of monocular vision I found it very difficult to locate birds and mammals when I tried to familiarize myself with them in the field. At the same time, the vision in the good eye was (and is) exceptionally acute, allowing me to read considerably finer print than that on the bottom line of ophthalmologist's charts. I am the last to spot a hawk sitting in a tree, but I can examine the hairs and contours of an insect's body without the aid of a magnifying glass.

In four continuous months in the rain forests of New Guinea, I collected over four hundred species of ants and studied many of them in some detail, but I never once saw a bird of paradise. My hearing has also been slightly impaired since early childhood, and I experience particular difficulty at higher frequencies. So, gone are the opportunities to hear the songs of many kinds of birds, frogs, and cicadas.

And finally, I have a peculiar inability to memorize poems, the lyrics of songs, and more than one or two sentences of prose. I never held a part in a chorus or school play, and never even tried—the very idea panicked me. As a result I was and still am virtually compelled to reconstruct most of what I learn, often with new images and ways of phrasing. I have no doubt that this slight learning disability added something to whatever originality of thought I have managed and helped to nudge me into part-time theoretical work—attempts, as it were, to cope with the world by reconstructing it into a form that I could more easily recall.

I was born in Birmingham, Alabama, on June 10, 1929. For four generations previously, all of my forebears had lived in Alabama, the first arriving in that state during the 1830s and 1840s. Almost all were of English origin. On my father's side, all lived in Mobile, where the men were principally shipowners and river pilots; my father's father was a railroad engineer who ran the train from Mobile north to Thomasville. On my mother's side, almost all were farmers in northern Alabama. Several fought for the Confederacy during the Civil War. One paternal great-grandfather, William ("Black Bill") Wilson, used his ship and skills as a pilot to smuggle guns past the federal blockade at Mobile, until he was captured by Admiral Farragut and sent to a prison in Tampa for the duration of the war.

So I enjoyed a sense of family roots even as my father pulled me in tow through his complicated odyssey. Of great importance, I benefited from the general approval and even esteem of my relatives for my interests in natural history. "Sonny" Wilson was thought to be a little strange but smart, and he was expected to make something of himself; what, nobody could quite figure out, but *something*. In the later years of my adolescence, my mother (who was then living in Kentucky) added her strong spiritual as well as financial support to the unusual career I was pursuing. Let me tell the rest of the story with the aid of several vignettes.

Washington, D.C., 1939

During a two-year sojourn in Washington, D.C., I was enthralled by visits to the Smithsonian Institution and National Zoo. The collections of insects and living animals were testaments to a magic world to which I felt I was owed special entry. The grounds of the zoo and nearby Rock Creek Park became the wilderness where, at the age of ten, I fantasized "expeditions" to collect insects. About this time I met a boy my age, Ellis G. MacLeod, who lived a block away and soon came to share my enthusiasm. Together we netted our first red admirals and fritillaries and sought the elusive mourning cloak. Armed with Frank Lutz's *Field Guide to the Insects* and poring over R. E. Snodgrass's *Principles of Insect Morphology*—which we could hardly begin to understand—we decided that we would devote our lives to entomology. In fact, Ellis is now a professor of entomology at the University of Illinois.

At this time, 1939–40, I also became fascinated with ants. I discovered a large colony of *Acanthomyops* in Rock Creek Park that seemed miraculous in the abundance, glittering yellow bodies, and citronella smell of the workers. I read the article "Stalking ants, savage and civilized," by William M. Mann, in the August 1934 issue of the *National Geographic*. In what was one of the more remarkable coincidences of my entire life, Mann was at that time director of the National Zoological Park. He became my hero from afar. In 1957, during the last year of his directorship, he gave me his large library on ants and escorted my wife and me on a special tour of the Zoo—for me a truly fulfilling event.

Mobile, Alabama, 1940

My little family returned to the South to commence a complicated sequence of residences in Mobile and other small towns and cities: Evergreen, Brewton, Decatur, and Pensacola. The woods and streams were always a short bicycle ride away. In 1942 I undertook a serious study of ants in Mobile, collecting and studying the tropical hunting ponerine *Odontomachus insularis,* the Argentine ant *Iridomyrmex humilis,* and a small species of *Pheidole.* At the age of thirteen, I made my first publishable observation, later to be used by William F. Buren and myself in scientific reports: that introduced fire ants were abundant in Mobile during 1942 and at that time included the reddish species, *Solenopsis invicta,* which has since spread throughout the southern United States.

My interests turned for a time to snakes. Little wonder—southern Alabama has one of the richest local faunas in the world. In 1942 I was appointed nature counselor at the Mobile area Boy Scout camp, primarily because the older, more competent young men had gone to war. I built a zoo with about ten species of snakes, which the other boys helped me to collect. Midway through the summer I was unfortunately bitten by one of my pygmy rattlesnakes and had to be rushed off for first aid. It was a bad day for herpetology at Camp Pushmataha. When I returned a week later, the senior staff had wisely disposed of the poisonous snakes, and the summer was concluded without further incident.

In 1944 I built an even larger collection of living snakes and frogs in my back-yard in Brewton. After capturing a specimen of the giant eellike salamander *Amphiuma means,* I wrote a letter of inquiry, with sketches and notes, to *Natural History* magazine. The editors responded by publishing an article on this remark-able animal, acknowledging the stimulus my letter had provided. I was thrilled! It was the first hint I had that the wondrous things in my private world might be valued outside Escambia County. Thus are Southern writers and artists born. I shared the feelings of Langston Hughes:

> When I get to be a composer
> I'm gonna write me some music about
> Daybreak in Alabama
> And I'm gonna put the purtiest songs in it
> Rising out of the ground like swamp mist
> And falling out of heaven like soft dew . . .
> In that dawn of music when I
> Get to be a composer
> And write about daybreak
> In Alabama.*

During the winter before entering college, at the now advanced age of sixteen, I decided that in order to realize my gathering ambitions I must specialize in a group of insects. Through competence in a specialty would come deep understanding of nature and a job in later years. After careful thought, I picked flies. These insects are abundant, extremely diverse, and with intriguing habits. Some are even beau-tiful! I was especially attracted to the metallescent blue and green dolichopodids. But this decision had to be quickly canceled. It was the last year of World War II and insect pins, manufactured mostly in Germany in that period, were unavail-able. I turned to my old favorites the ants: they could be collected in rubbing alcohol and stored in little medicine bottles purchased in pharmacies. With the generous help and encouragement of Marion R. Smith, the ant specialist of the U.S. National Museum, I set out to build a collection and prepare a monograph on the fauna of Alabama. The task was never completed, but in the course of pursu-ing it I learned a great deal about the classification and biology of this fascinating group of insects.

Tuscaloosa, Alabama, 1948

In my second year at the University of Alabama, I fell into the company of a remarkable group of young entomologists who had recently come from Cornell University: Ralph L. Chermock, with a new Ph.D., beginning an assistant profes-sorship; and George E. Ball and Barry D. Valentine, who enrolled in order to join Chermock as undergraduate students.

These three brought a professionalism, national perspective, and almost reli-

---

"Daybreak in Alabama," by Langston Hughes, from *Selected Poems of Langston Hughes,* (New York: Alfred A. Knopf, 1959). Reprinted by permission.

gious excitement about entomology and evolutionary biology. They greatly expanded my horizons. We pored over Ernst Mayr's *Systematics and the Origin of Species* as holy scripture. With Herbert Boschung, a fellow native of Alabama, we traveled all over the state, collecting and studying insects, reptiles, amphibians, and mammals. Our little band of zealots descended into caves to search for troglophilic crayfish and beetles, seined streams and ponds for fish, and drove along the highways on rainy nights looking for migrating tree frogs.

In 1948 I contacted William L. Brown, then a graduate student at Harvard University. Brown was (and is) a fanatic on ants, one of the warmest and most generous human beings I have ever known, and the single greatest influence on my scientific life. He fueled my already considerable enthusiasm with a stream of advice and urgings. What you must do, Wilson, he wrote, is to broaden the scope of your studies. Never mind a survey of the Alabama ants; start on a monograph of an important ant group. Look, you have the great advantage of living in the deep South, where there are a great many dacetine ants. No one has worked out the ecology and food habits of dacetine ants. There is an opportunity to do some really original research. See what you can come up with, and keep me posted.

I plunged into the project at once, tracking down one species after another, dissecting nests and culturing colonies in the laboratory. The dacetines are slender, ornately sculptured little ants with long, thin mandibles and bizarrely formed hairs covering most of their bodies. They also exhibit a striking form of behavior. The workers hunt collembolans and other soft-bodied arthropods with very slow, deliberate movements. They approach the prey with extended sensitive labral hairs and seize them with a convulsive snap when the hairs touch. Each dacetine species has a distinctive repertory directed toward the capture of a particular range of prey.

Several years before I heard of ethology, I prepared a detailed report on the comparative ethology of the Dacetini (Wilson, 1953). In 1959, Brown and I published a synthesis of dacetine biology (Brown & Wilson, 1959), in which we traced the evolution of food habits in correlation with social organization and biology. In brief, the anatomically most primitive species forage above ground for larger, more general insect prey; they form large colonies and often have well differentiated worker castes and a marked division of labor. As specialization on collembolans increases across the spectrum of species, body size correspondingly decreases, along with colony size, the nests become more subterranean and hidden, and the workers lose their polymorphism and caste-based division of labor. This was the first such study in the evolution of socioecology of which I am aware. It preceded the work by J. H. Crook and others in the 1960s on primate socioecology and in some respects was more definitive. But it remained known to only a few entomologists and had little effect on the subsequent development of behavioral ecology and sociobiology.

In 1949, while a senior at the University of Alabama, I was hired by the Alabama State Department of Conservation to conduct the first full-scale study of the imported fire ant. This formidable pest had been accidentally introduced into the port of Mobile by shipping from South America and had subsequently spread deep into Florida, Mississippi, and central Alabama. Much of my research that year was directed at the behavior of the ants (Wilson & Eads, 1949). At this time, for example, I discovered the phenomenon of queen execution by which colonies eliminate supernumerary queens (reported in Wilson, 1966).

In the fall of 1950, after completing a masters degree in biology, I moved to the University of Tennessee to begin work on the Ph.D. under the direction of the ant specialist Arthur C. Cole. But Brown intervened again like the good fairy. The dacetine and fire ant work was very promising, he said. You should come to Harvard, where the largest collection of ants in the world is kept. Get a global view; don't sell yourself short with entirely local studies. I also received encouragement from Frank M. Carpenter, the great authority on insect fossils and evolution, who was later to serve as my doctoral supervisor.

Cambridge, Massachusetts, 1953

I sat in the main lecture room of Harvard's Biological Laboratories listening to a lecture by Konrad Lorenz. At that moment things were going exceedingly well for an aspiring young entomologist. In the spring I had been elected to Harvard's Society of Fellows, which provided three years of completely unrestricted study anywhere in the world. In the summer I had traveled to Cuba and Mexico for two months of intensive study on tropical ants. A whole new world of ecology and behavior had been opened up, and I was primed for new ideas.

Now I heard Lorenz explain the basic principles of ethology in his unique, hortatory style. The importance of the central concept struck me like a thunderbolt. If birds and fish were guided to such a remarkable extent by auditory and visual releasers, ants and other social insects must be guided to an even greater degree by chemical releasers! Up to that time it was widely felt that ants were governed to a substantial degree by scent, such as nest odors, alarm substances, and trail secretions, but the sources of these substances and their chemical nature remained wholly unknown. Also, the notion that complex, stereotyped responses might be triggered by single chemical stimuli was quite new and strange.

The more theoretical work on ant behavior in 1953 was dominated by Theodore C. Schneirla, whose writings stressed the traditional ideas of comparative psychology, learning theory, and generalized concepts such as attraction/ repulsion. Even while working out the life cycle of those ultimate instinct machines, the *Eciton* army ants, an accomplishment that may well stand as the major contribution of his life, Schneirla still sought ways to explain behavior on the basis of the simplest possible schemes of stimulus and response. This behaviorist viewpoint was not fundamentally wrong in the relatively few conclusions it was able to draw, but it turned attention away from the kind of heuristic theory and experimental approaches that characterized the new ethology.

So it was fated that I would go in search of the chemical releasers of ants. However, I postponed this pursuit in order to take advantage of an opportunity of a wholly different kind. In 1954 the Society of Fellows and Museum of Comparative Zoology awarded me grants to work in New Guinea and surrounding areas of the South Pacific. The ant fauna of this part of the world was still very imperfectly known. There were opportunities to discover many new species and to explore the ecology and behavior of major groups of ants rarely ever seen alive by entomologists.

For ten months, from November 1954 to September 1955, I worked on Fiji, New Caledonia, the New Hebrides, New Guinea, Australia, and Sri Lanka. On the way home I visited the leading ant collections of Europe to include taxonomic research. I had experienced the great expedition at last: New Guinea was Rock Creek Park writ large. Camus was quite right when he observed, "A man's work is nothing but this slow trek to rediscover, through the detours of art, those two or three great and simple images in whose presence his heart first opened."

Every young evolutionary biologist should have such a *Wanderjahr*. The impulse to explore the physical world and discover hidden places, to see marvels and nameless creatures, and finally to return to the tribe with stories of the adventure is primitive and deep. It provides a source of information and ideas on which years of creative work can be built, and a lifetime of memories. "Something hidden," Kipling captured the thought for all. "Go and find it. Go and look behind the Ranges. Something lost behind the ranges. Lost and waiting for you. Go!"

In the fall of 1955 I married Irene Kelley, a Boston girl. We were fortunate to enjoy a relationship that was to last and grow in strength and pleasure for us both through the years ahead, and the raising of our daughter Catherine. I accepted an assistant professorship at Harvard. It was time to take stock. At the age of twenty-six, I had successfully completed a diversity of studies on ants: the behavior of the dacetines; the history of the imported fire ant; the evolution of caste systems based on allometry and frequency distribution analysis; a monograph on *Lasius*, the most thorough revision of any genus of ants undertaken to that time; and a variety of special studies of the ecology and behavior of hitherto little-known tropical forms.

With William L. Brown I had written an influential critique of the subspecies concept and explored the phenomenon of "character displacement" in species formation; we coined the term and popularized the study of the process. Now an embarrassment of opportunities lay before me. Modern biology was catching up with the social insects, yet only a handful of investigators worked on the group worldwide.

The result of this exceptional circumstance was that I commenced work in several fields simultaneously. I asked, what can the ants bring to modern biology? For the next twenty years my research comprised a thick tangle of studies spanning a large part of organismic and evolutionary biology. Immediately after my return from the Pacific, I began a series of taxonomic monographs on the ants of New Guinea and the surrounding Melanesian archipelagoes, describing new species, putting the classification of various groups in order, and incorporating my field notes of ecology and behavior. I published my way through about one-third of the fauna, which is one of the largest and most complex in the world. Later I completed a monograph of the ants of Polynesia with one of my students, Robert W. Taylor, now chief curator of the Australian national insect collection in Canberra, as well as special studies of other oceanic faunas.

In a second major effort at socioecology, I reconstructed the evolutionary history of the army ants. Relying in part on my own observations of the tropical Cerapachyini, which display the rudimentary elements of legionary behavior, I showed how mass raids and frequent nest change originated as adaptations to predation on other social insects and particularly large, formidable arthropods such as giant beetles and centipedes. Relying on the many species that comprise the mass raiders, I pieced together a reasonably full story of the steps leading to the origin of the advanced *Eciton* army ants of the New World tropics and *Anomma* driver ants of Africa. The procedure followed was similar to that used by previous workers to explain the origin of the balloon flies of the family Empididae. I was especially pleased with this accomplishment, because it added a much-needed evolutionary perspective to the splendid research of Schneirla on *Eciton* and Albert Raignier and others on *Anomma*.

Inspired by the theories of William D. Matthew and Philip J. Darlington on the origin of faunal dominance, I turned about the same time to the question of interspecific competition and the spread of groups of animals around the world. In his 1957 book *Zoogeography,* Darlington had identified the Old World tropics as the headquarters of major taxon evolution and the springboard for the evolution of dominant vertebrate groups.

As I sifted through the data from my taxonomic revisions and field notes, I detected patterns at the *species* level. I could see in finer detail the process of species multiplication as groups spread from tropical Asia into Australia and the Pacific islands. For the first time it was possible to correlate the passages of stages in geographic speciation with alterations in ecology and behavior. I invented the concept of the "taxon cycle," which has since been documented in modified form in some groups of birds, reptiles, beetles, and other organisms.

In essence, the cycle commences with the inter-island spread of species that live in ecologically more marginal habitats such as forest borders and savannas, and possess certain traits associated with this specialization, including larger colony size, soil-nesting, more frequent occurrence of caste systems and odor trails, and others. As the colonizing populations penetrate the more species-rich habitats of the inner rain forests, on islands such as New Guinea and Viti Levu, they tend to fission, while at the same time shifting toward wood-nesting and the associated social and behavioral traits.

The discovery of the taxon cycle in 1959 was an exercise not only in biogeography but also sociobiology, and it was to influence my thinking in the later attempt to synthesize that complicated subject. I also worked out area-species curves for the Pacific region, a prelude to later work with Robert MacArthur on the theory of island biogeography.

In the midst of these eclectic endeavors I returned to the idea of chemical releasers in ants. The story of how I hit on the first glandular origin of pheromones in ants may be of special interest. I was keeping imported fire ants *(Solenopsis invicta)* in culture. This species had always been a favorite of mine since my early work on it in Alabama. Fire ants have a dramatic odor trail system, and so I decided to try to find the source of the chemical scent. Could the key substance come from one of the organs in the posterior region of the abdomen? I painstakingly dissected out the rectal sac and two principal glands of the poison apparatus and washed them individually in insect Ringer's solution—not an easy procedure, since these organs are barely visible to the naked eye. I then crushed each in turn on the tip of an applicator stick and smeared it in an artificial trail across the glass plate being used by the fire ants as a foraging arena. At best I expected to find that some of the ants would follow the line when they were later stimulated by the presentation of food. But when I tested Dufour's gland, an insignificant finger-shaped organ located at the base of the sting, an astonishing thing happened. Worker ants poured out of the nest by the dozens, ran the length of the artificial trail, and milled around in confusion at its end. The same response occurred when I used an extract of the gland's contents in ether or ethanol. Dufour's gland, I saw at once, contains a chemical substance that not only guides the ants but also summons them out of the nest. The contents of one gland are enough to activate a large group of foragers. Stretched out in a line, the pheromone is not just the guidepost, but the entire message.

That night I couldn't sleep. I envisioned accounting for the entire social repertory of the ants with a small number of chemical releasers. Each of the substances might be produced by a different gland and stored in special reservoirs, to await release by the ant according to the message the insect wishes to transmit. I proceeded to discover alarm substances in the mandibular glands of fire ants and harvester ants. Independently in the same year Martin Lindauer and his coworkers demonstrated alarm substances in the mandibular glands of the major caste of leaf cutter ants. They also characterized the substances chemically, the first such identification made.

I went on to discover the "necrophoric substances" by which ants identify their own dead. When a corpse has decomposed for two or three days, it accumulates enough oleic acid and related esters to cause workers to remove it to the refuse pile. I was able to get workers to treat small ant-size wooden dummies as corpses by painting the objects with minute quantities of these substances. Even live nestmates were converted into "corpses" when contaminated with the necrophoric chemicals. They were carried off to the refuse, live and kicking. Only after they had cleaned themselves thoroughly were they allowed to return to the nest.

As this work proceeded, I felt confident enough to write the following:

> The complex social behavior of ants appears to be mediated in large part by chemoreceptors. If it can be assumed that "instinctive" behavior is organized in a fashion similar to that demonstrated for the better known invertebrates, a useful hypothesis would seem to be that there exists a series of behavioral "releasers," in this case chemical substances voided by individual ants that evoke specific responses in other members of the same species. It is further useful for purposes of investigation to suppose that the releasers are produced at least in part as glandular secretions and tend to be accumulated and stored in reservoirs. (Wilson, 1958)

Events have proved this prediction correct. In general, ants appear to communicate by approximately ten to twenty signals, most of which are chemical. In 1959 Peter Karlson and Adolf Butenandt first used the expression "pheromone" in a general review of chemotactic communication, and the word was quickly adopted by those of us working on the social insects.

The Sixties and Seventies

For several months in 1961 I suffered from a state of mild depression, and taking my first sabbatical leave, I headed, that spring, for a new round of fieldwork in Surinam and Trinidad-Tobago. The first several months went well, with the discovery of new forms of behavior and chemical communication in previously unstudied tropical ants. Then I transferred to Tobago to study and consolidate my work—still in a mood of deepening discontent. It just didn't seem enough to continue enlarging the natural history and biogeography of ants. The challenges were not commensurate to the forces then moving and shaking the biological sciences.

But doubts about my work were only a part of what was troubling me. A deeper cause of my malaise was the accumulated tensions stemming from nearly a decade of personal conflict and academic rivalries that had plagued Harvard's biology department.

Let me explain. In the 1950s the molecular revolution had begun, and vast sums of money and the best young talent were being committed to that end of biology. One major advance after another occurred at seemingly monthly intervals. I was in a particularly sensitive position at Harvard, because one of my colleagues was James D. Watson, one of the architects of molecular biology. In 1958 he and I were assistant professors in the Department of Biology. He openly expressed contempt for evolutionary biology, which he saw as a dying vestige that had hung on too long at Harvard. The Department of Biology was deeply split, and committee meetings were tense and often hostile. It pained me to see men of the stature of Ernst Mayr and George Wald at loggerheads. I was the only younger professor in evolutionary biology and seemingly the only heir to the imperiled tradition.

It must have come as a shock to Watson, therefore, when Stanford University offered me a tenured associate professorship in the spring of 1958, and Harvard quickly countered with an equivalent offer. Although Watson was a year older, the avatar of the new order, and clearly only several years from a Nobel Prize (he received it in 1962), I was in the process of being promoted ahead of him. He soon received an offer of a tenured position at another university and obtained his own promotion.

Feelings were very mixed on the Harvard faculty concerning the ultimate relative value of the various endeavors within biology. No one seriously questioned the great future of molecular biology, but there were other prospects, and only a zealot could delude himself into thinking that all further advances would be at the chemical level. About this time Harlow Shapley, the great astronomer and a friend, stopped me in the faculty club and said, "I have just come from a meeting with Mr. Pusey [Harvard's president] in which I told him you are the most important assistant professor at Harvard." The Stanford episode did not improve my relationship with Watson and his close allies, and I was the object of even colder rudeness from then on.

It was a double pity, because like most scientists of my generation I greatly admired Watson's achievement. I saw in its style and outrageous success the sword that might cut one Gordian knot in biology after another. As a graduate student in the early 1950s I had been taught that the gene was an immensely complicated tangle of protein and nucleic acids whose chemistry would not be worked out for generations. It was a thrill to learn that the underlying molecule was in fact quite simple and that straightforward, readily understood chemical principles could be translated upward into an all but limitless biological complexity. Could we look to the same form of mapping in the equally complex realms of population biology and behavior? The search for such a procedure was the logical task of the next generation of evolutionary biologists.

But in the late 1950s, there were exceedingly few such persons around; they seemed almost an endangered species. At Harvard lived some of the leaders of the earlier generation, including Philip J. Darlington, Ernst Mayr, Alfred S. Romer, and George G. Simpson. My respect for them bordered on awe. They also shared my distaste for the new molecular triumphalism that denied the value of a great deal of contemporary biological research at the organismic and population level. But these great men were not my allies. A full generation older, they were in the consolidation period of their careers. They were more concerned with pulling together the remaining loose ends into the Modern Synthesis of evolutionary theory that they had engineered in the 1940s and 1950s. At Harvard I felt squeezed between the younger generation of molecular biologists and the older generation of evolutionary biologists. It was impossible to identify with either.

Evolutionary biologists of my own age group with a similar outlook and ambition were nevertheless to be found in other universities. I encountered Lawrence Slobodkin at the University of Michigan. He had just completed a brief textbook on population ecology that expounded the model-building approach to evolutionary problems developed by his teacher, G. Evelyn Hutchinson, at Yale. Slobodkin introduced me in turn to Robert H. MacArthur, a charismatic genius who was destined to have the greatest impact on theoretical ecology of any single person during the 1960s and 1970s. The three of us agreed to write a monograph covering all aspects of population biology including, in at least a nascent form, sociobiology. But Slobodkin soon developed an intense dislike for MacArthur, even though he had befriended and encouraged him during their graduate student days. MacArthur in turn felt wounded by the sharpness of Slobodkin's criticisms, and the book project fell apart.

It was at this point that I went to South America and Trinidad-Tobago. The brief association with Slobodkin and MacArthur (and the looming presence of Hutchinson beyond) had persuaded me that much of the whole range of population biology was ripe for synthesis and rapid advance in experimental research; but this could only be accomplished with the aid of imaginative logical reasoning strengthened by mathematical models. I am at best a mediocre mathematician and in 1961 had no training beyond algebra and statistics.

So brooding in Tobago that summer, I realized that I would have to make a major effort in order to gain a minimal competence for the intellectual effort to come. I set a new goal: to lift myself to mathematical semiliteracy, to gain a sufficient competence to collaborate with those better gifted and trained. For the remaining months of the sabbatical I taught myself elementary calculus and probability theory. Returning to Harvard that fall, at the age of thirty-two, I began two years of undergraduate mathematics courses. As a result, I did acquire the level of competency I needed, and the collaborations I had envisioned were achieved.

The first joint effort was with MacArthur. Soon after we met in 1960 we had become fast friends. By comparing notes and establishing common interests, we realized that biogeography was an area in which rapid progress might be made in both the theoretical and empirical domains. From my studies on the Pacific ant fauna, we were aware that provocatively orderly patterns occurred in the diversity and distribution of faunas on islands. Still more data were available on birds, and we pieced those together. From all this, in 1962, emerged the first quantitative theory of species equilibrium, which was published in article form in *Evolution* in 1963. A full-scale monograph, *The Theory of Island Biogeography,* followed in 1967 (MacArthur & Wilson, 1967).

This work launched a whole array of similar studies, using various groups of plants and animals, and carried out in many parts of the world, and by the early 1970s became a substantial and generally successful part of both biogeography and ecology. From the time of our collaboration to his death from cancer in 1972, MacArthur referred to himself primarily as a biogeographer. He concentrated on patterns of distribution as well as the fundamental problems of demography and population interaction, and his last book, written while he was very ill, was entitled *Geographical Ecology.*

For my part, I collaborated with Daniel S. Simberloff, then a graduate student at Harvard, in conducting the first full-scale experiments on island colonization in the Florida Keys. This work, extending from 1966 through 1969, established the existence of species equilibria and tested some of the key assumptions of the MacArthur-Wilson models.

At the same time that the crucial work with MacArthur began, I collaborated with William H. Bossert in producing the first general theory of the chemical and physical design features of pheromones. Bossert was then a graduate student and is now Gordon McKay Professor of Applied Mathematics at Harvard, and a distinguished theoretical biologist. His skills in mathematics and physical theory were perfectly complementary to my own knowledge of the new but burgeoning field of pheromone research. Together we made the first distinction between primer and releaser pheromones, created the concept of the active space, measured the first Q/K ratios (that is, the ratios between emission rates of molecules and the response threshold), demonstrated the adaptive significance of the physical properties of pheromones, and devised techniques for the estimation of molecular response thresholds. This work had an important impact on later research on chemical communication, including the all-important role of pheromones in the organization of insect societies.

In July 1964 MacArthur and I met at his summer home at Marlboro, Vermont, with Egbert Leigh, Richard Levins, and Richard C. Lewontin, to discuss the future of population biology. To a certain extent we divided the subject up. We discussed what major problems lay ahead and how each of us, and others among the small number of colleagues interested at that time, might contribute to the push forward. The seeds of my own synthesis of sociobiology were now solidly planted. I saw that if MacArthur and I could make some sense out of biogeography, hitherto the most sprawling and disorganized of all biological disciplines, and Bossert and I were able to progress so quickly with the theory of pheromone evolution, the study of social organization should also be open to a rigorous theoretical and experimental approach.

That the Marlboro meeting should strengthen this conviction is ironic, for Levins and Lewontin were later to be among the bitterest opponents of sociobiology—not because of the failure of the dream we shared in 1964 but rather as a concession to their total commitment to Marxism, in a form incompatible with any notion of genetic determinism of human behavior.

The ideas of two other persons were decisive in leading me to a synthesis of sociobiology. The first was Stuart Altmann, who came to Harvard in the fall of 1955 just as I accepted an assistant professorship to begin the next fall. Altmann intended to work on the sociobiology (he used that name) of monkeys. At first he could not find a sponsor. Donald R. Griffin was a logical possibility but felt that Altmann's interests were too far from his own. And indeed, Altmann was an odd fish at that time. He was up to nothing less than the revival of primate field studies, which had lain fallow since C. Ray Carpenter's burst of research in the 1930s and early 1940s. Altmann had the courage, imagination, and vision to see the potentially great importance of the field. However, none of the senior Harvard faculty shared his view. They had difficulty seeing how the field study of primates could be made "scientific." It was suggested that because I at least worked on a *social* group (ants), I might serve as Altmann's thesis adviser. I realized that Altmann probably had more to teach me than I him, and I gladly accepted, even though my faculty appointment was not to begin for another eight months.

In January 1956 I joined Stuart at Cayo Santiago, Puerto Rico, and we spent several days on that hilly little island examining the free-ranging population of rhesus monkeys that was to be the object of his study. It was an illuminating experience, a firsthand look at an animal society radically different from those of the social insects. As we strolled among the chattering and gesturing monkeys, Altmann and I talked about the ways in which insect and primate societies might be critically compared, in effect how one might develop a unified sociobiology. There were few concepts on which to build such a framework in 1956. Altmann introduced me to the techniques of information analysis, which he was later to use with significant effect in his reports on rhesus behavior. I in turn employed similar methods in 1962 when I compared the accuracy of the fire ant odor trail to that of the honeybee waggle dance.

Other fundamental concepts for the construction of sociobiology were assembled during the 1960s. One of the most important was the application of kin selection theory to the social insects by William D. Hamilton in 1964. Others before Hamilton had conceived the idea, including Darwin, who used the notion to account for sterile castes in ants, and J. B. S. Haldane, who had devised the elementary calculus by which the genetic impact of altruism is discounted according to the coefficient of relatedness.

What Hamilton achieved of great and unique importance was the recognition that the haplodiploid method of sex determination, found only in the Hymenoptera (bees, wasps, and ants) and a very few other groups of organisms, makes sisters more closely related to each other than mothers are to daughters. He pointed out that a great deal of what is peculiar about social insects seems to flow from this cardinal datum: the almost complete restriction of advanced social life to the Hymenoptera; the restriction of worker castes to females; the short, rather solitary lives of the drones within the colonies; and other, finer points. In one stroke Hamilton had made kin selection convincing, while providing a powerful unitary theory of the origin of higher social life in the Hymenoptera.

I first read Hamilton's article in the spring of 1965 while traveling from Boston to Miami to start work on the Florida Keys project. My first reaction was admiration blunted by incredulity. It was brilliant, but evolution just couldn't be that simple! Why, this fellow was proposing to change our whole way of thinking about the origin of the social insects with a numerical exercise that anyone could do in three minutes on the back of an envelope. I tried to put the whole thing out of my mind. I could not. I tried to find the fatal flaw in Hamilton's reasoning that would allow me to dismiss him and rest easy. I could find none. Within twenty-four hours, as the train pulled into Miami, I was a convert. I had to admit that Hamilton, who knew far less about social insects than I did, had made the single most important discovery about them in this century. He had done so with a mode of reasoning that probably would have otherwise escaped me all my life.

That September I journeyed to London to deliver a paper on the behavior of social insects at the annual symposium of the Royal Entomological Society. I looked up Hamilton, still a graduate student at the University of London, and we talked at length about the many subjects we held of common interest. I was shocked to discover that almost no one, including his advisers, appreciated the value of his work. When I gave my talk to the Royal Entomological Society, I devoted about a third of it to Hamilton's main argument. And sure enough, the sachems of British entomology present at my lecture—Vincent Wigglesworth, J. S. Kennedy, and O. W. Richards—experienced the same difficulty I had on the train to Miami. They each stood up to dismiss the Hamilton theory, using various of the counterarguments that I myself had tried out and finally reluctantly abandoned. Speaking in turn, Hamilton and I easily answered their objections and, to use an appropriately British expression, carried the day.

We are in the midst of a great disproportion in the allocation of scientific effort. The social insects are among the great wonders of the living world. Aside from their colonial structures, far more intricate than we have any reason to expect from the limitations of their minuscule brains, they are ecologically dominant elements of the land fauna. About one third of the animal biomass of the Amazon forest consists of ants and termites. In most climatic zones these insects also exceed the earthworms in the amount of earth and humus they excavate and turn over. The ants are the principal middle-level predators, while the termites rank among the foremost decomposers of wood. In South and Central America several species of leaf cutting ants *(Atta)* are the dominant consumers of fresh vegetation and the leading agricultural pests.

Yet, despite the great significance of the social insects for mankind on these and other grounds, less than a hundred specialists were active around the world in the 1960s, if we exclude the large force of apiculturists working on honey production and crop pollination. On mainland China there were just three such experts—approximately one per 300 million people! I perceived that one of the principal reasons for this underrepresentation was the extraordinarily diffuse nature of the technical literature, comprised of thousands of articles in many languages scattered through often highly specialized, sometimes obscure journals. There had been no general English-language review of the subject since William M. Wheeler's *The Social Insects* in 1928.

I decided to try to put the matter right by conducting a comprehensive review of all aspects of the biology of the social insects. In doing so I integrated as many of the available data as possible into a framework of modern population biology. After all, colonies of insects are populations. They can be better understood by reference to the principles of demography and the mass effects of caste determination and communication within large groups. The book, entitled *The Insect Societies,* was published in 1971 and received highly favorable reviews. I entitled the last chapter "The prospect for a unified sociobiology." In it I argued that a single body of theory and vocabulary in the study of all kinds of social organisms is feasible:

> When the same parameters and quantitative theory are used to analyze both termite colonies and troops of rhesus macaques, we will have a unified science of sociobiology. . . . In spite of the phylogenetic remoteness of vertebrates and insects and the basic distinction between their respective personal and impersonal systems of communication, these two groups of animals have evolved social behaviors that are similar in degree of complexity and convergent in many important details. This fact conveys a special promise that sociobiology can eventually be derived from the first principles of population and behavioral biology and developed into a single, mature science. The discipline can then be expected to increase our understanding of the unique qualities of social behavior in animals as opposed to those of men.

In 1968, as the writing of *The Insect Societies* was underway, I faced a major decision: whether to continue sociobiology as a central, perhaps wholly consuming activity, or whether to proceed more fully into biogeography and ecology. Both enterprises seemed extraordinarily interesting and promising. I chose sociobiology, because it was *more* interesting and promising. Sociobiology entailed large domains of behavior and population organization that had never been subjected to analysis in the mode of evolutionary biology. Sociobiology also offered the prospect of connecting biology to the social sciences, which is intellectual high adventure beyond the limits of the conventional biological sciences.

So in 1971 I began to study literature and films on the sociobiology of vertebrates and the colonial invertebrates. This work went with surprising quickness. The literature was less extensive and technical than that dealing with social insects, and it had been more frequently synthesized in the recent past. Also, I received the encouragement and active help of specialists, including Irven De-Vore, John Eisenberg, Richard D. Estes, Sarah Hrdy, Peter Marler, Robert L. Trivers, and many others. I was able to lean on the exceptional skills of Kathleen M. Horton, who has played a key role in library research and manuscript preparation through all my research efforts since 1965. But an additional factor was that my devotion to the effort was total. For two years I averaged ninety hours of work a week in order to complete the book while meeting my duties at Harvard.

Sociobiology: The New Synthesis, containing 698 double-columned pages and over 2,000 references, appeared in the spring of 1975. The reviews, like those for *The Insect Societies,* were almost unanimously favorable, at first. Some were extremely generous in their praise, proclaiming the book a landmark and the start of a new discipline. There is no question that *Sociobiology* has had a major impact on the development of sociobiology, as well as on many domains of the social sciences and humanities, and that at this time of writing (1982) its influence continues to spread.

It also became the center of a major controversy, which proceeded at several levels. At issue was not general sociobiology, at least not the treatment of animal social organization and the integration of behavior with population biology, but rather the application of the basic ideas of the new discipline to human behavior. The bulk of *Sociobiology* was noncontroversial, but in the first and final chapters (1 and 27) I recommended the importation of the sociobiological program into human behavior, using a deliberately provocative style:

> Let us now consider man in the free spirit of natural history, as though we were zoologists from another planet completing a catalog of social species on Earth. In this macroscopic view the humanities and social sciences shrink to specialized branches of biology; history, biography, and fiction are the research protocols of human ethology; and anthropology and sociology together constitute the sociobiology of a single primate species.

My intention in these two chapters was to call attention to the relevance of biology to human social behavior in a direct and forceful manner. What I anticipated was that some social scientists and humanists, their interest pricked, would then begin to absorb evolutionary theory and the techniques of population biology into their thinking. The ultimate result, I felt confident, would be a basic alteration in the foundations of social theory. It seemed inevitable that biology and the social sciences will eventually be united, and that many of the bridging ideas will be contributed by sociobiology.

The response, both positive and negative, was far greater than I had anticipated. The first level of criticism came from a wholly unexpected direction. A Marxist-oriented group called Science for the People, one of the last such organizations still active on American campuses in the post-Vietnam years, saw human sociobiology as a major ideological threat. To them it legitimizes "genetic determinism" of social behavior, and any form of genetic determinism can—and will—

be used to justify the political status quo, IQ tests, racism, sexism, imperialism, and in fact the whole congerie of demons against which the far Left raged.

No concession was made to the possibility that a deeper knowledge of the biological basis of human nature might in fact be used to speed social progress in the direction desired by the radical Left. The reason, I suspect, lies in the precepts of traditional Marxist belief itself: that there is no human nature, that man's character is wholly the product of his political and economic practice, and that history is moving toward a dialectically achieved utopia in which human biology plays little or no part.

Sociobiology does seem to point in a direction wholly different from that perceived by the Marxist world view. To the extent that it can be made precise and subjected to verification, it appears to pit science against the dogmatic assumptions of Marxism and hence the main legitimation of revolutionary socialist change as it is being pressed worldwide.

In 1975–76 I took a crash course in political philosophy with the aid of Daniel Bell and a few other members of the Harvard faculty who could see I needed help. I succeeded in defending sociobiology on both scientific and philosophical grounds (e.g., Wilson, 1976). But the experience was made painful by the fact that the leaders of Science for the People were other members of the Harvard faculty, including Jonathan Beckwith, Stephen Jay Gould, Ruth Hubbard, and Richard Lewontin. Forming a special Sociobiology Study Group, they devoted large amounts of their time to the preparation of documents, including the famous letter to *The New York Review of Books* ("Against sociobiology," November 3, 1975) and the organization of lectures and meetings. The attacks were often personal in nature. The critics implied that I and others working in this area were promoting racism, sexism, and other political evils, either deliberately or else as a side product of being enculturated by a capitalist-imperialist state.

A somewhat less cerebral radical group, the International Committee Against Racism, took up the campaign and picketed or actively disrupted some of my lectures—despite the fact that none of my talks dealt with any subject other than the unifying traits of the human species. At the American Association for the Advancement of Science meetings of February 1978, held in Washington, D.C., INCAR protesters seized the stage as I was about to commence speaking and dumped water on my head. In time I gave up open public lecturing, confining myself to talks and seminars to universities, colleges, and professional groups.

Fortunately, few people in the academic community believed the charges of Science for the People and INCAR, especially when it became more apparent that my colleagues in the radical Left were promoting a political philosophy and not just defending society from genetic determinism. As intellectuals and the American public at large shifted more toward conservatism in the late 1970s, the purely ideological opposition to human sociobiology diminished to near insignificance.

But more enduring critiques had developed at a deeper level. To put the matter in a nutshell, it was observed that while biological reasoning could illuminate the central tendencies of such human behaviors as altruism, aggression, and parent-offspring bonding, it had nothing to say about the mind, free will, and cultural diversity. Sociobiology might help to explain the psychobiological traits human beings hold in common with animals, but not the mental qualities that distinguish the species. Psychology had not been incorporated into evolutionary theory.

At a still deeper level, some authors argued that mental life and cultural diversity can *never* be given a conventional scientific explanation. This most basic disagreement was between what Loren Graham (1981) has called the expansionists, those who believe that the natural sciences can be extended to all forms of mental activity and social phenomena, and the restrictionists, who believe that scientific investigation is intrinsically powerless to go that far.

In *On Human Nature* (1978) I took a strongly expansionist view. While reviewing human sociobiology more fully than in my earlier writings, I argued that even religious dogma and moral reasoning are based upon biological processes. They and other mental processes can be fully understood only by means of population biology and evolutionary reconstruction.

But the question of the linkage of heredity, mental activity, and culture remained unresolved. In July 1978 Charles J. Lumsden, who was then a young lecturer at the University of Toronto, suggested that he come to Harvard to pursue a collaborative effort in sociobiological theory. The prospect was attractive. Lumsden was obviously a highly creative scientist, filled with new ideas, and he had a strong background in mathematics. At first I declined. I was battle fatigued from the controversy over human behavior and wanted to draw more completely back into research on social insects. But Lumsden was very persuasive. He pointed out that large domains of sociobiological theory were unexplored and, at that time in the early development of the field, relatively easily penetrated.

Soon after Lumsden's arrival at Harvard in January 1979, we gravitated toward the key problem of human sociobiology: the nature of the linkage between genetic and cultural evolution. During the next three years of hard work, we constructed as complete a picture as possible of gene-culture coevolution, incorporating the main findings of cognitive and developmental psychology into evolutionary theory and sociobiology. The result was a series of articles, culminating in the book *Genes, Mind, and Culture*, published in 1981—and a whole new controversy. Whether or not the theory and techniques we proposed constitute the dreamed-of breakthrough remains to be seen. The logic and evidence has so far held up under close and sometimes hostile scrutiny, but the critical tests needed to establish the ideas solidly within the corpus of the natural sciences remain to be devised. A shorter book by Lumsden and myself, *Promethean Fire,* explains gene-culture coevolution in less technical language and attempts to trace the evolution of human cognition.

While pursuing human sociobiology, I did not neglect the study of social insects, which I consider my beginning and ultimate life's pursuit. In 1968 I had constructed a general but still rudimentary theory of caste evolution based on the techniques of linear programming. Through the 1970s I studied caste systems of one ant species after another in considerable detail, adding to the natural history of this subject and my personal knowledge of it. In 1977 I was joined for a year by George F. Oster, one of the foremost theoretical population biologists in the world. Oster had a special interest in caste development and the mathematical abilities to advance the subject fundamentally. After a year's close collaboration, we produced the book *Caste and Ecology in the Social Insects* (1978). This monograph laid down the theory of optimization in the evolution of insect castes and identified a series of previously unrecognized research problems that have occupied me and other entomologists since.

Cambridge, Massachusetts, 1982

I realize that while still not especially old at this time of writing (fifty-two), most of my principal scientific contributions have probably been made. I consider myself fortunate beyond any reasonable expectation to have been a young evolutionary biologist in a period of great opportunity, and I wish the same experience for younger generations of scientists as they each in turn cut more deeply into the phenomena of evolution and social behavior.

Although I and my family have suffered at times from the more unreasoning episodes of the sociobiology controversy, which has been one of the most intense and divisive disputes in the recent history of ideas, we have been sustained by the warmth and understanding of many friends and colleagues. I have also received more than my share of academic honors: the National Medal of Science, from President Carter in 1977; the Pulitzer Prize for *On Human Nature;* membership in the National Academy of Science, American Philosophical Society, and other academies; honorary degrees; and many prizes, awards, and endowed visiting lectureships. I cannot deny that this recognition has meant a great deal to me, especially because of the chancy and perilous nature of some of my investigations and the self-doubt and insecurity that such efforts inevitably engender.

I hope to continue research and writing in humanistic scholarship, pursuing the leads suggested by advances in sociobiology over the past ten years. But every creative person carries an innermost image of the routes of imaginative pursuit to which he returns for a strength independent of praise and human influence, and mine lies elsewhere. My ultimate retreat is in the natural world through which we are privileged to travel an endless Magellanic voyage. Each species of organism can consume lifetimes of fulfilling endeavor, and out of the millions that exist and the immense tangled histories they culminate comes the sense that no matter how sophisticated and intense our efforts may be in the future, the voyage will still have only begun, and however wise we believe ourselves to be we will never lose the feeling that the world is infinite, unfathomable, and filled with wonders—only a bicycle ride away.

References

Brown, W. L., and Wilson, E. O. 1959. The evolution of the dacetine ants. *Q. Rev. Biol.* 34: 278–94.

Graham, L. 1981. *Between Science and Values.* New York: Columbia Univ. Press.

Lumsden, C. J., and Wilson, E. O. 1981. *Genes, Mind, and Culture: The Coevolutionary Process.* Cambridge, Mass.: Harvard Univ. Press.

———. 1983. *Promethean Fire.* Cambridge, Mass.: Harvard Univ. Press.

MacArthur, R. H., and Wilson, E. O. 1967. *The Theory of Island Biogeography.* Princeton: Princeton Univ. Press.

Oster, G. F., and Wilson, E. O. 1978. *Caste and Ecology in the Social Insects.* Princeton: Princeton Univ. Press.

Wilson, E. O. 1953. The ecology of some North American dacetine ants. *Ann. Ent. Soc. Am.* 46:479–95.

———. 1958. A chemical releaser of alarm and digging behavior in the ant *Pogonomyrmex badius* (Latreille). *Psyche* 65:41–51.

———. 1966. Behavior of social insects. In *Insect Behaviour,* ed. P. T. Haskell, pp. 81–96. (Symposium of the Royal Entomological Society, no. 3). London: Royal Entomological Society.

———. 1971. *The Insect Societies.* Cambridge, Mass.: Harvard Univ. Press.

———. 1975. *Sociobiology: The New Synthesis.* Cambridge, Mass.: Harvard Univ. Press.

———. 1976. Academic vigilantism and the political significance of sociobiology. *BioScience* 26:183, 187–90.

Wilson, E. O., and Eads, J. H. 1949. A report on the imported fire ant *Solenopsis saevissima* var. *richteri* Forel in Alabama. Special Report to the Alabama Department of Conservation, mimeographed, 58 pp.

Vero C. Wynne-Edwards (b. 4 July 1906). Photograph by A. Lucas in August 1982

19

Backstage and Upstage with "Animal Dispersion"

Vero C. Wynne-Edwards

"The child is father of the man"

Though I have spent half my life in Scotland, my surname is Welsh, and I was born in England in 1906. I was the last but one in a family of six. My father was headmaster of Leeds Grammar School, and we lived in a house belonging to the school; but within a few years he had bought a stone-built house at Austwick, a small village in the Pennine hills forty-five miles northwest of Leeds, where we spent the school holidays. It was in limestone country, now included in the Yorkshire Dales National Park, and there were caves, potholes, fells, gills, scars, becks, and mosses, enough to give endless adventure for boys.

It used to be told that when I was just five, and we were out at a favorite picnic place, I disappeared from the party for a considerable time, but at last came running back shouting "I touched it." I had followed one of the innumerable rabbits to its hole and waited, crouched above the entrance, until it cautiously reappeared, only a foot from my face.

My father's subject was mathematics; but from one of his own boyhood schoolmasters he had gained a familiarity with British flora. Not surprisingly therefore, my own first scientific venture was learning to know the plants, especially those in the local limestone hills. I entered the Grammar School as a pupil in 1914, and in 1918 won the school's junior botany prize for a collection of named wildflowers and ferns, brought daily or weekly to be checked by one of the masters. As a prize book I received the eighteenth-century classic, *The Natural History of Selborne* by Gilbert White. In the summer holidays the same year I composed in pencil, most of it by candlelight in the very early mornings, the first twenty-two pages of a "Flora of Austwick," copying the sequence and nomenclature from a book, belonging to my father, on the same district, J. Windsor's *Flora Cravoniensis* (1873), but recording my own localities for the plants listed.

I have the penciled manuscript still; but, regrettably, my astronomy notebook of the same date got thrown out when my parents finally retired and went back to live at Austwick in 1937. In it there were several pages, each with several rows of circles to represent the sun's disc, on which I sketched the configurations of sunspots each fine morning. I used a 2½-inch astronomical telescope belonging to

487

one of my uncles. It stood on a tripod and had a dark filter, and I directed it out of the open nursery window. Big sunspots often last long enough to travel round the back of the sun and reappear, to cross the disc a second time, making it easy to confirm the approximately twenty-five-day rotation period, which I did. The notebook also had diagrams of the day-to-day positions of Jupiter's visible satellites, and of the slowly changing array of the four bright planets, all visible that winter in the evening sky. In the days when streets were lit with gas-mantle lamps, the stars shone bright on clear nights even in the city sky.

At about the same time, I began a diary that continued, with some lapses, for ten years. At first it was almost all about natural history—temperatures and rainfalls, birds seen, fossils and minerals found, animal tracks in the snow; but occasionally there were short descriptions, of a heron seen catching a fish, a hedgehog drinking, the plumages of garden birds, and my attempts to catch them under a sieve (which were illustrated).

In January 1920, aged thirteen, I was sent away to boarding school at Rugby, and my horizons immediately expanded. The school actually had a real observatory, with a nine-inch equatorial telescope; also a natural-history museum with library and cabinet collections, including a fine herbarium, and there was a flourishing natural-history society. We had extracurricular art and music schools (though I let my chances slip with the last); we had school and house libraries and reading rooms, and continual processions of visiting preachers, musicians, and lecturers. Some of the latter got two- or three-page writeups in the diary! I remember especially Sir Ernest Shackleton, on the eve of his last Antarctic expedition in the *Quest,* on which he was to die in January 1922; and Ernest Thompson Seton, whom I was privileged to meet in the headmaster's study afterwards. The diary says "he was a strange looking individual, with Paderewski hair and brown wrinkled face, the eyes deep sunk and sharp." Another visitor who lectured "awfully well" was Julian Huxley, of whom I was soon to see more.

Whether in Leeds, at Austwick, or at Rugby, our local transport was generally on bicycles. There was not much motor traffic and it went at a sober pace, especially before 1920. Fifty-mile trips, even when one went alone, were not unusual. My knowledge of the flora, both montane and lowland, rapidly grew. I made first visits to Snowdonia (North Wales) and the Lake District in the holidays of 1921, and to upper Teesdale in 1922. The last was on a cycling and camping trip from Austwick, with my father and two brothers. Upper Teesdale, on the northern boundary of Yorkshire, is a Mecca for British field botanists. The river Tees rises in the highest part of the Pennines and is soon plunging over an enormous dyke of intrusive dolerite, which has cut through the limestone country rock. The volcanic heat from the dyke metamorphosed part of the adjacent materials into "sugar limestone," which now carries a wealth of alpine plants unequaled elsewhere in England.

My father retired from Leeds in 1921 and, being a parson, became rector of a delightful country parish called Kirklington, in the Vale of York thirty miles north of Leeds. Simultaneously the Austwick house was let to a tenant, and we acquired a car, which meant we could make day visits to places like Austwick or Teesdale, and often did.

A highlight of 1922 was the finding of a small lime-loving plant that I could not identify on Ingleborough (723 m), the nearest "mountain" to Austwick. It was a sandwort, low growing and with white flowers a centimeter across. It was easy to overlook it because there were two superficially very similar flowers growing in the same communities. I convinced my father that it was distinct and that there were three look-alikes, not two. None of our quite numerous reference books gave a hint of its existence. It turned out to be *Arenaria gothica* (now often "lumped" with the equally disjunct *A. ciliata* and *A. norvegica*), first discovered in England ten to fifteen years earlier at the same place, which is still its only known station west of Sweden and the Jura. In 1922 only one British plant handbook had been revised recently enough to include it, and I promptly got someone to give it to me as a birthday present.

At Rugby there were kindred spirits in the School House. On our own initiative we collected plants and Lepidoptera, found birds' nests, hunted for fossils in the local cement pits, "fished" in ponds for aquatic life, made drawings of "scratch dials" on medieval church walls, and "excavated" for pottery in a Roman camp on Watling Street. Where possible we became technically proficient and possessed or borrowed the books we required. My closest friend was later to become a professional artist, and I myself spent a lot of time drawing, and particularly painting wild flowers, learning techniques at first hand from the headmaster's talented wife. In school we were expected to work hard, and in after years I often looked back in astonishment at the variety of things I had learned there—from Milton's poetry to military proficiency ("Certificate A," so called), and from theology to thermodynamics. At fifteen I entered the "Science Side" and thenceforward spent much of my time in laboratories. Physics and mathematics were exceptionally well taught; and for about a year at the end I attended an advanced biology class in which I was the only pupil.

Undergraduate

When I left the school in 1924 my ambition was to go on Himalayan expeditions, not necessarily to try to climb Mount Everest (this was in the period of Mallory and Irving), but more to study the superb alpine flora and fauna. Dr. W. W. Vaughan, the headmaster, advised me to go in for medicine so that I could double as an expeditionary doctor; but my father also sought advice on his own, and in the end I entered New College, Oxford, to read zoology, in the hope of having Julian Huxley as my college tutor.

I went there in October 1924; and having virtually covered the first-year curriculum at school, I took and passed the four "prelim" exams in two terms, and was allowed by the Linacre professor, E. S. Goodrich (1868–1946) to sit in at the advanced zoology class for the third term of my first year (April–June). The Oxford science degree—a B.A.—was of course extremely specialized; all the instruction for the second and final years was confined to a single department and discipline. Looking back fifty-five years later, I know how often I have been glad to have studied every phylum of the animal kingdom in detail, and thus acquired enormously more straight zoological knowledge than comes the way of students nowadays, when there is so much new to learn, especially in cellular and molecular biology, and when in most universities they have to carry a second or subsidiary subject to degree level as well. The apparent narrowness of the Oxford degree never seemed a handicap in my career in university teaching and research. The biological revolution has of course been running at flood level all my life, and we have never stopped having to study new subjects, often in widely diverse fields, just to keep up with our work.

Goodrich was the last of the great comparative anatomists. He lectured three to five mornings a week on a two-year rotation, talking in a quiet, matter-of-fact voice while gradually building and perfecting the most beautiful colored diagrams on six slate blackboards (he was a gifted landscape artist on his vacations). Often it took him two hours to finish the day's quota. He was a small tidy person, unfortunately shy of students so that very few of us penetrated his defenses and came to know the kind and cultured man inside.

I cannot recall ever being bored, and of course we greatly revered his erudition. He had made at least some important contributions to the anatomy of every metazoan phylum, especially on the subject of nephridia, coeloms and coelomoducts; but his tour de force was the cephalization of the chordates. The culmination of his many published works was his *Studies on the Structure and Development of Vertebrates* (1930), based on the Oxford lectures in the "vertebrate" year. When later I was at McGill University, Goodrich's teaching was the foundation for a vertebrate zoology course of my own, which I gave and developed with enormous enjoyment for twelve years.

Our other lecturers were notables too. Those who were there when I arrived were all sooner or later elected fellows of the Royal Society. We had Gavin de Beer for embryology and experimental zoology; he afterwards became the director of the Natural History Museum, South Kensington, and was knighted for services to science (as was Julian Huxley). E. B. Ford gave the genetics courses, John R. Baker the cytology, and Charles Elton the ecology. In my "borrowed" summer term I wrote a weekly essay on some topic chosen by Huxley, and presented it and discussed it with him, along with two fellow undergraduates, in his book-crammed attic study at 8 Holywell, where he then lived. A voracious reader and frequent traveler, in personal touch with much that was going on in biological laboratories in Europe and North America, he was not often at a loss in an argument. He had an extraordinary memory and a remarkable gift for putting two and two together and coming out with some new idea.

Sadly these encounters were not to last: Huxley accepted the chair of zoology at King's College, London, before 1925 was out. His successor as my tutor was Charles Elton, whose influence on me was to be much more specific and enduring. Charles (who had been one of Julian's pupils himself) had gone on a Spitzbergen expedition led by Julian, in the summer of 1921 while he was still an undergraduate, and again on a follow-up expedition to the same area in 1923. On his way home the second time he called in at Tromsö, and saw and bought a copy of Robert Collett's *Norges Pattedyr* (1911–12). From it he learned of the lemmings' extraordinary migrations in Scandinavia. He was greatly intrigued by their apparent periodicity; and further reading soon put him on the trail of the celebrated periodicities in the Canadian fur trade, recorded over many years by the Hudson's Bay Company and already publicized by, among others, Ernest Thompson Seton. Elton's own pioneer paper, "Periodic Fluctuations in the Numbers of Animals: Their Causes and Effects," came out in 1924; and by 1925 he was already launched on his own field researches in Bagley Wood near Oxford, systematically trapping voles and wood mice in order to find out if there were periodic rodent cycles in England.

The same year the Hudson's Bay Company retained Elton as a biological consultant for five years; and thus the seeds that grew and blossomed, in the early thirties, into the Bureau of Animal Population began to germinate. It gave Elton a scientific niche for forty years, until he retired in 1967. After ringing his doorbell for my first tutorial at the age of nineteen, I was caught up in no time in the whirl of excitement, and greatly enjoying it. Thereafter our tutorials no doubt dutifully ranged across the wider pastures of zoology, but their outstanding effect on me was to light an interest in population ecology that has burned for the rest of my life.

Elton never did solve the riddle of periodic fluctuations in the populations of small herbivores. Were they primarily due to environmental factors such as weather, acting through plant production to increase or diminish the consumers' food supplies? Were they predator-prey oscillations? In either case, why did they seemingly latch on to either a three-to-five year or a nine-to-eleven year cyclic period?

One hypothesis after another was investigated and rejected. A later student of Elton's, Dennis Chitty, who later joined the Bureau staff and eventually moved to Vancouver, has pursued a hypothesis of his own. In the 1950 to 1970 decades he suggested the cause of the cycles might be autonomic, or internal to the populations themselves; and that when the numbers of voles increased and their populations reached high densities, natural selection would favor genes for larger body size and greater aggressiveness (he had already found that voles averaged heavier when they were abundant, and lighter when the population had subsequently declined). He assumed that the aggressive types had lower reproductive rates, and poorer resistance to mortality from environmental causes, and that these factors progressively thinned the population down, until selection swung back the other way, to favor smaller, gentler, and more fecund phenotypes once more. The cycle was assumed to take three to five years to work itself out. The supposed advantage of the phenomenon was that it kept the populations in check, and prevented their numbers from increasing until they destroyed the habitat, and declined to extinction as a result.

Chitty's hypothesis has attracted a considerable following, but it has not been, and perhaps cannot be, convincingly proved or disproved. I mention it because in the last twelve months I have written down my own conclusions on the cycle phenomenon in the form of an entirely different hypothesis, which ought before long to appear in print.

I must mention one other initiation I owe to Elton. On his advice I read A. M. Carr-Saunders' seminal book, *The Population Problem* (1922), and bought myself a copy in 1927. Its author had formerly been an Oxford zoology student, and was incidentally a member of Huxley's party on the 1921 Spitzbergen expedition. Soon after graduating, his interests had veered away toward biometrics and social problems. His unusual gifts and versatility are apparent from his being at one time "called to the bar" (becoming qualified as an attorney), and later serving from 1937 to 1956 as the director of the famous London School of Economics and Political Science, during which he was knighted for public services.

In the 1922 book Carr-Saunders applied his zoological and evolutionary insight to human societies and populations, and came to the quite unexpected and revolutionary conclusion that, contrary to popular Malthusian concepts, the primitive tribes that had survived into modern times had all, virtually without exception, practiced population control in one form or another. They had lived within their own tribal lands at or near an optimal density, attuned to the resources their territories could produce. Within their social groups the people shared and managed their resources for the common good, and normally enjoyed good health.

It was only when I was about halfway through writing my own first book, *Animal Dispersion in Relation to Social Behaviour* (1962), that I had occasion to take *The Population Problem* down off the shelf, and suddenly realized I was applying to animals almost exactly the same ideas and principles that Carr-Saunders had found true of primitive humans (of this, more later).

For a time I was secretary of the Junior Scientific Society, and later of the Oxford Ornithological Society. Bernard Tucker, who did much to advance the field identification of birds in Britain, returned to Oxford as a member of the zoology staff in my last year, and we became friends and correspondents for the rest of his life (sadly he died before he was fifty). In 1927, a student contemporary of mine, Max (E. M.) Nicholson, organized the "Oxford bird census," in which I took part. This, in effect, introduced bird ecology to amateur ornithologists in this country. Such was the drive behind his project that, with Tucker's help, the census and its staff were officially taken under the wing of the university zoology department, to serve as a national center, and organize collaborative field studies throughout the country. Thus Oxford became a focus for ornithological research. From it arose the British Trust for Ornithology in 1932 and the Edward Grey Institute for Field Ornithology in 1938. The Institute's renowned director David Lack took over in 1945 and remained there until he died in office in 1973. Nicholson had a scarcely less seminal role in founding the Nature Conservancy, a government-funded agency charged with acquiring and managing national nature reserves, and carrying out (or promoting externally) research related to conserving the fauna and flora.

Marine Biologist

In my first "long vacation" (1925), I got a professional job (not an easy thing to do in those days) in fishery research. I wrote a letter to the chief fisheries scientist in the Board of Agriculture and Fisheries, and he replied by inviting me to lunch in London, and soon afterwards found me a two-month post with the Lancashire and

North-western Sea Fisheries Committee, under Professor James Johnstone of Liverpool. Most of my work was based at the Port Erin Marine Biological Station on the Isle of Man. I obtained regular samples of herrings from the fleet of "drifters," small gillnetting steamers, mostly from Scotland, that migrated each summer to exploit the huge spawning shoals round the island's coasts. I made simple measurements in the laboratory, aging the fish by scale reading, and estimating fat contents and stage of maturity. I spent one week at sea on a trawler, bottom fishing, and catching mostly hake and flatfish; and I finished the summer by returning to the mainland at Fleetwood, north of Liverpool, and going out at dawn each morning in a minute sailing beam-trawler, fishing for prawns. My job there was to search for fingerling-size cod among the catch.

I learned a lot in those nine weeks. At the lab I was among people working on plankton, physiology, fish, crustacea, and bottom faunas. Afloat I talked for hours to fishermen, watched them at work, and found my sea legs. From then on I felt a strong urge to make a career in marine biology. In each of the two succeeding springs I went to the Marine Biological Laboratory at Plymouth to take the students' "Easter Class" under J. H. Orton, so that when graduation from Oxford came in 1927 I considered I was well launched on my future career.

Happily I had a letter from Plymouth that July, asking if I wanted to apply for the "student probationership" they awarded annually at the Laboratory. It lasted two years and paid £200 a year, but as New College had just given me a senior scholarship, I would be well-off. I accepted, and moved to Plymouth in October 1927. The Laboratory was not connected with any university, so no question arose about entering for a doctoral degree, and anyway it was not considered a necessary qualification at that time.

I was given as a research project the investigation of a curious phenomenon in a small marine crustacean, an amphipod called *Jassa falcata* or *marmorata*. Like other members of its genus it had two types of adult male, a small and a large, which were morphologically very distinct (Wynne-Edwards, 1962, p. 260). I soon found the problem complicated by the fact that there were also two sibling species, overlapping in similar habitats, each with a "high" and "low" male; but I never got my supervisor, who seemingly lacked the taxonomist's eye, to accept the fact. Also I ran into trouble with trying to cut sections through their chitinous exoskeletons and was thus thwarted in trying to follow their sexual development. Sad to say, my project ran into the ground; and as far as I know no one has yet explained why "high" and "low" males exist in these and some other amphipods.

Part of the trouble may have been that I was developing a strong rival pursuit unconnected with marine biology. I got interested in the nightly roosts of starlings, birds whose numbers ran into millions, largely migrants from northern Europe coming to winter in that mild southwestern peninsula of England. They collected at dusk into teeming roosts, mostly in young conifer plantations. At the time I was engaged to be married to one of my former Oxford classmates, Jeannie Morris, whose home was in Exeter, only forty-five miles from Plymouth. I often went there for weekends, and before many weeks, I had decided to find all the starling roosts in between the two cities, and eventually surveyed the whole large county of Devonshire. It was an entertaining study. One went out from base to a selected observation point by the first bus or train in the morning and took up a stance with an unobstructed view. Shortly before sunrise a broad "wave" of starlings, radiating out from a roost, would fly overhead, often followed within minutes by one or two following waves. Their flight direction was checked by compass bearing. At dusk, if the weather held, one went to some other point, usually five to ten miles away, more or less at right angles to the direction of the morning flight, and watched for a returning flock. The two flight lines were then laid out on a map and their point of convergence gave the predicted site of the roost.

Before long I bought a small car. I spent the "dull" part of the day at the lab, and extended the survey at morning and night, trying in the evenings to locate the actual roosts, one at a time. I reckoned the biggest of them contained somewhere near a million birds, the most mobile individuals of which flew out and back each day to feed as much as twenty-four miles away. The chorus of a million voices made a noise like sausages sizzling in a pan, magnified at close quarters to a roar like a waterfall.

It turned out that the starling population partitioned the county into enormous territories, each of which drained back at night into its own particular roost, usually located somewhere near the middle of the territory. There was some overlap at the boundaries, where different flocks could occasionally be seen heading away in opposite directions. There were also territorial boundary changes during the winter, usually when one territory was split by the formation of a secondary roost, or when two previous territories amalgamated.

I did of course keep some marine biology going, and picked up a great deal from my colleagues, who included C. F. A. Pantin, F. S. Russell, Marie Lebour, W. H. Harvey, and W. R. G. Atkins. I made routine studies of planktonic fish larvae, and wrote a paper on my Manx herring, showing that although they went on growing right to the end of their lives, at a relative rate that diminished with increasing body size, the relation between body weight and gonad weight remained almost linear throughout. That was in the early days of statistics, and I had a helpful correspondence with R. A. Fisher about testing the significance of mean weight differences between the sexes. But I have to confess that on "extracurricular" subjects I wrote no less than five ornithological papers.

When my two years were up, I moved to Bristol University as an assistant lecturer in zoology. I had a short but eventful time there. Having now obtained a steady job, I got married to Jeannie Morris. A month or two later I received a surprise invitation to go to McGill University, Montreal, as an assistant professor. The venture strongly appealed to us, and we emigrated to Canada in September 1930.

On the voyage from Southampton on the Canadian Pacific liner *Empress of Scotland* I spent much time on the lookout for birds and kept a numerical log. The data threw new light for me on seabird ecology. For example, I found the gull family are mostly left behind as one steams out over the ocean; only the kittiwake is at home over the deep water. Petrels and shearwaters, on the other hand, are essentially pelagic when they are not breeding, and keep well clear of the land. On arrival I quickly wrote a popular article for the magazine *Discovery,* in which the basic pattern of inshore (coastal), offshore (out to the edge of the continental shelf), and pelagic (deep-water) zones of seabird distribution were first outlined.

The voyage also coincided with the autumnal migration, and I saw parasitic jaegers, arctic terns and (twice) a solitary golden plover winging their way south over the trackless sea. The distribution of shearwaters and storm petrels was conspicuously regional or patchy; and in fact our five-day transect from Land's End to the Straits of Belle Isle left a host of questions answered.

It was clear that if one could travel repeatedly backwards and forwards every couple of weeks over the same track, as the steamship liners did, a picture might begin to emerge of how the different species shifted their distributions with the changing seasons. My teaching commitments would allow this to be attempted only in the vacation, but even this might provide a useful start. Remembering my earlier success with approaching the man at the top, I wrote to Sir Edward Beatty, chairman of the governors and chancellor of McGill, who was also president of Canadian Pacific. I outlined my project, stating that I had no source of research funds, and politely asked whether there was any chance that his steamship company might concede me a reduced fare on one of their ships for a series of voyages between May and September 1933. Within forty-eight hours, the university principal, General Sir Arthur Currie, a great man, summoned me to his office. My letter to Beatty was in his hand. It had a comment by the chancellor scored across the top asking the principal, I could only suppose, to let me know how outrageous and impertinent my request appeared.

It was a pity I had been so impulsive, Sir Arthur said. The result had been to get the Canadian Pacific door slammed in my face. If I were to succeed now, a completely new plan would have to be developed. There were only two companies with passenger ships plying between Montreal and England, the second being the Cunard line. He advised me to go and have a word with the dean of the medical faculty, whose wife's brother was the sole Canadian director on the Cunard board, and see if I could sell my project to him.

To cut a long story short, it worked. I made four round trips in one of Cunard's small "A" boats, the *Ascania*, in May, June, August, and September, staying ashore for July, the month I judged to be most likely to provide a midsummer lull as far as bird movement was concerned. The project succeeded far beyond expectation. I abundantly confirmed the reality of my three ecological zones. I found the greater shearwaters, which breed on islands in the subantarctic and come north to escape the southern winter, arriving in hundreds and thousands on the North American side of the Atlantic in May, many of them gradually drifting east on the prevailing westerly winds as the northern summer proceeded, and these same birds migrating south again in the fall on the eastern side. A similar loop movement was later discovered affecting southern-breeding shearwaters in the North Pacific. I saw arctic terns migrating northwest in spring and southeast autumn, to reach Baffin Bay and Greenland via West Africa and vice versa; unless they can rest on a floating object they remain airborne night and day because, as I was to learn years later, terns become waterlogged if they stop even a few minutes on the water. And of course there was much more to be told.

I wrote up my paper, after combing the literature for seabird observations made at sea by other authors, anywhere in the Atlantic longitudes, and in the end I felt I had broken new ground with my study; in fact many others have amplified it since. Wondering where to publish a rather long manuscript, I happened to see an advertisement for the Walker prize of the Boston Society of Natural History, for an essay on "any subject in the field of ornithology," so I sent it along. They gave me the prize and published the paper (Wynne-Edwards, 1935).

Incidentally, the captain of the *Ascania*, James Bisset, was a tremendous character who afterwards became captain of the *Queen Mary* and commodore of the Cunard-White Star fleet (and "Sir James"). Passenger ships were sailing more than half empty in 1933, in the aftermath of the Great Depression, and he and I become close friends and confederates, at least for the time being. To everyone on board I was the "bird man," with a permanent place at the captain's table for dinner. He was an uproarious raconteur; and over a preprandial drink in his cabin each evening we decided which of his stories he should tell that night. All I had to do then was give the conversation a nudge in the right direction. How we laughed afterwards! He had never had it so good.

A Botanical Digression

I have grown accustomed since 1962 to being associated with controversial ideas, but an earlier episode of the same kind, much longer ago, is now almost forgotten. It was exciting enough at the time. I had not left my botanical interests behind in England. In 1932, my second summer in Canada, I made a trip to the Shickshock mountains in the Gaspé peninsula of eastern Quebec, in order to study, among other things, the unusually rich alpine flora on Mount Albert (1150 m). In 1934 I gained more experience in Newfoundland and southern Labrador; and I made my first real Arctic trip in 1937, as the Canadian "official" on Commander Donald B. MacMillan's cruise that year, to Labrador and southern Baffin Island.

On that occasion, instead of using his own vessel the *Bowdoin*, MacMillan chartered the Gloucester schooner *Gertrude Thebaud*, then the champion of the New England fishing fleet and current rival of Canada's *Bluenose*. We had a crew of thirty-six college boys from homes in the eastern states, and a professional master and mate, plus several scientists. My job was to collect material for the National Museum of Canada and see that Canadian law was duly observed. We sailed into Frobisher Bay, one of the two great inlets in southeast Baffin. Its mountainous south coast had only been roughly sketched long ago in 1862 by Charles Francis Hall, who drew in what he could make out from the opposite shore of the bay, across twenty-five miles of water. I was able to make a much more detailed sketch map as we sailed along, or viewed it from vantage points ashore at three places where we stopped.

One of the boys and I managed to reach the top of the Grinnell Icecap (850 m), which buried about fifty square miles of mountains under a flattish dome of ice, and discharged by glaciers, four of which reached the sea. Interestingly, it was possible to identify correctly for the first time, and locate on the map, the York Sound and Jackman Sound described and named by Martin Frobisher in 1576–78. In 1945, when the first aerial survey maps of the region appeared, both Hall and I had small bays named after us nearby. We nearly wrecked the *Thebaud* in another little fiord a few miles further west. She was stranded when at anchor in the night by the falling 10-meter tide, lay over on her side, and shipped 100 tons of water before righting herself as the tide returned. We pumped her dry by hand before the tide went out a second time.

In addition to these excursions, my wife, two children, and I were lucky enough to rent a cottage for eleven consecutive summers near Bic on the south shore of the Lower St. Lawrence, about 180 miles northeast of Quebec city. The place was another botanists' honeypot, already known as such from visits made by Professor Merrett Lyndon Fernald of Harvard in 1905 to 1907. Within two or three years I had myself added another dozen rare species to the local list.

Fernald's name is still a household word among North American field botanists because of his masterly centennial revision of *Gray's Manual of Botany* (1950). He was long the director of the Gray Herbarium. At the time he first appeared on the scene, the botanical exploration of the continent appeared to be nearly complete. It had begun, naturally, in the east, and in the course of the nineteenth century had been progressively extended west and northwest to the extremities of the continent. The largely circumpolar flora of the Arctic was fairly well known, and so was the mountain flora of New England, which had been a happy hunting ground for Boott, Tuckerman, and Bigelow. But there were still large mountainous and maritime tracts in eastern Quebec, Newfoundland, and Labrador that were virgin ground; and these young Fernald, with great foresight, set out from Harvard to explore. To his amazement, he and his companions began turning up species only described and known from the Far West, particularly from the Rockies and Pacific coastal ranges, and completely new to the east side of the continent; or else they were undescribed endemics whose nearest relations were known from the western cordillera. The discoveries made them redouble their efforts, and the list grew until it exceeded 100 such species.

Fernald found, in short, sandwiched between the familiar boreo-alpine outliers of the Appalachians, and the true arctic flora of the high north, a region bordering the Gulf of St. Lawrence that held a substantial number of plants separated by 1500 or 2000 miles from their nearest known representatives in the west. What could the explanation be?

In a famous paper, "Persistence of Plants in Unglaciated Areas of North America," Fernald (1925) offered his hypothesis. The Pleistocene ice ages had, as everyone accepted, largely obliterated the previous flora and fauna, driving them south as the huge ice caps grew and advanced. When the last ice age ended 10,000 years ago, both flora and fauna spread north again as the land ice melted away. Fernald had been particularly struck by how extremely sporadic and local in their distribution some of his "cordilleran" species had proved to be; a few of them have still only been found in a single eastern station. He postulated that *they* must be the true relics of the preglacial flora, which had held out for 100,000 years (or however long it was) on ice-free nunataks, either on the coast or on unglaciated mountain tops. At the retreat of the ice, the robust circumpolar flora had rushed in, spreading north, east, and west, and up to the cool mountain tops (as was also generally accepted). But the "old" cordilleran species had been thinned out and genetically impoverished by their rigorous experience, and were unable to compete, and now survived only on or close to their ice age refuges. That was why they were so localized to this day.

When I read this, it seemed wholly implausible and full of special pleading. I was brought up in the knowledge that rare plants at the limits of their ranges, as at Teesdale, are often exceedingly dependent on optimal soil conditions. Follow arctic species north to the Arctic and they can grow on many different substrates; alternatively, follow the alpines south to the Alps, and exactly the same is true. At Bic we had limestone cliffs, even lime-infused bogs, isolated like an island in a sea of acid Precambrian rock. On Mount Albert there was an isolated mass of ultrabasic serpentine. Almost all of Fernald's localities for cordilleran isolates could immediately be pinpointed similarly as base-rich "islands."

Furthermore at Bic there was abundant evidence not only of glaciation (which most likely buried the hilltops), but also of deep submergence beneath the sea after the ice burden had melted. The rare plants there *must* therefore have been colonists; and they belonged to both of Fernald's categories, the cordilleran and the arctic, and not just to the former. As at Teesdale, they included northern outliers of southern species as well as southern outliers of northern ones.

My explanation (Wynne-Edwards, 1937, 1939) was that, historically, the flora has all been one. It was all displaced southwards and has returned to the north again. However, by accidents of geography there happen to be no large mountains in the center of the continent. Instead there is a mountain fringe that can be likened to an arch or bow, extending northward up the western cordillera, eastward over the arctic archipelago, and southward into Greenland, Labrador, and the Appalachians. The highest arctic species are now confined to the crown of the arch; the more eurythermal species come more or less far down the two sides as well, but the specialized alpines are found today only in the two limbs of the bow, typically on special rock or soil types that have confined them to a sporadic distribution. There are few calcareous or ultrabasic mountains rising above the timber line in New England, but the suitable habitats that do exist, like the Smugglers' Notch cliffs in Vermont, can harbor great rarities too (e.g., *Saxifraga oppositifolia*).

My interest in this subject led to my supervising a graduate student, Homer H. Scoggan, in a study of the flora of Bic. Later he published a 400-page book, *The Flora of Bic and the Gaspé Peninsula, Quebec* (1950). It also led to my election to the Royal Society of Canada (1940); and, almost incredibly, to another Walker prize from the Boston Society of Natural History!

Freshwater Fishes

Going to Canada did put an end to active marine biology, apart from the seabird transects in 1933, but I transferred my interest in fishes to the remarkably rich freshwater fauna of the St. Lawrence and its tributaries. The reason for its diversity, compared, say, with those of European rivers, is that the Pleistocene ice sheets never seriously affected the great Mississippi basin, nor was the basin submerged by the sea at any time after the Cretaceous period, 60 million years ago; but the glaciation did cause changes of drainage in the Great Lakes region, and allowed a sizable part of the Mississippi fauna to spill into the St. Lawrence headwaters.

I made a faunal survey, partly for the Quebec provincial government and partly on my own, lasting several summers, from the southwest corner of the province right to the Gaspé peninsula, 500 miles to the northeast. The survey was still in progress when I eventually left Canada, and my species maps and locality records were turned over to the Quebec ministry of tourism, hunting, and fishing, and have been incorporated into publications by their biologists. During the Second World War, the Fishery Research Board of Canada sent me on two long journeys, with Dr. Ronald Grant as companion, to report on the fishery resources of the Mackenzie and Yukon rivers. It was part of a reconnaissance commissioned by the federal government of the resources of the far northwest, and of the potential there was for settlement there of men discharged from the armed forces after the war. In the summer of 1944, Grant and I went 1300 miles down the Slave-Mackenzie system in a freight canoe, from Fort Smith to Aklavik, and thence across the delta to the "Reindeer Station"; it took eleven weeks. In 1945 we spent two months in the Yukon Territory, traveling by truck on the new Alaska highway and its feeder roads, and descending 400 miles down the Lewes-Yukon river in a 27-foot flat-bottomed boat belonging to an Indian named Frankie Jim. In both summers, especially the second, chartered float planes flew us to some of the more remote lakes and rivers; and at the end of both seasons we flew out on scheduled air services.

There were some memorable incidents, such as shooting the Five-Finger Rapids; going into the small Eldorado mine on Great Bear Lake when they were working night and day to win the ore for a hush-hush weapon that was soon to erupt over Hiroshima; helping to carry the week's gold dredgings from the re-worked Klondike river gravels into the bank at Dawson City; or watching the panic stations in the cockpit when our engine went dead, 4000 feet above desolate forested foothills and seven miles from the nearest water.

The Mackenzie runs for the most part over vast alluvial plains, often in several channels separated by willow-covered islands. One can seldom see further than the forested shores—often more than a mile apart—on either side, and the length of the reach ahead and the reach astern. It carries a vast volume of water, warmed in summer by the hot Alberta sunshine, down to the Arctic Sea, and when it gets there three or four weeks later it is still much warmer than the innumerable ponds in the permafrost muskeg on either side. We were the first biologists to interest ourselves in the small kinds of fish as well as the large, and we found that several temperate-zone species extended all the way down to the delta, thus enlarging their previously known ranges by more than a thousand miles. Food productivity in the river is remarkably high.

The upper Yukon is entirely different. Many of the headwaters are in snowy mountains; the rate of descent is much faster and the speed of flow about 6 knots, exceptional for a great river in a mature bed. The cold water swirls and "boils" as it goes. Not far above Dawson City, it receives from a tributary an enormous load of white pumice in suspension, and the mixed water looks like café au lait and makes a curious low hissing sound. The scenery is majestic, and I have little doubt the Yukon is one of the world's most beautiful rivers. King salmon migrate up from the Bering Sea, some of them reaching 1750 miles from the mouth; and because of the current against them, even though they hug the banks, it presumably means swimming two or three times that distance through the water. They eat nothing en route. It seems an almost incredible mechanical achievement on a kilogram or two of stored fat! (Wynne-Edwards, 1952)

Vicissitudes

When I first went to McGill, the head of the zoology department was Arthur Willey, FRS, a distinguished Cambridge zoologist still remembered for his studies on *Amphioxus* and *Nautilus.* Unfortunately he was about to retire, and the university made a sad mistake in the choice of his successor, who eventually died in office six years later. After that, no immediate replacement was made. Instead the university appointed my friend and colleague N. J. Berrill and me as joint acting heads, presumably because there was no academic authority decisive enough to choose between us. To us it presented no problem at first, and the department soon returned to prosperity. The arrangement was allowed to ride until 1939, when Canada joined forces with Britain in the Second World War and it became perfectly clear that nothing more would be done to fill the Strathcona chair of zoology until hostilities ceased. Berrill and I made up our minds that the best way to break the deadlock would be for me (three years the younger) to move elsewhere, and leave him with an uncontestable claim to the chair; but that too would now have to wait.

To a man with three brothers, one of whom had been an infantry soldier decorated twice for bravery in France and Belgium in the 1914–18 war, a second who was decorated for bravery in the Royal Navy at Dunkirk in 1940, and the third killed in active service in the Royal Air Force in 1941, my own contribution to the war effort seemed second-class, to say the least. I retained my post at McGill but devoted most of my teaching effort to a crash course in electronic physics for radar mechanics in the Royal Canadian Air Force, which was repeated over and over as one draft of trainees succeeded another. I enjoyed learning the physics; it seemed an elegant and tidy discipline compared with the intricacies or zoology. The four biologists who had been pressed into service to swell the physics staff, and were accustomed to making complex matters in biology understandable to students, were conspicuously more successful as teachers of difficult physics to military personnel than the physicists were. In the event, the troops obtained their grasp of the principles and practice of radar very largely from the biologists in tutorials and practicals, where we were left to put them wise to what the terse and often mathematical lectures meant.

I also enlisted for basic training in the Royal Canadian Naval Reserve, in order to take command of a university naval training squadron consisting of McGill students. I enjoyed that too, especially the classes in which I taught navigation and the history of naval warfare! Just at the time of V-E Day, which ended the war in the Atlantic in May 1945, I was given a trip to the United Kingdom in a frigate, shepherding what turned out to be the last convoy of merchant ships from Halifax to Londonderry. I had a week's leave there, and was overwhelmed with nostalgia, and with the bond of comradeship that made everyone so warm-hearted. When I got home and was telling my wife and children about all that had happened, I said, "Well, I know where we are going when the war's over: to England!" The children quickly objected: "There wouldn't be any skiing", and I replied, "Then we'll have to go to Scotland instead." We had close friends at McGill, Dr. and Mrs. David Thomson, who were both Aberdeen graduates; his father had actually been professor of natural history there. We also knew that of the Scottish university cities, Aberdeen had the best access to mountains and snow for skiing. That was the summer I went off to the Yukon; and by a striking coincidence, I heard while I was there in a letter from my wife that the selfsame chair of natural history had fallen vacant, and was advertised in *Nature*.

Of course I applied as soon as I got home; and, after an interminable wait during which I had obtained, as a second option, a marine biology post in British Columbia, I heard that I had been appointed. I resigned from McGill, and in next to no time John Berrill was offered the Strathcona chair!

Scientific Administration

This heading may look as if it has nothing to do with animal behavior; but administration in a wide sense has occupied a notable part of my professional life, and often enough I have found that sitting on committees, especially the ones that fund and manage research, draws one's attention to useful scientific knowledge that would otherwise have gone unnoticed. The same is true of editing and refereeing for scientific journals and examining doctoral theses, in each of which I have done my full share.

Soon after going to Aberdeen, I was put on a London-based national advisory committee on fisheries research, chaired by Professor James Gray of Cambridge. We reviewed the annual programs submitted by the government fisheries laboratories at Lowestoft, Aberdeen, and elsewhere, and advised on the funding and programs of all the grant-aided laboratories in the country devoted to marine and freshwater biology and oceanography.

That kept me in close contact with scientists working at sea and at the bench, and added very opportunely to my intellectual stock in trade in the important decade of 1952–62, as will shortly become apparent. At the same I was riding another rather similar horse, as a council member of the Nature Conservancy, mentioned earlier, most of whose research staff were terrestrial ecologists.

I do not wish to stress this committee sitting unduly, and I need only record that a big reshuffle in the funding of government-sponsored science in Britain took place in 1964–65. A new Natural Environment Research Council (NERC) was formed, incorporating the Nature Conservancy en bloc, and the functions of Professor Gray's committee together with the grant-aided laboratories themselves, and similar institutions dealing with the earth sciences and hydrology. I was not altogether surprised at being asked to be a council member, but astonished when, three years later, the first chairman retired under the rules and I was asked to succeed him. So from 1968 to 1971 I became the NERC's halftime employee, commuting 1000 miles a week in order to give the university a crack of the whip as well. Never have I worked harder; but the council members were as clever a bunch of people as I have ever had the fortune to work with, and my term of office also coincided with what we look back upon now as a golden age of expansion and prosperity. Most of the projects the council planned or approved were eventually realized, and our efforts well rewarded.

The same expansive period allowed the zoology department at Aberdeen to grow. For many years I took every opportunity of giving the department a strong ecological bias and by degrees we acquired a distinctive cachet. More and more students came from all parts of the United Kingdom, and many, especially graduate students, from overseas. There was a most effective research unit attached to the department working on the population ecology and behavior of the red grouse from 1956–68, after which I turned it over to the NERC (who were already funding it); and I also established a substantial field station, located twelve miles up the coast, on the edge of a beautiful estuary and a national nature reserve. In 1970 a new zoology building was opened, to house what had become the university's largest science department in terms of student numbers.

Until I retired in 1974, the university kept to the Victorian tradition that the professor should be the permanent head of his department as long as he remained in post. A professorship of "civil and natural history" had existed in name since the eighteenth century, but the first zoologist appointed (at which time the word *civil* was dropped) was William MacGillivray in 1842.

Among the foreign academic connections I have enjoyed was a semester at the University of Louisville, Kentucky, in 1959, as Visiting Professor of Conservation. Another was a visit to New Zealand in 1962 as a British Council Commonwealth Interchange Fellow. The latter gave my wife and me a trip round the world, and the chance of seeing some of the zoological wonders I had quoted at second hand from other authors in *Animal Dispersion*. Among them were the synchronously flashing fireflies in Thailand, life in the canopy of the rain forest in Malaya, swiftlet and bat caves and the spawning of the palolo worm in Fiji, hibernating monarch butterflies at Pacific Grove, California, and acorn woodpeckers at Stanford University. I visited eastern and southern Africa in 1964, Nigeria in 1974, and the West Indies in 1967. In all, I lectured in approximately fifty universities and research institutes between 1962 and 1973, including a particularly memorable occasion at Lorenz's institute at Seewiesen, Bavaria, in 1967, when Leyhausen, Eibl-Eibesfeldt, and Lorenz himself were all present.

The Royal Society of London elected me a fellow in 1970, and I became a Commander of the Order of the British Empire in 1973.

"Animal Dispersion"

My own chief contribution to science is a 650-page book published in 1962, *Animal Dispersion in Relation to Social Behaviour.* The interest Charles Elton had awakened in me in the population biology of animals was still very much alive when David Lack produced his pioneer book, *The Natural Regulation of Animal Numbers* (1954). I read it from cover to cover, and went back over it carefully a second time. I also wrote a dissenting review of it for the magazine *Discovery*. The train of thought and events it set off have in fact been dominating my scientific life ever since, and are only now nearing their conclusion. In the pages that follow I have tried to sketch for the reader the course of my own intellectual journey.

As one might expect, my background experience was considerably different from Lack's. In particular, I had repeatedly come up against the "overfishing" problem when in Canada, as it affected sport fishing, hunting, and fur trapping; and as soon as I moved to Aberdeen, a major sea-fishing port, I began living with it almost on my doorstep. Since *this phenomenon of overexploitation is fundamental to all that follows* in my scientific story, I must try to put it in perspective at the start.

Suppose a commercial venture is established to exploit a hitherto virgin fish stock. By their fishing, the exploiters add to the mortality that the stock has previously experienced; and normally they will find that the stock's recruitment rate quickly rises in response, and compensates for the extra toll. Commonly the oldest, biggest fish are removed and not replaced, so that the average size of individual fish caught tends to drop. The total quantity or biomass of the population may also diminish a little, but the catch per unit effort is maintained. The operators can continue indefinitely to exploit the resource at this rate without risk to future annual harvests of fish.

Exploitation of such a naturally regenerating resource is only possible up to a certain limit, however, which varies with the species and locality and is not therefore predictable beforehand. If the exploiters were progressively to increase their rate of fishing, by using more vessels and nets, they would sooner or later reach a threshold where the annual mortality on the fish equaled the highest recruitment rate of which the stock was capable. Any further increase in the fishing rate would then begin to eat into capital, by depleting the biomass of the stock itself; and when that happened the fishermen's catch per unit effort would also fall.

Historically, most profitable fisheries have tended to overstep the threshold, at least a little, though not enough to do serious damage to the productivity of the resource. But if a fishery is extremely profitable, or if the recuperative powers of the stock are extremely low, or both, as with whales, overexploitation can eventually damage or destroy the resource beyond recovery.

These developments were confirmed in the North Sea by what happened in the two world wars. The fishermen in the warring nations left their nets and joined their respective navies, and the neutrals were precluded from using the main fishing grounds because they were studded with mine fields. The fish stocks, more or less heavily overfished beforehand, benefited by four to five years respite, and at the end of each war the size and weight of the average fish, and the catch per unit effort, had increased dramatically, in some species to several times what they had been when war was declared. On both occasions the high productivity was eroded again by renewed overfishing in the years that followed.

In the last decade, because of ineffectual quota systems and failure to get international agreement on conservation strategy, overfishing has been aggravated so desperately that investment and manpower in the North Sea industry have rapidly been evaporating, and disaster looms ahead. Translating this into the language of animal population ecology, the predators have overexploited their prey to the point at which their own survival is threatened.

With this much background we can return to the theme of Lack's book. His conclusions about the regulation of animal numbers were based of course on classical Darwinism, that is to say on the principle of natural selection and the survival of the fittest. The units in which to measure who the fittest are, appear to be the number of offspring that individuals contribute to the next generation. The fitter ones hand down more genes to posterity, and the less fit fewer, or perhaps none at all. Evolution then proceeds through the changes in gene-frequencies that result, as generation succeeds generation.

If one assumes, with Lack, that this is how natural selection works, the next logical step is that, if a new hereditary adaptation were to develop that enabled individuals to produce a still larger number of successful progeny, the adaptation would be bound to spread through the population in the course of time; and this leads one to the conclusion that the fecundity of every species must rise until it reaches the maximum level attainable.

A fundamental problem unconnected with this, but also stressed by Lack, is that, although the populations of organisms often fluctuate strongly in numbers from time to time, progressive changes in one direction, that is, population explosions or extinctions, are the exception and not the rule. Normally if fluctuations occur, they are not of enormous proportions, because a trend in one direction eventually sets off some compensatory process that drives the numbers back toward the norm. In other words, one has to infer that normally there are stabilizing forces at work.

Putting these two propositions together—maximum reproduction plus stabilizing forces—it would appear that the kind of stabilizing force in most demand would be one that curbed excessive population growth. Darwin recognized this in the *Origin of Species*. "We must never forget," he said, "that every single organic being may be said to be striving to the utmost to increase in numbers." It appeared to him that, for animals, there were four natural "checks" that would prevent this from continuing indefinitely. They were:

the food supply,
predators,
climatic extremes, and
epidemic disease.

Lack, and I as I followed his train of thought, were well aware that there are some animal species that have no predators, notably the top predators themselves. Second, some habitats are very equable, so that extremes of climate rarely influence mortality. Third, tolerance and immunity between parasite and host are continually being increased by selection, so that mortality from infectious disease cannot be a reliable check either, in the long run. One is therefore obliged to fall back on the amount of food available as the only dependable factor in limiting population growth, at least for some species.

It was here that I began to have misgivings about the soundness of the arguments. If food supply were the only effective check, the population would go on rising unhindered until some individuals were starving and dying. Even if the whole population did not immediately suffer the same fate, we have here the classical preconditions for overexploitation, a misfortune that applies with no less harmful consequences to many kinds of living food resources other than fish. Carnivores may overexploit their prey to their own detriment, and herbivores may overgraze. Both result in driving food productivity down, so long as the demand for consumption is allowed to exceed the rate at which the food supply can regenerate. The golden rule we learn from farming is to keep the land "in good heart," producing high but *sustainable* yields; and the same precept is just as true for the majority of animals and their habitats.

However, the overfishing principle does not apply universally. There are ecological niches in which the food resources are not susceptible to damage by overexploitation. Scavenger species, for example, eat dead organic matter, and do no harm to the source of supply by eating every scrap of it. And, especially in the tropics, many other animals subsist on nectar (and pollen), or on fruits, freely given away by plants in exchange for cross-pollination or seed dispersal. But every one of these foods is still in finite supply; and if the number of consumers increased to the level at which there was no longer enough to go round, the tendency would be for many individuals to feel undernourished before any significant number actually died and thus reduced the demand. The relief of consumer pressure by mortality would typically be too little and too late. Fruit eating in temperate latitudes, incidentally, tends to provide a transient extra resource, at least to the vertebrate consumers, which afterwards return to their staple diets, most of which would be prone to overexploitation, were this not being actively prevented.

I can sum up these thoughts by saying that, in the wild, there are some types of food of which only a proportion can be eaten with impunity, if there is to be enough stock left to ensure the full renewal of the crop the following year; and there are some types of which all that is accessible (or available) can be eaten. But for either type there is always a finite rate of food production per unit area and unit time (often a year); and in an ideal world, the consumers' demands should never exceed the available productivity. In other words, the number of mouths to be fed should not be allowed to exceed the number of appetites the available production can satisfy. The limited amount of food available cannot therefore be used with impunity as a "check" to prevent further increase of consumer numbers.

A Hypothesis Takes Shape

When this much had become clear in my mind I saw almost in a flash, not only that something closely resembling this "ideal world" does normally exist in nature, but also what the mechanisms are that bring it about. For thirty years I had been reading books and papers on territoriality in birds and other animals, and studying the phenomenon in the field. Birds tend to be most strongly territorial before and during the breeding season; and whereas for some species the territory is large and serves as a complete microcosm, providing shelter, a breeding site, and a food supply for parents and young, in others, especially in colonial breeders, it may consist of nothing more than a cramped nest site. Controversy had long reigned about the apparently varied character and functions of territories, although, in fact, two of the early pioneers of the territory theory, each of whom "discovered" the phenomenon for himself, had both assumed its primary function was to limit the number of breeders in a given habitat, and space them out so that they could enjoy sufficient food.

With this conception I could now completely agree. Even in a bird colony, quite artificial restrictions commonly appear to be placed on the area or volume of space the tightly packed colony fills. My experience with gannet colonies, for example, made me suspect that the compact group of nonbreeding adults often present, standing apart from the breeders in a customary area of their own nearby, were surplus adults for whom no room could be found within the conventional perimeter of the colony. I began thinking of territory as a possible mechanism for *excluding the surplus* of potential breeders, over and above the quota that the habitat can carry. Possession of different forms of personal real estate at other times of the year, such as roosting sites, might perform the same function, of allowing an established quota in, and keeping extras, surplus to the habitat's capacity, out.

Of course not all the higher animals by any means are territorial, at least in this simple manner. But there was a second, parallel mechanism, as effective as the first, namely personal status or rank in a group hierarchy. Social hierarchies or pecking orders had been known and speculated about since 1922, but no universal biological function had been agreed upon, any more than it had for territory. They were first discovered by Schjelderup-Ebbe when he was studying the behavior of domestic ducks and fowl competing for food; and it became clear that there were birds of higher rank who possessed a prior right to take any desirable morsel of food, in the presence of contenders of lower rank. Normally the latter acquiesced and avoided a dispute.

If food were short, this could mean that high-ranking individuals would be able to satisfy their appetites while lower-ranking birds starved—another alternative exclusion mechanism for the same purpose, namely squeezing out surplus consumers! Since brevity is of the essence here, let it suffice to say that territory and hierarchies alike can each give an individual the right to, or can exclude it from, the same set of biological goals, namely a domicile (or citizenship), food, and reproduction.

The next "revelation" was even more exciting. When animals are competing thus for basic rights, they usually abstain from violent combat and make use of innocuous forms of gamesmanship instead; and they compete not for the actual food, for example, but for substitute prizes, either a piece of real estate, or personal rank, or both combined. These are artificial, token prizes, serving like season tickets to the real necessities of life. Even the competition itself can be remarkably sublimated. Thus many birds defend their established territories mostly by singing, and alpha peacocks acquire their rights to polygamous matings by displaying the unbeatable magnificence of their trains.

Pondering these things, I suddenly realized that this looking-glass world, so long seem blindly and taken for granted—of artificiality, make-believe, status symbols, obedience, and conformity—was the image of society itself: it was in purest essence the social state. Within it, the number of "tickets" per unit area of habitat or per gregarious group, could be artificially manipulated, so as to prevent the population from exceeding the carrying capacity. Its primary biological function could be expressed as simply as that!

Sociality was of course another of these general phenomena to which, hitherto, no universal biological function could be attributed, though here it found a natural, obvious place. According to this hypothesis, all animals that competed by conventional methods for conventional prizes would, by definition, be social. It would not matter at all whether, as individuals, they were gregarious ("sociable") or solitary in their way of life. Solitary hawks or foxes could acquire property rights or personal status through intermittent confrontations with their neighbors, no less than the members of a herd of antelopes.

In the wild, sociality usually finds its fullest expression in small neighborhood groups, often of less than fifty individuals, the majority of whom are natives to the district, and all of whom are personally known to one another. It appears to be centered on competition for property and status almost as much as on brotherhood and companionship; but the latest development of the hypothesis has shown that mutual cooperation can confer a large extra survival value on the group, and increase the members' reproductive output. Demonstrations of brotherhood are in fact frequently apparent, and often accompanied by expressions of enjoyment among the participants. I shall return to this later. But for the present we, as human beings, can easily see that the hypothesis implies little if any violation, and much revelation, of sociality as we ourselves experience it.

When we look back toward the probable course of the evolution of sociality, we should bear in mind that all animals are consumers of organic matter that was originally synthesized by plant or microbial producers; and that the problems inherent in excessive population must therefore be as old as the animal kingdom itself. The development of sociality as a means of beating the problems was no doubt gradual, but its rudiments are recognizable today even in the lower metazoan phyla (Wynne-Edwards, 1962). The successive appearance in evolutionary time of one vertebrate class after another, from the fish to the mammals, suggests that effective sociality, already present in the paleozoic fish, would have been handed down, like the notochord and vertebrae, to every subsequent species of vertebrate. Its mechanisms must have diversified and no doubt grown more elaborate on the way. A similar long history most likely unfolded in parallel in the arthropod phylum as well. And though I have not had time yet to think it through, I suspect a counterpart has also evolved in plants, in order to allocate a sufficient quantum of resources (light, nutrients, water) to individuals to let them attain their full or most effective development.

A Controversy at Last Resolved?

Animal Dispersion aroused controversy from the moment it appeared. Reviews of it ranged from extolling its Darwin-like vision to deriding its credulous author. In the course of the ensuing twenty years, the majority of professional opinion has hardened against it, in the belief that it is founded on the sands of delusion. Nevertheless it gives some idea of the interest it aroused, that an article outlining these ideas, published in the *Scientific American* in 1965, has sold 350,000 offprints.

The flaw in the argument in the eyes of the critics was that it flew in the face of accepted Darwinism; or, to be more exact, it contradicted traditional tenets about fitness and natural selection. It postulated (1) that animals collaborate socially for the benefit of the group, because in no other way can their numbers be matched to the carrying capacity of their habitats; (2) that they compete conventionally for property and status, rather than for the real necessities of life, and the losers patiently accept their lot; (3) that animals are not, as Darwin supposed, always striving to increase their numbers; instead they are programmed to regulate them.

If the last postulate is true it implies that selection does not favor selfish individuals that maximize their own posterity, regardless of the long-term consequences, as conventional neo-Darwinists assume it is bound to do. Instead, somehow, the self-interest of individuals must have been subordinated to the long-term interest (i.e., survival) of the group and species to which they belong; so that selection, instead of favoring the individuals that leave the most offspring, favors the small groups that maximize their collective export of emigrants, who go off and establish themselves (and their genes) in other groups and localities.

When I wrote *Animal Dispersion*, and for a long time afterwards, there was a serious gap in the argument. I could not explain how these ideal cooperative societies could have evolved in the first place—how they could have achieved their supremacy over a primitive and unscrupulous individualism. The traditionalists therefore assumed that, as with all previous prophets of "evolution for the good of the species," my ideas were illusory, and the hypothesis wrong. Few evolutionists admitted to themselves that the evidence for self-regulation of numbers in the higher animals was already overwhelming, and that such homeostasis could not have evolved without some form of group selection. In 1956, when the gist of the argument was clear in my mind and writing the book already under way, I initiated a major research project on the population ecology of the red grouse in the vicinity of Aberdeen. It has run for twenty-five years and has tested experimentally in the field virtually all the premises on which the hypothesis is based. The red grouse has proved to be an almost ideal research animal; and the premises of population homeostasis have not only been greatly clarified, but have proved to be entirely correct.

The research teams that carried out the investigations and published their results was led in succession by Doctors David Jenkins and Adam Watson. They were able, among other things, to convince David Lack that the grouse regulate their own numbers through the territorial system. The grouse are herbivores, subsisting on heather; and by means of experiments in which an agricultural fertilizer was applied to the habitat, it was shown that their mean territory size varies inversely with the amount of heather protein present per unit area. Yet though the population density thus varies with the plane of nutrition, it rarely if ever rises to the threshold at which their food consumption would begin to injure the renewability of the heather crop. A third important prediction confirmed, was that if an established owner of a territory is experimentally removed, his place is normally taken over by one of the surviving outcast males, who had previously been excluded from the territorial system; that is to say, there is nothing wrong with the outcast birds except that there happen to be enough birds who rank above them in the social hierarchy to fill up all the territorial space available.

For the last four years, I have been able to devote myself to writing a new book, factually founded largely on the grouse work. As before, writing has deepened my insight, with the result that early this year (1982) I realized what the missing factor was, and can now give a complete, and I hope acceptable explanation.

Briefly it is this. Through their ability to match their numbers to the carrying capacity, the social animals generally manage to keep their habitats in excellent condition. In particular, food productivity is virtually as high as good consumer husbandry can make it. Most individuals are consequently born in, and spend their lives in, the most productive conditions procurable in that locality. *Their own average reproductivity is far and away superior to what would be possible, if numbers were allowed to exceed the habitat's carrying capacity, or if food productivity were driven down through overexploitation, as a result of the excesses of selfish individualism.* This is not the place to spell out the full details, and it must suffice to say that, if a selfish mutation could and did appear on the scene, it could never become common, because it would confer a frequency-dependent disadvantage (that is, the commoner the gene, the worse the damage), and its increase would only hasten its own demise.

This is the missing link in the argument, but it is not the only new light on the hypothesis. I have been postulating since 1962 that animals achieve population homeostasis in part by regulating natality and recruitment, and in part by excluding any surpluses that appear and thus piling on extra mortality. As a result of social competition, some individuals are normally barred from the opportunity of breeding at all, as I have just implied above; and they may also be driven out of the habitat, to end their days on marginal or hostile ground. Social competition not only sharply divides the candidates for admission (or readmission) to the social group into "passes" and "failures," but in order to do the job it is normally necessary to let the contenders grade themselves, through one-to-one encounters, into a single linear pecking order from top to bottom of the group.

The pass level varies with circumstances, and is determined by feedback from the current feeding conditions that influence the individuals' homeostatic responses. All those whose qualifications put them above the cutoff point win their right to live and feed or breed; but typically the social dominants at the top of the hierarchy *do* enjoy a greater personal "fitness" than the pass-grade candidates below them. An illustration of this is again seen in species where a few dominant males perform most of the matings with females in their group. Both individual and group selection can thus work in with, and complement, each other; and the former no doubt contributes much toward maintaining the high genetic quality of local groups.

Another new factor to emerge is that many if not all the highly socialized species that have been closely enough investigated so far, appear to have elaborately structured populations, by which I mean that their dispersion over the habitat is not only controlled as to population density, but is structured genetically as well. Members of both sexes are programmed to control the degree of panmixis (outbreeding) at mating time; and in many species (including the red grouse) we know the males and females are programmed differently, on the average, in terms of the closeness of their attachment to the exact site of their birth. The difference may serve to minimize the chances of incestuous matings. Programmed dispersal gives the local gene-frequencies a nonrandom distribution, and tends to foster genetic differences between local groups. This contributes enormously to the effectiveness of group selection.

In some species of the higher animals, one can see groups of individuals that habitually gather to take part in ceremonial displays, which appear to emphasize their social bonds and at the same time give hilarious enjoyment. Birds are well endowed by flight in this respect; and here in Scotland at certain times of year I can see parties of common swifts, dashing in close formation at top speed and screaming as they go; or piping parties of oystercatchers. In early spring, flocks of rooks often rise in a loose spiral to a considerable height, "cawing" noisily, until individuals begin to peel off in sudden spectacular tumbling dives, which continue until the flock disperses or resumes its spiraling. Social groups, even by popular definition, experience cohesive ties; and one can now appreciate how vital the members of such groups are to one another, and the extent to which each group makes up an evolutionary team. Typically, each member was individually programmed so as to locate itself in that neighborhood, and was screened by its peers before gaining admission to the group. Each carries its personal contribution to their usually excellent and diverse genetic heritage. Their contributions to each other's fitness, and to the posterity of their group in the longer term, are immense and indivisible. Small wonder if their cohesion is reinforced by feelings of delight when they demonstrate their solidarity and mettle.

In 1966 I took part in a conference at the Rockefeller University in New York on genetics and the social sciences. The distinguished geneticist Theodosius Dobzhansky was the leading speaker, and in private afterward, while we were discussing the implications of my own paper on population control and social behavior in animals, he put a specific challenge to me. "Don't you think you could sort this thing out once and for all?" he said. "This thing" was how group selection (which has vexed geneticists for half a century) could take place, and lead to the evolution of collaboration and altruism, and of statistical kinds of genetic adaptation as well.

Now, too late I regret to say for Dobzhansky, I believe I have the answer. Sociality, which enables consumers to collaborate in regulating their demands for food, has involved the subdivision of large populations into smaller, partially isolated, self-perpetuating groups. Each group differs from the rest in the exact constitution of its gene pool; and partly as a result of this, their net reproductivities will also differ over any given run of generations. The groups as units are small and individually expendable, and it benefits the survival of the rest that a proportion of them should be weeded out or supplanted on account of their poorer reproductivity. The relative fitnesses of the individuals in a group in their own lifetimes get submerged within a very few generations afterward into the collective fitness of the group as a whole.

This structuring of populations into social units provides a mechanism by which selection can promote *any* group-advantageous trait, even though the trait has nothing to do with sociality and controlling consumer demand—for example, traits affecting the number of chromosomes, or the frequency of crossing-over at meiosis.

All along I have expected the missing piece in the puzzle would turn out to be something simple, staring one in the face like the rest of the animal dispersion hypothesis; and so it appears to be. I should have tumbled to it long ago. No mathematical models are necessary. If the hypothesis as it now stands were to prove acceptable to biologists generally, it would broaden our ideas about how natural selection can operate in a profound and constructive way, by its validation of group selection. It would resolve a great controversy that began fifty years ago, between the advocates of group selection, especially Sewall Wright, and its adversaries Haldane and Fisher. It would remove the theoretical difficulties that have surrounded the evolution, not only of population homeostasis, but of cooperation and altruism as well. It would establish the primary biological functions of sociality, which would promptly appear to rate it among the most valuable general adaptations the higher organisms possess, because it promotes the process of adaptation itself.

Finally I must emphasize again that this has been a very brief summary of my conclusions on a complicated subject, without the benefit of full discussion or supporting evidence. If in spite of that the argument rings true, well and good: I shall have achieved my aim. But if to some readers it sounds dubious, I ask them to be patient and await the appearance of the full text before a final judgment is passed.

Reflection

My new book is not yet finished and is unlikely to appear in print before 1985. It is the culmination of thirty years of my scientific life, and that is why there has seemed no choice but to anticipate its findings here.

Although sociality is outwardly a behavioral phenomenon, I regard myself as an ecologist rather than an ethologist, though the borderline between the two is not always distinct. To a varying extent, all biologists are also evolutionists, but in that respect I have probably given more attention than most ecologists to evolutionary problems. We all accept evolution as an established fact, and the knowledge illuminates our approach to, and understanding of, our own specialist field, whatever it happens to be. But though we can be certain that evolution has occurred, its mechanisms are still in part conjectural, simply because the processes are so slow that we cannot investigate them by experiment.

Consequently evolutionary theorists are necessarily in some part speculators. My own speculations are backed to an important extent by twenty-five years of research on a single species of animal in the wild, the red grouse. The research was begun after the animal dispersion hypothesis had been formulated, when I knew what predictions I wanted to test. The answers were duly obtained and thoroughly tested; and as I build on this narrow but secure foundation, I can feel virtually certain in my own mind that the hypothesis is correct, and may legitimately be extended on circumstantial evidence to cover other species far and wide, at least in the animal kingdom. The pages of *Animal Dispersion* were already laden with such circumstantial evidence, and vastly more has accrued since 1962.

I am hopeful therefore for the future, although experience warns that the reactions of biologists to new theoretical ideas tend to be cautious, and are liable to be delayed. As far as science is concerned, therefore, what I hope most for myself is to stay alive long enough to see what judgment is passed on "Evolution through Group Selection," if that is what I finally decide to call the new book!

References

Wynne-Edwards, V. C. 1935. On the habits and distribution of birds on the North Atlantic. *Proc. Boston. Soc. Nat. Hist.* 40:233–346.

———. 1937. Isolated arctic-alpine floras in eastern North America: a discussion of their glacial and recent history. *Trans. Royal Soc. Canada* (section 5, series 31): 1–26.

———. 1939. Some factors in the isolation of rare alpine plants. *Trans. Roy. Soc. Canada* (section 5): 1–8.

———. 1952. Freshwater vertebrates of the arctic and subarctic. Fish. Res. Board Canada, Bull. 94, 29 pp.

———. 1962. *Animal Dispersion in Relation to Social Behaviour.* Edinburgh: Oliver & Boyd; New York: Hafner.